About Island Press

Island Press is the only nonprofit organization in the United States whose principal purpose is the publication of books on environmental issues and natural resource management. We provide solutions-oriented information to professionals, public officials, business and community leaders, and concerned citizens who are shaping responses to environmental problems.

In 2003, Island Press celebrates its nineteenth anniversary as the leading provider of timely and practical books that take a multidisciplinary approach to critical environmental concerns. Our growing list of titles reflects our commitment to bringing the best of an expanding body of literature to the environmental community throughout North America and the world.

Support for Island Press is provided by The Nathan Cummings Foundation, Geraldine R. Dodge Foundation, Doris Duke Charitable Foundation, Educational Foundation of America, The Charles Engelhard Foundation, The Ford Foundation, The George Gund Foundation, The Vira I. Heinz Endowment, The William and Flora Hewlett Foundation, Henry Luce Foundation, The John D. and Catherine T. MacArthur Foundation, The Andrew W. Mellon Foundation, The Moriah Fund, The Curtis and Edith Munson Foundation, National Fish and Wildlife Foundation, The New-Land Foundation, Oak Foundation, The Overbrook Foundation, The David and Lucile Packard Foundation, The Pew Charitable Trusts, The Rockefeller Foundation, The Winslow Foundation, and other generous donors.

The opinions expressed in this book are those of the authors and do not necessarily reflect the views of these foundations.

About the Federal Highway Administration

The Federal Highway Administration (FHWA) is a part of the U.S. Department of Transportation, with headquarters in Washington, D.C., and field offices across the United States. The mission of FHWA is simply to create the best surface transportation system in the world for the American people through leadership, innovation, service, and technical excellence. The agency provides expertise, financial resources, and information to continually improve the nation's highway system and its intermodal connections, while striving to protect and enhance the country's environmental and ecological resources, economic vitality, and quality of life.

About the California Department of Transportation

The California Department of Transportation (Caltrans) is responsible for the design, construction, maintenance, and operation of the state's highway system and portion of the Interstate Highway System, and in addition it provides support to local transportation partners in meeting their goals. Caltrans' mission is to improve the mobility of people and goods across California while protecting the environmental values that are important to the state's residents and economy.

About The Nature Conservancy

The Nature Conservancy is a nonprofit organization with headquarters in Arlington, Virginia, and with offices across the United States and in many parts of the globe. The mission ofThe Nature Conservancy is to preserve the plants, animals, and natural communities that represent the diversity of life on Earth by protecting the lands and waters they need to survive.

Road Ecology

Road Ecology

Science and Solutions

Richard T. T. Forman • Daniel Sperling
John A. Bissonette • Anthony P. Clevenger
Carol D. Cutshall • Virginia H. Dale
Lenore Fahrig • Robert France
Charles R. Goldman • Kevin Heanue
Julia A. Jones • Frederick J. Swanson
Thomas Turrentine • Thomas C. Winter

ISLAND PRESS
Washington • Covelo • London

Library of Congress Cataloging-in-Publication Data
Road ecology : science and solutions / Richard T. T. Forman ... [et al.]
p. cm.
ISBN 1-55963-932-6 (hardcover) — ISBN 1-55963-933-4 (pbk.)
1. Roads—Environmental aspects. 2. Automobiles—Environmental aspects. 3. Roadside ecology. I. Forman, Richard T. T.
TD195.R63 R62 2002
577.5'5—dc21

2002010027

British Cataloguing-in-Publication Data available.

Book design by Brighid Willson.

Printed on recycled, acid-free paper

Manufactured in the United States of America
10 9 8 7 6 5

Contents

PART II. VEGETATION AND WILDLIFE

PART III. WATER, CHEMICALS, AND ATMOSPHERE

Foreword

For a generation, North Americans have been in simultaneous pursuit of twin goals that are inherently in conflict. On the one hand, they seek to harvest the manifold benefits of an expanding road system, including a strong economy, more jobs, and better access to schools, friends, family, recreation, and cheaper land on which to build ever-larger homes. On the other, they have growing concerns about threats to the natural environment, including air and water quality, wildlife habitat, loss of species, and expanding urban encroachment on rural landscapes. They pursue the former goal by increasing their use of larger and larger vehicles and their demand for more roads to accommodate them. They pursue the latter by demanding more regulation of vehicles, policies to discourage auto use and increase use of mass transit, and stricter controls on local land development. Not surprisingly, these conflicting demands clash wherever transportation decisions are made, whether at the federal, state, or local levels. Thus analysis paralysis and stalemate often result.

Enmeshed in this gloomy scene, some choose to curse the darkness. But others seek to light a candle. Richard Forman, Daniel Sperling, and their colleagues have chosen the latter course. Assembling a team of experts from all sides of this tangle, they have neatly sidestepped most intractable parts of the struggle by accepting that there are already many cars, trucks, and roads and that, given continuing growth in population, there are likely to be more. Then they consider what can be done to mitigate some of the weightier problems, whether caused by the existing network or by future additions. The authors describe the tentacles of the road system as wrapping themselves around the land in an "uneasy embrace," in which nature affects the roads while the roads influence the land in countless ways.

For more than a century, the transportation community has been increasing its knowledge of how to guard the road system against nature's assaults, through better planning, design, materials, and construction. But we are just beginning to recognize the many ways that roads assault nature, and consequently realize our need to understand these phenomena so as to mitigate negative outcomes.

This book proclaims this need and elaborates a clear call for a new field of study, which is identified by the authors as "road ecology." For some time, existing requirements to assess environmental impacts before new road projects are undertaken have resulted in the development of a group of environmental experts skilled in the conduct of independent ad hoc studies of proposed projects. Their work has produced a process and a body of literature and has doubtless improved the design of many poorly conceived schemes. But this book makes clear that ad hoc environmental analysis has left many gaps in our understanding of effective mitigation for individual road projects and is unlikely to ever lead to effective mitigation of the macro effects of a growing system of roads.

By looking at problems associated with vegetation, wildlife, aquatic ecosystems, wind and atmospheric effects, and flows of water, sediment, and chemicals, the authors have described the issues and provided a target for researchers in many fields to focus their efforts. Until now, the fields of opportunity in road ecology have been ripe but the workers few. Let us hope that this book will provide the incentive and direction that will lead to a new generation of leaders and specialists dedicated to finding answers to these pressing problems.

THOMAS B. DEEN
Executive Director (retired)
Transportation Research Board, National Academy of Sciences
Member, National Academy of Engineering

Preface

Humans have spread an enormous net over the land. As the largest human artifact on earth, this vast, nearly five million mile (8 million km) road network used by a quarter billion vehicles permeates virtually every corner of North America. The network is both an engineering marvel and an economic success story. Indeed, it provides unprecedented human mobility, greatly facilitates the movement of goods, and stretches the boundary of social interactions. In effect, roads and vehicles are at the core of today's economy and society.

These roads are superimposed on mountains, valleys, plains, and rivers teeming with natural flows. Streams and groundwater flow through the land. Wind carries and deposits seeds, spores, and sediment. Wildlife forage and disperse and may migrate. Fish do too. In effect, nature's never-ending horizontal flows and movements mold the land mosaic and create its patterns of biodiversity.

These two giants, the land and the net, lie intertwined in an uneasy embrace. The road system ties the land together for us yet slices nature into pieces. Natural processes degrade and disrupt roads and vehicles, requiring continuous maintenance and repair of the rigid network. Conversely, the road system degrades and disrupts natural patterns and processes, requiring management and mitigation for nature. Both effects—nature degrading roads and roads degrading nature—are costly to society. They also increasingly gain public attention.

The road network was largely in place well before Earth Day 1970 and the emergence of modern ecology. It was built in an era when transportation planners focused on providing safe and efficient transport, with relatively little regard for ecology. That is changing. Today, the transportation community and ecological scientists increasingly seek to undo major mistakes of the past and prevent new ones in the future. With travel on the rise everywhere and roads expanding at urban edges, often intruding into ecologically important areas, the call for new knowledge and skills is stronger than ever.

Perhaps just in time, a solution appears to lie before us. Its underlying foundations include knowledge in transportation, hydrology, wildlife biology, plant ecology, population ecology, soil science, water chemistry, aquatic biology, and fisheries. Fitting these fields together should lead to a science of road ecology, bulging with useful applications. However, landscape ecology has emerged as a key ingredient, or glue, elucidating spatial patterns, ecological flows, and landscape changes over large areas. Although its principles have been mainly incorporated into such fields as forestry, conservation biology, and landscape architecture, they are ideal for transportation planning. Road networks, vehicle flows, biodiversity patterns, and ecological flows are distributed and operate at the same scale. Indeed, integrating landscape ecology principles with road and automotive engineering, travel behavior, and transportation planning should provide a treasure chest of new solutions for both the transportation community and society.

Thus the core objective of this book is to begin integrating the dispersed theories, principles, models, and concepts important to road ecology in order to build a coherent state-of-the-science framework and body of principles useful to transportation planning, practice, and policy. In the process, we identify a range of examples, applications, and case studies—in effect, an illustrative array of possible solutions for the reader to consider. A North American focus (the USA and Canada) is complemented by worldwide examples. Although the atmosphere is briefly included, we focus on land, water, plants, and animals. Bringing the field of road ecology together as a compelling subject should open up obvious research frontiers in both science and engineering. Rather than becoming an exhaustive treatise, the book uses principles, illustrative applications, and plentiful references to open a door to further thought and discovery by the reader.

The transportation community of engineers, planners, environmental specialists, economists, and social scientists represents the primary audience. Ecological scientists are the second audience. We also write for policy makers, public interest groups, and informed citizens. Professionals, practitioners, and planners in state and province departments of transportation, federal agencies, nonprofit organizations, and equivalent entities in other nations should find the text indispensable. Most important, the book should stimulate students to explore its ideas and change the world.

In essence, road ecology uses the science of ecology and landscape ecology to explore, understand, and address the interactions of roads and vehicles with their surrounding environment. The book has four areas of focus: (1) roads, vehicles, and transportation planning; (2) roadsides, vegetation, wildlife, and mitigation; (3) water, sediment, chemicals, aquatic ecosystems, and the atmosphere; and (4) road systems, major landscape types, and further perspectives.

Two giants in uneasy embrace—road system and land—represent an especially useful metaphor for road ecology. But other perspectives are also valuable. For example, the journey along a road, like life, encounters a series of openings, surprises, forks, joys, and sorrows. The traveler follows a drumbeat of dashed lines, punctuated by solid-line stretches of predictability or uncertainty. The less-traveled road taken at a fork. Route 66, or Route 1, as a strip of culture. Reading the landscape en route. The turtle-eye's view at the precipice of a busy road to cross. The rectilinear network grid enclosing and controlling. Road as a catalyst for development. Road as cold concrete. Road as mirage. Road corridor as noisy, dangerous strip to avoid. Road as creator, and destroyer, of communities. Road as environmental injustice. Road as mobility and freedom. Each perspective highlights an area where engineering, science, and society come together. Indeed, such perspectives also suggest many of the scientific topics in road ecology.

With 14 coauthors collaborating (four transportation specialists, one hydrologist, and nine ecologists), writing this book was a complex yet synergistic 27-month process. Initially some of us considered the road system as overwhelmingly "bad," providing access with negative environmental effects, while others considered it to be "good," essential to society despite some environmental problems. Twice, we met as a group (in 2000 at the annual meetings of the Transportation Research Board in Washington, D.C., and the Ecological Society of America in Utah), each time including presentations and discussions with leading transportation specialists. Small groups of authors met for planning sessions at Lake Tahoe in 2001, the International Association for Landscape Ecology 2001 meeting in Phoenix, and the Ecological Society of America 2001 meeting in Madison, Wisconsin. The authors also presented symposia on road ecology at the 2001 International Conference on Ecology and Transportation in Colorado and the 2002 annual meeting of the Transportation Research Board in Washington, D.C.

Individual authors generally wrote brief first drafts of sections, which were reviewed by other authors, revised, and thereafter melded into the book manuscript. All authors reviewed and interwove the material into the final tapestry. Almost all chapters were reviewed by outside specialists. Richard Forman shepherded the process of writing, reviewing, revising, and incorporating artwork, and Daniel Sperling facilitated the book development process and helped engage the transportation community at every step. We plan to deposit key materials elucidating the book development process in the archives of the Ecological Society of America.

As collaborating transportation specialists and ecological scientists, we are awed by both the present and the imminent environmental challenge. With an additional 60 million North Americans anticipated in 30 years, how will society accommodate their desire for space and travel? Can society halt, or even reverse, the environmental deterioration caused by more roads and vehicles? Yet, as collaborators on this book, we sense a promising solution. The emergence of a science of road ecology, with its applications by transportation specialists and ecological scientists, now provides society, faced with a burgeoning population, a unique opportunity to enormously improve the road- and traffic-related environment.

We envision a transportation system that provides effectively for both (1) natural processes and biodiversity and (2) safe and efficient human mobility. Without road ecology, a successful meshing of nature and people will never occur. Wise solutions lie within grasp. The window of time remains just wide enough. Society can see the benefits, a gentle roadprint on the land.

RICHARD T. T. FORMAN, DANIEL SPERLING, JOHN A. BISSONETTE,
ANTHONY P. CLEVENGER, CAROL D. CUTSHALL, VIRGINIA H. DALE,
LENORE FAHRIG, ROBERT FRANCE, CHARLES R. GOLDMAN,
KEVIN HEANUE, JULIA A. JONES, FREDERICK J. SWANSON,
THOMAS TURRENTINE, THOMAS C. WINTER

All 14 coauthors gratefully acknowledge the Federal Highway Administration, California Department of Transportation, and The Nature Conservancy for sponsoring preparation of this book. The tapestry of nature and roads thoroughly engages these organizations. Their vision and leadership, hand in hand with the public, will strongly mold landscapes of the future. We offer *Road Ecology* to the effort.

The contents of this document reflect the views of the authors, who are responsible for the accuracy of the statements and data herein. The contents do not necessarily reflect official policies of the U.S. Department of Transportation, the Federal Highway Administration, California Department of Transportation, or The Nature Conservancy.

Acknowledgments

We warmly thank many friends and colleagues in the transportation and ecological communities who helped us accomplish this book, especially the following. Robert E. Skinner Jr. and Jon Williams (both of the Transportation Research Board, National Academy of Sciences) encouraged us at the outset, and four Transportation Research Board committees and groups aided us as the book evolved. Fred G. Bank (U.S. Federal Highway Administration), David G. Burwell (Surface Transportation Policy Project), Michael Cameron (The Nature Conservancy), Mark A. Delucchi (University of California, Davis), Judith M. Espinosa (University of New Mexico), Gary L. Evink (formerly Florida Department of Transportation), Martin Lee-Gosselin (Université Laval), and Leslie M. Reid (USDA Forest Service) shared their transportation expertise at our authors' meetings. The following colleagues kindly provided critical reviews of chapters: Fred G. Bank, G. J. Hans Bekker (Dutch Ministry of Transport), James R. Brandle (University of Nebraska), David G. Burwell, Steven S. Cliff (University of California, Davis), Thomas R. Crow (USDA Forest Service), Gary L. Evink, Kerry R. Foresman (University of Montana), David R. Foster (Harvard University), Anna M. Hersperger (ETH-Zurich), Louis R. Iverson (USDA Forest Service), Scott Jackson (University of Massachusetts), James R. Karr (University of Washington), Wayne W. Kober (AASHTO, formerly Pennsylvania Department of Transportation), Kenneth Kurani (University of California, Davis), P. Spencer (Sam) Lake (Monash University), Martin Lee-Gosselin, Thomas Linkous (Ohio Department of Transportation), David Montgomery (University of Washington), John F. Morrall (University of Calgary), Barbara Petrarca (Rhode Island Department of Transportation), John Poorman (Capital District Transportation Committee, Albany), Leslie Reid, Hein van Bohemen (Dutch Ministry of Transport), and William E. Winner (Oregon State University). Fred G. Bank and Cynthia Burbank (U.S. Federal Highway Administration), Jeff Morales (California Department of Transportation), and Bruce Runnels and W. William Weeks (The Nature Conservancy) enthusiastically arranged sponsorship for the book. Richard Forman gratefully acknowledges the special aid of Lauren E. Alexander, Lawrence Buell, Wayne Franklin, John H. Mitchell, Kent Ryden, and Barbara L. Forman.

Taco Iwashima Matthews kindly and efficiently prepared the illustrations, which add lucidity, consistency, and pleasure to the presentation. Finally, we deeply appreciate Fred G. Bank, G. J. Hans Bekker, Michael W. Binford, David G. Burwell, Gary L. Evink, Wayne W. Kober, Martin Lee-Gosselin, C. Ian MacGillivray, and Hein van Bohemen, who made special contributions to the book as a whole.

The Metric System in North America

Questions and Answers

What's your stride? Probably 2 feet (women) or $2^1/_2$ feet (men). How about a long step forward? 1 meter.

How long is a meter? Three feet plus three healthy inchworms.

How big is an acre? Enough for a 1-acre house lot with corners about 200 feet apart; in other words, four-tenths of a hectare.

What's a hectare (abbreviated "ha")? Football-field size, so players have $2^1/_2$ acres to run around.

How long is a mile? A typical 20-minute walk, or 1.6 kilometers, or a one-minute drive at 60 miles per hour (mph). Or simply the distance between milestones.

How about a kilometer? A typical 12-minute walk, equal to six-tenths of a mile on the odometer. Also, the distance between kilometer-rocks.

How big is a square mile? Same as the big checkerboard squares across the U.S. Midwest. They can hold either 640 acres or $2^1/_2$ square kilometers.

Helpful Conversions

1 meter (m) = 3.28 feet (ft) — 1 foot = 0.304 meter

1 hectare (ha) = 2.47 acres — 1 acre = 0.405 hectare

1 kilometer (km) = 0.621 mile (mi) — 1 mile = 1.609 kilometer

1 square kilometer (km^2) = 0.386 square mile (mi^2) = 247 acres

1 square mile = 2.59 square kilometers = 640 acres

1 kilometer = 1000 meters — 1 square kilometer = 100 hectares

1 hectare = 10 000 square meters (m^2)

1 km/km^2 = 1.609 mi/mi^2 — 1 mi/mi^2 = 0.621 km/km^2

Road Ecology

PART I

Roads, Vehicles, and Ecology

CHAPTER 1

Foundations of Road Ecology

> What is the use of running when we are not on the right road?
>
> —German proverb

> . . . great technical advances occurred in the technology of pavement structures and surfacings during the nineteenth century. Almost in their entirety, these advances predated the development of the motorcar. . . . Communities at last saw an alternative to a life full of mud, stench, dust, and noise.
>
> —M. G. Lay, *Ways of the World,* 1992

Transportation lies at the core of society. It is what links us together. Both businesses and individuals depend on safe and efficient mobility. In the past century in North America, roads and vehicles have enlarged the spiderweb of our interactions and activities. Now we routinely use vehicles on roads to visit a friend, go shopping, travel to school, or dine out.

Unfortunately, with this dependence on roads and vehicles comes deep and widespread environmental damage.[674,675] As a result, environmental protection now plays a key role in transportation policy and decisions. Ever-increasing resources are devoted to minimizing the adverse impacts of roads and vehicles on species and ecological systems.

Environmental protection is viewed and approached from many perspectives. The engineer seeks technical solutions and designs technical devices to abate damages. The economist seeks the best use of societal resources and identifies actions that yield the highest return. Legislators and lawyers craft sharply defined rules to preclude certain behaviors. Ecologists emphasize that we are too human centered in our responses and seek to elevate the importance of plants and animals. They seek to maintain the diverse characteristics and services of intact, or undegraded, nature and to maintain or reestablish relatively natural

What's Nature Like Near a Busy Highway?

Consider taking a leisurely stroll or nature walk in the edge of woods by a busy two-lane highway.[675] The sense of leisure quickly evaporates in the face of traffic noise. Speeding vehicles evoke a sense of danger. You may be confronted underfoot with society's refuse. Busy roads and a bucolic outdoors seem incompatible.

So you move back into the wooded edge to look more closely. Many of the native forest birds seem to be missing—even for quite a distance into the forest; apparently it is too noisy. Indeed, few other forest vertebrates—mammals, frogs, turtles, snakes—are seen; it must be a road-avoidance zone for them, too. If you had ventured to walk along the roadside, you might have seen road-killed animals, though carcasses disappear quickly where road-kill scavengers hunt. The combination of road-avoidance zone and road-kill strip makes you realize what a barrier the busy highway is, dividing large natural populations into small ones that may be prone to local extinction. Also, wildlife movement corridors that connect distant patches across the landscape may be severed. You wonder whether this is an inadvertent collective assault on biodiversity.

Unlike the adjoining forest interior, your forest edge seems to be full of generalist "weedy" plants, some of them non-native exotics, all persisting next to the open environment of a frequently mowed roadside. The roadside vegetation growing on earth that was homogenized and smoothed during road construction seems monotonous, largely devoid of its natural heterogeneity and richness. A few grasses, plus some non-native plants, tend to dominate at the expense of a diversity of native wildflowers. Open, straight roadside ditches carry warmed water, alternating with pulses of rainwater, into a narrow, wooded stream that lost its valuable curves during road construction. A specific set of invisible chemicals has reached the roadside and perhaps the forest—nitrogen oxides, hydrocarbons, herbicides, roadsalt, and heavy metals such as zinc and cadmium are typical. Entering the streams, wetlands, and groundwater around you, they inhibit all kinds of natural processes and are toxic to some of the species.

What is it like next to a busy road? No place for a neighborhood walk. Or a path in a park. Or even a nature reserve. Here nature is both severed and impoverished. Road ecology is needed.

ecological systems in human-imprinted areas. Meshing this goal with the economic and social activities of our busy highways remains a daunting challenge.

In market economies, prices are a primary mechanism for allocating resources and guiding behavior. Environmental impacts remain largely outside the marketplace.[200] When we drive a car, we degrade the quality of everyone's air. But we do not pay for damage to health or vegetation. If we did, we would probably pollute less. Although conceptual models exist, no effective mechanism in society ensures a proper balance between supply and demand for clean air. The same basic problem exists for noise and water pollution, climate change, aesthetics, loss of wetlands, and loss of biological diversity. The absence of a pricing mechanism has led to regulatory approaches.

Environmental protection is complex yet more easily regulated in transportation than in most other sectors of society because transportation networks are mainly in the public domain. Governments at various levels build and maintain most roads and largely own, operate, and subsidize transit services. Governments also own and manage many ecologically important lands where public roads exist (Figure 1.1). Entwinement of transportation with

Figure 1.1. Government land, where a state- and federally funded road and nongovernment vehicles interact with a sequence of species and ecosystems in a national forest. Design of the road included varying the edges of tree lines to eliminate straight lines; reducing slope angles and varying cut-and-fill slopes to eliminate flat planes; creating rock outcrops similar to native ones; and planting only erosion-control grasses so that natural plant succession follows. State Route 410, Snoqualmie National Forest, Washington. Courtesy U.S. FHWA.

the public domain means that public goals, such as environmental protection, play a more direct role in investments and institutional behavior. Public pressure can translate directly into action by elected leaders and public officials.

Environmental protection came to the forefront of public discourse in the 1960s. Such environmental disasters as London's "killer smog," which killed scores of people, and the Cuyahoga River in Ohio (USA), which caught fire, galvanized worldwide attention. The realization that newly developed and widely used chemicals could decimate ecosystems and poison humans on an extensive scale, a discovery highlighted by Rachel Carson's *Silent Spring*, catalyzed public action.

In the transportation sector, air pollution proved the initial and most compelling call to action, first in California and then elsewhere.[674] Widespread pollution in the USA culminated with the federal 1970 Clean Air Act, which accelerated the process of eliminating lead from gasoline and reducing vehicular pollution. This law was followed in the mid-1970s by fuel economy rules and "gas guzzler" disincentives. Japan pursued roughly the same track in reducing emissions and fuel consumption, as did Western Europe somewhat later.

In a larger sense, many nations were becoming more environmentally conscious as the 1960s ended. Rules and laws were passed to reduce noise and decrease air and water pollution. Greater concern for aesthetics was emerging. In the USA, the National Environmental Policy Act (NEPA) became law, which required that environmental impacts be documented for new projects using federal funds. By the 1970s, environmental and aesthetic concerns were beginning to play an important, if not always well informed, role in the design, construction, and operation of roads.

But even as environmental consciousness evolved, knowledge and political will lagged. As one concern was addressed, another would emerge.[675] As roadside aesthetics received greater emphasis, concerns about non-native and invasive species grew. A phalanx of new rules and institutions emerged to control a carefully specified set of air and water pollutants. But new threats from new pollutants kept appearing. As four-wheel-drive and other high-clearance vehicles tended to replace cars, remote natural areas became accessible to recreational vehicles, and telecommuting from rural areas gained appeal, the threat to ecologically sensitive land increased.

Furthermore, environmental impacts have become global in nature, through the cascading accumulation of ecological stresses and altering of ecological interactions of the earth system itself.[428, 674] The pervasiveness of roads and their cumulative effect on the environment are now of increasing concern for habitat fragmentation, rare species, and aquatic ecosystems.

What Is Road Ecology?

In 1994, a lone ecologist slowly drove a long, winding road up a canyon in the Rocky Mountains. Front views, like an ancient movie, flashed back and forth from towering granite cliffs to precipitous forest slopes. The road, an engineering marvel, crept over old landslides, and the car sidled past avalanche tracks. The destination was a conference of the Ecological Society of America, where 2500 ecologists were packed into the canyon. Upon arrival, the driver, who sensed that road ecology might be important but had never heard of a meeting on the subject, studied the printed program of over 2000 presentations. The word "road" appeared in only one title. He talked with people, from world authorities to promising students. Everyone could speak knowledgeably, even passionately, about the unusual birds around, how to measure the vegetation, the water flows, erosion patterns, wildlife trails, and mathematical models. No one mentioned that omnipresent road running through the canyon.

The ecologist then walked up the canyon to look more closely. The serpentlike route through the heart of the valley was the organizing force for almost all human activities. Hordes of early miners had used it as access to their dreams of wealth, and tens of thousands of sheep must have been shepherded up and down the canyon every year along this solitary route. Bandits and predators lined the route. Today, hotels and tourists, ski areas, homes, and everything else human depend on the condition of that lifeline. But what about those birds and vegetation and water flows and erosion patterns and wildlife movements? Do they affect the road? Or does the road affect them? Indeed, how does life change for plants and animals with a road and traffic nearby?

Answering these questions, and similar ones from local spots everywhere, leads inexorably to road ecology. Indeed, a handful of key concepts and terms here helps bring the big picture into focus.

A *road* is an open way for the passage of vehicles,[1015, 532] and *ecology* is the study of interactions between organisms and the environment.[856, 776] Therefore, the combination describes the essence of *road ecology,* namely, the interaction of organisms and the environment linked to roads and vehicles. More broadly, traffic flows on an infrastructure of roads and related facilities form a *road system*. Thus road ecology explores and addresses the relationship between the natural environment and the road system.

Let us delve into that concept to learn more. Roads come in many varieties, from multilane highways to suburban streets, from logging roads to farmers' lanes (Figure 1.2). All are the focus of road ecology. Sometimes the term "road" or "roadway" refers to the roadbed area between roadside ditches. Other times, *road* or *road corridor* refers to a wider strip where the land surface has been altered by construction, maintenance, or management regimes. Com-

Figure 1.2. An urban collector-distributor roadway linking major arterial highways with downtown streets. This six- to eight-lane, median-divided roadway carrying high volumes of traffic through changing residential and commercial areas has curbside and median plantings, brick splashblocks, sidewalks, screening walls, and quality light fixtures. Harbor City Boulevard (State Route 720), Baltimore, Maryland. Courtesy U.S. FHWA.

monly, this wider strip includes the road surface, shoulders, ditches, and outer roadsides. Where cutting through the side of a slope, the road corridor typically includes a cut surface on the uphill side and a filled area on the downhill side.[2] Various engineered structures to control, for instance, water flow or snow accumulation may be included in this wider road strip. A *highway corridor* usually also includes the strip of cultural structures, as in strip development, associated with the highway.[995]

These attributes, and many more that will be discussed later, are useful in describing a road location or site. Road ecology also focuses on a *road segment,* the stretch of road between two points, such as between two intersections or towns (Figure 1.2). A road segment thus slices through a heterogeneous landscape, so that the pair of adjoining local ecosystems or land uses on opposite sides of the road keeps changing along the segment. The sequence of pairs is little studied but may be ecologically important in a road segment.

In addition, road segments are linked together to form a *road network,* which, with its moving vehicles, we call a road system, as mentioned earlier.[568, 302] The road system connects *nodes,* or important locations such as cities, at a broad, or regional, scale or schools and clusters of shops at a fine, or local, scale.

Across the network, traffic may vary from nearly zero to over 200 000 vehicles per day and change markedly through the day, week, or season.

In effect, the road system ties almost every piece of land together for society. Yet the same road system slices nature into pieces, like little polygons enclosed in the mesh of a network. This network produces major ecological effects in a landscape.

Road density, the average total road length per unit area of landscape (i.e., kilometers per square kilometer, or miles per square mile), is a handy overall measure of a road network or the amount of roads in an area.[898, 754, 310] Many ecological phenomena, from wildlife to flooding to biodiversity, have been related to road density.[288, 320] *Mesh size,* the average area or diameter of the polygons enclosed by a road network, as in a fishnet, is proportional to road density but focuses on the enclosed parcels rather than the roads.[302, 450] Just as average mesh size for a fishnet is of limited use to a fisherman, since all fish could escape through a single large hole, average road density, or mesh size, provides only an overview. The combination of average mesh size and variability of mesh size is ecologically more informative.

However, *network form,* the explicit spatial arrangement of roads and intersections (linkages and nodes), is still better. Network form determines the relative sizes, shapes, and arrangement of enclosed patches.[304] Like the fish example, plants and animals are probably less affected by averages and variability and more sensitive to size, shape, and arrangement of habitat.

Earth, Fill, and Soil

Ancient peoples lived with, though also feared and revered, the four basic elements of the universe: earth, water, air, and fire. The first two—earth and water—form the core of road engineering in action. Earth is moved and molded to create a road that will persist through the vicissitudes of both daily and heavy water flows. Highway engineers routinely deal with earth and fill, while ecologists deal with soil. This section ties the two perspectives together.

Rock, either by weathering in place or being mechanically splintered by machine, forms various smaller rocks, gravels, sands, and finer particles. This *earth* or *earthen material* may be transported and deposited as *fill* in road construction to form much of the roadbed beneath the road surface and shoulders, as well as to cover roadside areas[2] (Figure 1.3). Fill areas are often covered by topsoil and then seeded. Roadside vegetation, whether originating naturally or by seeding, helps create a thin layer of soil, which contains blackish organic matter in the upper portion of the fill.[455]

The particle sizes in fill vary widely, from boulders to clay.[455, 186] *Gravel* (2 to 75 mm diameter) is particularly useful because of the large pore spaces between particles, which permit relatively copious and rapid water flow. *Sand* (0.05 to

Figure 1.3. Earth, fill, soil, landslides, and roads. The earthen material in the road cutbank (upper right) has a fairly stable, steep surface with a thin soil layer containing roots on top. The fillslope below the road (center) consists of earthen fill and is a less-steep, less-stable surface, in this case showing erosion channels and a mudslide. The gully on the far left had a long, shallow landslide, so roads at three levels were stabilized with retaining walls using a blend of steel piles, creosote timbers, and soil/rock anchors. By working from the top of a slope, no excavation was required. State Route 226 by the Blue Ridge Parkway (visible at top), North Carolina. Courtesy U.S. FHWA.

2.0 mm diameter), *silt* (0.002 to 0.05 mm), and *clay* (<0.002 mm) are the fine particles. The *texture* of earth or soil refers to the relative proportions of sand, silt, and clay.[455, 186, 516] Sand has good (rapid) water drainage, silt is intermediate, and clay drains poorly (slowly). Puddles and mud tend to form on clay.

The *roadbed* supports the road surface and shoulders and typically is sandwiched between ditches. To reduce flooding problems for both the roadbed and the traffic on it, usually the roadbed is higher than the surrounding land surface. In addition, various base layers of sandy or gravelly material are commonly laid down in the roadbed to support traffic on the road surface.[184, 2] When water penetrates the roadbed, these porous layers facilitate water drainage into deeper levels or adjacent ditches.

In contrast, ditches may accumulate silt in their bottom or be lined with impermeable material in local spots to facilitate rapid, unimpeded runoff of surface water horizontally. In flat terrain, the outer roadside beyond the ditch is often covered by a material with mixed particle sizes, which is intermediate in porosity and water drainage. In hilly and mountainous terrain, roadsides are more variable in porosity because some natural earth surfaces are not covered by fill.

Where a road surface is more than about a meter above the surrounding land, ditches may be absent, since road water runoff simply runs down and away on the outer slopes of the roadbed. These outer slopes or surfaces of a roadbed composed of deposited earthen material are *fillslopes* (Figure 1.3). A fillslope tends to be highly erodible because its particles have only partially self-compacted over time.[516]

In hilly and mountainous terrain, fillslopes may be quite long on the downhill side of a road.[2] On the uphill side, in contrast, a *cutbank* surface remains where earth, rock, or both were removed for constructing the road. Normally, a ditch to catch and drain water separates the cutbank from the roadbed. A *roadcut* has cutbanks and ditches on both sides of the roadbed. Cutbanks range from a complete earthen bank with particles naturally compacted over time to a rock face that is often topped by a thin mantle of soil. Cutbanks are conspicuous to travelers, whereas fillslopes are rarely noticed.

Along a road segment, the roadside soil tends to be much more constant than the heterogeneous sequence of soils in the adjoining land. In road construction, rock particles originating from different nearby sites tend to be intermixed, averaging out differences, or fill is trucked in from a single site. Furthermore, the fill is deposited and contoured to a relatively smooth surface. Analogously, the chemical constituents of the earth material are averaged in the mixing process. The net effect ecologically is that microhabitat heterogeneity is sharply reduced in roadsides. Earth-moving equipment normally works at a broader scale than the natural processes of soil and plants. Consequently, the range of plant species and natural communities on roadsides is truncated.

At the same time, cutbanks, open ditches, road shoulders, and fillslopes are novel microhabitats in a natural landscape. They increase overall habitat heterogeneity in a natural landscape, whereas they may add little heterogeneity to a suburban or agricultural landscape.

As the upper portion of earth altered by plants and other organisms, *soil* is a rich, dynamic combination of mineral particles, roots, air, water, dead (blackish) organic matter, bacteria, fungi, and soil animals[455, 186] (Figure 1.3). Roots grow, porosity increases as worms move about and roots die, and the composition of air in the pores changes. Water moves up and down, as do tiny soil animals, mineral nutrients, and organic matter. Earth and fill seem almost inert compared with the action in soil.

Over time, soil formation normally produces conspicuous soil *horizons,* or layers.[455, 186] An upper *A-horizon* has leaf litter and humus (both dead organic matter) on top and mineral soil (rock particles) mixed with blackish organic matter beneath. Rain water percolating through the A-layer dissolves or leaches out chemicals. Under the A is the *B-horizon,* composed of mineral soil. Some of the chemicals leached from above, such as aluminum and iron, accumulate here, though overall the B is relatively nutrient poor. Since the black organic matter is rich in nutrients, a thick A-horizon indicates a highly productive soil (assuming it is not extremely acid, cold, or wet). Grasslands with at least moderate rainfall tend to have deep, thick, rich, and blackish A-horizons.

Four basic principles of environmental engineering may help guide roadside construction to a worthy ecological result.[184,267]

1. Mold cutbanks and especially fillslopes to minimize erosion.
2. Leave no severely compacted areas.
3. Minimize the release of chemicals.
4. Create diverse roadside microrelief with rocks and fill.

Introducing Ecological Concepts

A century of research has provided a broad foundation and stimulus for the growth of ecology. The field greatly accelerated in the past four decades, after ecology was discovered to be the primary field of theory and principles for solving environmental issues.[691,856,620,776] Road systems intersect almost all areas of ecology.[674] Yet, at this stage, a limited number of phenomena and concepts from ecology, which we introduce below, are emerging as central to road ecology understanding and solutions. Just as transportation specialists would probably skim through the preceding section, many ecologists will doubtless skim through this section. Yet, if a handful of important ecological concepts are absorbed here, the chapters ahead will provide deeper insight.

The following conceptual foundations of road ecology begin with water and water flows, followed by microclimate, wind, and atmospheric effects; vegetation and biodiversity; populations and wildlife; and, finally, landscape ecology and habitat fragmentation.

Precipitation water, either running off the road surface or directly falling onto roadsides, basically follows one of three routes.[231,856] Plants pump some water vertically to the atmosphere via *evapo-transpiration.* Some water *percolates,* or drains down through the soil, and may run diagonally downward to groundwater or a stream. Finally, some water, especially in heavy rains, may flow over the soil as *surface runoff.* Only this last water flow can cause the *erosion,* or removal, of particles from the soil or earth surface.[455] Eroded material then may be carried by water as *sediment transport* for some distance, to be followed by *sediment deposition,* which occurs where water velocity drops, as in a relatively flat ditch, a lake, or the quiet pool of a stream.

Hydrology refers to the quantity of water present or flowing.[231] Hydrologic flows are mainly driven by gravity. *Groundwater* fills the spaces between rock particles, with its upper surface of saturation called the *watertable*. Groundwater in sand or porous rock is called an *aquifer*.[231] A watertable persisting for long periods at or above the soil surface forms a *wetland* (Figure 1.4), whereas a watertable may also be many meters below the surface.[641, 856] *Surface water* includes streams and rivers, as well as lakes and ponds in topographically closed depressions. The surface runoff of water over the soil transports not only sediment but also a wide range of chemicals, both natural and produced by humans. Most such chemicals are in solution in the water and move invisibly.

Water quality describes the physical, chemical, and biological characteristics of water.[1022, 423,856] Temperature, turbidity, and velocity are physical attributes. Chemical attributes include nitrogen and phosphorus, pH (acidity), oxygen

Figure 1.4. Lagoon estuary with intertidal wetland ecosystems after mitigation. Tidal ebb and flow had been restricted by an earthen causeway (bottom) for residential development, a small landfill (center), and roadway slide debris and fill (left and center). Considerable earthen material and sediment was removed (lower left), lead-containing and hydrocarbon-contaminated material from the landfill was removed, alder *(Alnus)* trees were planted (lower left), and upland slopes (left) were seeded with native plants to reduce erosion. Vegetation areas on heterogeneous slopes and wetland contain diverse plant communities, numerous species, and wildlife movement. The road through the land mosaic plays several important ecological roles. State Route 1, Marin County, California. Courtesy U.S. FHWA.

levels, and organic toxic substances. The biological attributes of water range widely from concentrations of greenish algae to rooted plants, aquatic insects, worms in the mud, and fish populations. *Aquatic ecosystems,* the array of organisms interacting with the environment in lakes, streams, rivers, and saltwater, are strongly affected by water quality. In addition, *habitat structure,* especially the complex of rocks, logs, gravel, mud, and slopes covering the bottom of a water body, helps mold aquatic ecosystems[1022, 856] (Figure 1.4). Native fish populations or the abundance of game fish is sometimes used as an overall measure of water quality and habitat structure, since virtually all physical, chemical, and biological attributes affect fish.[470, 159] Indeed, in some places, the abundance of fishermen is a reasonable indication of water quality, though many fishermen simply catch introduced rather than native fish.

The *microclimate* (the local climate near a surface) surrounding a road usually differs markedly from that of adjoining areas[343, 794, 11] (Figure 1.1). A black road surface absorbs considerable heat from solar radiation and on cool days may get warm enough to attract animals, including amphibians and snakes. Hot air rises from the road, sucking in cooler air from adjacent areas. *Solar angle,* determined by the height of the sun relative to the ground surface, also sets up wide temperature variations within a road corridor, as in roadsides on north versus south sides of an east-west road.[241] Also, *road orientation* or direction relative to wind direction exerts major effects on wind speed, turbulence formation, relative humidity, and soil desiccation patterns. In effect, a wide range of microclimatic conditions is typically concentrated in the narrow zone of a road corridor. This pattern helps create diverse microhabitats for plants and animals.

Wind and atmospheric effects related to transportation also operate at global and regional scales. *Greenhouse gases* (those that absorb infrared radiation emitted by the earth, thus warming the atmosphere), including carbon dioxide (CO_2) from fuel combustion, accumulate in the stratosphere and are associated with global climate change, which in turn produces numerous effects on plants, animals, and ecological systems.[11, 428] Analogously, ozone and nitrogen oxides (NOx) from exhaust accumulate in the lower atmosphere and cause diverse ecological effects on plants. Pollution particles and aerosols (tiny liquid particles suspended in air) accumulate at scales from global to local. *Wind erosion* of soil and other particles, as well as deposition of soil and snow, has many linkages with roads and vehicles.[110]

Vegetation generally refers to the types, density, and arrangement of plants covering an area (Figure 1.4).[483, 856] Somewhat more specific is the concept of a *natural community,* an assemblage of predominantly native animal or plant species, such as an avian or herbaceous community.[653] Change in vegetation or natural communities over time is called *succession* or ecological succession.[723, 59, 856, 923] *Biological diversity* or *biodiversity* refers broadly to the variety of life forms, including genetic types, species, and natural communities, present.[434, 620, 776] However, *species diversity* or *species richness,* which refers more specifically to the number of species present, is the predominant measure of biodiversity used by most ecologists. *Non-native species* (sometimes called "exotics," "introductions," or even "aliens") are purposely or inadvertently introduced species that are native elsewhere. Some non-native species are *invasive,* meaning they successfully invade and become widespread in a natural community.

The concept of a *population* refers to all individuals of one species, such as a butterfly species or humans, living in a particular place.[148, 856, 653, 776] Births, deaths, and movements are key to population rises and falls and are especially important for small populations wavering on the brink of *local extinction,* disappearance from a place. When an entire species in a large area is on the brink of extinction, we designate it a threatened or endangered species, which is listed by a state or province or national government. *Road-kills* (faunal casualties or road mortality) are animals killed by vehicles and, like predation, represent an abrupt way to decrease population number.[248, 250, 251, 139] However, the overall effect of road-kills has to be balanced against birthrate, or baby production, as well as other causes of mortality.

The linkage between wildlife and roads is deeply ingrained in North America, and indeed the term "wildlife" has many meanings, ranging from game or hunted animals to all living organisms in nature. This book takes a middle ground by considering all nondomesticated terrestrial vertebrates (plus "showy" invertebrates, such as colorful butterflies and large snails) to be *wildlife.*[434, 148] Such a concept excludes plants, microorganisms, domestic animals, and most insects, and emphasizes that nongame species constitute the vast majority of wildlife. Fish are considered in their own right, separate from wildlife.

Wildlife tend to have a *home range,* an area of daily movements where foraging for food occurs. Animal *dispersal* is movement well beyond the home range to locate and establish a new home range. *Migration* is cyclic movement between different areas that generally avoids cold or dry seasons. Wildlife species tend to kindle interest in the public and therefore are often an important basis for land-use planning.

Finally, from landscape ecology comes the view of a *land mosaic,* whereby any point in an aerial photo or the view from an airplane window is in either a patch, a corridor (strip), or a background matrix[299, 302, 935, 441, 675, 1033] (Figures 1.3 and 1.4). This simple *patch-corridor-matrix model* has stimulated ecological analysis and impressive understanding of highly diverse landscape types. *Habitat fragmentation,* the breaking of a habitat type into pieces (with consequent loss of connectivity), and *stepping stones,* a row of disconnected habitat patches through which animals may move, are two important examples of our understanding of land mosaics.

Patch characteristics, including size and shape, are especially important ecologically.[816, 434, 302, 620] *Edge species,* which mainly live near the perimeter of a patch, tend to be high in density and diversity (known as the *edge effect*). In contrast, *interior species,* which generally avoid the edge area, depend on large habitat patches and are commonly of conservation interest (Figure 1.1). Corridors come in many forms, from wide to narrow and from straight to curvy, but all exhibit five general *corridor functions:* barrier, conduit, source, sink, and habitat[815, 73] (Figure 1.4). The *barrier,* or *filter effect,* is of particular interest in road ecology. In this case water, animals, and people commonly encounter, and may be partially or entirely blocked by, the road corridor.

In addition, the spatial pattern of land mosaics changes over time, with major consequences for both nature and society.[1074, 720, 257, 935] Finally, the land mosaic perspective has greatly enhanced natural resource management, biological conservation, and land-use planning.[815, 571, 130, 199, 379, 93, 417, 697, 560]

The Roots and Emergence of Road Ecology

When a subject or an effort begins is often unclear, since antecedents always exist and so-called discoverers "stand on the shoulders of giants." Thus, highlighting the 1980s as the beginning of road ecology is really based on a perception that, until then, research and government programs had mainly reflected individual interests and specific or local road problems. The 1980s continued this trend, but some coordinated and sustained programs also emerged that largely continue today. In this overall evolution, "hot issues," open "policy windows," and "critical masses" of effort have doubtless propelled road ecology forward.

The Early Period

Perhaps the deepest root of road ecology is buried in mud. Erosion, sediment transport, poor drainage, and muddy roads were a central part of daily life from earliest times into the early twentieth century in North America.[532] Rain and snowmelt water drained poorly from compacted soil deeply rutted by wagon wheels. Roadcuts and fillslopes facilitated mudslides and landslides (Figure 1.3). Safety hazards, intense and frequent road maintenance, and spiraling costs were chronic problems demanding study and solution. Systematic research on road surface design beginning in the nineteenth century by the French, the Scotsman MacAdam, and others focused on finding solutions for the ever-present mud, erosion, and sediment flow.[532] The relatively recent arrival of paved hard surface eliminated mud as a central problem of transportation.

Road-kills are also a root of road ecology, perhaps especially in Canada, the USA, and Scandinavia, where large mammals are abundant. Measurements and calculations of road-kill rates abound since at least the 1920s, though often studies lacked the methodological rigor of modern ecological research. Collisions with deer *(Odocoileus)* and other large mammals, such as moose and elk *(Alces, Cervus)*, have usually driven such research. Human fatalities and injury, vehicular damage and repair, medical costs, and insurance costs have held the public eye and fostered research. Proposed solutions, including a range of animal deterrents, have so far proven to be of limited success (Chapter 6).

In the 1920s and 1930s, many scenic roads and parkways were constructed. The Blue Ridge Parkway in North Carolina and Virginia, Skyline Drive in Virginia, roads on Mount Desert Island in Maine, and the Going-to-the-Sun Road in Montana were essentially built for people who had cars to enjoy vistas and recreation in extensive natural surroundings (Figure 1.3). A network of parkways in the New York City area combined traveling and commuting with vistas and recreation.[532] Such projects had massive ecological impacts but occurred well before the explosion of environmental public consciousness and the rise of modern ecology.[691] Ecologists took little note of these rural and urban parkways. Surprisingly, even after the emergence of modern ecology, the vast 74 500-km (46 300-mi) interstate highway network constructed in the USA mainly during the 1960s and 1970s generated little professional interest from ecologists.

In the 1960s, expanding road construction in France ran afoul of French hunters. A pioneering agreement was reached whereby *game bridges* were constructed for game animals to cross over highways[41,135] (F. G. Bank and G. L. Evink, personal communication). Some 150 were built, many of them narrow, 5–10 m (16–33 ft) wide structures containing local roads or farmers' lanes. These game bridges were probably an inspiration for the later construction of wildlife overpasses, or "green bridges," in several nations.

The 1970s saw considerable research on deer relative to highways, especially along the Pennsylvania Turnpike, America's trend-setting, high-speed, multilane highway.[142] Also, North America's first wildlife overpass appeared in 1978–79. A narrow one (8 m wide) in southwestern Utah enhanced migratory deer movement along a ridge. In 1985–86, two wider ones (each about 30 m wide) were built in northern New Jersey to maintain connections for horseback riders and deer where a multilane interstate highway sliced a park in two.[513] These are still the only wildlife overpasses in the USA. Canada has built one in British Columbia (5 m wide) and two recent ones in Alberta (each 52 m wide).[611,359] In addition, the design of an occasional bridge over a stream, river, or floodplain has doubtless been altered to enhance the movement of terrestrial wildlife.

Also in the 1970s, environmental consciousness in society reached full swing. Many environmental issues made headlines and were tackled. Yet, except for air pollution, surface transportation largely escaped the concerns of environmentalists.

Air pollution as a new dimension of road ecology became an important issue in the 1970s in several industrialized nations, including Germany and parts of the United States. Acid precipitation, mostly the result of coal burning, was an early concern, particularly in the northeastern USA. But the major role of transportation, especially emissions from vehicles, was becoming clearer and was demanding research and solution.[674] A variety of gases and particles was threatening human health, damaging crops, and blighting urban areas with dense brown clouds. Established in 1970, the U.S. Environmental Protection Agency was empowered to set ambient air quality standards and require reductions in new-automobile emissions. A decade later, Americans could breathe more easily and see the landscape more clearly.

The Recent Period

A broadening of topics in road ecology, an accelerated research effort, and greater information dissemination followed in the 1980s. In Germany, a leading ecologist spearheaded perhaps the first overview of the diverse themes in road ecology.[241] In The Netherlands, the establishment of a road ecology unit within the Ministry of Transport, Public Works and Water Management was an especially important development.[63, 952, 339] Indeed, the background for this action is interesting.

The Dutch have literally created a third of their land by building huge dikes and pushing back the sea. In 1953, an enormous North Sea storm breached a dike and quickly drowned 2000 people. In the subsequent restoration period, the linkage between nature protection and roads and dikes became especially evident. Later, in 1979, a proposed highway, which would have razed a batch of ancient trees, generated a national public outcry. These two events, in a culture that has long manipulated, even created, the nature that people experience, helped catalyze the government into establishing and funding a group of ecologists and related specialists to work within the appropriate ministry (H. D. van Bohemen and H. Bekker, personal communications). That road ecology group is still one of the world's leading units and does research, pilot projects, mitigation, compensation, monitoring, and public education.[1,63,953] It recommends new solutions and plays an active role in all major highway projects. The result is a long record of accomplishments in applied road ecology.

Germany, Switzerland, and several other European nations also have a tradition of addressing the interactions among roads, vehicles, and wildlife.[476, 722, 474] Germany has a strong set of laws, especially to protect motorist safety. Since many highways have no speed limit, mitigation along highways to prevent wildlife-vehicle crashes has been of major importance (F. G. Bank and G. L. Evink, personal communication). Investment in fencing, signs, under-

Figure 1.5. Landscape-connector overpass. This "green bridge" contains continuous vegetation strips used for movement by wildlife and a small road for local vehicle usage. The five such overpasses on this highway generally have grassland and planted savanna on the surface plus shrubs and small trees planted around the fences on both edges of a crossing structure. Route B31, Nesselwangen, Germany. Courtesy A. M. Hersperger.

passes, many green bridges, and land conservation is widespread (Figure 1.5). Although these accomplishments mainly benefit the traveling public, they also enhance landscape connectivity. In addition, nature and wildlife benefit from widespread public support in Germany. In Switzerland, the transportation focus has been not only on motorist safety but also on maintaining landscape connectivity for wildlife and rural human uses (A. M. Hersperger, F. G. Bank, and G. L. Evink, personal communications). The scientifically based programs have produced a wide range of mitigation approaches. These include wildlife overpasses, which have been found to provide the best landscape connectivity for many species. Such "landscape connectors" have been built in other European nations, including Italy and Spain (see Figure 6.4 in Chapter 6). France apparently built the first green bridges and has built the most.

The 1980s also saw a swarm of studies and projects on amphibians and roads.[522] Hundreds of amphibian tunnels in several nations of Western Europe funneled the animals under roads in their seasonal reproductive migration. Massive squishings and associated messy auto accidents were reduced. In the preceding decades in Western Europe, research focused on human concerns of safety, efficiency, and cost. But by the 1980s, research, technology, and government were also enhancing nature protection.

In North America, perhaps all states and provinces had environmental transportation programs and some successes in the 1980s. However, Florida certainly had the highest profile, as it continued to accomplish an impressive array of road ecology projects, underpinned by continuing research.[250, 251, 248] Today, Florida continues as the flagship program, though a handful of other programs may be growing faster.

Especially notable is a series of 23 underpasses and 13 bridge widenings built in the 1980s and 1990s along Alligator Alley (now Interstate Highway 75) to facilitate water movement into the Everglades National Park and to reduce road-kills of the threatened Florida panther (*Puma concolor coryi*).[324, 251] This ambitious project along a 64-km (40-mi) road segment contrasted with most previous study and work at individual locations. It required much joint federal and state collaboration, planning, and funding. The Florida underpasses caught and held the interest and imagination of the American public, even though hardly anyone has ever seen a Florida panther. Even among the 14 authors of this book, only one has seen a panther, and another followed panther tracks through a wildlife underpass beneath a highway. The group of environmental managers in the Florida Department of Transportation, like the Dutch road-ecology unit, has addressed a wide range of ecological issues in a coordinated, sustained manner.

Widespread wildlife-vehicle collisions, plus some high-profile cases, have helped catalyze road ecology interest in Canada.[1054, 535, 170] Many moose-car crashes result in a human fatality. Much of Canada seemingly has moose around every bend, but also elk, deer, and bear (*Ursus*) are frequently hit. Understanding the population dynamics and the movements of these impressive animals is still of great interest, not only on the palette of provincial transportation departments but also in the public eye.

In the 1990s, many trends continued and gathered steam. In the USA, sediment transport, wetlands, air pollution, and climate change all were high-profile transportation concerns.[674] In spite of significant growth in population, vehicles, and travel, some key components of lower-atmosphere air quality improved overall (Figure 1.6). Violations of the national standard for carbon monoxide and ground-level ozone (or smog) were reduced. However, these pollutants, especially ozone, continue as significant problems in certain geographic areas (Figure 1.6). Nitrogen oxides (NOx) and particulate matter (PM-10, PM-2.5) from vehicular use were increasingly recognized as both ecological and human-health problems. Meanwhile, the transportation focus increasingly turned to the unmistakable increase in vehicular emissions of carbon dioxide and other greenhouse gases, which were accumulating in the stratosphere and threatening global climate.[674]

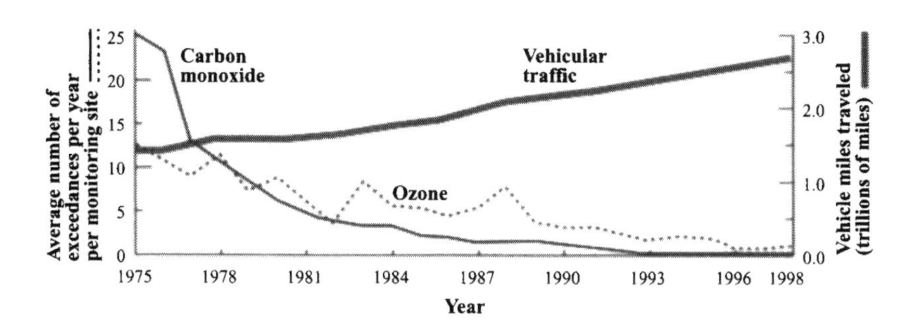

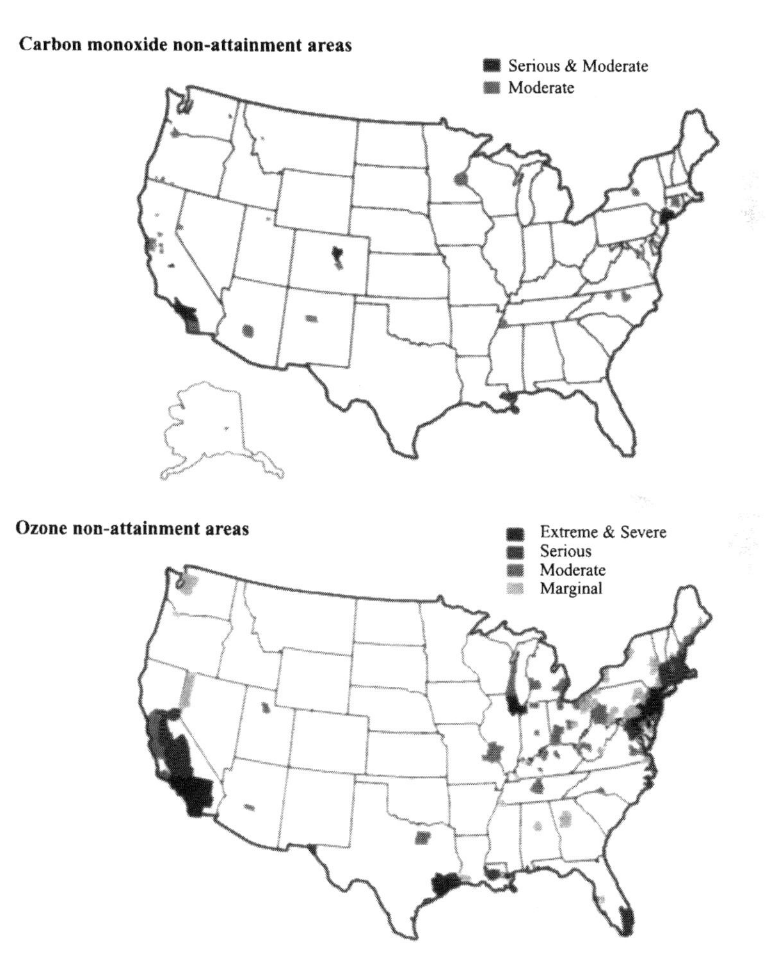

Figure 1.6. Air quality trends and spatial patterns. Data are annual vehicle miles traveled on U.S. roads and number of times carbon monoxide and ozone air concentrations have exceeded federal one-hour air-pollution standards (FHWA 2000a). Violations of Clean Air Act standards are graded from Marginal to Extreme; mapped non-attainment areas are for 1995 (FHWA 1996c).

The Dutch focused on enhancing habitat connectivity by making road corridors permeable for the movement of wildlife.[952,139] The Australians managed their giant green network to protect biological diversity.[815,887] New roads constructed through rainforest in tropical regions were increasingly recognized as a primary cause of deforestation and ecological degradation.[260,198,525,750] In the USA, headlines and public concern about the environmental effects of roads and vehicles increased noticeably in the 1990s, along with not-in-my-backyard and quality-of-life concerns.

Although roads are often a conspicuous feature in a landscape, they tended to escape research notice as the field of landscape ecology mushroomed through the 1980s and early 1990s.[317,816,934,302,695] Finally, in the late 1990s, roads and road systems received scrutiny, not just as objects themselves but as key components of landscape pattern and functioning.[674,760,310,463] The importance of vegetation and native plants in roadsides was highlighted.[1,393] Articles reviewing pieces of the road ecology literature began to increase in frequency.[72,310,930,862,944]

Also at this time, in Canada, two 52 m (171 ft) wide wildlife overpasses were built with federal funding to reduce road-kills and enhance connectivity for movement of elk, wolf, black bear, grizzly, and many other species (*Cervus elaphus, Canis lupus, Ursus americanus, U. arctos*) (Chapter 6).[611] Meanwhile, after a half century of building forestry roads, the USDA Forest Service began to also close roads on a significant scale in certain areas and to identify remote areas that should remain roadless.[376] The economic reasons included eliminating the costs of road maintenance and human-access impacts. Ecological reasons, such as reducing sediment problems in streams and protecting large remote areas for wildlife, were also important.

Meanwhile, the late 1990s and early 2000s witnessed five important international conferences on road ecology, held in The Netherlands, Florida (twice), Montana, and Colorado.[139,250,251,248,150] Symposia on road ecology were held at annual meetings of the Ecological Society of America and the American Geophysical Union in 1998, and collections of articles were published in the journals *Conservation Biology* (2000) and *Earth Surface Processes and Landforms* (2001).

Interactions among European colleagues especially catalyzed by the Dutch conference[139,953] led to a loose but productive Europe-wide organization, Infra Eco Network Europe. This has in turn spawned a planning process for road systems (including a "COST" program), all sorts of research and subsequent research papers, and an accelerated recognition by governments that ecological sensitivity may be a key part of transportation.[725] Elsewhere, World Bank efforts and Indian scholars helped stimulate a book on roads, habitats, and wildlife for South Asia.[750] In the USA, two National Research Council (NRC) Transportation Research Board committees highlighted road ecology in National Academy of Science books linking surface transportation, environment, atmosphere, research, planning, and policy.[674,675]

Conclusion

What is the take-home message for ecological research, planning, and mitigation of road systems today? Ongoing research projects address a relatively wide range of important road-ecology topics and issues, though many others remain unaddressed. What is known comes mostly from a variety of scattered projects—geographically and scientifically—though a few places have multidimensional research efforts. The total amount of research is minuscule considering the magnitude of the road system and its ecological effects, as well as the many environmental controversies flagged by the public.

To illustrate the challenge, consider the indexes of 100 ecological books (1973–2002) on the first author's shelf: 28% contain the word "road"; 11% roadside; 10% transportation or transport; 5% road-kill or road casualty; 3% automobile; 3% vehicle; 3% traffic; 2% truck or trucking; 2% street or streetscape; 1% car; 0% roadway, verge, right-of-way, road mortality, auto, and lorry. Or, consider a recent computer search on the word "ecology" that yielded 82 757 "hits," but in which the combination "road and ecology" resulted in only 0.03% of that number. Research gaps and frontiers lie in wait at every turn. Ecologists and the transportation community can no longer avoid stepping forward and taking the lead for society.

Pioneering research often occurs in unexpected places. For example, an academic has a research hypothesis of some sort and tests it on organisms near roads. As a by-product of this process, we learn much about a piece of road ecology. Such inquiry, lurking in the vast research community, is limited but keeps enriching the field and periodically produces quantum jumps in understanding. Although many specific concepts, principles, and models have emerged, encompassing paradigms, such as transport node-and-linkage-network theory and sequential-process population-reduction theory,[144,139,304] have only begun to receive attention in road ecology.

Multiyear research projects and programs in major research institutions or organizations typically have greater cumulative impact and are mainly funded by state, province, or national governments. These programs often have a mitigation component related to specific road or vehicle projects, and they tend to be well known by the transportation and ecological communities through meetings and publications. However, today such programs are scarce.

An important benefit of these multiyear projects, with associated mitigation programs, is broad dissemination of knowledge in local communities. However, only a small amount of information on ecological planning for roads is being published and has thus become widely available.[139,247,851,603] Mitigation and planning solutions, at least in the USA, have typically been on a case-by-case basis to solve specific local problems. Slowly, we see funded multiyear programs emerging to address issues of road ecology policy.

This book, *Road Ecology: Science and Solutions,* attempts to bring the field of road ecology together from its scattered origins. The book highlights major patterns and trends and articulates principles. Although an array of possible applications and solutions enriches the pages, no specific solutions are recommended by the authors. This is not a how-to guidebook. Rather, the pages explore the research that has been done, and that needs to be done, and presents the findings in forms useful to the reader. Society should benefit from the synthesis. Ultimately, though, transportation and the land and water around us should be the main benefactors.

CHAPTER 2

Roads

> A National Road . . . ran from the navigable limit of the Potomac . . . [nearly] to St. Louis. Travel along the road peaked between 1820 and 1840. . . . The funds available for maintenance were not sufficient. . . . Local inhabitants . . . even stole broken stone from the road bed. . . . It was by far the best road that had yet been built in North America.
>
> —M. G. Lay, *Ways of the World,* 1992

> . . . connecting everywhere to everywhere.
>
> —Mel Webber, *The Joys of Automobility,* 1992

Roads facilitate the movement of people and goods, play a central role in urban and economic development, and enhance countless social interactions (Figure 2.1). Without roads or other forms of transport, mineral resources would have no value, farm production would be limited to areas adjacent to consumers, and workers would be rooted near their jobs.

The first roads were footpaths and trails for humans and pack animals. The earliest manufactured roads date to 4000 B.C. in cities in the Middle East, India, and Britain.[532] To accommodate and guide heavy carts, early road builders commonly employed artificially rutted roads, often with two lanes. Road building was expensive and motivated primarily by the need to move armies, slaves, and other conscripted labor. The Greeks, Persians, Assyrians, Romans, Chinese, Japanese, and Incas all invested heavily in roads. By A.D. 200, the Romans had built an 80 000-km (50 000-mi) network of first-class cement and stone roads around the Mediterranean.

Figure 2.1. Space beneath busy highway used to knit adjoining neighborhoods. Landscape architects used wooden retaining walls to create distinct areas for court games and passive recreation, and the "forest" of pillars was painted with festive colors. State Route 53, Duluth, Minnesota. Courtesy U.S. FHWA.

Roman technology, using cement and roadbed designs, was later lost for a millennium.

Well into the eighteenth century, Europe's post-Roman roads were unimproved dirt paths. It was railways, steamships, telegraphs, and telephones that first overcame distance.[111] In the late 1700s, several countries began building higher-quality roads, sometimes charging tolls. Still, it was not until the advent of motor vehicles in the twentieth century that significant investments were made in the regional and interregional road systems outside of cities.

Large numbers of motor vehicles came sooner to the USA than elsewhere. With its wide-open spaces and dynamic urban and rural economies, the USA was ready for small personal vehicles. Henry Ford responded by mass-producing a small, simple, low-cost vehicle that could transport goods and people. The Model T Ford went into full mass production in 1914, and 400 000 vehicles selling for US$440 each were made in 1915.[840] By 1920, when motorcars equaled horse-drawn vehicles in number,[532] there was one car for every 13 Americans.[681, 265, 273] By 1929, there was one car for every five Americans. Europe would not reach that level of ownership until the 1960s.[553] The car would reshape every activity of American life, from friendships and family to place of worship, work, and recreation. The demand for car travel would also reshape and expand the U.S. road system.

History of the U.S. Road System

When cars burst upon the American scene, an extensive road system providing access to virtually every parcel of land in the country was already in place. But the system consisted of only simple roads, especially outside cities. Most had dirt and gravel surfaces. Roads served farmlands, with a few interregional roads facilitating travel between states. Only 4% were paved.[262] Roads were narrow and closely followed the topography of the land. Road surfaces generally were not designed for drainage of water. Thus, during rainy periods, wheels would sink into mud and roads often became impassable.

The U.S. road system has less than doubled in length since 1900.[262] What has changed is the quality and capacity of that system. With improved pavements, controlled access (freeways), and other innovations, the primitive pre-auto road network was transformed into an efficient personal and commercial network—in the words of Melvin Webber (1992), "connecting everywhere to everywhere."

To this day, virtually all urban and rural roads in the USA are built and operated by government—local (Figure 2.2), state, and federal. Exceptions

Figure 2.2. Covered bridge built by local government in 1879 in a period of major stream flooding. The structure features oak flooring, barn-red siding, and a slate roof, which keeps dangerous winter ice off the bridge surface. Brattleboro, Vermont. Courtesy U.S. FHWA.

include some private subdivision and shopping mall roads, logging roads on private forestland, home driveways, ranch roads, and some modern toll roads. In this chapter, and elsewhere in this book, we especially focus on roads from suburbs to remote areas, where roads and vehicles strongly affect the natural environment.

Colonial Roads and Early Turnpikes

Today's roads mostly follow routes and patterns laid out a century or more ago. The colonial economy of the USA was primarily resource extractive and export oriented. Thus the early road system radiated outward from eastern coastal cities. Most roads were little more than natural surfaces and pathways, improved just enough to allow passage of wagons. Such roads were built and improved with local conscripted labor from landowners or sometimes with slave and indentured workers.[262]

In debt and in recession following the War of Independence, the new states of the USA offered concessions to private companies to build toll roads (and canals).[262] These early roads and canals complemented each other, with the roads providing local access to the longer-haul segment of a trip or placed where canals were not viable. Roads and canals rarely competed. Toll roads, called *turnpikes* in the USA (because of turnstiles), were made to specifications developed in France, the acknowledged leader in road engineering and maintenance protocols.[262] On a rolled crowned bed, large 18-cm (7-in) stones were laid on edge, set tight with smaller stones, and then overlaid with several centimeters of smaller, hand-crushed rock, interlocked and compacted. This method was superseded in the 1820s by methods developed by MacAdam and other Scottish engineers.[262] MacAdam's technique emphasized the use of local soils that were compacted and overlaid with a 15–25 cm (6–10 in) layer of angular, small stones. This pavement-like surface was compressed to interlock, drain rainwater, and protect the sub-base against water so that little mud could form.

By 1808, Connecticut had chartered 50 companies to build 1240 km (770 mi) of roads.[262] New York had chartered 67 companies for 5000 km (3110 mi) of roads and 21 companies for toll bridges. By 1812, the road from Philadelphia to New York was paved its full length with hand-crushed stone (the first mechanical crusher was not patented until 1858). Toll companies focused primarily on heavily traveled routes between large cities.

The new U.S. government became interested in roads to assist development of a national postal service, develop interior lands, and defend remote territories.[262] The government had bought most of the lands to the north and west of the Ohio River in 1803, and it built roads into those areas to encourage immigration. Land purchases generated revenue for the cash-strapped federal government and became an important revenue source for road building and maintenance in sparsely populated areas.

In 1806, the federal government began its most ambitious project, constructing the National, or Cumberland, Road from the Potomac River at Cumberland, Maryland, inland to the Ohio River at Wheeling.[262] The right-of-way was cleared 20 m (66 ft) across and ditches were built for drainage. The roadbed was 9 m (30 ft) wide; the middle 6 m was built with the French method, using 18-cm base stones. The road cost US$1.75 million, about $8700 per kilometer ($14 000 per mile).

This period of federal interest in road building was short lived. By 1831, the completed sections of the Cumberland Road were turned over to the states (see quote at beginning of chapter). The federal government halted all road funding. Because states had no interest in roads, local governments oversaw all road building from the 1830s to the 1920s (Figure 2.2). Not until the early 1900s, prodded by federal incentives, did states begin to organize highway departments and seriously engage in road construction.

This period from 1830 to the early 1900s was dominated by railroad expansion. Railroads often laid tracks alongside turnpikes (toll roads), in direct competition. By 1850, hundreds of turnpike companies were driven into bankruptcy, affecting thousands of kilometers of roads. The turnpike roads fell into disrepair, and the companies usually turned over their assets to local governments, mostly serving local farm traffic. The railroads carried most long-haul passenger and freight traffic until after World War I.

Good Roads Movement

As indicated, most roads in the USA were rural, under local control, and in place by 1900. In general, the roads were unsurfaced, dusty (especially once autos and trucks began to use them), heavily rutted, and often impassable after heavy rains. They were used to carry farm goods to market or to the nearest grain elevator, with elevators spaced along rail lines according to the distance a horse could travel from farm to elevator and back in one day. Roads were little used in winter.

Interest in improving roads began in the late 1800s with the proliferation of bicycles.[262] Cyclists began to call for "good roads" with hard, smooth surfaces for country rides. As cars became more common and touring clubs were organized, this *good-roads movement* strengthened and expanded. Car drivers disliked the dust raised by high-speed travel and sought smooth, dust-free roads that were passable year-round.

Initially, farmers opposed demands by city drivers for better roads, but as farmers acquired their own cars, they also joined the call for improved roads. The first response was oiling of roads, later followed by use of asphalt surfaces.

Auto clubs soon began to promote between-state travel and began formulating "trails" across the USA. These routes were essentially ad hoc sequences of roads that could be linked to allow long-distance travel. The trails took on greater importance during the two world wars, when factories were converted to production of military goods and delivery to ports became urgent.

The USA engaged in a flurry of road building after World War I. The first goal of the construction, politically supported by farmers and the touring clubs, was to "get the farmer out of the mud." The essential funding mechanism was the gas tax. Beginning in 1919 with Oregon, states began imposing gas taxes (and vehicle registration fees) to finance the huge cost of building and improving roads. Gasoline taxes were preferred because they functioned as user fees. Thus the people using roads paid for road improvement and expansion. By 1923, 33 states had imposed a gasoline tax, and by 1929, all states had.

Another important element in the expansion and institutionalization of road building was the *city beautiful movement.*[553] Stimulated by the Chicago World's Fair and by travel to Paris and other European cities, many Americans began to view roads differently. Well-designed boulevards began appearing in most large American cities around the turn of the century, followed by parkways in the 1920s. During the same period, the U.S. government built a number of scenic highways to provide attractive access to the emerging National Park System and to forestlands. These included the Going-to-the-Sun Road in Montana and the Blue Ridge Parkway and Skyline Drive in the mountains of North Carolina and Virginia. Such national parks as Yosemite in California and Zion in Utah also built access roads. Funding for these projects was conceived in light of increased leisure driving but was often justified as depression-era public work investments. Such parklike road project ideas were halted by World War II and were never reestablished on a significant scale.

Federal Involvement and National Highways

In 1893, with the creation of the Office of Road Inquiry in the U.S. Department of Agriculture, the federal government began to renew its interest in roads. The first serious embrace of roads came in 1916 when the U.S. Congress created the Federal-Aid Highway Program, still the basic program for federal support of highways.[353] This law specified that states serve as recipients of federal funds subject to various conditions. Reliance on the states as the senior partners in the federal highway program continues to the present, although local officials in counties and municipalities have played a stronger role since the early 1960s. The 1916 law encouraged the creation of a statewide agency in each state that could bridge the jurisdictional fragmentation among county governments. An important feature of these state highway departments was that they were technically oriented, staffed by engineers, and managed in accordance with the principles of scientific management and administration.[353] The law required states to create a state highway department to serve as a partner in the federal program and have sufficient authority to supervise the expenditure of the funds. Federal funds could not simply be passed to localities. It also required the states to match federal funds dollar for dollar.

The federal government enacted its first national gasoline tax in 1932, later converting the tax into a "dedicated trust fund" for highways. Federal involvement in road building was legally based on facilitating postal service and serving military needs. But the tax was only 0.1 cent per gallon. The flurry of rural road building slowed during the 1930s economic depression and World War II, except for government make-work programs and militarily important roads.

Military planners also took note of the need to improve roads.[262] The strategic importance of roads had been well understood, dating back to Roman times and before. During the War of 1812, as the British navy blockaded ports, the newly finished U.S. coastal highway became essential to transport goods northward and southward along the Atlantic states. Afterward, the military supervised the building of many roads into the western frontier. But most important in the twentieth century was the need for well-made roads to move heavy equipment. The nation's experience during World War I was pivotal. American railroads proved inadequate to move equipment and military materiel to ports for overseas transport. After the spring thaw in 1918, trucks were used extensively to transport military equipment. The roads were unsuited to the heavily loaded trucks and were seriously damaged.

In 1922, after having observed the difficulty of moving military equipment and soldiers to Atlantic ports during World War I, General Pershing unveiled a plan for a system of national highways of strategic value.[353] This plan was the precursor of the interstate highway system built later in the century. It also laid the foundation for broad federal involvement in road building.

The Interstate Highway System

After World War II, auto ownership and suburbia flourished in concert. Public transit use declined, and congestion in urban areas became severe. In 1956, the U.S. Congress enacted an extraordinarily ambitious plan to build and finance the National System of Interstate and Defense Highways, now known as the interstate highway system.[262]

The *interstate highway* program evolved from demands for a national system to serve auto, truck, and strategic military needs. It was inspired in part by the New York parkways built by Robert Moses in the 1930s, the limited-access German autobahns, and the Pennsylvania Turnpike built after World War II.[262] Early studies of limited-access highways showed that they carried three times more volume per lane than a conventional highway and that the accident rates were only a fraction of the rates of conventional highways. The Pennsylvania Turnpike provided a real-life example of a modern freeway, but it was argued that it could not serve as a national model because many parts of the country did not have enough traffic to amortize the tolls.

Finally, after many years of debate, the plan was enacted in 1956. The road system was to be 68 400 km (42 500 mi) in extent, with four lanes divided by a median strip, with limited access in its entirety, and with the federal government paying 90% of the cost. Standards included bridge clearances and pavement strengths to support military vehicles, such as trucks carrying tanks. Financing was made possible through taxation of gasoline, tires, and vehicle parts. Considered complete in 1990, the system may be enlarged only if a state uses its own funds to build a road to interstate highway standards and petitions the federal government to have the route added.

The interstate system was initially envisioned as an intercity network. But with growing urban traffic congestion, cities demanded to be part of the program. Organizations such as the League of Cities and the Conference of Mayors lobbied successfully to include metropolitan areas in the program. The final system has a major urban (including suburban) component, with urban ring

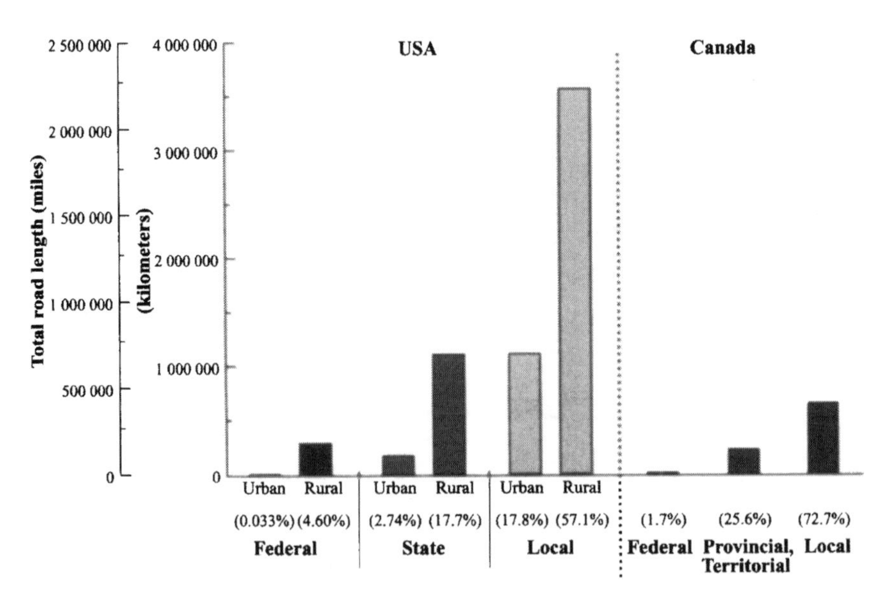

Figure 2.3. Jurisdictions for roads in the USA and Canada. Urban = municipalities with ≥50 000 people; rural = all other areas. USA: federal = 291 055 km (180 745 mi); state = 1 286 625 km (798 994 mi); local = 4 706 359 km (2 922 649 mi) (FHWA 1995b). Canada data from Transportation Association of Canada (2000).

roads and many additional spurs within metropolitan areas. Over 28% of the current 74 551 km (46 334 mi) is in urban areas. Although the interstate highway system accounts for only 1.2% of the country's total public road length (Figure 2.3), it carries 22.8% of all travel in the USA.[276]

Post-Interstate Era

Following the completion of the interstate highway system but with population and travel demand still increasing, new programs and visions were needed. The post-interstate era began in 1991 with passage of the Intermodal Surface Transportation Efficiency Act (ISTEA) by the U.S. Congress. This law set a new course for highway and transit decision making. Responsibilities and funding shifted from national roads and federal direction to local needs and greater state and local government authority. Fully half the funds were eligible to be spent on either highways or public transit—a major shift from the past, when transit was largely not eligible. However, highway expenditures predominated overwhelmingly. Although only a small part of the total effort, environmental mitigation and enhancement received a higher priority. New funding was made available for bikeways and pedestrian facilities as well as for restoration of historic transportation facilities, such as canals and towpaths, abandoned rail stations, and urban waterfronts. In the subsequent major transportation law of 1998, the Transportation Equity Act for the 21st Century (TEA 21), Congress reaffirmed its commitment to the principles and programs of ISTEA, making only minor changes and retaining the basic program structure.

U.S. roads are owned, financed, and maintained by different governments and management agencies in urban and rural areas (Figure 2.3). Each responsible entity has different laws, finance structures, and visions of the road system.

The Canadian Road System

While Canada shares a similar car culture and a long permeable border with the USA, the geography, governing, and history of its roads are distinct. Canada is a federation of 10 provinces, with a population of about 30 million, one-tenth that of the USA. Three major metropolitan areas, Montreal, Toronto, and Vancouver, each have around three million residents, and the rest of the population is spread thinly over an area slightly larger than the USA. However, since 1975, the growth in area of all large metropolitan regions has outstripped the urban population growth (Figure 2.4). Canadian and U.S. cities both are sprawling outward at their fringes. Like their U.S. neighbors to the south, Canadians enjoy a high standard of living, which is based on an extractive export economy tied

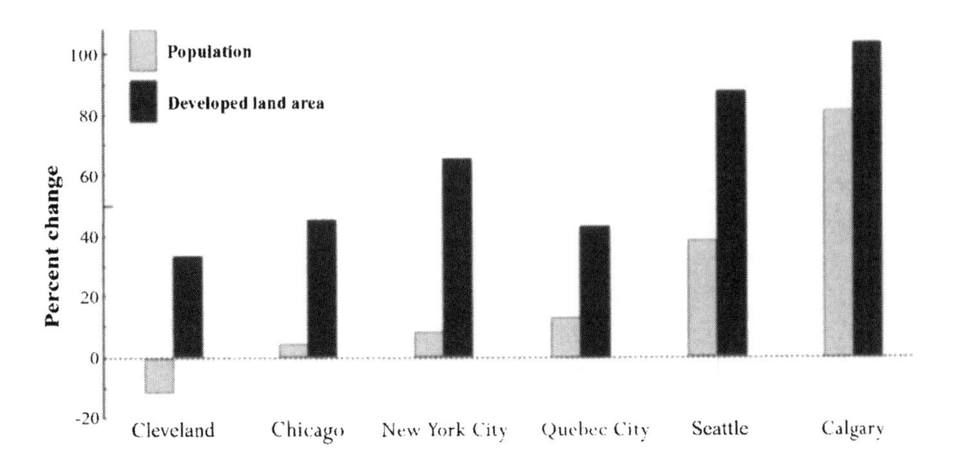

Figure 2.4. Change in population size and developed land area for metropolitan areas over two decades. Data for USA, 1970–90 (Diamond and Noonan 1996). Canada 1976–96: population data from Statistics Canada; estimates for area of Québec City (Statistics Canada) and Calgary (courtesy City of Calgary).

to the Commonwealth and to the USA. There are more miles of roads per person in Canada than in any other nation on earth.[271]

Development of the Road System

In the early days of Canadian settlement, most roads in Canada were the financial responsibility of the localities they served (Transport Canada website 2001).[415] Their primary function was to meet the short-haul requirements of these settlements, and roads seldom extended much beyond community limits. The roads basically provided access to the major transport modes of the time—waterways and, later, railways. Road development was undertaken by a system of "statutory labor," under which each settler was expected to maintain the road adjacent to his or her property or to work for several days a year on road maintenance. Over time, this labor was commuted into the payment of a fine in lieu of actual work, which formed the first source of funds for road expenditures. As road requirements developed beyond the capabilities of this local activity, toll roads were introduced.

The advent of the automobile changed the future of roads. The roads that had proved acceptable during most of the year for travel by horse and wagon now proved to be insufficient in width, composition, and strength for the motor vehicle. Between 1902 and 1912, the number of motor vehicles rose from 2131 to over 50 000. As automobile use expanded, the interregional importance of roads grew. Road construction increasingly became a provincial undertaking, with the provinces of Ontario and Quebec being the first to establish highway departments. Major highway construction efforts followed, and between 1929 and 1969, the provision of paved road surfaces across Canada increased tenfold.

The Canadian Constitution assigns responsibility for highways to the provinces, so the federal government has played a more modest role in developing and financing highways than is the case in the USA[415, 924] (Figure 2.3). Provinces and cities control 98.3% of the highway system in Canada, and federal highways comprise 1.7% of the total (J. F. Morrall, personal communication). Also, almost all road funding comes from the provinces and cities.

Federal highway-related activities can be categorized into three areas:

1. *Military roads.* These include a road between Prince Rupert and Cedarvale in British Columbia built in 1942–46, as well as the Alaskan Highway connecting Dawson Creek, British Columbia, with Fairbanks, Alaska, built in 1942–43 by the U.S. government at U.S. expense under Canadian authorization, with the Canadian portion (about 80%) turned over to Canada in 1946.
2. *Federal highway ownership.* This comprises 15 080 km (9365 mi) of roads on native reserves, on federal properties such as national parks and the Alaskan Highway, and in the National Capital region (Figure 2.2).
3. *Financial contributions* to the provinces and territories for highway construction.

Federal assistance for provincial road construction began with the passage of the Canada Highways Act in 1919.[415, 924] In the 1930s, province-federal cost sharing led to increased road building associated with mining industries, tourist highways, and some sections of a projected highway across Canada. In 1949, a law was passed establishing a *Trans-Canada Highway* program to connect the Atlantic and Pacific Oceans with a highway[654, 611] (Figure 2.5). The provinces carried out the actual construction using federal standards. Although it was originally envisioned as a two-lane highway, extra lanes were added near major urban centers and elsewhere at province or municipality expense and at federal expense on federal land. No previous Canadian project matched the extent or cost of building a 7821 km (4860 mi) long highway from Newfoundland to Vancouver Island, through all 10 provinces. It is the world's longest paved highway.

Figure 2.5. The Trans-Canada Highway, the longest paved highway in the world, shown here in Banff National Park, where the highway is under federal control. Because the highway bisects this flagship national park of Canada, major wildlife research and mitigation efforts are centered here (see Chapter 6). Photo by A. P. Clevenger.

The Road Network

The distinctiveness of Canada's road network is evident in a comparison with those of many other leading nations (Figure 2.6). Canada has the lowest road density (0.10 km/km^2, or 0.16 mi/mi^2), only slightly lower than Mexico (0.13 km/km^2).[271] However, when measured in light of the country's population, Canada has the greatest length of roads per 1000 people (19.93 km), nearly 25% greater than the USA, the next closest country. On a per capita basis, Canada's road network is twice that of France and almost four times greater than that of Great Britain. On average, only 9 km (5.6 mi) of road exist for every 100 km^2 (39 mi^2) of territory.

The underlying explanation is clear. Canada has an immense landmass of nearly 10 million km^2 (3.9 million mi^2) that covers six time zones. Spread across the land are just over 30 million people (fewer than in California), and most are compressed along the southern margin of the country within 400 km (250 mi) of the USA. A far-reaching road network is needed to connect the dispersed towns and cities. Canada has 901 903 km (560 082 mi) of roads measured on a two-lane equivalent basis. A *two-lane equivalent* is a length of road measured as if there were only two lanes. Just over one-third of the Canadian road network (317 920 km) is paved (Figure 2.7). The network of paved

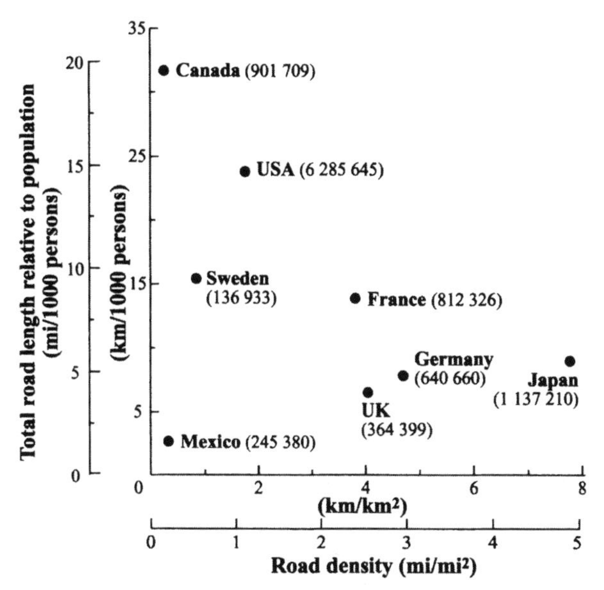

Figure 2.6. Road density relative to population in various countries. Total road lengths (km) in a country are in parentheses. Adapted from FHWA (1996c).

roads, which have a surface pavement of asphalt or concrete, has grown considerably, from just over 100 000 km (62 000 mi) in 1959 to over 300 000 km in 2001. Interest in transportation and environmental sustainability is beginning to emerge.[654, 611]

Of the total road network, about 2% is under federal jurisdiction, 25% is under provincial or territorial jurisdiction, and the remaining 73% is under local jurisdiction (Figure 2.3). Nearly half of Canada's road network is made up of gravel roads (49.1%), 19.6% is paved rural roads, and 13.8%, paved urban roads (Figure 2.7). The *winter roads* (almost 0.6% of the network) listed in Figure 2.7 are roads built of ice or snow in the winter over frozen lakes, rivers, and muskeg (Chapter 13).

The road network varies considerably by province or territory, from a low of five people per kilometer in Saskatchewan to a high of 66.3 in Ontario. The number of vehicles per two-lane equivalent kilometer is lowest in Saskatchewan, at 3.5 registered vehicles for every kilometer. British Columbia has the most densely used road network, with 40.5 registered vehicles per two-lane kilometer of road. The total vehicle kilometers of travel (VKT) in Canada are estimated to be 215 to 250 billion kilometers annually.

In 1989, Canadians devoted CD$103.1 billion to transportation, equivalent to 16% of the gross domestic product (GDP). This is substantially more than is allocated to health care (9% of GDP).[800]

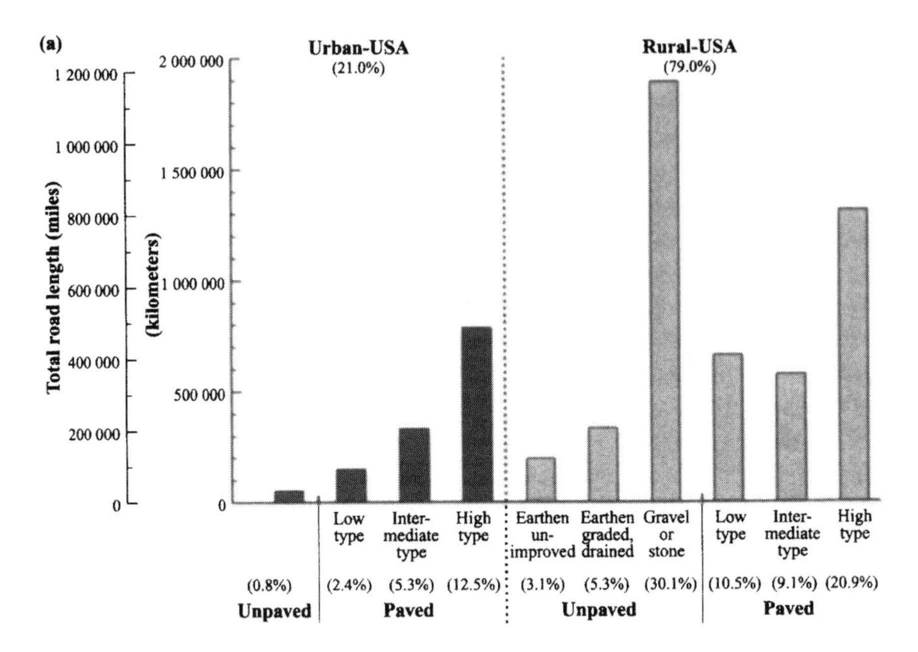

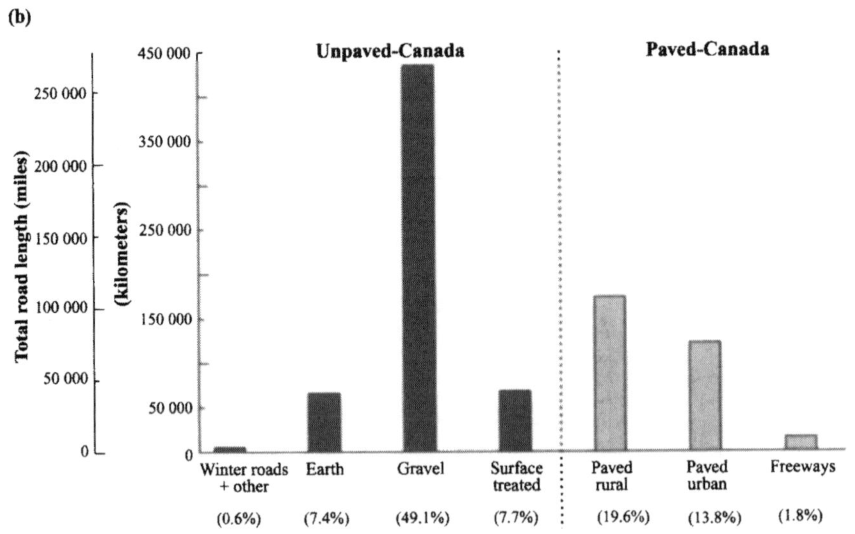

Figure 2.7. Road-surface types in the USA and Canada. USA: urban = 1 318 907 km (819 041 mi); rural = 4 975 865 km (3 090 012 mi) (FHWA 1996c). Canada: unpaved = 574 400 km (356 700 mi); paved = 312 000 km (193 750 mi); winter roads and other = 5400 km (3350 mi); freeways = 16 000 km (9940 mi) (Transportation Association of Canada (2000).

Size, Growth, and Pattern of the U.S. Road Network

The USA has 6.3 million km (3.9 million mi) of public roads (Figure 2.7).[265, 127, 730] These provide 13.2 million lane kilometers (8.2 million lane miles), reflecting the fact that roads are overwhelmingly two-lane, although multilane highways are common. Most of the roads are in rural areas: 80% when measured in terms of center-line kilometers and 76% when measured in lane kilometers.

The road network is expanding slowly in relative terms, with only 88 000 new kilometers (55 000 new miles) added in the decade from 1987 to 1997, an increase of 0.2% per year. In terms of road capacity for vehicular travel, most growth is through road widening—about 80% of all expansion. Lane kilometers (or miles) increased 0.3% per year during the decade. In 1997, 98 000 new lane kilometers (61 000 lane miles) were added.[126]

In addition to these public roads, a considerable number of private roads (not open to the public) exist, including military roads, many forestry roads, roads on cattle ranches, roads in shopping malls, and driveways. Also, some roads under federal control, especially on USDA Forest Service land, are not open to the public. The total length of non-public roads in the USA is unknown. A current study of aerial photographs for the northern third of Wisconsin (with much private and some government forestland) finds the density of visible roads to be about 60% greater than that in official Federal Highway Administration statistics (T. J. Hawbaker, personal communication), suggesting that in certain landscapes the length of private or unofficial roads may be considerable. The increased usage of sport-utility vehicles, four-wheel-drive vehicles, and off-road vehicles since the 1990s (Chapter 3) may function as an active road-establishment process.

Ecologically, a particularly important subset of the total road network is forestry roads. USDA National Forests have an approximately 600 300-km (373 000-mi) system of roads serving 192 million acres (77.8 million ha).[175, 273, 93] Of this total land area, 34 million acres (13.8 million ha) are in "wilderness areas" with few or no roads, 117 million acres are in "general forest," and 40 million acres are currently "under review." Of the total roads, 138 000 km (86 000 mi) are maintained for passenger cars, and 339 000 km are for high-clearance vehicles. An additional 122 000 km (76 000 mi) are for high-clearance vehicles but are closed to the public. (Federal Highway Administration statistics for public roads do not include these closed roads.[271] Thus, FHWA statistics indicate that only 554 400 km [343 200 mi] of public roads are under federal control.[273] These include Park Service, military, Bureau of Land Management, and USDA Forest Service roads open to the public.)

Most of these USDA Forest Service roads have been built for logging purposes and some for mining and recreation. Temporary *skid roads* or *trails* created by off-road logging vehicles, typically among the trees, are not included in the preceding totals. Also, off-road-vehicle routes and other illegal or unofficial roads, perhaps totaling several tens of thousands of kilometers in length, are excluded. One estimate suggests that there are 96 500 km (60 000 mi) of "non-system roads" (not maintained as part of the USDA Forest Service road system),[175] alternatively described as "ghost roads that have escaped the government's inventory."[93] In addition, roads on privately owned forestlands are not included due to the scarcity of data. Road densities on private forestlands are often much higher than on National Forest land.[175]

A key statistic suggesting the continuing ecological impact of roads is quantity of land consumed by metropolitan growth. Metropolitan areas have been spatially spreading considerably faster than population growth (Figure 2.4). This growth is also spreading to rural and small-town areas. According to research at the U.S. Department of Agriculture, Americans in the 1990s converted open space to developed land at a rate of 0.89 million ha (2.2 million acres) per year, or 102 ha (252 acres) per hour—a rate of conversion 50% greater than in the 1980s.

Although the USA is in no danger of "running out" of land given the immense size of the continent, much of the country is desert or mountain terrain suited for only low-density settlement. The land most affected by the extent and pace of expansive development is that close to existing cities, towns, and farms—the land most likely to affect transportation trends. Sprawl development threatens the effective functioning of highways as well as ecosystems, farmland, scenic and historic resources, and even sense of place and community.

Road Density

Road density is an important but crude measure to assess the potential impact of roads on local environments. The USA devotes about 0.45% of its land area to roads, based on the average road density of about 0.75 km/km^2 (1.2 mi/mi^2) for the nation.[126] Adding the right-of-way of these roads would roughly double the amount of land devoted to roads. This crude estimate is probably a significant underestimate because it excludes private roads in suburban areas as well as driveways and parking areas.

Not surprisingly, road densities vary greatly between nations and regions (Figure 2.6). Japan has a road density about four times that of the USA, and Germany, France, and England have road densities 2.5 times greater. But these aggregate measures mask large variations. For instance, Japan and New Jersey have about the same road density, and in densely populated areas of Germany paved-road density exceeds 3.6 km/km^2.[584]

Types of Roads

Roads are classified according to a variety of legal and functional criteria, but no single method is well suited for analyzing landscape ecology concerns. Highway planners in many parts of the world use *functional classifications*.[2] In the USA, the classes in this functional typology are limited-access highways (mostly interstate highways), arterials, collectors, and local roads. This classification relates to the service provided to people by a road. *Local roads* serve primarily a land access function, while the limited-access interstate highways are designed to serve long-distance traffic. This functional classification scheme also distinguishes between urban and rural roads (Figure 2.7). *Urban roads* typically have abutting residential, commercial, or industrial development. *Rural roads* typically abut farms, forest, or nonforested mountain or desert terrain. It may be useful to develop a new, more ecologically based road-classification scheme, but it should be noted that the existing functional classification is widely used in the transportation field and has a rich base of data.

The highway planning classification scheme had its origins in the state highway needs studies of the 1940s and 1950s.[262] To establish a more rational basis for investment, many states undertook comprehensive assessments of their road systems. The studies consisted of a complete road inventory, including critical dimensions, an assessment of the condition of the road, *traffic volume* (level of vehicle use), and ownership. The U.S. Congress undertook several national needs assessments in the 1960s. In 1968, it required that a report on the "condition and performance of the nation's highways" be prepared biennially.

Overlaying road type on topography provides a clearer picture of the potential impacts of a road system on ecological conditions. This approach also indicates where opportunities lie for mitigating existing impacts and for designing features that enhance ecological functions. Almost 70% of all roads are local, and thus their function is primarily land access[3] (Figure 2.2). Local roads primarily provide access to residences, farms, businesses, or adjacent property rather than serving through traffic. They are designed to serve local or repeat drivers who are familiar with the road. Nearly 80% of U.S. roads are "very low-volume roads" with traffic volumes of ≤400 vehicles per day.[3]

Collector roads account for 20% of the nation's road length and, in addition to providing some access, deliver traffic from local roads to arterial roads and limited-access highways.[2] Major and minor *arterial roads* (mainly for through movement or long-distance travel between cities and major towns), together with the interstate highway system, constitute only 11% of the nation's road length but serve 72% of all travel. Travel on arterial roads grew 39% between 1983 and 1993. In 1993 a National Highway System, composed mainly of major arterial roads and interstate highways (about 4% of the nation's public roads, which carry about 42% of the vehicular travel miles), was adopted by the U.S. Congress.[265]

The functional classification system camouflages considerable variation, especially across regions of the USA (Figure 2.8). The eastern coastal states have rather dense road systems, which were built to serve the small farms of colonial America. The southern states, early populated by large plantations, have less-dense road systems. The plains states have roads that fit *mile-square sections* (2.59 km²), forming an immense grid. Every section-sized farm was completely surrounded by roads. Even quarter-section farms (65 ha) commonly have roads on two sides representing half their perimeter. These roads are rather dense given the sparse population. The western mountain and desert states have leaner road systems, with an even sparser population and few settlements. Many western roads surround or serve as connectors between urban areas.

The arterial road system also has different characteristics in different parts of the country.[2] Eastern arterials serve high volumes of traffic because of the den-

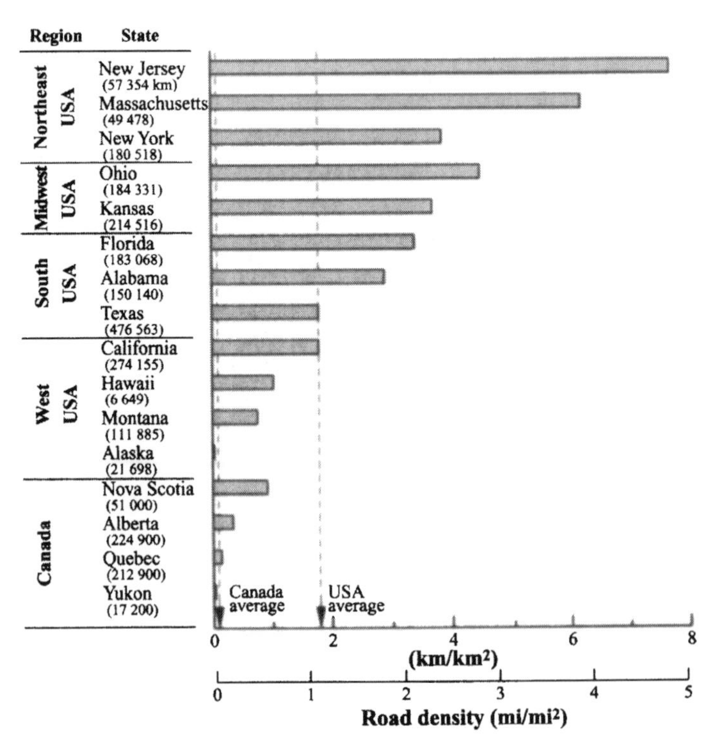

Figure 2.8. Road densities of selected states and provinces in North America. Total road lengths (km) in a state or province are in parentheses. Adapted from FHWA (1996c).

sity of overall development. Western arterials serve less traffic on average than their eastern counterparts. The higher traffic speeds of western arterial roads tend to overcome the vast distances needed for personal and commercial travel.

In general, rural networks are less dense than urban networks, carry lower traffic volumes, and have weaker connectivity and higher average speeds. In the rural eastern USA, rural roads tend to follow natural boundaries, whereas in much of the West, rural roads follow legal boundaries set by surveyors. Road systems in the section grids of midwestern and western agricultural lands tend to be extensive and geometrical and to serve minimal traffic. Relatively wide roadside rights-of-way were granted along the grid lines, which in places may be significant wildlife corridors.

New low-density roads in some sprawl areas (e.g., serving five-acre ranchettes and hobby farms) often have different features, including wider roadbeds, greater density (compared with older rural developments), and better connectivity to higher-speed arterials and multilane roads. These public roads may also connect to private roads. Suburban road networks are characterized by dense traffic, wide arterials, and limited-access freeways.

Road Surfaces

Of the 6.3 million km (3.9 million mi) of roads in the USA, about 3.65 million km (2.3 million mi) are paved (Figure 2.7). About 5% of the paved roads use Portland mix concrete. Virtually all of the rest use asphalt (bituminous) pavement, which typically is a mixture of 92% aggregates and 8% asphalt glue. The mixture is heated to about 175 degrees F (79.4 degrees C) and then rolled and cooled.

Initially, most asphalt was harvested from lakes in Trinidad and Venezuela, but since the early 1900s it has been produced at oil refineries as a petroleum product. About 500 000 tons of asphalt are produced each year in the USA. In addition, 100 000 tons are removed annually from roads, usually for repaving surfaces, of which 80% is recycled.[268, 274] Roughly 65% if interstate highways have an asphalt surface.

Roadsides, Bridges, and Other Engineering Structures

Roads are built within legally defined rights-of-way, with paved surfaces normally placed in the middle of the right-of-way. Most of the U.S. lands surveyed by the federal government west of the original eastern lands provided for a 66 ft (20 m) right-of-way along section lines. These rights-of-way became the paths of least resistance for most road construction in the Midwest and flat farmland areas of the West, and still account for a significant portion of the roads in these regions.

In the West, the Washington State Department of Transportation reports that it manages 100 000 acres (40 000 ha) of roadsides along its 7000 mi (11 300 km) of state highways (out of a total road mileage of 79 428 in the state). This is about 14 acres of roadside per highway mile (3.5 ha/km). The California Department of Transportation indicates that it maintains 230 000 acres (93 000 ha) of right-of-way along 15 000 mi (24 000 km) of highways, a ratio of about 15 acres per highway mile (4.2 ha/km). In contrast, roadsides are normally narrower along smaller roads, in mountainous areas of the West, in rural eastern parts of the USA, and in urban areas.

Apparently no detailed calculations are available for the total area of road surface, roadside, road right-of-way, road corridor, or simply road in the USA. The National Research Council (1997) reported that U.S. highways, streets, and adjacent rights-of-way (sensu roadsides), which included paved ways, roadsides, and medians, totaled roughly 20 million acres (about 8 million ha), based on an average 5 acres/mi (1.4 ha/km) times 4 million miles (6.25 million km). That is equivalent to the surface area of South Carolina, or 1% of the total U.S. land area. The figure "12 million acres of highway corridor" in the USA has also been mentioned.[393] An estimate that about a fifth of the U.S. land area was directly affected ecologically by the road network used the 1% figure for total area of roads.[306]

To make a rigorous estimate of total road surface and roadside areas, the following information should be useful: (1) total length of public roads = 3 950 000 mi (6 250 000 km), (2) 20.5% of road length is in the West (New Mexico to Wyoming and Alaska), 41% in the Midwest (Texas to Nebraska and Michigan), 17% in the South (Florida to Arkansas and Virginia), and 18% in the Northeast (Kentucky to New York and Maine),[280] (3) over 75% of U.S. roads are in rural areas,[674] (4) government land surveys in the Midwest and in large areas of the West provided 66 ft (20 m) width for the road surface and roadside, (5) minimal two-lane road width = 20 ft (6 m),[268,2,3] (6) rural roadsides in the West and Midwest are considerably wider than those in the South and Northeast, and (7) roadsides in rural areas are considerably wider than those in urban areas.

Using this information, and making some rough estimates of the percentage of urban and rural roads in different regions, and the width of rural and urban roadsides in different regions, provides some insight into the total area of roads in the USA. There may be about 15 million acres of road surface and 12 million acres of roadside, totaling nearly 27 million acres of (public) roads or road corridors. This suggests that roughly 0.6% of the USA is road surface, 0.5% roadside, and 1.1% road or road corridor. While these rough numbers are generally consistent with the estimates cited above, they are no more robust than the estimates. None of the estimates includes private roads and unofficial and non-public roads on public land.

Barriers along Roads

In addition to roads and roadsides, the road system includes many engineering structures. These include concrete barriers, guard rails, and noise barriers, considered in this section, and bridges, culverts, and pipes, discussed in the following section.

Barriers separating lanes on roads, usually constructed of concrete, are called *Jersey barriers* in the USA.[2] Some are used for traffic separation on freeways and multilane interstate highways, and many are temporary, portable units used to enhance safety in construction projects. The original height of Jersey barriers was 1 m (32 in), but in special situations barriers as high as nearly 1.5 m (57 in) have been used to block headlight glare from opposing traffic. Jersey barriers hooked together to form a long wall are formidable obstacles to movement by most wildlife. Animals may become trapped or disoriented by traffic or headlights on the road surface and therefore are subject to high roadkill rates. Apparently, only limited testing of various Jersey barrier designs with holes for wildlife and water movement, reminiscent of scuppers on shipboard, has been done, so such barriers are scattered and rare on North American highways (Chapter 6).

Horizontal *guard rails* of various sorts on posts along roadsides and median strips are widespread[2] and are readily crossed by small, medium-sized, and large animals. Heavy-duty "thrie-beam" or three-ridged guard rails used to constrain errant trucks also normally have space beneath and above the rail that facilitates ready movement by wildlife. Right-of-way fences along multilane highways also tend to block large mammals from moving in roadside areas and crossing highways (Chapter 6).

A more recent feature of roads is the *noise barrier*, or sound wall.[272] Noise barriers are designed primarily to insulate residential areas from sound intrusion. Required on some new highways and retrofitted on some existing highways, almost all sound walls in the USA have been constructed since 1986. Two-thirds of the barriers have been constructed of precast concrete and block, and the rest of wood, earth berm, and metal. The typical height is 3 to 5 m (10 to 16 ft), although 4% are over 7 m (23 ft). The federal government has funded 2120 km (1317 mi) of barriers, almost 30% (701 km) of which are in California. Another third are in Virginia, New Jersey, Minnesota, Colorado, New York, Pennsylvania, Washington, Oregon, and Michigan in descending order.[272]

Noise barriers are mainly built in suburban settings. Although walls are major barriers to movement by many wildlife species, wildlife readily use *soil berms* (low smooth earthen ridges), which are generally considered aesthetically more pleasing to the traveling public. Various combinations of soil berms, walls, and below-grade roads can be used. For example, to provide the multiple benefits of aesthetics, wildlife movement, and noise reduction, in suitable areas a roadway 2.5 m (8.2 ft) below grade might be combined with narrow 2.5-m-high soil berms on roadsides.

Bridges, Culverts, and Pipes

The road system in the USA includes 582 976 bridges longer than 6 m (20 ft).[278] These structures principally serve to separate a road from an intersecting waterway, railroad, or other road. Of this total, 83%, or 481 000 of the structures, are over waterways (Figure 2.3) and thus generally help protect the waterways from road crossings, though the bridges still have many specific effects on the waterways. Nearly 30% of the bridges in 1998 were classified as "deficient" (16.0% structurally deficient and 13.6% functionally obsolete).[278] Construction addressing deficient bridges offers a unique opportunity to address and mitigate important ecological issues.

The U.S. Federal Highway Administration estimates that there are an additional 12.5 million smaller structures, mostly box culverts and pipes and mostly in rural areas, which provide drainage from one side of the road to the other (FHWA Office of Bridge Technology, personal communication). This amounts to a structure on average every quarter mile (400 m). Pipes serving as drainage structures generally have a minimum diameter of 60 cm (24 in).

While drainage structures and pipes were generally designed strictly with drainage in mind, many structures also serve as passageways for fish and wildlife. On certain streams, the structures have been specifically designed to facilitate the passage of fish as well as water (Chapter 8). Designing for the passage of wildlife is still uncommon in the USA, though highway structures represent an excellent opportunity to do so. At present, often inadequate information is available regarding habitat areas and species movements near and across roads. With better information and mapping, the existing 13 million road structures could be adapted to the movement of fish and terrestrial animals and new facilities could be designed to greatly enhance fish and wildlife passage.

Conclusion

The lattice of roads is extensive and pervasive. It was built to provide humans with access to every piece of farmland, forestland, park space, and other resource location, and to every home. Overall, the road network is growing slowly, except on urban fringes, where it rapidly expands. In some places—especially in urban and suburban communities, farmlands, and forestry lands—the network is dense.

Many roads are imposing barriers to various forms of life. On the other hand, road networks can include expansive roadsides that serve as habitats, as well as overpasses, bridges, culverts, and pipes, which can facilitate, and even channel, water and species movement. In the chapters ahead, we examine the effects of these roads on ecological systems. We also offer frameworks for making road systems work better with ecological systems.

CHAPTER 3

Vehicles and Planning

> Route 66 expresses who we are, where we live, what we do, and whom we can become.
>
> —Michael Wallis, *Route 66: The Mother Road,* 2001

> Everything in life is somewhere else, and you get there in a car.
>
> —E. B. White, "Fro-Joy," *One Man's Meat,* 1944

Roads and driving are so entwined that the appearance of one leads to the other. More and better roads encourage more driving. Indeed, vehicle and travel saturation is nowhere in sight. We have not come close to reaching the upper limit of travel demand. Suppose we could travel at 10 000 miles an hour and driving were nearly free. Wouldn't we travel much more—to have dinner in Bali from time to time, to see children or parents in other cities more often, to attend a play in New York or London on a whim? Travel is highly valued and highly desired. Individuals limit their travel, though, because it consumes time and money or, in some circumstances, may be uncomfortable or unsafe. Society further limits travel in a variety of ways, including speed limits, fuel taxes, pedestrian zones, and restraints on road construction. Such limitations by society and its leaders reflect widespread recognition that adverse impacts are associated with roads and vehicles.

We do not explore the dynamics of travel behavior in this chapter. Rather, we provide some background on travel trends and behavior and relate those trends to road investment decisions and their ecological impacts.

Automobility and Its Challenges

The proliferation of automobiles is the primary challenge to planners.[674] Once a prized luxury of the rich, autos are available to almost everyone of driving age in the USA. There are now more than 1.1 vehicles for every licensed driver.[280] The automobile has facilitated a lifestyle in which homes, shopping, work, and recreation are dispersed over a wide region. Licenses and vehicle ownership have become modern rites of passage.

Road Driving and VMT/VKT

Americans own 230 million motor vehicles and use those vehicles for 89% of all daily travel and 79% of trips over 100 mi (161 km).[126] Travel by cars continues to grow faster than population or the economy, with more cars, more drivers, and more miles per person each year (Figure 3.1).[674, 675] Travel distances continue to increase for all major categories of highway travel. Between 1983 and 1995, the average commute distance grew from 13.7 to 18.7 km (8.5–11.6 mi), in concert with increasing suburbanization.[204] Leisure and visiting travel are growing even faster. There is no evidence that increasing suburbanization, and the travel growth associated with it, will be soon reversed or replaced by an alternative.

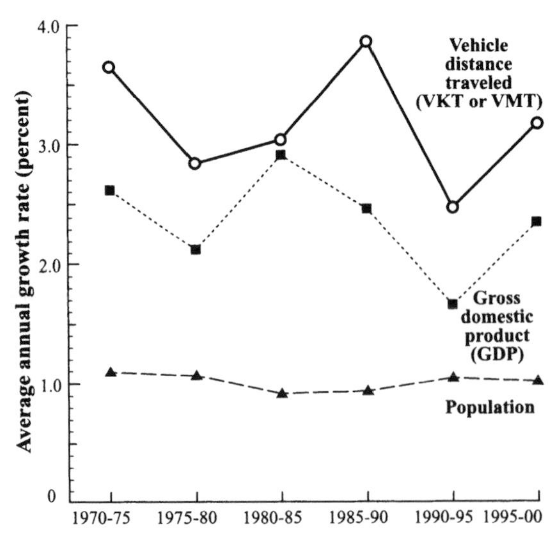

Figure 3.1. Growth rates of population, gross domestic product, and vehicle distance traveled in the USA. VKT = vehicle kilometers traveled; VMT = vehicle miles traveled. Adapted from Davis (2000) and FHWA (2002).

This automobile-facilitated transformation of culture and lifestyle is being replicated worldwide. Though the USA leads the world in such *automobility*, others are following the same pattern. In middle-income industrializing countries, such as China, India, and Mexico, vehicle growth is particularly rapid. For every 1% increase in income in these countries, there is a corresponding 1% increase in vehicle purchases.[202] Freight trucks exhibit similar growth trends.[1052] If trends continue, we can expect considerably more vehicles and travel, in the USA and everywhere else in the world.[178,867]

Travel is commonly measured as distance traveled by vehicles and passengers. In the USA in the 1950s, vehicle travel increased about 4.5% per year.[273] The rate of growth has noticeably declined since, with a growth in the 1990s of <3% per year (Figure 3.1). Despite the declining growth rate, the absolute increases continue to be huge, since the base is so large. For example, between 1950 and 1960, annual vehicle travel (measured as *vehicle miles traveled*, or VMT) grew by 260 516 million mi (419 170 *vehicle kilometers traveled*, or VKT). But between 1990 and 2000, the absolute increase was much larger, over 660 000 million mi (1 062 000 km), even though the rate was declining.

VMT has been increasing faster than population growth in the USA and in virtually all other countries since the advent of cars (Figure 3.1). This increase is the result of people shifting from transit to personal vehicles, driving their vehicles more, and carpooling less. These increases in travel and car use are explained by the following factors.[273,674,204,127,730]

- *Increase in migration to travel-intensive regions.* In the USA, people have been migrating from the East and Midwest, where vehicles are driven less, to the South and West, where VMT per person is about 5% greater.
- *Drop in vehicle occupancy.* With the dispersal of jobs and homes and the shift away from fixed-schedule jobs, carpooling and conventional transit services become less convenient. With smaller families and more cars, single-occupant vehicle travel becomes more common. Carpooling has fallen from 15% of all car trips in 1977 to 10% in 1995, and the number of people per commute trip dropped from 1.18 in 1970 to 1.15 in 1980 and 1.09 in 1990.[728]
- *Lengthening of commute trips.* With suburbanization, commuter trips have lengthened, from 13.7 km (8.5 mi) in 1983 to 18.7 km (11.6 mi) in 1995. Despite these longer distances, travel time is not increasing much, since an increasing share of travel is in less-congested suburban areas rather than in traffic-clogged cities. Average commuter speed increased from 45 kilometers per hour, or km/h (28 mph), in 1983 to 55 km/h (34 mph) in 1995.
- *Drop in fuel costs.* Fuel costs (1996 US$) dropped from 10 cents per kilometer (16 cents per mile) in 1936 to less than 3.7 cents per kilometer today.
- *Increase in number of vehicles per household.* There has been a decline in the number of persons per vehicle as well as an increase in the number of vehi-

cles per household. There are now more vehicles than licensed drivers in the USA.

- *Increase in number of licensed drivers.* The number of licensed drivers continues to increase, due to more women driving, people living and driving longer, more people gaining affluence, and relatively fewer people living in dense city centers.

Future travel growth in the USA is likely to slow. Most women now have driving licenses, and a growing proportion of the population is over 65 years old, an age when people drive less. But because the population and the economy continue to expand, absolute increases in travel will continue into the foreseeable future. This continuing growth in vehicle travel results in more air pollution, road congestion, noise, and water pollution, as well as traffic congestion. The continuing increases in travel tend to undermine technological improvements in vehicular emissions, energy efficiency, and safety.

For many years, rapid increases in travel went hand in hand with rapid expansion and improvement of the highway system. A central focus of transportation was *meeting demand* by the public for more travel, which was accomplished mainly by increasing the capacity of the road network and secondarily by facilitating movement on the system. That no longer is the case. The period of rapid road-capacity expansion is over. Since the mid-1970s, the focus of the transportation community has gradually shifted away from road construction and toward demand management and reduction of adverse impacts of vehicle use.

Public policy now aims to reduce vehicle emissions, encourage more compact suburban development, and restrain peak travel.[675] While the most powerful influence on demand is income, other factors also play a role. Among these other variables are income distribution, status, economic productivity, societal trends such as integration of women into the economy, economic development patterns, presence of domestic energy resources, presence of a domestic auto industry, transport infrastructure investment patterns, commitment to environmental protection, and loan policies of multilateral lending institutions. Governments can and do influence many of these factors.

Despite the growth in U.S. travel, safety has improved. The percentage, and even absolute number, of human fatalities in vehicles has dropped steadily over many years due to stricter safety standards for vehicles, safety improvements on roads, and harsher drinking-and-driving laws. In 1997, there were 42 013 fatalities, 3 383 000 injuries, and 6 624 000 accidents in passenger vehicles.[126] The fatality rate of 4.7 deaths per million miles driven in 1960 dropped to 1.4 in 1997. However, human injuries due to collisions with wildlife are increasing.

Recreational travel particularly concerns road ecologists. Like other types of travel, it has increased rapidly. Automobiles, especially sport-utility vehicles, allow easier access to the seaside, mountains, and other recreational areas, changing both the length and the frequency of such travel. For instance, in Yosemite National Park, where the availability of overnight accommodations has been capped, the number of visitors continues to increase—the result of day-users commuting in cars and tour buses from nearby regions. In fact, weekend and day use for the entire Sierra Nevada region of California has been growing rapidly, in concert with increases in the population of the adjacent Central Valley. Vehicle travel is increasing, even though the capacity of roads serving the Sierra region is not expanding. Traffic is becoming more congested at peak times and is spreading to off-peak times and the "shoulder seasons." Moreover, the increase in all-wheel- and four-wheel-drive vehicles facilitates travel into the mountain areas off the main roads in winter. Portable cell phones and GPS navigation systems also encourage visitation to remote areas.

Off-Road Vehicle Use

Of particular concern is the use of *off-road vehicles* or off-highway vehicles for travel and recreation (Figure 3.2). The use of off-road vehicles threatens sensitive habitats and ecological processes. Unfortunately, data for such use are sparse. Existing data come from state governments, which are required to report to the federal government the amount of fuel used off road. These estimates are used to determine the amount of fuel tax revenue to be provided to government agencies that manage the public lands where this travel often takes place. The estimates of off-road use do not consider private roads, for instance, in private parks and forests. There are four categories of off-road vehicles[205]:

- Management vehicles, such as USDA Forest Service vehicles
- Street-licensed vehicles that go off road
- Registered off-road vehicles, such as trail bikes, dirt bikes, All Terrain Vehicles, dune buggies, specialized four-wheel-drive jeeps and trucks, and snowmobiles
- Unregistered off-highway vehicles in all three preceding categories.

Estimates of the number and use of these off-road vehicles are based on indirect measurements, few direct observations, and many assumptions, and thus are not accurate. Estimates in an Oak Ridge National Laboratory study suggest that privately owned pickup trucks and sport-utility vehicles (SUVs) traveled a total of 36.2 billion km (22.5 billion mi) off road in the USA in 1997.[205]

The U.S. motorcycle industry estimates that about 2 319 500 motorcycles, one-third of the U.S. total, are used off road at some time. These include registered and unregistered dirt bikes, competition dirt bikes and dirt motorcycles, dual-use motorcycles, and highway motorcycles. The Oak Ridge study

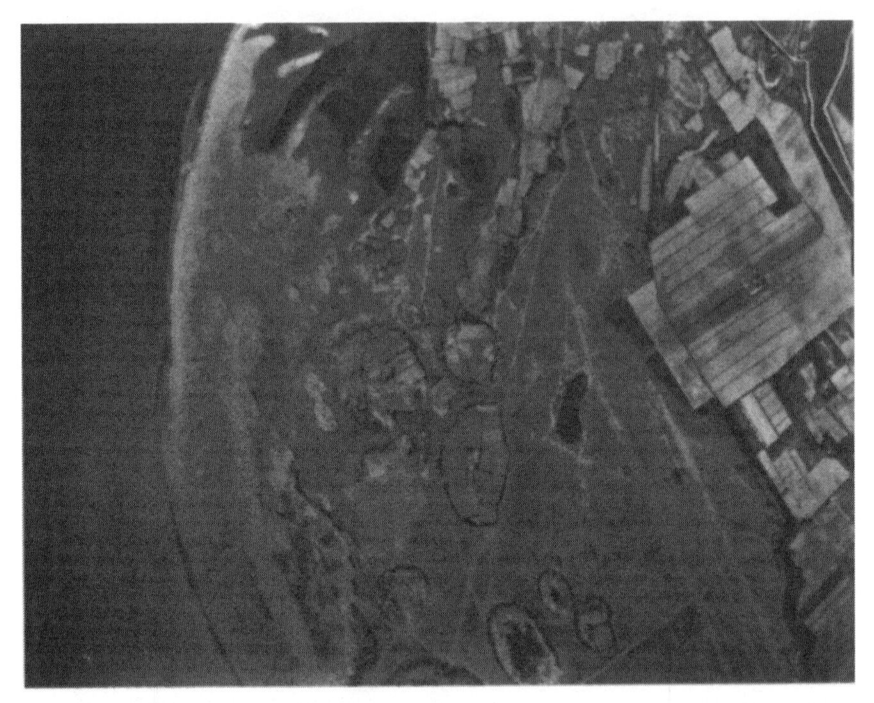

Figure 3.2. Network of off-road-vehicle tracks in marshland habitat. Powerful Mediterranean currents and storms mold the large sand dune area (left); vineyard farmland (right); vineyards (center); and shrub-and-pine habitat (bottom) on isolated dunes surrounded by salt and brackish marsh area. Petite Camargue area, southern France. Courtesy Government of France.

estimates that the average annual miles driven off road by motorcycles was 1321 km (821 mi) over 50.4 days per motorcycle. These data suggest that motorcycles drove some 3.1 billion km (1.9 billion mi) off road in 1997.[205]

More evidence of off-road use comes from the U.S. Consumer Product Safety Commission. Based on a 1989 survey, the Commission estimated that 3.9 million all-terrain vehicles (ATVs) (both three- and four-wheel vehicles) were owned in the USA (60% in the South and the Midwest) and that 27% were used exclusively on private lands. The average number of hours of use per year per vehicle was 170.9, but the distance driven was not estimated.[205]

Snowmobiles are another type of off-road vehicle. There are about 370 000 km (230 000 mi) of groomed snowmobile trails and 1.56 million snowmobiles in the USA. According to a 1992 survey by the International Snowmobile Association, snowmobiles are used 80% of the time for recreation, 15% for ice fishing, and 5% for work. The association estimates that the average snowmobile is driven about 2446 km (1520 mi) per year, which makes for total snowmobile travel of 3.81 billion km (2.37 billion mi) per year in the USA.[205]

Vehicle Size, Fuel, and Technology Trends

Vehicles have become more obtrusive since the early 1980s. Not only are they more capable of operating off road, but they also are larger and more powerful. For all these reasons, they consume more fuel. With oil prices steadily declining in real terms since the 1979–80 petroleum price spikes, and with the federal Corporate Average Fuel Economy (CAFE) standards unintentionally providing vehicle manufacturers with an incentive to switch from cars to light trucks, the car market has been transformed.[674] First, minivans were introduced in the early 1980s, and then in the 1990s, SUVs became popular. In 2002, sales of light-duty trucks (<6000 lb) including SUVs essentially equaled car sales in the USA. Sales and use of heavy-duty trucks to serve freight transport needs also have grown, further increasing total fuel consumption.

The combined effect of bigger, more powerful vehicles and more travel has resulted in large increases in fuel use, virtually all of it petroleum. Today, transportation as a whole accounts for 96.9% of petroleum use and 27.4% of all energy use in the USA[204]. Automobiles, light-duty trucks, and buses consume 60% of all transportation fuel (Figure 3.3). Large trucks use 16.0%, and non-highway modes, such as air, rail, pipelines, and water, use 21.1%. Use of off-road vehicles in agriculture and construction consumes 2.9%. For all transportation modes, gasoline consumption represents 15 303 trillion BTUs (British thermal units) and diesel fuel 5275 trillion.[204] Heavy trucks use

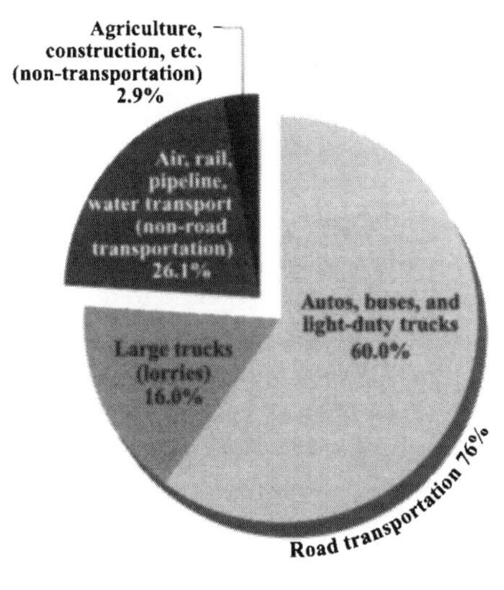

Figure 3.3. Annual petroleum use by road and other types of transportation in the USA. Adapted from Davis (2000) and Bureau of Transportation Statistics (2001).

3619 trillion BTUs compared with 1656 trillion for rail, agriculture, construction, and big ships.

Petroleum fuels retain their dominance in the market because of many attractions, including high energy density and high portability. But they also have major downsides. Increased fuel consumption means that more fuel is being transported in ships, pipelines, and trucks and that more is being stored in large tank farms and at local service stations. More fuel movement and storage means a greater likelihood of spills. Petroleum fuels create serious health and ecological threats when spilled.

Petroleum combustion produces large amounts of carbon dioxide, the principal greenhouse gas, and large amounts of air pollution. Increasing production of carbon dioxide and other greenhouse gases is causing greater climate change threats[428] (Chapter 10). Air pollution is a serious problem, though in the USA government regulations and massive automotive engineering investments in emission control exist. Overall, these efforts have been effective at more than offsetting increased travel and fuel use. However, some pollutants continue to increase, and those that have decreased at the national level remain problems in certain geographic areas (see Chapter 10 and Figure 1.6 in Chapter 1).

The enormous increase in petroleum fuel use, with all its adverse impacts, has created pressure to introduce cleaner alternative fuels. Canada, Italy, Argentina, Russia, and others have made massive investments in natural gas-powered vehicles.[866] Brazil went even further and switched almost totally to ethanol (made from sugar cane) in the 1980s, though it has backed off considerably since. Even the USA has gradually expanded its use of ethanol (made from corn) to about 2% of gasoline sales in 2000 and increasing. But in none of these cases was environmental impact the motivating concern.

Perhaps the more pivotal event occurred in 1990, when California adopted a *zero emissions vehicle* (ZEV) mandate. ZEVs have no combustion products and no evaporative emissions at the car's location. If powered by electricity, they will cause emissions to be generated at the site of the power plant. In most cases, these emissions are less damaging because they tend to be generated remotely and at night (fewer people exposed) and can be controlled more easily in centralized facilities.[865] Resulting emission effects tend to be concentrated rather than widely dispersed. *Battery electric vehicles* also reduce or eliminate many coolants and lubricants, a source of ubiquitous leaks and pollutants. A disadvantage is the need to dispose of and recycle large amounts of often-toxic materials used in the batteries. The principal impact of the ZEV mandate, however, was to ramp up interest and investment in inherently low emitting vehicles. Battery electrics were the first target, but the high cost of batteries has limited their prospects.

Two new technologies now receive the most attention.[865, 674, 675] *Hybrid-*

electric vehicles have a small internal combustion engine mated with an electric motor and battery. *Fuel-cell electric vehicles* have a fuel cell that creates electricity on board the vehicle from hydrogen or some other fuel. Toyota introduced the first mass-market hybrid-electric car in 1997 in Japan. In 2000, Honda and Toyota began mass marketing of hybrid-electrics around the world, and most other major automotive companies have announced plans to introduce their own hybrid-electric models by 2005. Initial hybrid-electric vehicles all operate on gasoline, but more efficiently than vehicles with only internal combustion engines.

The more dramatic change is likely to come from fuel cell vehicles. The vision, articulated in many books and now espoused by major car companies and some U.S. government leaders, is that vehicles will one day be powered by fuel cells that operate on hydrogen produced from water using solar energy, and that emit only water vapor. That vision is far off, though Toyota and Honda have announced that they will sell limited numbers of fuel cell cars in Japan beginning in 2003, and DaimlerChrysler has announced that it will begin selling fuel cell buses worldwide in 2003. Fuel cell vehicles, which have the potential to be the dominant technology, are more environmentally benign. They have zero or near-zero emissions, are more energy efficient than hybrid-electrics or vehicles with internal combustion engines, and provide a pathway to the use of solar hydrogen.

In addition to changes in the fuels and drive systems, additional efforts are being made to alter the size of vehicles for local use. In a number of resort locations as well as urban areas, planners are testing and encouraging the use of smaller vehicles.[865] These range from micro-electric (neighborhood) vehicles in Swiss cantons and U.S. retirement communities to modestly sized two- and four-passenger city vehicles in many cities. These vehicles have special infrastructure needs and designs, such as special right-of-ways and parking privileges.

Many programs exist to encourage walking, biking, carpooling, and public transit to reduce car use in urban areas of the USA and Europe. The infrastructure for pedestrians, transit, and bikes has eroded over many decades in the USA, but has enjoyed some new congressional funding since the early 1990s. After many decades of decline, *transit* (public transportation) provided more than 9 billion trips in 1999, representing the highest level of ridership in 40 years and a growth rate higher than that for VMT on roads.[16] Indeed, since 1995, transit ridership has grown by 20%.

The Planning and Project Development Process

Facilitated by vast investment in roads, travel growth has been explosive almost everywhere in the world. Even today, with slowing expansion in road construction, the USA still spends about US$100 billion per year on roads for new construction, upgrading, maintenance, and operations. Most of these funds are collected from vehicle users through fuel taxes and are placed in a transportation trust fund.

Each country has developed a different approach for determining how much money to devote to roads and how to prioritize the spending of those funds on particular roads and other transportation facilities. In the remainder of this chapter, we examine methods and planning processes used to select and fund road projects in the USA. High points of this complex subject important to road ecology are briefly introduced, but exploring the underlying laws, regulations, and procedures is essential for preparing documents and understanding the processes.

Evolution of the Process

Highly elaborate and increasingly decentralized, the U.S. planning and project selection process is the result of a centralized funding mechanism being imposed on a decentralized system of government. Five "players" interact in diverse ways to accomplish the federal-aid highway program: federal government, state governments, metropolitan planning organizations, local governments, and the public.

As indicated in the previous chapter, the federal government has played a central role in the development of the U.S. road system. Federal involvement ramped up in three steps. First, in the early twentieth century, it became clear that coordination was needed among states since many federally financed roads crossed state boundaries. Then, after World War II, as the federal government became more involved in financing highways that passed through urban areas, it quickly saw the need for coordination by the many local governments involved. For instance, today the San Francisco Bay Area encompasses at least 100 cities, 9 counties, and more than 400 special-purpose local governments such as fire and transit districts. Finally, beginning in the late 1960s, the federal government became more attentive to its role in preserving environmental quality and began to link transportation planning to a broader set of goals.

Federal involvement in transportation planning began in 1921.[353] In 1916, the U.S. Congress had appropriated US$25 million for highway construction and created a formula to apportion those funds to states according to population and postal route mileage.[262] It soon became apparent that states were building roads to varying standards, without coordinating their efforts at state lines. Congress responded with the Highway Act of 1921, which limited federal aid to a designated road system coordinated at state lines and ordered that pavement on new projects be at least 5.5 m (18 ft) wide. In 1934, Congress stepped up its involvement in planning by setting aside 1.5% of highway funding for states to conduct planning and research.[1016] The result was the development of highway mapping, traffic counting, and truck-weighing techniques. Standard practices were adopted that were tied to research and established an analytical basis for designing and selecting projects.

In the 1930s, the U.S. government began providing road financing to cities in dire financial straits suffering through the economic depression.[262] For the first time, federal funds were made available for the construction of urban roads. This initial engagement in urban roads was greatly expanded with the enactment of the Interstate Highway Program in 1956. This act provided 90% of funds for limited-access intercity and urban connector highways (Figure 3.4). The urban highways were particularly controversial. To make room for them, huge swaths of urban land were condemned, causing a storm of opposition in many cities. "Freeway revolts" successfully blocked construction in Boston, San Francisco, and elsewhere and led to new federal requirements for planning and public involvement that have been applied across federal programs and activities. For example, they led directly to an important 1962 act passed by Congress.

Figure 3.4. A four-level highway interchange that replaced a 1950s cloverleaf interchange and resulted from transportation models of a metropolitan planning organization. The design is expected to reduce traffic delays and accidents and to respond to future traffic demand. Interstate Highways 85 and 285, Atlanta, Georgia. Courtesy U.S. FHWA.

The 1962 Highway Act was a landmark event for regional planning generally.[1016] Concerned that cities and counties in metropolitan areas were not working together in planning and building the new (federally funded) interstate highways, Congress required for the first time that each metropolitan area create a regionwide entity to conduct "continuing, comprehensive and coordinated" transportation planning. This "3-Cs" requirement applied to every urban area with a population over 50 000 and included public transit as well as highways. Cities rapidly created *metropolitan planning organizations* (MPOs) for this mission.

Since 1962, Congress has repeatedly strengthened the metropolitan planning process. First, through the 3-Cs requirement, it engaged local elected officials in planning, gradually reducing dominance by the states. Later, it expanded the scope of analysis beyond the initial focus on travel demand to incorporate a wide range of social, economic, and environmental concerns. Then, culminating in the Intermodal Surface Transportation Efficiency Act of 1991 (ISTEA), Congress delegated decision-making authority downward to the metropolitan level. Local officials, coordinating with others in their metropolitan region, now have effective control over transportation spending decisions in their region. Previously, funding was based on formulas and was highly constrained. Now federal highway and transit funds are largely integrated into the same process, and local officials have considerable discretion for how they spend funds.

The process by which major transportation projects come into being in the USA may be characterized in three steps. First, a regional transportation plan is approved that contains a particular road project. Second, the project is scheduled for funding as part of a group of projects. Third, if a project involves a major federal action, approval of the appropriate environmental documentation is secured.

The process has two distinct components—rural and urban. The rural process is state oriented but must involve a broadly based analysis, views of local officials, and public involvement through public hearings or other forums. Typically, states hold annual public hearings at various locations statewide requesting public input on future transportation investment programs. Most often, more requests are made for new or improved facilities than available budgets can fund. State transportation commissions—or in some instances, state legislatures—make the decisions regarding which proposals will be advanced to project development. Ecological considerations are not typically a major consideration at the planning stage.

The transportation planning process created since 1962 in response to federal government requirements now dominates, so that virtually every major transportation infrastructure investment passes through this planning process. All transportation improvement projects with federal funds, which include almost all roads except neighborhood streets, must originate in the detailed transportation planning process described below. But even those transportation projects without federal funding also generally pass through the process, because the projects are in regions violating air quality standards (requiring conformity determinations) or because the sponsors (local or state governments or private companies) want the projects to remain eligible for federal assistance should the need or opportunity arise.

Local neighborhood roads generally do not pass through this elaborate regional process of metropolitan planning organizations. These roads are built on the outskirts of existing areas or as in-fill within the areas. Typically, developers apply to local governments for approval of their subdivision plans, of which road proposals are a significant component. The approval process is entirely within the purview of the local government. However, if wetlands are involved, the U.S. Army's wetlands permitting process comes into play, not only for the roads but also for any part of the development that would involve the filling of wetlands. The wetland provisions are so strict that most developers will avoid wetlands if the process will be invoked. Exceptions occur, for example, in coastal areas where upscale resort development is proposed and the high costs of mitigation can be absorbed.

The urban process, which originated in the 1962 Highway Act, is highly formalized and requires the concurrence of local elected officials for every urban transportation project (highway and transit) proposed for federal assistance.[1016] Local officials act through metropolitan planning organizations, which are designated by a state's governor for every urbanized area (areas over 50 000 in population) in the country. There are currently over 380 MPOs in the USA.

In this process, proposals for additions to the highway and transit systems of an area are analyzed and examined (Figure 3.4). If found to have merit, proposals must be formally approved by the MPO as a part of a long-range (20-year) system plan, which includes all of the existing and proposed major routes in an area. At a point well before actual construction (generally six years), an individual major project must be programmed by inclusion on the official statewide and, if appropriate, metropolitan area program of projects. Major project proposals are subject to much more rigorous evaluation in the project development process, administered under the provisions of the National Environmental Policy Act.

The metropolitan planning organizations typically have an executive director and staff who report to the local elected officials on the board of the MPO. The MPOs monitor highway and transit travel conditions; track population, employment, and land-use changes; and model and forecast future land use and travel under varying assumptions. Since passage of the Clean Air Acts, especially the 1990 amendments, MPOs in areas not meeting air quality standards have been deeply involved in air quality analysis.[1016] As part of the conformity requirements established in 1990, the MPOs must ensure that regional transportation plans are consistent with the region's plans to attain ambient air quality standards. No analogous requirement yet exists to ensure that transportation plans and projects are consistent with regional quality for other key environmental concerns, including hydrologic conditions, water quality, wildlife movement, and biodiversity.

The metropolitan transportation plans and programs developed and approved by the MPOs include all projects, both capital and operating, proposed for federal funding. Plans typically cover a 20-year forecast period and include 6-year investment schedules for projects. The approved plans and programs of projects must be financially constrained to funds shown to be available from those government bodies or agencies making transportation investments.

In other words, locally elected officials, acting through the metropolitan planning organization, determine which federally assisted projects will be implemented. This single provision of federal transportation law that delegates responsibility to regional metropolitan planning organizations, dating from 1962 and strengthened in subsequent legislation, has greatly reduced transportation controversies. The formalization of environmental reviews further reduces transportation controversies by ensuring a process for considering environmental impacts. The 1990 requirement for conformity between transportation and air quality plans, together with federal requirements for environmental impact statements (or EIS; discussed later in this chapter), provides a process and framework for environmental reviews. It allows locally elected officials, serving on the MPO board or not, to act through established MPO procedures in reviewing and participating in project designs and investment plans. Road ecology issues in a metropolitan region would normally be addressed in the planning and NEPA/EIS process. If the environmental consequences of a project are found to be significant, locally elected officials can exercise their voting power to block projects at least until the time when approval to acquire right-of-way land for roads occurs.

In many U.S. metropolitan regions, decisions on the locations of most major highway projects have essentially already been made. Thus road ecology concerns have much to do with mitigation of environmental impacts, management of the existing infrastructure, local land-use planning and projects, sprawl outside the urban fringe, and regional planning that includes the sprawl and nearby rural areas. These critical activities are largely beyond the focus of MPOs. Consequently in these arenas, state environmental quality laws plus the environmental interests of local governments, in addition to state departments of transportation and state natural resource agencies, loom large. This milieu lends itself to significant progress in some states but to roadblocks in others.

Transportation Planning and Air Quality Models

A vast amount of effort is invested in planning and analyzing transportation projects. More than perhaps any other public planning process, transportation planning relies on an elaborate set of standardized mathematical models, complemented by considerable detailed survey data.[1016]

In rural areas, however, the process is less formalized and elaborate, relying on vehicle counts collected by various means along roads. Such data, collected since the 1930s in many areas, provide a database of travel growth trends by highway type and land use. To estimate future patterns for rural areas, highway planners review historic trends in traffic volumes, work with county officials to estimate land-use trends, and estimate possible induced travel based on the nature of a proposed project. Planners then make calculations of traffic levels for the forecast period, generally 20 years.

In contrast, elaborate planning and forecasting tools are used in urban areas. These models, which date back to 1950s Detroit and Chicago, were created to provide an analytical basis for forecasting demand and determining where to site new freeways being planned for those metropolitan areas, and have since become institutionalized.[980] The tools are now used by practically every major city in the world, including those in developing countries. How this came about is rather simple: when the U.S. 1962 Highway Act imposed the 3-Cs requirement (discussed earlier in this chapter), the models developed in Detroit and Chicago were the only credible ones available to forecast travel demand in corridors of a metropolitan area and to analyze the travel impact of proposed road projects.

The models became known as the *urban transportation planning system* (UTPS). They comprised four sets of models—trip generation, trip distribution, mode choice, and route assignment—which were usually run in that sequence. Travel data for the models came from extensive household surveys. Household members were asked to record the details of all trips for a previous day or sequence of days. The travel data elicited in the home interviews were assembled and incorporated as inputs into the series of four UTPS models. This four-step modeling system is used to forecast travel within corridors of the urban area based on anticipated changes in land use and population. The evolution of this elaborate modeling and decision-making process corresponded, perhaps fortunately, with the growth of the computer age.

Land-use inputs are essential to these modeling exercises. In practice, the transportation demand models were constructed and operated quite separate from land-use analyses. Forecasts of land-use patterns were conducted independently and then incorporated into the transportation forecasting models. This process essentially treated the relationship between land use and transportation as cause and effect. In the 1990s, public interest advocacy groups began questioning this approach, arguing that transportation plans and projects cause sprawl. Some of the larger metropolitan areas (with greater transportation problems and more resources for planning and modeling) have begun to integrate land-use models with the transportation models so that potential interactions can be highlighted and analyzed.

In addition to the problems of modeling the connections between land use and transportation, state-of-the-practice models also are relatively insensitive to impacts of small- and medium-scale projects and to most transport-policy initiatives. The fundamental problem is that the models were developed initially to site and analyze a major new highway project, and thus are not well suited to the types of transportation projects and policies now under consideration in most areas. Research initiatives are under way to develop a new generation of transportation modeling procedures, but significant change is not likely in the near future.

The shortcomings of the current urban-transportation-planning-system models have become more glaring with the new federal requirements that regional transportation plans be in conformity with regional air quality plans. The conformity requirement was first enacted as part of the U.S. Clean Air Act Amendments of 1977 and was greatly strengthened by subsequent amendments in 1990 followed by ISTEA legislation. Conformity is the mechanism by which federal funding of road projects can be formally linked to air quality goals. The conformity process requires that new transportation facilities or activities not lead to new violations of air quality standards, not increase the frequency or severity of existing violations of standards, and not delay the attainment of standards. The U.S. Environmental Protection Agency and Department of Transportation jointly administer the conformity process.[947,281] In conformity analyses, vehicle emissions and air quality models used by air quality planners are linked with the UTPS models.

Many argue that the models are not accurate enough to estimate the air quality impact of new transportation facilities. In fact, the air quality and emissions models are known to be subject to large errors, just as the transportation models are. As a result, the conformity process is founded on a shaky analytical framework. Recognizing the validity of the modeling concerns, analysts have devised approaches that recognize the modeling limitations. For example, the pollution burden is calculated for a specific proposed future transportation system. Then a limited set of changes is made to the transportation system model that characterizes a limited number of projects to be evaluated. The models are run again with everything else held constant and the pollution burden again calculated. In this manner, the change in the predicted emissions associated with the projects can be crudely estimated.

No regional requirements or modeling approaches yet exist for road ecology issues other than air quality. Thus hydrologic conditions, water quality, wildlife movement, biodiversity, and other key factors at the landscape or regional scale currently must be addressed piecemeal, project by project, through the NEPA/EIS process. Furthermore, most of America's road system is rural (Chapter 2), outside of the 380 or so municipalities with 50 000 people or more and an MPO. Indeed, the bulk of the U.S. road system (80%) has very low traffic volumes of 400 veh/d (vehicles per day) or less, and 70% of the system is on local roads. These roads would mainly fall through the cracks of the current planning process, even if the process could be extended to encompass the extensive road system covering rural lands. New planning and project development approaches may be needed to effectively address road ecology and landscape ecology issues for society.

Project Development, the Environment, and NEPA

The National Environmental Policy Act is the key legislation available to address road ecology issues other than air quality. NEPA-related provisions are much broader than transportation alone and are administered by the U.S. Departments of Agriculture, Commerce, Defense, Interior, Justice, and Transportation, plus the Council on Environmental Quality, Environmental Protection Agency, and Federal Emergency Management Agency.

Fifty-four federal laws related to NEPA provisions have been identified by the U.S. Department of Transportation and were printed in the May 25, 2000, *Federal Register* (pp. 33978–33979) of the U.S. Congress. These spell out the federal responsibilities that must be addressed in the NEPA process whenever applicable to a proposed project or action. The laws are grouped into five categories: (1) 6 laws in the category of individual rights; (2) 13 in communities and community resources; (3) 13 in cultural resources and aesthetics; (4) 10 in waters and water-related resources; and (5) 12 in wildlife, plants, and natural areas.

Thus NEPA does not exist in isolation. Many laws, federal agencies, state agencies, and legal precedents combine to determine the effectiveness of the process. An introduction to these important components here precedes a closer examination of the NEPA process and its relationship to local land use.

Transportation Projects and the Environment

The enactment of the National Environmental Policy Act in 1969 and the Clean Air Act Amendments of 1970, along with subsequent amendments, requires that transportation plans and project designs take into account environmental factors.[1048] All projects coming from either the statewide or the metropolitan planning process that involve a major federal action are subject to NEPA.

The National Environmental Policy Act guidelines require early coordination between project developers and state and federal environmental resource agencies. NEPA guidelines and agency regulations also require that public notice be given in the *Federal Register* when a project that is expected to have environmental consequences begins the project development stage. The lead federal agency must file a *purpose and need statement,* which specifies the area in which the project is to be built, the project's general alignment, and the purpose for which it is being proposed.

The enactment of the Intermodal Surface Transportation Efficiency Act in 1991 placed further importance on the purpose and need statement by making the majority of federal-aid funds available for either highway or transit investments, with local officials acting through the metropolitan planning organization having the responsibility for initiating the option. The purpose of an urban project generally cannot be to fulfill a highway or transit need but, rather, must be to fulfill a transportation requirement. This assures that the transit option is examined as a part of every project proposal. The environmental analysis, building on the planning process, must demonstrate that the selected project best meets the transportation requirement. For example, this transportation planning process in the Portland, Oregon, metropolitan region led to a project that combined light rail, highway, and local road development (Figure 3.5).

Figure 3.5. Shops and offices in a new planned community surrounding a light-rail stop on the outskirts of a city. The residential, "environmentally friendly" community begins just behind the photographer, and the easily walkable light-rail platform is near the trees visible in the distance. Portland, Oregon. Photo by R. T. T. Forman.

In addition, ISTEA specified that, in air quality non-attainment areas, transportation plans must be in conformity with air quality plans and new transportation investments cannot be made if they will lead to a violation of air quality standards. As part of this conformity deliberation, ISTEA does not allow new lanes to be considered unless it can be demonstrated that *high-occupancy–vehicle* (HOV) lanes will not work. ISTEA not only leveled the playing field between highways and transit but initiated a preference for high-occupancy–vehicle lanes over general-purpose highway lanes.

The early coordination of the NEPA process usually involves a meeting of federal and state resource agencies. A state's department of transportation (DOT) begins the meeting by describing its proposal, including an outline of the purpose and need for the project and, to the extent developed and known, the expected environmental issues. The state DOT staff also describes the inventories and analysis that will be undertaken for each environmental resource (human and natural) that might be adversely affected by the proposal.

The NEPA process requires that all feasible alternatives be examined along with a "no build" option. This typically translates into four to six options being subjected to full and equal analysis after a much greater number have been screened for possible consideration. The environmental resource specialists consider the adequacy of the methodology proposed by a state, and plans are made for periodic review of the work as it progresses. The methodologies for evaluation of some environmental resources, such as air quality, are well defined. Other resources, such as endangered species, require field inventories to determine whether there is a problem. If there appears to be a problem, a mitigation plan must be developed for approval by the appropriate state resource agency. Departments of Transportation in large states have teams of environmental specialists, whereas smaller states often rely on consultants. Even in large states, however, it is not unusual to employ consultants to work on complex environmental issues.

In 1966, the U.S. Congress enacted a strict standard for the use of parkland in highway construction. The Transportation Act states that any proposal using federal-aid highway funds may not take parkland if there is "a feasible and prudent alternative." A series of court decisions has resulted in a very strict interpretation of this phrase, so that the only circumstance in which parkland may be used is where a proposed highway must cross a linear park. The same law also restricts the use of historic and archaeological sites for highway construction.

The NEPA Process Up Close

The *NEPA process* is the most important process currently in place to ensure that ecological impacts of roads are considered. The process for reviewing environmental impacts of road projects is based on NEPA's requirements plus joint regulations of the Federal Highway Administration (FHWA) and the Federal Transit Administration (FTA).[947,1048,281] Over 90% of transportation projects nationwide are determined by a state department of transportation to have minor or no impacts and are called *categorical exclusions.* These projects are not reviewed. Another 3% to 4% are labeled *environmental assessments,* for which brief reports are filed by a state department of transportation with a *finding of no significant impact.* The remaining projects (5% or less), which are determined to have a potentially significant impact on the environment, must have an environmental impact statement prepared before they can be funded.

The review process for a project subject to an EIS has six key steps: (1) notice of intent, (2) scoping and coordination, (3) draft EIS, (4) public hearing, (5) final EIS, and (6) record of decision.

When the decision has been reached to prepare an environmental impact statement, the FHWA and FTA (both parts of the U.S. Department of Transportation) work with the respective state department of transportation to prepare an official *notice of intent* for publication in the *Federal Register.* This notice advises the public and other agencies that an EIS is being contemplated for a particular project. The FHWA works with state and resource agencies in a *scoping and coordination* process to develop an appropriate scope of work for the environmental analysis.

A draft *environmental impact statement* may follow the notice of intent by six months to a year or longer. The development of the project to solve a particular transportation problem follows a process prescribed in part by federal laws and regulations. In this process, the state department of transportation—working with the FHWA or the FTA (or both), environmental agencies, affected local jurisdictions, and the public—investigates alternatives to solve the problem and analyzes how well the alternatives would satisfy the defined project purpose and need. The evaluation considers the affected environment (both human and natural) and analyzes and balances the various resources affected, along with any potential impacts to those resources. The evaluation also looks for ways to minimize the impacts to the environment and analyzes the potential impacts of those mitigation measures.

When the FHWA division office in a state is satisfied that all of the information needed to meet federal requirements is presented in the draft environmental impact statement, it approves the draft for circulation for comments by the public and other agencies. The state department of transportation then publishes an availability and public hearing notice and, after a 30-day period, conducts a public hearing. The draft EIS itself is available for comment for a minimum of 45 days. The process uses the draft EIS and the public hearing to provide interested parties with the latest information regarding the project and also to give the public, interested parties, and other government agencies the opportunity to provide information that can aid development of the project.

The state department of transportation, working with the FHWA/FTA, analyzes and considers all substantive comments received on the draft environmental impact statement as well as comments from the public hearing. The department uses this input to select a recommended alternative, which is described in the final environmental impact statement. This recommendation is made in compliance with applicable environmental laws, taking into consideration that the various environmental laws protect different resources. Balancing resources and impacts to those resources must be considered. Other decisions to be made include identifying mitigation measures planned for the proposed project. This information is reflected in the final EIS. A final EIS can be prepared no sooner than 15 days after the public hearing has been held and must be approved by the FHWA/FTA. Once approved, it is made available for review through a notice published in the *Federal Register.* The availability notice will indicate where the final EIS can be obtained or examined.

A *record of decision* (ROD) is the official document presenting the federal agency (FHWA/FTA) decision. Only after the final environmental impact statement has been available for 30 days can a record of decision be approved, and only then can any future project actions or approvals be made. The record of decision (1) states the basis for the decision; (2) identifies all of the alternatives considered, the resulting selected alternative, and whether or not the selected alternative is the "environmentally preferable alternative"; and (3) states whether all practicable means to avoid or minimize environmental harm from the alternative selected have been adopted and, if not, why they were not. The record of decision also indicates what environmental mitigation measures will actually be incorporated into the proposed project; those measures then become a legal obligation of the state department of transportation.

The NEPA process also encourages coordination with other environmental agencies that need to issue an environmental permit or grant some kind of approval for the project. In this way, information necessary for such a permit or approval can be addressed and included in the environmental documentation. This helps avoid duplication when gathering and presenting decision-making information.

Local Land Use and NEPA

The 1962 Highway Act, in mandating a "continuing, comprehensive, and coordinated" transportation planning process, required that transportation plans be coordinated with local land-use plans. This much-amended requirement still recognizes the impact of transportation policy on land use as well as "the consistency of transportation plans and programs with the provisions of all applicable short- and long-term land use and development plans." Local governments, acting through metropolitan planning organizations, provide the land-use, demographic, and employment data plus the forecast statistics. States then rely on the land-use plans of metropolitan areas as the basis for all metropolitan transportation planning.

A thorny issue often arises. Arguing that transportation plans and projects cause sprawl, environmental organizations increasingly challenge the purpose and need statements. Federal natural resource agencies also argue that federally assisted transportation investments serving new suburban development are inappropriate. Transportation planners respond that they have no other choice according to the law than to serve the approved land-use plans of metropolitan areas.

Today, transportation-planning funds are the only federal resource helping metropolitan areas consider the impact of future development. When the infrastructure on which regions grow—water, sewer, utilities, roads, and so forth—is considered, only transportation facilities receive federal investment (Figure 3.4). Thus transportation projects become hot buttons—and NEPA, the judicial mechanism—for those unhappy with the way urban areas are developing. Because NEPA is a process law, most successful challenges to transportation occur because of process violations. There is no body of law or legal standard to judge the quality of development or of ecological conditions on the land.

At the same time, federal aid for transportation planning can assist state and local officials in sound planning. Some states have established a degree of oversight over local development, through such measures as adequate public-facilities (water, sewer, roads, schools) ordinances, smart growth policies, urban growth boundaries, bikeways and walkways, and so forth (Figure 3.5). The plans for federally assisted transportation therefore must be consistent with, and may enhance, these local land-use plans and policies.

Conclusion

Vehicles and roads continue to expand. In the USA, road expansion has slowed even while vehicle use continues to increase. The result is heavier use of roads over longer periods of the day and vaster expanses of the road system. Increasing vehicle use at the urban periphery and in rural and off-road areas is creating greater ecological degradation.

An elaborate set of institutions and planning methods has been created to determine exactly how, where, and when transportation funds will be spent. In the USA, the planning process explicitly considers environmental impacts via federally required environmental impact statements. Overall, these analyses are haphazard and fragmented with respect to the road ecology issues highlighted in this book.

The models and analyses are neither haphazard nor fragmented with respect to air pollution. As a result of increasingly aggressive requirements, culminating in the 1990 rules requiring conformity between metropolitan transportation and air quality plans, an elaborate set of protocols and institutions was put in place to ensure that air pollution is fully considered in any transportation expansions.

The development of air quality models and the associated conformity process is perhaps a useful approach or "model" for addressing other key ecological objectives—in essence, to create an effective mesh of transportation and ecology for society. For over 20 years, transportation planners and air quality specialists have worked together in both system planning and project development to ensure that transportation investment and air quality requirements are not only coordinated but mutually supportive. This accomplishment also highlights the huge technical challenge to create a solid analytical and scientific framework for decision making.

Many ecologists are working in the transportation field, but few work in areas of transportation planning besides air quality. Most ecologists focus on project development, where the specific impacts of transportation proposals can be and are assessed. This crucial work underpins the transportation project–related NEPA process. Others work on topics related to the management and operations of transportation facilities, where appropriate practices to minimize ongoing negative impacts or to specify practices that benefit the environment are important.

Transportation planning and road ecology share many parallels. Both approach their disciplines broadly. Transportation planning considers all modes—existing and proposed—and also encompasses land use, travel behavior, and social, economic, and environmental consequences. Such planning occurs when the options and the opportunities to recommend transportation policies and practices most beneficial to the environment are the greatest.

Like transportation planning, road ecology is, by definition, interweaving. Multiple systems—such as watersheds, air sheds, species habitats, migration routes, and species protection—are treated in an integrated or holistic manner. To the degree that the ecology of an area—be it a watershed, landscape, or political jurisdiction—is understood, an informed dialogue between transportation planners and ecologists can occur and offer the most potential for positive interaction.

Pilot ecological management efforts undertaken in the 1990s under federal sponsorship have demonstrated that significant opportunities exist for positive interaction and that major benefits to transportation and the environment are possible. Transportation planners and ecologists need to work together to foster more rational and sustainable land-use practices. Land consumption for development, particularly on the urban fringe, drives the demand for auto-oriented transportation. It also devours valuable natural resources.

PART II

Vegetation and Wildlife

CHAPTER 4

Roadsides and Vegetation

Bumping across the ruts . . . without benefit of filling stations, or numbered roads, or even "Blue Books," those "horseless buggies" drew behind them, across the face of the prairie, the first string of a network of roads whose strands were to increase in width, number and strength, and enmesh an ever-increasing number of foreigners, both plant and animal.

—May Theilgaard Watts, *Reading the Landscape of America,* 1975

THE STORY OF A ROAD. *First—a foot path* with sugar maple, lady ferns, trilliums, ovenbirds, foxes. *Soon—a country lane* with hawthorn, wild plum, bracken fern, violets, brown thrasher, woodchuck. *Presently—a country road* with apple trees, white pine, elderberry, rabbits, wild strawberry, meadow larks. *Then—the edge of the country* with silver maple, Norway maple, Norway spruce, dandelion, robin, deer mouse. *Then—the edge of the city* with cottonwood, catalpa, box elder, plantain, starling, house mouse. *Finally—the city* with tree of heaven, English sparrow, Norway rat, pigeons, German cockroach, African violet.

—*Reading the Landscape of America*

Most car drivers stare blankly through a trapezoid of glass to reach a destination safely and efficiently. The narrow bottom and wide top view through a windshield mainly frames the highway surface beneath an expanse of sky. To the left and right are slivers of roadside almost touching up front, and above them loom the adjoining land uses. These triangular slivers normally fade out of the driver's mind, as a void against which everything else is sharpened. Roadsides seem constant, endless, the most boring objects on earth. No one looks clearly at a roadside because, it seems, there is nothing to see.

Yet passengers in the car can see the world differently. Rather than seeming like two little triangles of void, side windows open up a panorama, indeed a moving roadside show. To understand the show, consider the lives of plants, the leading characters. These prominent green forms reveal complex lives, growing, waving, making colored flowers, attracting bees and butterflies, competing with neighbors, being sucked by bugs, getting eaten by hairy mammals, absorbing mineral nutrients from the soil, pumping up water to the sky, and dying in place to become black organic matter, an ending that benefits the plants that replace them.

Reading the Roadside

So let the show begin. Rather than reading the surrounding landscape,[1007] let us read the *roadside* (verge) with this first field guide. Clues to understanding nature abound in roadsides, as illustrated in Figure 4.1. Indeed, a landscape detective or roadside detective sees many more, including the following:

- *Line of low scrawny plants within 0.5 or 1 m (2 or 3 ft) of highway pavement.* "Tough," disturbance-resistant species that survive coverage by chemicals and particles, soil compaction by tires, and scraping by maintenance equipment.
- *Variegated roadside with varied colors and tones.* Management is for nature, with fertilizer, herbicides, and other chemicals absent or nearly so. Probably plenty of colorful butterflies and other pollinators present.
- *Wildflower patches with distinct borders and one or two species in flower.* Planted.
- *Bare sandy or rocky patches.* Contain a relatively different set of plant species.
- *Ovoid depressions in the outer roadside.* Probably ephemeral pools (e.g., vernal), with a distinct set of species that thrive because the pool dries up for part of the year.
- *Scattered shrubs or shrub patches.* An area with relatively high bird density, diversity, and nesting, plus cover for ground animals.
- *"Transparent" tree planting or vegetation* (without a shrub/sapling layer). Much lower wildlife density and diversity than in natural woodland.
- *Older trees present.* Contain mosses, lichens, bark insects, and birds that feed on bark insects.
- *Holes in older trees.* Hole-nesting birds and mammals probably present.
- *Bird boxes on posts or the back of signs in the outer roadside.* Bluebirds, kestrels, or other species enhanced by roadside management may be present.

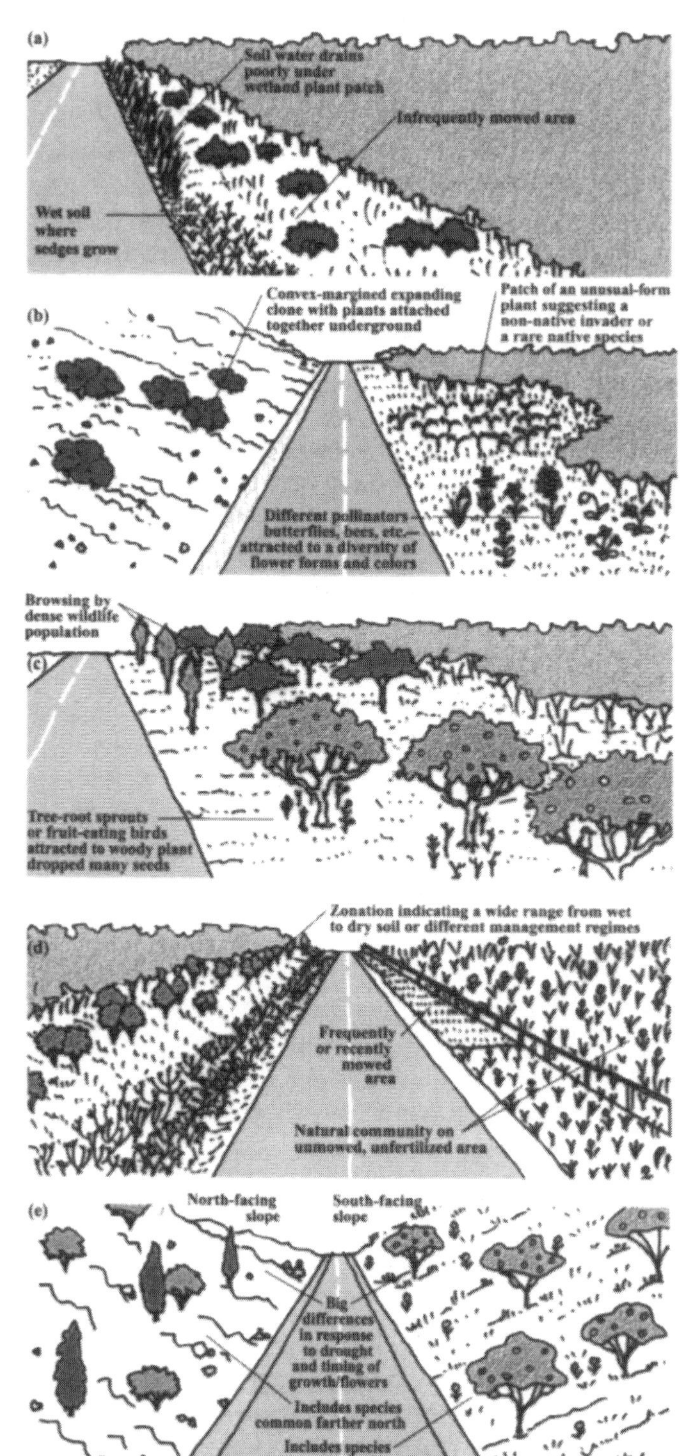

Figure 4.1. Reading the roadside for ecological clues.

- *Surrounding landscape of intensive agriculture or built area.* Roadside may harbor almost the only natural vegetation.
- *Tall plants lining a fence in open rangeland.* Species that escaped the mowers on one side and livestock on the other.
- *More luxuriant growth of a species next to the road in dry country.* In part due to rainwater irrigation from the road surface.
- *Spherical dead-looking plant rolling across the road in dry country* (and often caught in fences and ditches). Tumbleweed, a non-native plant spreading its seeds (see book cover).

Read the roadsides on your way to work. For longer trips, have a passenger take this roadside field guide along. How many of the clues listed, and how many new ones, can be detected? (Be sure to drive extra cautiously; road dangers lurk where roadside clues abound.) Landscape detectives delight in the cascade of discoveries invisible to everyone else.

Types of Plants

Now let us look more systematically at the vegetation. What are the main types of plants typical of roadsides? How does the vegetation change over time? And how is it spatially arranged, both at a location and along a road segment? Of course, variations are infinite over millions of roadside kilometers or miles, most more or less open to the sun, so here we highlight only predominant patterns.

Botanists begin by labeling most roadside species as *flowering plants,* which means that they have an internal tube system for transporting water and chemicals and that they produce flowers with visible reproductive organs.[941,942] Most species are *herbaceous,* that is, without woody stems, and have growing cells at the tip of the stem, so cutting the plant is serious. *Grasses,* on the other hand, have growing cells at the base of the stem, so mowing the plant simply stimulates more growth. *Annuals,* which go from seed to flower to death in a season, are often present, but *perennials,* which grow year after year, dominate roadsides.

Ecologists see *generalist species* rather than *specialists* in roadsides. These plants have a wide genetic tolerance to environmental conditions, such as moist to dry or cool to hot.[59, 856, 923] Most of the species are "light demanding" and would grow poorly or get outcompeted in dense shade. Roadside species are commonly *disturbance tolerant,* which reflects a history of surviving or benefiting in the face of repeated disturbances such as mowing.[724, 1]

Natural *plant communities,* assemblages of mainly native plants that commonly grow together, are usually recognizable in roadsides.[241,815,942] However, in heavily polluted or intensively mowed sites, or where non-native (also called exotic, introduced, or alien) species are prevalent, natural communities are degraded or unrecognizable. Non-native species are widely distributed in roadsides, especially near sources such as built areas. Although many exotic species present in roadsides are uncommon there, a few—including kudzu-vine (*Pueraria thunbergiana*) in the southeastern USA and purple loosestrife *(Lythrum salicaria)* farther north[1036]—are invasive and form conspicuous roadside *monocultures*, in which one species covers the ground.

Natural communities differ widely in the typical number of species present. In comparison, plant communities in roadsides may be less diverse because of the loss of sensitive native species that are inhibited by roadside conditions. On the other hand, non-native species that have colonized or been planted add to the total number of species present. However, non-native species may have limited integration into the food web. The extreme alternative to the typical natural community is a monoculture. Most monocultures in roadsides are invasive exotics, planted or oft-mowed grasses, or native wetland species.

Roadsides are packed with common species. Rare species for a town or county are often present in roadsides, though few studies exist.[337,356] State- and nationally listed rare plant species are sometimes present yet are probably scarce in roadsides (Figure 4.1). Roadsides by unpaved roads may harbor rare species, whereas common species predominate by paved roads bathed in nitrogen from traffic.[1077] In short, roadsides cover an enormous area and contain a large percentage of the total flora of a landscape.[1008,897] Nevertheless, a distinct subset of plant types dominates the roadside view from a car.

Plant Succession over Time

Road construction creates a long bare-earth surface where *plant succession*, the sequence of vegetation over time, starts from scratch.[1064,856,923] The surface is commonly planted with annuals to reduce erosion. Irrespective, spores or seeds of algae, mosses, lichens, and flowering plants are promptly blown in by wind. Except for local spots, flowering plants with a root network and overlapping leaves normally form a green cover within weeks. During that period, however, the earthen fill remains especially subject to erosion.

Seeds germinate, sending down anchoring roots to absorb water and mineral nutrients from the fill, which essentially has no *humus*, fine black organic matter, to hold nutrients near the surface. Sun and wind can easily dry out the surface layer, or runoff water can easily wash it away. The plant dies either way. If the seedling survives, a rosette of leaves may form just above the surface. Photosynthesis traps solar energy, and roots grow deeper. Stem and leaves then shoot upward, forming dense vegetation, which sharply reduces the chance of raindrop energy causing erosion.

In natural succession, most of the germinating seeds make tiny rosettes of leaves, so initially plant species richness may be high.[59, 923] When expanding rosettes bump into one another, competition favors the fittest plants. Species richness (number of species) drops, so at the first stage of essentially complete plant cover, often only a few species predominate. These are rapidly growing, "sun-loving" plants. Most tend to be weedy annuals that quickly produce flowers and wind-dispersed seeds, the kinds of plants that spring up in almost any disturbed spot of bare earth.

Plant succession in the roadside commonly proceeds to a cover of perennial herbaceous plants.[941,942] In dry climates, these may then mix with or be replaced by shrubs. In moist climates, the perennial herbaceous cover is often replaced by shrubs, tree saplings, or both, and then by trees. However, individual spots within the roadside progress at different rates, so that at any point there tends to be a fine-scale mosaic of plants in different successional stages.[316]

From the initial low-diversity stage of annuals, plant species richness typically increases progressively.[856,59,723] More species colonize, the mosaic of stages becomes richer, and species disappear slowly. Diversity occasionally remains constant or even drops a bit, when one persistent species, such as a particular perennial grass or woody plant, tends to predominate. Irrespective, soil formation and the buildup of dead organic matter during the early stages of succession provide significant stability to the roadside ecosystem. Also, shrubs and trees provide greater shade and a less variable microclimate.

Roadside management activities such as mowing alter this successional sequence in diverse ways, as discussed later in this chapter. For example, to reduce erosion, new roadsides are often planted with a monoculture of rapidly growing grass or a grass-and-legume mixture. This shortens the window of time when raindrop energy can directly affect the soil and cause erosion of surface particles.

The cover of planted grass produces shade and cools the earthen surface. The plants accelerate the soil formation process by shedding dead leaves and roots to produce blackish, dead *organic matter,* like humus in the garden. In turn, this soil organic matter is especially effective in holding nutrients that plants absorb while growing. The presence of shade ensures that of all the seeds blowing in, only the *shade-tolerant species* thrive in colonization. Also, colonizing species must be able to compete with the root network that becomes established first. If the planted grass is an annual, its roots and stems die within months, so the colonizers that survive the initial grass covering have a great head start on later colonizers.

Animals form important *feedbacks* with plants during succession.[856,653,923,776] For instance, fruit-eating birds and other animals drop seeds of woody plants, which colonize and add richness to roadsides. Thus, the more berry plants there are, the more the fruit-eating birds come, which in turn may lead to more berry plants. Or masses of showy roadside flowers attract hordes of bees and butterflies. More bees and butterflies pollinating more flowers means that more flowers and more seeds appear, which in turn may lead to more roadside flowers.

Other herbivores eat leaves and stems. Unlike a florist shop, nature's vegetation has holes in its leaves, a "stamp of approval" by chewing insects. More insects may attract more birds and other roadside animals, which in turn relish the insects. Fewer insects may then mean fewer birds, which in turn could lead to more insects.

More conspicuous feedback effects on succession are produced by some mammalian herbivores. Nothing eats at random: all herbivores select preferred foods.[653, 776, 923] So deer faced with a cornucopia of roadside plants target the delicious or nutritious species, a practice that favors the spread of less palatable species and may also change species richness. If herbivores mainly feed on a dominant species, plant diversity may actually rise as other species colonize the newly available space. However, if feeding targets delicious rare species, diversity may drop. Of course, if herbivores are extremely dense, much vegetation will be eaten and some plant species may be eliminated.

Spatial Patterns of Vegetation

Perhaps most conspicuous is the *vegetation zonation*, which appears as a series of zones or strips[1026, 653] laterally from a road surface to the adjoining land-use type.[941, 942] Next to the road surface, soft shoulders are rare. Shoulders atop the roadbed commonly get compacted by vehicles and scraped by heavy maintenance machinery. Herbicides may be used to kill colonizing plants. Thus the typically sparse vegetation cover on shoulders is usually composed of tough, drought-resistant, low-growing plants.[640]

In contrast, the roadside ditch is irrigated by rainwater running off the relatively impervious road surface. Except on a downslope, water also runs into the ditch from the outer roadside and from the adjoining land beyond the roadside. Ditch vegetation therefore is usually dominated by moisture-demanding plants, such as grasslike sedges (Figure 4.1). Wetland plants, including cattail (*Typha*), may form distinctive lines along ditches (Figure 4.2). In dry climates, roadside ditches often may contain rare plants and larger plants.[422] Common plants, such as sagebrush (*Artemisia tridentata*) in the western USA, tend to be more luxuriant.

The outer roadside beyond the ditch, which normally has soil conditions intermediate between the shoulder and the ditch, may have vegetation more typical of the well-drained open sites distributed elsewhere over the landscape. This assumes that mowing has been infrequent, because mowing equipment tends to compact the soil and mowing itself markedly changes the vegetation.[356]

Figure 4.2. Tiny, elongated wetland marsh with cattails *(Typha)* in highway median strip. Note the strip of median trees (top), the extensive area of monotonous mowed grass subject to high levels of winter roadsalt, and the diverse structures associated with highways, including drainage ditch, guard rail, various signs, and cellular-phone microwave tower. Also note the incongruity of a family enjoying nature and cattails by a busy highway (see Chapter 1). Route 2, Phillipston, Massachusetts. Photo by R. T. T. Forman.

Roadside zonation beyond the ditch may be more evident on longer or steeper slopes, reflecting, for instance, a greater range of soil moisture conditions (Figure 4.1d). The type and height of vegetation in the adjoining land also exert a major effect. Surrounding vegetation type affects the kinds and abundance of seeds, animals, wind, and fire that impact the roadside, especially in its outer portion next to the adjoining vegetation.[815] If adjacent vegetation, such as forest, is higher than roadside vegetation, the amount of solar radiation is affected. Forest on the *equatorward side* of an east-west road shades and cools the roadside, whereas with forest on the *poleward side,* the heat level on the roadside tends to be greater (Figure 4.1e). In the outer roadside next to forest, vegetation usually looks quite different than that near the ditch.

Often superimposed on this lateral zonation are small, distinct vegetation patches (Figures 4.1 and 4.2). Some are due to special soil conditions at a microsite, such as inhibited drainage creating a patch of cattail or reedgrass *(Phragmites)* or unusual plants associated with a patch of sand or buried concrete. Other distinct vegetation patches often occur where a single roundish or oblong plant *clone* with attached underground stems, such as sumac *(Rhus)*, remains for a protracted period. Animal activity, fire, or human disturbance at a spot can also produce a small distinct patch of plants within the roadside zonation.

Now consider the roadside vegetation pattern as you drive along a road through a landscape. Large patches pass sequentially, reflecting differences in slope angle, solar exposure, surface length, successional stage, substrate type, adjoining land use, and so forth. Like a movie, these strikingly diverse scenes usually rapidly replace one another along a route. Moreover, linear strips appear like arrows within some patches and sometimes cut through several patches.

Finally, at an even broader regional scale[302] (see quote at beginning of chapter), such as in climbing a mountain or proceeding from city to countryside, astute observation often reveals a *vegetation gradient.* Here the repeated cluster of patch types gradually changes during a trip.

Vegetation

Roads slice through most of the ecosystem or vegetation types in a landscape.[974] Each vegetation type is composed of a relatively distinct set of species and serves as a source of species "raining down" upon the roadsides. Not surprisingly then, the roadside of a road segment may contain a large number of native species. Roads also cut through most human land-use types. These include not only patches, such as farmland and residential areas, but also corridors, including powerlines, railroads, and other intersecting roads. These land uses also have somewhat distinct sets of species, often rich in non-native species, that may colonize roadsides. The net effect is to produce species-rich road segments cutting across the land.

Thus roadsides in the United Kingdom were found to contain 870 of the 2000 plants in the total U.K. flora.[1008] The roadsides of one multilane highway in The Netherlands were found to have nearly 800 plant species, which was half the total flora of the nation.[897, 640] These Dutch roadsides included "very rare" species. Twenty percent (160) of the species were uncommon elsewhere because few if any suitable habitats remained. In comparing samples along the highway with the same number of samples in comparable surrounding agricultural land, the total number of species was slightly greater in the roadside (245 versus 200 species).[897, 640]

However, the fine-scale pattern was quite different. Comparing plant diversity in an average 100-m^2 roadside plot with a comparable plot in surrounding land showed that local diversity was three times higher on roadsides. In other words, an average roadside plot had 32 species compared with 11 species in a non-roadside plot. Thus, while diversity in a road segment parallels, or is marginally greater than, that in the surrounding landscape, local diversity in the roadside appears to be much higher than in the surroundings. (This means high alpha diversity in roadsides and high beta diversity in the surroundings.[1026])

Pieces of this overall picture that are available from several other studies of roadside plant diversity in various countries generally mesh to reinforce this broad pattern.[72, 626, 972, 955, 941, 942, 547, 944] We now need to look more closely at the pattern of (1) vegetation and then (2) plants.

Perhaps the simplest description of roadside vegetation types is visually based: woodland, shrubland, herbaceous vegetation, and wetland. The first three are on well-drained soil, with *woodland* dominated by trees; *shrubland* by shrubs, small trees, or both; and *herbaceous vegetation* by non-woody plants, often up to a meter or so in height. *Wetland* may be dominated by plants of any height but has water at or above the soil surface for an extended time during most years.

The intensively managed areas of roadsides are overwhelmingly covered with herbaceous vegetation. As discussed later in this chapter, this is mainly for driver safety. Shrubland commonly follows herbaceous vegetation in succession, so shrubland is often present on steep slopes, the outer parts of wide roadsides, and where management is less intensive. Woodland tends to be most common in these three situations as well as along roads with low traffic volume. Woodland is also common in locations along roads where *errant vehicles* (those that unexpectedly run off the road surface onto the roadside) are unlikely to reach.

The specific types of vegetation or plant communities vary widely by region or climate. Although no nation has done an exhaustive study of its roadsides, information on roadside vegetation types is available from many countries: Japan,[469] New Zealand,[1043, 941, 547, 942] Australia,[521, 71, 18, 701, 815] the Middle East,[53] Europe,[108, 940] Czech-Slovakia,[813, 435, 504] Germany,[915, 241, 807, 665, 943, 884] Finland,[916] Sweden,[488] Denmark,[384, 385] Great Britain,[54, 717, 1008, 796, 828, 23] Belgium,[903, 1076, 1078] The Netherlands,[1021, 897] France,[410] and Canada.[530] In the USA,[393] studies include Hawaii,[1020] Alaska,[530] California,[337, 123] Oregon,[702, 703] Montana,[937] and the Midwest.[394]

Many other roadside vegetation studies exist, especially from Germany and surrounding regions. Nevertheless, reading a sample of the listed studies will underline the limited and mainly local nature of the research. Systematic studies of whole road networks, or even road segments, are relatively hard to find.[1043, 941, 942]

Woody vegetation normally occurs in narrow strips, as in the outer portion of a roadside, along a fence, and along a median strip. Moderately wide patches are mainly limited to special locations, including highway interchanges, rest areas, and wide median strips. Thus, overall, roadsides include few forest-interior species that require large forest patches. The forest edge is typically wider on the equator-facing edge[987, 317] than the pole-facing edge, mainly because of solar angle. The *mantel* (shrub/sapling layer of the forest edge) is denser on the equator-facing side. Also, the type of vegetation adjoining a roadside doubtless affects the roadside vegetation.[684, 973, 974, 944] Shrub strips are often used to minimize the glare of oncoming headlights and occasionally to absorb the energy of errant vehicles.

Wetland communities, including open marshes and wooded swamps, take many forms in roadsides. Ditches may be thin strips of wetland. Tiny inhibited drainage spots may appear almost anywhere across the outer roadside, for instance, caused by buried clay, rock, or concrete. Tiny wetland spots also often occur around the ends of culverts and pipes (Figure 4.2). Rivers, streams, and intermittent channels frequently cross roads and may have associated wetlands. Streams, rivers, lakes, and estuaries with wetland often border roadsides.

In addition, tiny distinct plant communities commonly develop by roadside engineering structures, ranging from culverts and bridges to runoff drainage structures, noise barriers, snowfences, and guardrails. Microclimatic conditions and sometimes the chemicals leaching from the structures help create these vegetation spots, which cumulatively raise the roadside biodiversity. Also, tiny distinct communities often appear on spots repeatedly missed by mowers.

Herbaceous roadside vegetation that survives mowing and soil compaction tends to be highly resistant to disturbance. It is also subjected to an array of chemicals from vehicle emissions, vehicle parts wear, roads, and such management activities as herbiciding and road salting (Chapter 8).[639] The presence of bare soil patches facilitates plant colonization. Ample sunlight combined with usually reasonable soil quality and the availability of moisture means that growth and succession are relatively rapid. Nitrogen from NOx emissions from traffic may fertilize the roadside.[23]

Two additional types of roadside vegetation are noteworthy: prairie corridors and roadside natural strips (road reserves). *Prairie corridors* in North America are strips of natural, tall, or mid-height grass vegetation dominated by native plants alongside roads and railroads.[541, 759, 655, 394, 827, 777] Relative to the currently widespread herbaceous roadside vegetation, the native prairie corridor appears to contain plant species that compete well against weeds, are effective as an erosion control cover, are tolerant of drought conditions, and require minimal mowing.[777]

Roadside natural strips, or *road reserves,* are strips of natural vegetation along roads. They are found predominantly in the agricultural landscapes of Australia[815, 701, 133, 302, 887] as well as in the fynbos area of South Africa.[207] Most are strips of woodland vegetation. An analogue may be the "beauty strips," or buffer zones, of trees left along certain roads largely to shield the view of clearcut logging behind them. Australian roadside natural strips typically vary from 10 to 40 m (33 to 131 ft) wide on each side of the road. Strips up to 100 m (328 ft) wide occur occasionally, and only in certain areas (e.g., New South Wales) do strips sometimes reach a few hundred meters wide, where there were major livestock-movement routes.[687] Detailed elegant protection and management regimes have been developed and are used by planners, maintenance personnel, adjoining landowners, and the public.

In the intensive agricultural landscapes of Australia, large and medium-sized patches of natural vegetation tend to be scarce, so the natural strips are especially important for maintaining biodiversity. Many native species and natural communities of the landscape are essentially found only in these strips of nature.[72] In fact, the *giant green network* formed by the roadside natural strips is a centerpiece of conservation.[815,302] These landscapes are the most biologically connected heavy-human-imprint landscapes on earth.

Several important ecological dimensions of roadside vegetation remain little known. What is the effect of roadside width? Could there be an ecological minimum or optimum width to guide planning and management? Similarly, what about connectivity? How connected along a road should the roadside be, or does it matter to plants? How much of the landscape ecology literature of the past 20 years on powerline, railroad, hedgerow, and stream corridors is applicable to roadsides?

Plants

An overview of the types of plants in roadsides leads to a more detailed consideration of native, non-native, and invading plants. This in turn leads to exploring the spread of plants, both along roads and from roadsides to the surrounding land.

Types of Plants

The flowering plants covering roadsides are generally dominated by native species but often have many non-native species present and in some sites pre-dominating. The native species tend to be common ones in the landscape, but rare species may also be present. *Wind-dispersed species* (with lightweight seeds carried by air movement) generally dominate the open herbaceous layer, and *animal-dispersed species* (mainly dispersed by birds and mammals) may be abundant in the woody vegetation.[488,971] For example, fruit-eating birds often feed on berries in a forest edge, fly to a nearby perch such as a shrub or sapling or sign, and defecate, thus dispersing the seed from forest to roadside.[610] Annuals are common in roadsides, but most species are perennials. Most perennials can spread not only by seed dispersal but also by vegetative reproduction, where horizontal roots or stems extending outward sprout, thus forming a clone with new attached shoots.

Nonflowering plants in roadsides, including conifers, ferns, mosses, fungi, algae, and lichens, are little studied.[626,942] Some are especially sensitive to chemical pollutants and would grow poorly near busy roads. Nitrogen enrichment of roadside soils would favor a few dominant flowering plants,[23] including grasses, at the expense of these mostly smaller species. Species that grow on bark are rare in the absence of mature roadside trees. In some intensive-agriculture landscapes many of the old trees, which harbor uncommon species such as certain arboreal mammals, mainly persist along roadsides.[994,71,72,73]

Native and Non-Native Species

While walking, bicycling, or driving along many roads, an ecologist often notices the abundance of non-native species in roadsides (Figure 4.3). These "exotic" plants, often spreading as "weeds," are common features of our roadsides.[751] *Exotics* are common in all microhabitat zones from road shoulder to outer roadside. An extensive study of New Zealand roadsides found no significant difference in total exotic species diversity from road shoulder, ditch, outer slope of ditch, and outer roadside, with about 100 to 110 plant species in each[942] (Figure 4.3). Rare non-natives (only one individual of a species recorded) composed about a quarter of the total diversity and also did not change significantly across the roadside. But where do these plants come from? How do they survive and spread? What will they do to the broader landscape?

Every region has its own set of predominant non-native species. In the U.S. Pacific Northwest, roadside vegetation in places is dominated by the non-natives Canadian thistle *(Cirsium arvense)*, false dandelion *(Chrysanthemum leucanthemum)*, or Scotch broom *(Cytisus scoparius)*.[212,702,703] Other introduced species flourish in roadsides in, for instance, California,[337] Florida,[252] Britain,[718] France,[410] and Australia.[357,563]

Residents of Australia, which has a history of non-native species invasions of major economic, ecological, and social importance, usefully differentiate between native and indigenous species. *Native species* are those originating in natural communities on the (Australian) continent, whereas *indigenous species* originated in the region. This seems to be a useful differentiation for plant restoration objectives where a climatically (and culturally) determined region, such as the Upper Midwest or New England, is important. Of course, for special purposes, various subdivisions within a region—such as states, counties, towns, and parks—could be added.[393]

Historically, people have carried certain plants with them deliberately or accidentally along trails and roads.[218] However, the proliferation of people and their expansion into every region on earth over several millennia, and espe-

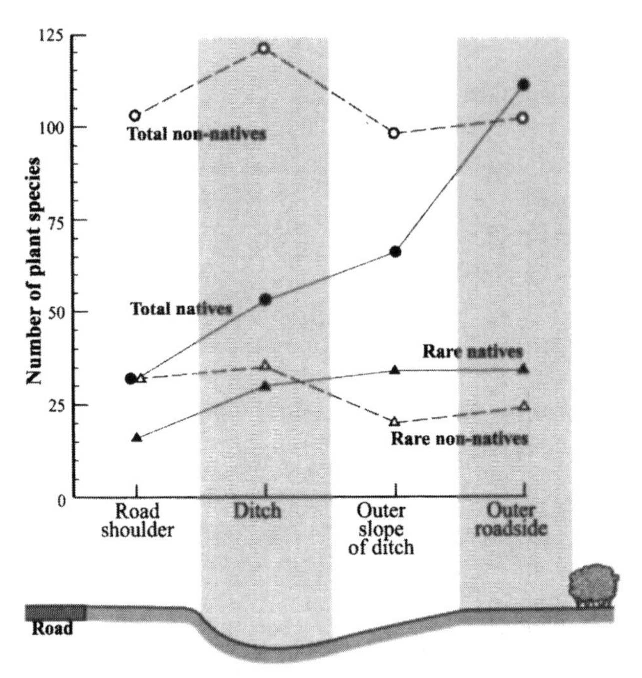

Figure 4.3. Diversity of native and non-native plant species in roadside habitats. Rare species = a single occurrence; the number of plots sampled in each of the four habitat types varied from 99 to 136. Based on sampling 50 km (31 mi) of roadsides along main roads in South Island, New Zealand. Adapted from Ullmann et al. (1998).

cially the past few hundred years, have greatly increased the transport and mixing of roadside plants. Many roadside plants are opportunists, whose dispersal mechanisms may allow them to be carried in the clothing, vehicles, or fur of animals belonging to travelers along roads. The human modification of land and natural disturbance regimes also has no doubt affected roadside plants. Other common roadside plants are escaped ornamental or food plants, whose seeds may be eaten and transported by insects, birds, and mammals.

Most of the common exotic roadside plants in Canada and the USA originate from Europe and Asia, reflecting human migrations and geographical patterns of commerce.[218] Roadsides are commonly seeded with vegetation, sometimes using non-native species, to reduce erosion. Species have been planted for cultivation or ornamental use and then have expanded their ranges. Scotch broom and gorse *(Ulex europaeus),* which proliferate along certain roads in the U.S. Pacific Northwest, are among the most striking of these "escaped ornamentals," having been originally planted in only a few locations.[702, 703] Himalayan blackberry *(Rubus discolor),* originally planted in the same area because it is more prolific and less thorny than the native blackberry, dramatically expanded its range from a few original plantings and now is common along field margins and roadsides.

Introduction also occurs when propagules (often seeds) of a species are included with a shipment of some economic product. Such is believed to be the origin of cheatgrass *(Bromus tectorum),* a common annual exotic grass of western U.S. rangelands, whose seeds were probably included in early shipments of hay or carried in the fur or hooves of imported cattle.[298, 581, 302] A fungus (Phytopthera) responsible for high levels of mortality of incense cedar *(Libocedrus decurrens)* apparently was introduced in shipments of imported wood. The list of such documented and suspected introductions goes on and on.

A successful roadside non-native species is one that can get there, get established, adapt to roadside conditions, and wait for the right conditions to spread.[648] Thus the classic invasive plant has many easily dispersed seeds or spores, has good seed viability, and is adapted to high light and variable soil moisture conditions. The roadside contributes to this process by serving as both a habitat for plants to grow in and a corridor along which the exotic plant may distribute its offspring. Many of the species produce numerous lightweight seeds with an appendage that facilitates dispersal by wind.[37] Some readers may recall their mixed sense of pride and dread at watching a favored child blow dandelion seeds *(Taraxacum officinale)* into the wind.

Roadside non-native plants also benefit from being able to wait in the soil until environmental conditions are suitable for their establishment and growth.[635] Buried seeds of weed species are known to remain viable for over 50 years.[911, 487] Although higher densities of seeds occur in roadside seedbanks, especially for seeds of non-native species,[702] seeds of exotic roadside plants are often found in the soil of the surrounding forest.[738, 479, 442, 391] Since the weedy annual roadside plants in forest sites may remain viable for several years, this increases the chance that environmental conditions will become favorable for plant germination, establishment, and invasion of the forest.

Roadside plants must be able to survive in the environmental conditions of roadsides. Roadsides in forested environments have higher light and more variable temperature and soil moisture than in the native forest. Ten of thirteen exotic species along roads in an old-growth forest in the U.S. Pacific Northwest were significantly more frequent under high or medium light than under low light conditions.[703] On the other hand, wall lettuce *(Lactuca muralis)*, a non-native species was more frequent along streams and under low light conditions in old-growth forest.

Although the total number of non-native species in a road network is similar from road shoulder to outer roadside (Figure 4.3), many more native species grow in the outer roadside. Thus the road shoulder tends to have a high proportion of exotics.[942]

Several ecological properties of roadsides seem to favor non-native plants. In Massachusetts, soil disturbance favored the presence of the non-native Norway maple *(Acer platanoides)* but not the similar native sugar maple *(Acer saccharum)*.[20] On infertile soil in California grassland, ungrazed roadsides had a higher proportion of exotic plant species than did grazed sites away from a road.[810] However, on fertile soil, the opposite was found. Therefore, both soil fertility and grazing level affected the prevalence of non-natives. The total amount of biomass (plant tissue) present and the amount of sunlight or shade also were significant factors. The ungrazed roadsides, irrespective of soil type, had less total plant diversity than did the surrounding grazed grasslands.

Pathways and roadsides with a distinct set of physical and biological properties have been common for a few millennia. Perhaps a set of plants has genetically adapted to roadsides, evolving along with the expansion of the road network globally[218] (E. Menges, personal communication). Irrespective, successful introduced species probably have escaped from many inherent limitations of habitat, predators, or diseases when they reach roadside environments. This implies that the current levels and types of disturbances to which ecosystems are being subjected, especially along roads, may be quite different from the disturbance regime under which native communities (including plants, insects, animals, and diseases) evolved. Indeed, it has been observed that there is nothing in nature that mimics a road.

The Spread of Roadside Plants

Two routes of species movement are considered here: (1) the spread along roads and (2) the spread from roadsides into the surrounding landscape.

Spread along Roads

Several studies indicate that once a non-native species colonizes a roadside, the species can spread along the road. Dispersal of adhesive seeds, such as those with hooks or hairs, is an effective mode of transport.[488] Favored by turbulence of passing vehicles, seeds of common milkweed *(Asclepias syriaca)* in the southeastern USA appear to move long distances to roadside "vanguard sites."[1058, 1059] From these sites, the species then apparently spreads to saturate appropriate open sites in the local surroundings.

In the U.S. Pacific Northwest, mature forest encompassing forestry roads with exotic species had no visible exotic species, but several such species were found in the soil seedbank.[702] The number of non-native species along high-use and low-use forestry roads was much higher than along abandoned forestry roads. Soil disturbance promotes the recruitment of seedlings. In New York State, the non-native purple loosestrife *(Lythrum salicaria)* appears to have spread along a multilane highway by short-distance movements, favored by ditches and culverts.[1036] In a sandy Florida scrub area, the introduction of limestone rock material for roadbeds changed the soil attributes of roadsides, which in turn supported more nonindigenous species (not characteristic of the surrounding scrub vegetation) and more plant cover than on roadsides without the imported roadbed material.[372] In Western Australia, fire in woodland next to a road enhanced the spread of two non-native grass species, which in turn were considered to make the area more fire prone.[634] In Victoria, Australia, wetter and more open forest along roadsides had more non-natives than did dryer and denser forest. Many other authors suggest that roads facilitate the movement of non-native species.[1020, 613, 393, 315, 930]

Based on a handful of studies, vehicles are known dispersers of seeds. In Ibadan, Nigeria, mud was collected from 75 vehicles and placed on moist sand, and the resulting seedlings were counted.[173, 547] Forty plant species were identified. By comparing the list with species found in vegetation sampling of roadsides of the area, most were common local roadside species. However, several, including fig (the epiphyte *Ficus*), were not found in the roadside samples. Seeds carried by vehicles were considerably smaller on average than those of plants growing in the roadsides. An analogous study of the area surrounding Gottingen, Germany, found 124 plant species in the mud from a car driven approximately 10 000 km (6200 mi) on paved roads and 5000 km on "partly unsurfaced" field and forest roads.[822] All but three of the species were known roadside species of the area. The car-carried flora included annuals, perennials, herbaceous and woody species, and species with various seed-dispersal mechanisms. Grass seeds and small seeds were most common.

In Kakadu National Park, Australia, 304 entering tourist vehicles (cars) were vacuumed, and only 10 non-native weed species were found.[563] The frequency of weed species on cars did not correlate with weed abundance (percent cover) in the park, though the car-carried weeds were at three times the number of sites within the park as for other weed species. This suggested that tourist vehicles contribute to weed spread across the park. Four-wheel-drive vehicles, which were more likely to have driven off road, carried more total seeds (presumably mostly native species) than did two-wheel-drive vehicles.

Finally, in Canberra, Australia, seeds were collected from the wastewater of a car-wash establishment and germinated to identify seedlings.[984, 985] A total of 18 500 seedlings representing more than 259 species was recorded. Species from the city and from surrounding croplands and woodlands predominated (Figure 4.4). As in the German study above, the species found exhibited a wide range of biological properties; no clear grouping of species was evident. Twenty species (7% of the total) were not known from the region, and most were considered to represent long-distance transport of species by cars along roads. The frost-sensitive species identified probably traveled at least 100 km (62 mi), coastal species at least 170 km (106 mi), and others from farther afield.

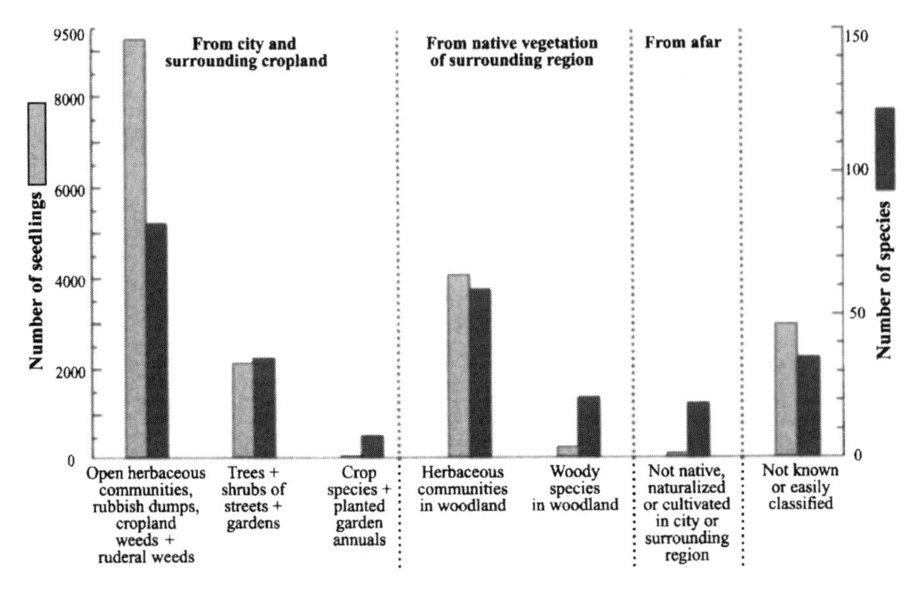

Figure 4.4. Source of plant seedlings and species derived from an urban car-wash facility in Canberra, Australia. For 26 months, seeds were filtered from car-wash water and planted on nursery soil; seedlings were recorded during the period and for 9 months thereafter. Adapted from Wace (1977).

Given the enormous number of vehicles endlessly moving, vehicle transport of seed and the consequent mixing of plants growing along roadsides are doubtless a major process of the road system.[916] The Kakadu park study[563] indirectly suggests that off-road vehicles[1013] may add native species to the roadside mix. The Canberra study[984] suggests that vehicles from homes and built areas add non-native species to the mix.

Other mechanisms facilitate the movement of plant species along roads. Plants with edible fruits may be widely spread by birds or mammals, which consume the fruit, often dropping seeds along forest edges by roads. Insects, including ants, may also collect and move seeds. Nevertheless, the transport of seeds by animals moving along roads or roadsides is probably of limited importance overall (Figure 4.5). Few animals move significant distances along roads by locomotion, and most of the animals that do are predators and scavengers.[302]

In cold climates, roadsalt applied to roads (Chapter 8) may facilitate plant movements of salt-adapted or saline plants.[828, 1, 201] In The Netherlands, salt-tolerant plant species have spread along roadsides as far as 150 km (93 mi) inland from the coast.[955] Also, water is channeled along roadside ditches and streams, so species with water-resistant seeds may spread in this way.[1036]

Figure 4.5. Livestock moving along and grazing in roadside. Note the livestock trails and the relatively high vegetation, which probably indicates low animal density. Northwestern Colorado. Photo by R. T. T. Forman.

Several aspects of wind facilitate the movement of lightweight wind-dispersed seeds along roads (Chapter 10).[18, 796, 1036, 72] High windspeed or the acceleration of wind can lift particles from a surface, such as seeds from a road. Turbulence tends to have an even greater lifting power. Airflow can be channeled along the trough of an open road in forest and accelerated when passing through narrows. Wind diagonally entering such a trough bends and accelerates along the downwind roadside-forest edge. Wind flowing perpendicular to such a road produces downdrafts and turbulence on roadsides. Wind often accelerates and is turbulent around valleys and bridges. Moving vehicles themselves produce turbulence, and a seed, like a paper cup on a busy highway, is sucked along in the direction of vehicle movement.[1058, 1059, 916] Even differences in heat absorption between the road surface and the roadside can lead to airflows in the corridor sufficient to lift lightweight seeds.

These microclimatic wind patterns in roadsides may explain, in part, why there is a close resemblance between the non-native species in roadside *seedbanks* (seeds stored in soil) and the non-native species growing in roadsides.[635, 702] In contrast, for native species, the seedbank and the growing plants in roadsides are dissimilar.

Spread from Roadsides to the Surrounding Land

The expansion of non-native plants in a landscape is a source of concern. Many fear that exotic plants, especially those able to spread along roads (Figure 4.4), may expand from the roadsides and "take over" the landscape. Near roads that penetrate much of the land, introduced plants might modify the way landscapes function, such as by altering water use, fire danger, or soil nutrients. Or they might facilitate invasion by other undesired pests, predators, or diseases. Although these fears are mainly untested speculation, some concrete answers are available. For example, landscapes with many barriers to invasion will be less susceptible than ones with few barriers.

The most likely direction for spread of non-native plant species from roads is into adjacent disturbed environments. A landscape with frequent severe disturbances is more subject to invasion than one lacking such disturbances. Both natural and human disturbances may facilitate exotic plant invasions from roadsides into the landscape as a whole. Streams, which are frequently disturbed by floods, cross roads in multiple locations at bridges and culverts, where introduced species may spread from roads to the stream network (Chapter 7).[703] In steep forested landscapes, many landslides originate from forest roads, and exotic plants may spread from the road along the landslide scar. Cultivated fields, suburban yards and gardens, construction sites, and harvested areas in forests are all examples of human disturbances that adjoin roads and may permit the spread of exotic plants. The spectacular widespread occurrence of Scotch broom in logged clearcuts and landslide scars in the Oregon Coast Range attests to the susceptibility of clearcuts to invasion under certain conditions.[702]

While non-natives are seen to alter roadsides and other local ecosystems, little is known about how they may affect whole landscapes. For instance, tamarisk *(Tamarix)* in streams in the southwestern USA is believed to use more water than do the native species. An introduced succulent plant *(Carpobrotus edulis)* increases soil salinity in certain coastal plant communities of California.[201] A dense shrub cover of Scotch broom and gorse may increase the susceptibility to fire or fire severity in conifer forests of the Pacific Northwest. Also in this region, exotic plants are generally less frequent on abandoned forestry roads,[703] which suggests that management of a road network in a landscape context may help control exotic plant spread.

An important but little-known question is how often and how far roadside weeds or non-native species spread into the surroundings.[1020, 937, 613, 372, 930, 315] In some areas, non-native species invade livestock rangelands, even resulting in weed laws and "weed police" to stop the spread (e.g., in Alberta). Roadsides with weeds are nearby, but little evidence is available that the weeds in the rangeland spread from the roadsides rather than from house lots or built areas.

Similarly, many farmers have weed problems in cultivated fields, and some nature reserves and parks have prominent invading non-native plants.

Available studies give some insight into how far species may spread from roadsides. In three forest communities (schlerophyllous) in Victoria, Australia, an elevated frequency of non-native species extended only about 5 m (16 ft) beyond the roadside of a graded earthen road.[18] A few species apparently showed a higher frequency, extending outward for at least 30 m (98 ft). In Western Australia, non-native weeds after fire in adjoining woodland were found at an elevated level up to about 2 m (7 ft) from the open roadside of a two-lane highway.[634] In Glacier National Park, Montana, species richness of non-native species was elevated in forest plots 1 to 2 m (3 to 7 ft) from the open roadside of a two-lane highway.[937] Forest plots 25 and 100 m (82 and 328 ft) from the roadside were generally low in non-native species richness.

Forest plots near graded earthen roads showed similar patterns. In soil samples taken at 0 and 5 m from roadsides along forestry roads in Pacific Northwest conifer forest, about 95% of the seeds present were exotic species, but at 50 m into the forest only about 3% were exotics.[702] Near a two-lane paved road lined by planted non-native Norway maples in Massachusetts, the seedling density was higher in roadside plots than in forest plots 50 m (164 ft) from the road.[20] An elevated maple seedling density was only present on the roadsides.

Along a 25-km (16-mi) segment of a multilane highway with 50 000 veh/d in the outer suburbs of Boston, 22 roadside locations were present where a planted woody non-native species of seed-producing size was adjacent to a natural forest ecosystem.[314,315] Half of the locations showed evidence of species invasion into the adjacent forest, with the farthest distance detected for invading plants ranging from 10 to 120 m (33 to 394 ft). Finally, dry Florida scrubland appears to be resistant to the invasion by roadside weeds.[372]

In short, in these studies of species spread from roadsides into adjoining land, the farthest distance reported was but 120 m (394 ft). Indeed, the bulk of the species invaded only up to about 10 m (33 ft) from a roadside. More research is warranted.

Roadside Habitats and Animals

A distinctive hawk is commonly seen by the side of logging roads in the tropical rainforest of the Sierra Imathaca in Venezuela. Said to be absent from the dense forest, this species is aptly named the roadside hawk (*Buteo magnirostris*).[88] Animals mainly limited to roadsides are unusual. Rather, in this section we introduce animal diversity in roadsides, types of animals using roadsides as habitat, and the movement of animals along roadsides. Chapters 5 and 6 provide further insight into animal movements around roads, especially for wildlife of surrounding land.

An early comprehensive study[1008] found that roadsides in the United Kingdom contained 20 of the 50 mammal species (40%) of the British fauna. Similarly, roadsides contained 20% of the 200 bird species, 100% of the 6 reptiles, 83% of the 6 amphibians, 42% of the 60 butterflies, and 32% of the 25 bumblebee species in Britain. In The Netherlands, 50% of the Dutch butterfly fauna (80 species) are common species in roadsides.[640] This includes 20 species resident in the roadsides, plus a few rare butterfly species. The corresponding percentages for grasshoppers, cicadas, bees, and ground beetles are higher than that for butterflies. Dutch scientists believe the connectivity of roadsides contributes to these high numbers. In short, where roadside management aims to enhance biodiversity, the overall diversity of animals in roadsides is surprisingly large.

In most nations, recognition of the conservation value of roadsides still barely exists. The Netherlands takes many measures to protect and enhance natural communities in roadsides.[63] In Britain, six roadsides are officially designated as Sites of Special Scientific Interest,[944] somewhat analogous to the most significant state-level nature reserves in the USA. In addition, one roadside has a higher designation as a National Nature Reserve, and two roadsides are Specially Protected Areas, a European Community designation for important species-protection areas.

As a habitat or series of habitats, the roadside is home to species for many different reasons.[137] Many bird species nest nearby and feed in roadsides.[162, 528, 88, 547, 77] Grain-eating birds, but not insect-feeding and other birds, are attracted to roadsides in Punjab, India.[216] Some bird species, including waterfowl, nest in roadsides, especially unmowed roadsides.[693, 779, 1000, 137] In Iowa, the rate of nest predation (eggs or young eaten by predators) in roadsides was related to the type of adjacent land use.[77] Predation was higher next to wooded habitat than by open habitat.

The often-dense herbaceous cover of roadsides is attractive to many small mammals.[225] Also, planting non-native shrubs among scattered native shrubs along Indiana highway roadsides changed habitat conditions for many animals.[779] Rabbit density increased slightly, bird density and species richness rose sharply, and bird nesting greatly increased. No change in road-kill frequency was observed. Another study found road-kill rates of small mammals to be proportional to mammal density in the roadside of a multilane highway.[8]

Designing roadsides for wildlife meshes somewhat conflicting goals. For example, animals want cover, escape cover (in case of danger), visibility, and the ability to cross a road, while drivers want efficient travel without hitting wildlife. In Figure 4.6, the open roadside (a) provides maximum visibility for both driver and animal, and would be best if we designed only for people. With roadside woody vegetation angled toward the oncoming vehicle, design (d) appears second best for the driver. However, the convoluted forest edge

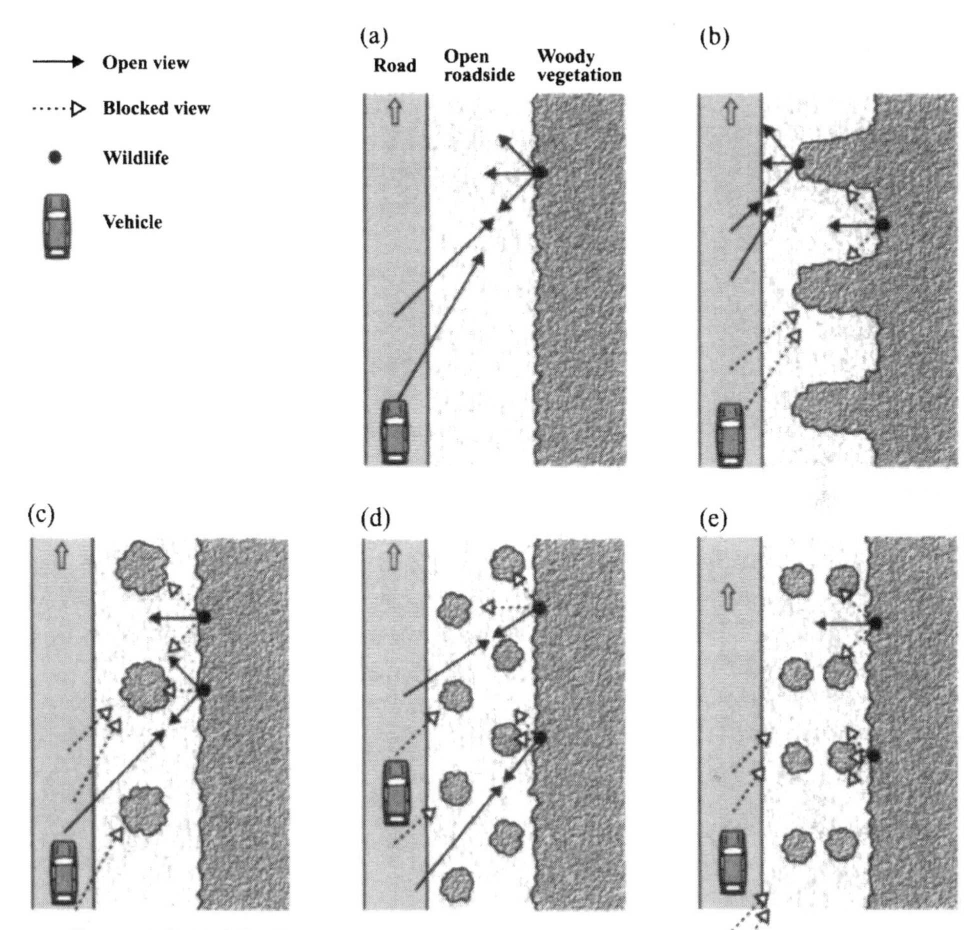

Figure 4.6. Visibility between wildlife and vehicles according to the pattern or design of roadside vegetation.

(b) seems best for all four of the animal's needs. The larger woody patches in design (c) may be second best for the animal. Still other roadside designs may optimize effectively among the societal goals of transportation and biodiversity.

Less is known about the many insect species in roadsides.[583,246,742,973,944] Some insects are observed to retreat to roadsides when nearby agricultural treatments occur.[583] In Iowa, butterfly diversity, density, and road-kill rate were all higher where roadsides were planted with diverse native prairie wildflowers compared with roadsides planted with a brome *(Bromus)* monoculture.[777] The volume of passing traffic seemed to have no effect on breeding moth and butterfly species in roadsides.[661,547] But lots of insects end up smashed on passing vehicles.

One study found that in median strips, small mammal density tended to be highest where unmowed grassy roadsides bordered wooded strips.[7] No difference in density was found between unmowed areas in the median strip and the outer roadside. Also, no difference in small mammal density was recorded between the woods in the median strip and woods up to 400 m (1300 ft) beyond the roadside.

The movement of animals along roadsides overall is limited[302] but presumably is enhanced by the high degree of connectivity of roadside habitat.[72,971,973] Movement at the broad landscape scale, such as ravens *(Corvus)* feeding on road-killed animals along roadsides[496] and large predators tracking at night along remote roads,[302] is discussed in Chapter 5. Rather, the focus here is on species living in or adjacent to roadsides. In one study, small mammals moved at least 28 m (92 ft) along a roadside in home-range movement.[774] The spread of small mammals over a few years has been described for a California desert roadside[431] and an Illinois multilane highway.[347] In the latter case, a roadside vole species *(Microtus)* that could not move through villages and towns was able to expand its range 90 km (56 mi) when a multilane highway with continuous roadside vegetation was built to bypass the towns and villages.

Simulation modeling using data for ground beetles and other data from roadsides adds insight into the dispersal process along roadsides of major highways.[973,974,959,983] The authors concluded that long-distance dispersal was more frequent on wider than on narrow roadsides because fewer animals were lost to adjacent areas (highway and adjoining land uses). If animals lost to adjacent areas are compensated by reproduction, then roadsides may be dispersal corridors for movements of a few hundred meters by these species. If periodic nodes of suitable open habitat are present, the movement of species along roadsides may be enhanced. In North Dakota, the harvester ant *(Pogonomyrmex occidentalis)* has expanded its range several kilometers in 12 years, apparently by moving along roadsides.[215]

Thus, despite the near absence of construction and management practices to enhance wildlife in roadsides, a fair number of species uses them. Many species reside in nearby habitats and feed in roadsides, while some occupy roadside habitats. Some move along roadsides in foraging home-range movements, and a few disperse and spread long distances along roadsides. Presumably, these dispersals occur mainly by walking and flying, though an occasional ride on a vehicle surely occurs.[72] Although woody vegetation in roadsides (Figure 4.6) could be arranged to enhance movement along a road, creating effective habitat and facilitating movement across a road are higher priorities.

Roadside Management

Roadsides in the USA are maintained according to local desires and demands. In the 1930s, it was declared that "roadsides should be maintained as if they were *our nation's frontyards.*"[393] Beautification was added as an objective in the 1960s. Since the 1970s, an ecological approach has gradually grown as an objective. In essence, the trend has been from lawn to grassy strip of beauty to attractive strip of biodiversity. The last phase, which has only begun, translates into replacing "blanket spraying, wall-to-wall mowing, and monoculture seeding" with "reducing herbicide use, protecting ground water, encouraging beautiful native plants, preventing weeds, and providing for wildlife."[393] The broad national trend of thinking sets the framework within which roadside designs and maintenance fit local desires and demands.

Management of roadside vegetation became especially active in the mid-1960s. For more than 30 years, Lady Bird Johnson, wife of former U.S. president Lyndon B. Johnson, promoted *roadside beautification*.[393] Mrs. Johnson was fond of saying, "Where flowers bloom, so does hope." Her efforts resulted in the Highway Beautification Act of 1965, which called for improving landscaping, removing billboards, and screening roadside junkyards.

Ecological goals for U.S. roadsides have in turn focused on reducing soil erosion and sediment flow and, more recently, on planting native wildflowers and reducing non-native invasive plants.[393] In The Netherlands, Australia, and some other European nations, a much richer array of ecological goals and practices exists.[63, 310]

Consider a representative array of broad objectives for managing roadsides:

- Minimize cost.
- Maximize motorist safety.
- Enhance visual quality while maintaining ecological benefits.
- Reduce erosion and sediment flow.
- Control non-native species.
- Enhance biodiversity.
- Enhance wildlife density and reduce the road barrier effect to improve animal crossing of roads.
- Reduce wildlife density and increase the road barrier effect to reduce roadkills and wildlife-vehicle crashes.
- Accomplish a multiple-use array of societal goals.

The first two are single objectives, which can be pursued without the others. The other objectives require optimization and trade-offs among conflicting goals and contain major ecological dimensions. With ecological expectations growing in America, these last seven objectives emerge as worthy challenges for roadside management.

Fortunately, the roadside planner and manager has a wide array of approaches and tools to accomplish goals, including mowing at different times and at different frequencies, cutting brush, pruning, applying herbicides, scraping or blading surfaces, clearing ditches, filling eroded banks, smoothing or increasing the surface heterogeneity of the outer roadside, repairing and painting structures, widening or narrowing the intensive-management zone, grazing livestock, haying, compacting and loosening soil, adding topsoil, fertilizing, planting annuals or perennials, native wildflowers, shrubs, or trees (plus other measures for erosion control), and much more. For many goals, management focuses on decreasing or increasing the results of natural ecological processes.

Washington State provides a useful example of roadside management today in the USA. The state department of transportation divides its highway system into 24 maintenance areas, each with 483 km (300 mi) of roads and two or three maintenance crews. The total maintenance staff for the state is 250 persons, with a budget of about US$9 million. The budget devotes approximately 24% for litter control, 27% for weed control, 33% for clearing, 13% for landscaping, and 3% for other activities.

Washington classifies its roadsides into three zones. Zone 1 is closest to the road and, until recently, kept bare and free of vegetation—often up to 4 m (12 ft) from the road. A herbicide regime is used for this, though it leaves no competitors against invasive weeds and therefore encourages undesirable plants. New guidelines allow for leaving vegetation that is neither a fire hazard nor damaging to pavements, with the goal of reducing the need for herbicides and weed control. Zone 2 is sometimes called the recovery zone, because it is the area in which vehicles are driven or pushed off-road due to vehicle or driver problems. The width of this zone varies according to the speed and volume of traffic, motorist visibility, drainage, and fire control issues, though typically it extends at least 9 m (30 ft) from the highway pavement.[2]

In Zone 2, plant stems should not exceed 10 cm (4 in) in height. While this zone has typically been mowed close to the ground, it is now the intention to mow the zone at a height of 30 to 61 cm (12 to 24 in). Zone 3, farthest from the road, may be intensively mowed or increasingly may be planted in a somewhat natural way, but it normally accommodates utility access, corresponds with adjacent neighbors' wishes, and forms a transition to adjacent lands. Although the state has a huge road network, the management changes encapsulated here could be rapidly implemented to produce a widespread, visible ecological benefit.

The delineation of roadside zones that receive different management regimes is widespread among states and provinces. We now introduce sections on mowing, controlling non-native and invasive species, plantings, and variegated roadsides.

Mowing and Vegetation Control

The prime purpose of mowing along roads is safety. Grass, weeds, brush, and trees can limit a driver's view of approaching vehicles and signs. Lush vegetation can hide pedestrians and bikers from drivers, and vice versa. Therefore, U.S. states have established a policy for the frequency and type of grass cutting to keep the vegetation below a height that would compromise safety. Brush and shrub control is usually performed on a regular basis as well.[370] In addition to enhancing visibility, in places vegetation is controlled to prevent the growth of trees, which could be a hazard to errant vehicles.

The ecological implications of roadside mowing and management are surprisingly little documented. Certainly the natural old-field succession from herbaceous cover to shrubs and trees is arrested, which is an intent of mowing.[707] Furthermore, the establishment, growth, and survival of certain species are either promoted or retarded by regular mowing. Uncommon disturbance-sensitive plants often disappear, whereas grasses tend to thrive.

Roadside mowing requires important decisions about time, space, and intensity.[528,796,707,1,638,596] As in the Washington State example above, mowing intensity varies according to how low the vegetation is cut. This is ecologically important for the relative success of seedling growth, grass dominance, snowdrift accumulation, wildlife habitat, and more.

The key temporal dimensions are *mowing timing* (time of mowing during the season or year) and *mowing frequency* (how often or how long between mowings). Regular year-after-year timing depends, for example, on normal periods of vegetation growth, nesting and denning of animals, flowering of plants, pollinators and pollination of flowering plants, frost periods, and the retention of vegetation cover for wildlife during nongrowing periods. In addition, flexibility is built into a mowing regime to allow for occasional wet periods, droughts, and other less predictable conditions that are relevant to the timing of mowing.

The frequency of mowing has a major ecological effect as well as a cost effect.[796,707,626] Mowing once or twice a year is much cheaper than mowing five to six times a year. It also greatly favors a richness of plants, whereas frequent mowing typically favors a few grasses that outcompete the diverse plant species.

Not surprisingly, the combination of timing and frequency has a major effect on plant diversity and offers the roadside manager a range of options. For example, one study of wide roadsides beside multilane highways found the highest plant diversity by mowing twice a year, early and late in the growing season (Figure 4.7). Mowing once a year either early or late in the season or mowing every other year early in the season resulted in fewer species. The lowest diversity was found with no mowing. However, these results are somewhat at variance with ecological expectations. Theory suggests that frequent human-created disturbance would reduce diversity, late-season disturbance would favor grass dominance leading to less diversity, and natural succession of the herbaceous community in the absence of disturbance would produce a species-rich community.[856,923,776] This example underlines the need for rigorous research in different regions and roadsides, and highlights how little we know about the ecology of roadside vegetation considering the decades of mowing and what we regularly drive past.

The spatial dimensions of roadside mowing focus on the zones across a roadside, sections along a road, and special locations.[637] On wide roadsides, two or three parallel strips can be mowed at different times or different frequencies, resulting in greater vegetation diversity across a roadside. Analogously, alternate sequential segments or stretches along a road can be mowed at different times or frequencies. Special locations, such as mowing relatively steep slopes or fillslopes with unstable soil, may produce significant soil erosion. But in cases with minor erosion, mowing may produce small disturbance spots

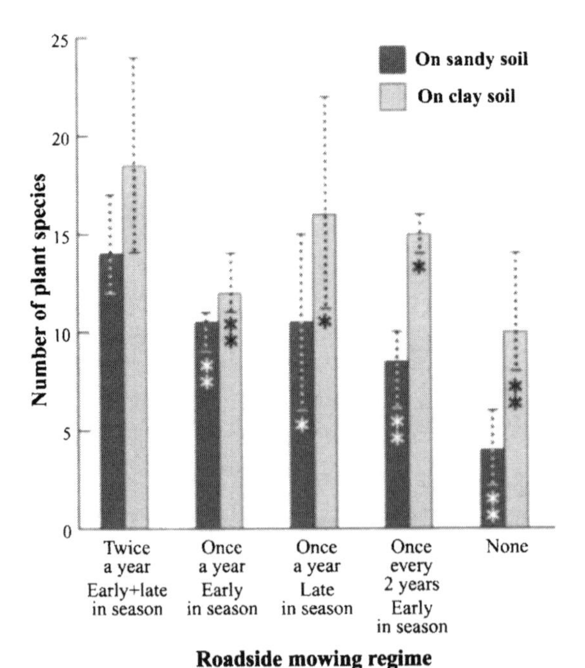

Figure 4.7. Plant species diversity related to roadside mowing frequency and timing. Early in season = mid-June; late in season = mid-September; ⋆ = significant decrease and ⋆⋆ = highly significant decrease relative to mowing twice a year. A few days after mowing, cuttings were removed for composting or animal fodder. Ten-year study in The Netherlands. Sandy soil by highway A1 at Voorthuizen; clay soil by highway A15 at Tiel. Vertical dotted line apparently indicates range of values. Adapted from Ministry of TPW&WM (1994b).

suitable for the colonization or persistence of uncommon species. In short, by combining the temporal and spatial options for mowing, the roadside manager has a rich palette indeed.

At Fort McCoy in central Wisconsin, the threatened and endangered Karner blue butterfly *(Lycaeides melissa samuelsis)* is obligated to spend its larval phase on wild blue lupine *(Lupinus)*, often found growing along roadsides susceptible to mowing (see Figure 14.3 in Chapter 14). Thus the locations of roadside populations of wild blue lupine are marked with stakes every year, and maintenance personnel are instructed to lift their mower blades as they come upon a marked population of lupine. Through this management process, a rare species persists in the roadsides.

Hay can be cut for fodder, compost, or "very clean compost."[344,639] During times of drought in the USA, farmers are sometimes allowed to mow roadsides for hay. The rules and regulations vary from state to state based on a state's own weather patterns, terrain, and legislation. During a summer drought in 2000, at least four states (Nebraska, Wisconsin, Montana, and Georgia) permitted farmers to engage in roadside mowing for a small fee. In others, the department of transportation defined the policy, generally on a county-by-county basis. In many nations, livestock grazes along certain roadsides, thus economically helping both the farmer and the transportation manager (Figure 4.5).

Roadside vegetation in different countries is managed according to differing guidelines. For example, many roads in Australia contain conspicuous roadside natural strips of woodland and large trees relatively close to the road surface.[72,815] The proximity of trees often provides connectivity overhead for arboreal animals, plus shade and a pleasant driving experience for the traveler. Roadsides with natural strips in the state of Victoria contain 25% of all endangered species and 45% of the remaining native grasslands (B. L. Harper-Lore, unpublished review). In Australia, these roadside reserves may be inventoried, assessed, and preserved by a Roadsides Conservation Advisory Committee, an interagency partnership formed in 1975. Roadside handbooks based on state and local efforts plus the committee's work call for many ecologically beneficial activities, including the "avoidance of tidying up vegetation." In some cases, the upgrading of paved roadways—even in a growing suburb—has been eliminated when roadside inventories revealed high conservation values.

Controlling Non-native and Invasive Plants

Mrs. Johnson continued to champion environmental initiatives by helping to add a native wildflower requirement as an amendment to the U.S. Surface Transportation Urban Relocation Authorization Act in 1987. Native plants are typically effective when used for erosion control, landscaping, and maintenance of highway rights-of-way. Different species are adapted to the varied climates and geology that roads cross. Native plants can prosper without fertilizer and, once established, can ward off the invasion of weeds, thereby reducing the need for herbicides.

Road transportation is a critical component in the fight against invasive species because roads can facilitate the spread of plants and animals in the landscape and outside their native range.[962] For example, the non-native invasive multiflora rose *(Rosa multiflora)* was planted along highways in Virginia and elsewhere and quickly spread into natural habitats to become a widespread infestation.[879]

Invasive plants can also be removed during spraying and mowing operations. Weed seed can be inadvertently introduced into roadsides on equipment during construction or through the use of mulch and imported soil, gravel, or grass sod. The use of plants native to North America, rather than those indigenous to a region (see discussion earlier in this chapter), has resulted in California poppies being planted along roadsides in North Carolina, and many other disjointing examples. In the 1980s, few U.S. native plant nurseries existed that could supply the large amount of seed needed for highway projects. But today such nurseries are common in different parts of the country, producing seeds and plants suited for each ecological region. As a result, the use of native wildflowers and grasses along roadsides has noticeably increased. Unfortunately, invasive plant species are sometimes still deliberately planted for erosion control, landscaping, or wildflower projects.

A 1999 U.S. executive order established to prevent and control the introduction and spread of invasive species targeted species that are likely to harm the environment, human health, or economy.[393] The order builds on the National Environmental Policy Act of 1969, the Federal Noxious Weed Act of 1974, and the Endangered Species Act of 1973 to reduce introductions of invasive species, control the species, and minimize their economic, ecological, and human health effects.

In some cases, it seems impossible to eradicate introduced invasive species. For instance, a tiny introduced insect (Balsam woolly adelgid) is slowly killing the Fraser firs *(Abies fraseri)*, a dominant of the spruce-fir subalpine forests of the southern Appalachian Mountains.[746] Apparently, the only effective control is to spray a soap solution on the trees, which is time consuming and expensive. In Great Smoky Mountains National Park, the Fraser fir majestically lined the scenic road to Clingman's Dome,[1007] and soap solution was sprayed from trucks for several years to protect the fir. Finally, because this roadside contained the only mature living specimens of Fraser fir in the park, it was determined that artificially maintaining the doomed trees was not realistic. The spraying was stopped, and now only dead firs line the scenic road.

The 1999 executive order means that federal funds cannot be used for construction, revegetation, or landscaping that purposely includes the use of known invasive plant species. It provides support to control and prevent invasive species. New construction and landscaping techniques and equipment have been designed to help meet the intent of the executive order. These include bio-control delivery systems, equipment cleaners, seeding equipment for steep slopes, burn-management equipment, geographic positioning systems to map existing invasive populations, and techniques to reduce soil disturbance during vegetation management activities to minimize opportunities for introductions. Coordinated research and training also help to develop integrated vegetation management principles. Such trends, plus research to evaluate progress, should help lead to ecologically sound roadsides in the future.

Plantings

Plants in roadsides may serve at least nine important roles[859, 394]: (1) control erosion, (2) lower maintenance costs, (3) provide aesthetic beauty, (4) control snow drifting, (5) reduce headlight glare, (6) reinforce road alignment, (7) serve as crash barriers, (8) reduce wind, and (9) provide wildlife habitat. Most planted plants serve more than one function.[779, 1, 954, 63, 393]

Planting of annual and perennial herbaceous plants to minimize soil erosion is especially common following road construction activities. Seed mixtures and monocultures, sometimes including non-native species, have been planted. Experiments comparing various seed mixtures of natives/non-natives and annuals/perennials have been carried out,[123] mainly to evaluate the ability of the plant to reduce erosion. Accelerating natural succession is also a goal, and using seed mixtures of native successional plants or successional plants with other start-up plants would be promising.[1064] Seedlings and mature herbaceous plants are also sometimes planted for erosion control, especially on more erosive surfaces. Wildflower plantings, usually using native species, are also becoming more common on established roadsides. Again, these could be done to increase native species richness or to accelerate natural succession. Plant communities typical of natural succession in a region are likely to provide the best habitat conditions for wildlife.

Shrub planting is also used to control soil erosion, particularly on steep slopes, such as fillslopes around overpasses. Shrub planting is used for reducing headlight glare and in some places for absorbing the energy of errant vehicles or crashes. Because planted shrubs mainly reach to the height of people in cars, planting and controlling shrubs is especially important relative to visibility, such as on the insides of curves and around intersections. Furthermore, shrubs are at a critical height to provide cover for almost all wildlife species. Thus shrubs serve as important habitats for breeding and as important cover for movement. In particular sites where wildlife should be discouraged for safety reasons, shrub planting may be limited, shrub growth controlled, or low shrub species planted.

Trees relatively close to road surfaces are the rule in most of the road network, 80% of which has less than 400 veh/d.[3] In sites with a high probability of errant vehicles, guard rails can be installed or mature trees either removed or not planted. Alternatively and more ecologically, a clump of shrubs can be carefully planted in front of an abutment, signpost, or tree to absorb some of a vehicle's crash energy. The force of impact on a solid object is decreased by one shrub. Two sequential shrubs further reduce the force, though the second shrub provides less gain than the first. A third and fourth provide additional gain, but at a progressively lower rate. Also, an engineered, energy-absorbing structure could be placed in front of a hard structure and hidden by shrubs. The results are less vehicle damage and cost, fewer injuries and less medical cost, fewer human fatalities, and more wildlife benefit.

Trees and woodland in roadsides relate to a still wider range of issues. In addition to the preceding ecological benefits, tree plantings can reduce windspeed and provide firewood,[129,317] fruit from fruit trees, holes for hole-nesting birds and mammals, habitats for bark insects, mosses and lichens, leaf litter and humus for holding soil nutrients, shade, visual quality benefits, screening from adjacent land uses, and carbon sequestration. Little research has yet been published evaluating tree species, locations, and these diverse important roles of roadsides for society.

Trees and shrubs play a special role in reducing the barrier and fragmentation effect of roads, thus providing connectivity for animal crossing of roads (Figure 4.6). Tree canopies over a narrow road provide for movement across a road by a range of arboreal animals. On a medium-width road, short stretches of roadside trees touching overhead at intervals along the road can provide a series of connections.[750] On wider roads, these stretches of roadside trees on both sides of the road do not touch. Thus animals must cross the road surface separating the trees where arboreal animals are subject to predation and roadkill. Alternatively, various canopy bridges, hammocks, or other structures can be added (Chapter 6). Planting the roadside stretches on opposite sides of a road—with shrubs and trees especially favored for cover, and perhaps food, for native wildlife—should increase the success rate of crossing by flying, arboreal, and ground animals.

Ecology and Visual Quality

Roadside managers may plant or remove vegetation (Figure 4.6) to create roadsides with high visual quality (VQ). Similarly, plantings and removals can block or open up views of the surrounding landscape from the road. Broadly speaking, *aesthetics* means beauty, while *visual quality* also includes views and scenic values, though they both are strongly tied to culture.[850, 670, 237] Methods for estimating visual quality or visual preference include surveys having different groups of people classify photos, take their own photos, or rank computer simulations of scenes. Several quantitative VQ-assessment models, especially developed by government natural resource agencies, use different key variables and therefore give sometimes similar and sometimes dissimilar results.[877]

Most agreement occurs for high visual quality, some for low quality, and typically little for intermediate quality. A high-VQ scene catches the eye, draws the viewer's attention, and therefore has unusual form, color, texture, location, arrangement, juxtaposition, or a combination of these. An individual object may be beautiful and the whole scene ugly, or vice versa.[485] Foreground, midground, and background are often important in judging visual quality.

A detailed visual quality study in Connecticut found that the presence of water or a church was most preferred, an old traditional farm or agriculture with animals was good, and a successional field area was lowest in visual quality (Figure 4.8).[481] In comparing these results from a fairly diverse set of people with results that might be expected from ecologists (Figure 4.8), an interesting relationship emerges. The three highest visual-quality ("best") scenes recognized by the public ranged from high to low in overall ecological quality, yet the three "worst" scenes also ranged from high to low ecologically. Visual quality and ecological quality seem not to be correlated (C. Steinitz, personal communication).[237]

If the Connecticut results are representative, the roadside manager can readily highlight the scattered locations with high values for both ecological and visual quality. But for ecologically sound management along a road segment, most locations would be classified as medium or low VQ. In designing roadsides that include views of surrounding areas,[333] the *view-from* and *view-of* dichotomy is important. People on a highway gain views of the surrounding landscape, while people in the surroundings may gain views of the highway (see Figure 14.4). Providing view-from and view-of visual quality for both sets of people is an important challenge.

To the undiscerning eye, the endless natural vegetation favored by ecologists lacks contrast and therefore is often not considered to be of high visual quality. Most people prefer contrast rather than monotony. Yet even a slight presence of human development is often found to lower VQ rankings, which continue dropping with additional amounts of development.

Two attributes seem to produce considerable agreement between ecologists and the public. Both the presence of water, such as a stream or lake, and the presence or suspected presence of wildlife in a forest increase visual quality.[877, 708] Yet even these can easily be considered to be low visual quality if the

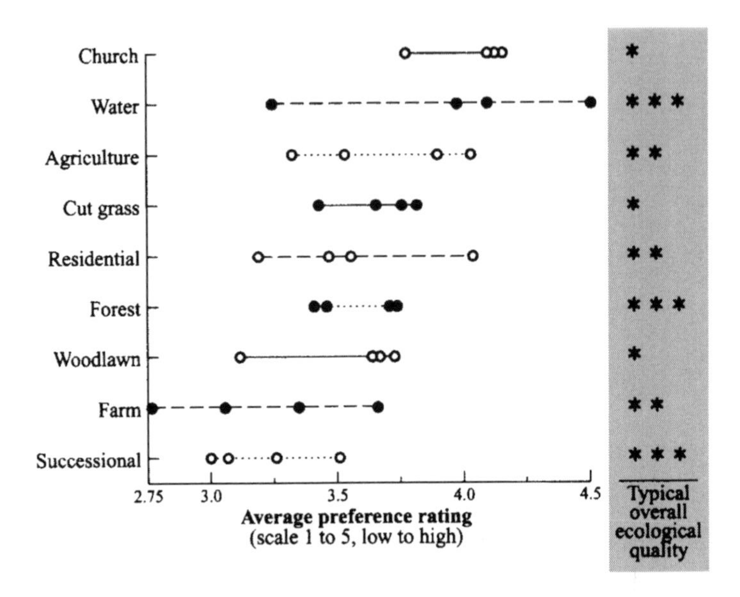

Figure 4.8. Visual preferences of people for nine types of scenes in southern New England. Four different scenes along roads for each of the nine types (36 color prints) were ranked by 249 people (95 highway residents, 27 transportation planners, 127 other citizens). Scenes were ranked on a relative attractiveness scale from 1 (unattractive) to 5 (highly attractive); the overall average ranking was 3.61. Average preference rankings are given for the 36 scenes. Adapted from Kent (1993). Typical overall ecological quality (* = lowest, *** = highest) is suggested by the authors of this book.

lake is polluted, the stream is in raging flood stage, or the forest harbors dangerous animals.

To some, a mowed area looks trim, tidy, and in control, whereas a successional field of tall grass and weeds looks unkempt.[671] To most people, fires, dead trees, or fallen logs are bad, and breaking up a large forest into accessible pieces is good. For most ecologists, the opposite is true.

In the view from a vehicle moving rapidly along a highway, roadside flowers as flecks may become streaks, and objects may dissolve.[333] In a vehicle (unlike a train), the landscape comes right at us. The road goes up and down. The car heads off in different directions. The driver has the sense of being in control. The windshield view is dominated by road and roadside, and perspective lines converge ahead at a point. Squint, and the road is a triangle and the roadside a diagonal wedge in the upper right or left. A side view is like the view from a train.

While driving, the overall visual quality one experiences is the combination of the quality of individual scenes and the order in which they are viewed.[859, 24, 824] Thus the first scene, early scenes, and the last scenes are considered the most important. A view exists in the context of preceding views as well as of the surroundings. Thus the roadside manager faces a quandary. A roadside, like a movie, should hang together, but for how long? 100 m? 1 km? 10 km? 100 km? Or for different scales? Making a road segment visually integrated without compromising essential ecological soundness is a challenge.

Thus, although the subject of ecology and visual quality is not considered further in this book, it remains an important research topic. Cases doubtless exist where the two variables are positively correlated and where they are negatively correlated.[498,360] The roadside manager would welcome solid information on both to benefit the traveling public.

Variegated Roadsides

The preceding trends in the ecological management of roadsides appear to lead to a pattern best described as a *variegated roadside,* recognized by a richness of natural pattern and process and providing diverse benefits to society. Bulging with native plant species, heterogeneous or patchy in surface microtopography, water conditions, and vegetation type, vertically diverse, abundant in animal habitats and movement routes, and serving a raft of societal benefits and objectives, the variegated roadside might be a vision for the future. But let us look more closely.

The book *Nature Engineering and Civil Engineering Works*[1] seems to have been the first synthesis that pointed in this direction. In a land of finite area and intensive agriculture, the authors recognized both that roadsides represent a large total area and provide rather few benefits to society, and that, with good management, valuable alternatives to the endless mowed grassland could be realized and visible to the public. Building on the diverse nature and engineering approaches introduced, we now can begin to visualize what a variegated roadside could look like and the benefits that could be provided.[852,63,613,640,777]

Reducing the frequency and intensity of mowing is a start. But using the range of temporal and spatial mowing regimes described above helps provide a template for open portions of a roadside.[637] The reduction in non-native species and the elimination of invasive species broaden the range of native plant communities in roadsides.

Soil conditions are especially important here.[596] Eliminating fertilization (and reducing the blanketing of nitrogen from vehicles), except in exceptional situations, helps prevent one or two plants from dominating, and minimizes the eutrophication of pools, ditches, streams, and adjoining water bodies. Building up the soil organic matter through natural succession helps retain soil nutrients and sustain a rich soil animal community. Keeping sandy, rocky, and other nutrient-poor spots in a nutrient-poor condition favors the rarer plants that thrive there.

Recuperating the natural surface heterogeneity lost during the former homogenizing and smoothing road construction process is more difficult. This goal can be accomplished in the upgrading or improvement projects and the more ambitious maintenance projects. Microtopographic differences in soil surface are fairly easy to create, and the vegetation is apt to respond rapidly and become remarkably diverse. A more marked manifestation of this goal is the creation or re-creation of pools, wetlands, and other watertable-related changes that produce major increases in habitat types and species richness.

Accomplishing these changes in the plant communities would go a long way toward providing cover and food for wildlife.[1064, 852, 137] Habitat heterogeneity can be readily increased with brush piles, rock piles, and the like. Shrub lines and stump lines (Chapter 6) help guide the movement of wildlife in directions desired in management.

Variegated roadsides can also provide diverse benefits to society. Along a road segment, the traveling public could point to a sequence of distinctive segments, each providing a different set of benefits. The list of possible benefits is long, from controlling stormwater pollutants and protecting rare plants to providing wood production, carbon sequestration, and cultural and nature education for the commuter and traveler.

Conclusion

"Reading the roadside" offers the traveler considerable insight into the nature surrounding a road as well as the history, management, and future of roadside ecosystems. The abundance of non-native species in roadsides is important as a potential source of invasives that may damage surrounding pastureland, cropland, and nature reserves. Yet, at present, the evidence for the linkage is weak. Roadsides contain rare native species, yet detailed inventories of their abundance and distribution are scarce and are needed to establish effective management and protection regimes.

Roadside native vegetation is relatively homogeneous and impoverished. Zonation across a roadside and patchiness at at least two spatial scales along a roadside are conspicuous. Widespread frequent mowing favors the spread of grasses at the expense of most native plants and animals. Changes in the arrangement and timing of mowing and other management activities can produce a natural mosaic of plant communities that form a variegated roadside. Roadside design and planting may visually enhance travel, but since visual quality and ecological conditions are often not correlated, care is required to establish high-quality ecological conditions using high- and medium-quality visual designs. Roadsides, which cover about 0.5% of the entire USA (Chapter 2), represent an enormous opportunity for new thinking, planning, and management to benefit society. Future North American roadsides could fit traveler safety solutions into a rich tapestry of natural ecosystems.

CHAPTER 5

Wildlife Populations

> The next moment all the rabbits leaped up in panic. . . . Hazel looked down at the road in astonishment. For a moment he thought that he was looking at another river—black, smooth and straight between its banks . . . "But that's not natural," he said, sniffing the strange, strong smells of tar and oil. "What is it? How did it come there?" "It's a man thing," said Bigwig. "They put that stuff there and then the hrududil [cars] run on it—faster than we can. . . ."
>
> "It's dangerous, then? They can catch us?" "No, that's what's so odd. They don't take any notice of us at all. . . . As a matter of fact, I don't think they're alive at all. But I must admit I can't altogether make it out."
>
> —Richard Adams, *Watership Down,* 1972

> Fragmented mammals: What does that mean?
>
> —Rob C. van Apeldoorn, title of an article in *Habitat Fragmentation & Infrastructure,* the proceedings of an early road ecology symposium in The Netherlands, 1997

Imagine roads as footprints. If our imagination is working at half speed, we see only the physical footprint. If we were able to look at the accumulation of footprints over time, we would see an environment with an ever-increasing number of "tracks." Now place your imagination at full throttle. In addition to this physical footprint, you will see a "virtual footprint" surrounding roads. This *virtual footprint* is the accumulated effect over time and space of all of the activities that roads induce or allow, as well as all of the ecological effects of those activities. The virtual footprint can have positive and negative aspects. Access to hiking trails, bike paths, and scenic vistas is an example of the positive benefits of roads.

Yet negative impacts in the virtual footprint, such as species declines near roads, disturbances caused by off-road vehicle access into roadless areas, disruption of hydrologic regimes, and fragmentation of habitats can be reduced significantly by intelligent action. The virtual footprint can be a baby-size 2 rather than a boot-size 13 with cleats. Incorporation of mitigation measures into the planning and development of road systems up front can keep the virtual footprint to a minimum. The goal of building "smart roads" is to minimize negative road effects.

To extend the metaphor, the goal is to "tread lightly" on the land. To the extent that this can be done, the quality of people's lives should improve. Mitigation measures designed to minimize the negative aspects of the virtual footprint, if done properly, maintain or enhance road services while restoring ecosystem services. The latter occurs when important functions of energy flow and nutrient cycling continue to operate, when hydrologic regimes are maintained, and when the natural movement of animals is functionally unimpeded.

Roads and vehicles affect wildlife in several important and interesting ways. Most of the ways are well documented[52,958,768,660] and have been described in the literature for over 50 years.[882,403,290] Roads can cause a direct loss of habitat, alter the quality of adjacent habitat, lead to road-kills, and impede animal movements. As roads are upgraded to accommodate greater traffic volume, the rate of successful wildlife crossing decreases significantly.[48,896,107,802] Thus, roads may effectively fragment habitats and otherwise continuous population distributions. Smaller populations typically result, with a greater potential of genetic problems and an increased chance of local extinction.

In this chapter, we focus on the main patterns, issues, and principles that link roads and vehicles with wildlife populations, be they beneficial or harmful. In addition, we touch on some of the commonly tried (and some tested) solutions to mitigate the harmful effects roads may have on wildlife. Although considerable information is widely dispersed in internal agency reports, the presentation here is mainly based on literature in technical and scientific journals.

Wildlife Mortality on Roads

Anyone who has watched a deer try to cross a busy road, or seen the rumpled remains of one that didn't make it, can begin to imagine the diverse effects of *road-kills* or *road mortality* (or faunal casualty). Although deer *(Odocoileus)* and other large animals, such as bear *(Ursus)*, moose *(Alces alces)*, and elk *(Cervus elaphus)*, are much more noticeable when killed on roads, it is possible that a greater biomass of small animals is road-killed. In some cases, a road-kill rate may be high enough to exceed natural causes of death due to predation and disease.[585, 134, 163, 92, 406]

Collisions between the two moving objects, animals and vehicles (with animals normally coming out second best), doubtless date to the origin of vehicles. In the mid-nineteenth century, Henry David Thoreau described the results of a direct wagon-wheel hit on a turtle, and in 1895 Barbour commented on bird kills caused by Nebraska railroads.[46] Many articles written early in the twentieth century reported vertebrate road mortality.[1023, 1024, 98, 99, 100, 164, 117, 55, 805] Most early reports were the product of a single automobile trip, often when the recorder was on vacation, and as a result are subject to large sampling error. A few notable exceptions were based on multiple and longer trips or on scheduled surveys.[830, 873, 219]

Similarly, in earlier reports of bird kills, most writers failed to consider more than "an exceedingly small portion of the questions involved."[555] Counts of dead birds were inadequate for understanding the relationship of roadways to birds. Several reasons were listed for why roads and roadsides were attractive to birds, thus placing the animals in proximity to fast-traveling automobiles. One especially meaningful insight suggested that the relevant question was not "How many birds are killed by speeding automobiles?" but rather "How many pairs of birds are prevented from successfully rearing broods?" Because lost mates are often replaced, destruction of roosting areas, hunting perches, and nest sites during road building and maintenance can influence local population welfare more than does the actual loss of individual birds to automobiles. Early observers recognized the importance of habitat and road variables that influence bird collisions with vehicles. Thus height of cover, land use, local hazards, weather conditions, and size and speed of automobiles may explain most bird-vehicle collisions.[290]

Although, as mentioned above, many early estimates of mortality were based on single-trip road counts, one paper provided an expanded analysis of at least six studies (Figure 5.1).[883] The author reported an average of about one vertebrate casualty (0.95) per 10 km (6 mi). He estimated that about 153 vertebrates per 1613 km (1000 mi) of "main traveled highway" were killed in the "northern" states. Several early authors calculated the percentage kill for different vertebrate groups. For example, one study counted 892 mortalities over 12 115 km (7529 mi) driven in Illinois; mammals represented 24%, birds 68%, amphibians less than 1%, and reptiles 7% of mortalities.[873] Another author counted 512 animals on the road in 3473 km (2157 mi) traveled between Albany, New York, and Iowa City, Iowa, and the return (Figure 5.1).[883]

Many other authors addressed multispecies road mortalities.[99, 381, 206, 228, 292, 1001, 805, 830, 873, 497, 599, 608] Most often, birds are listed as killed more frequently than are mammals. Reptiles and amphibians appear less often in the record.

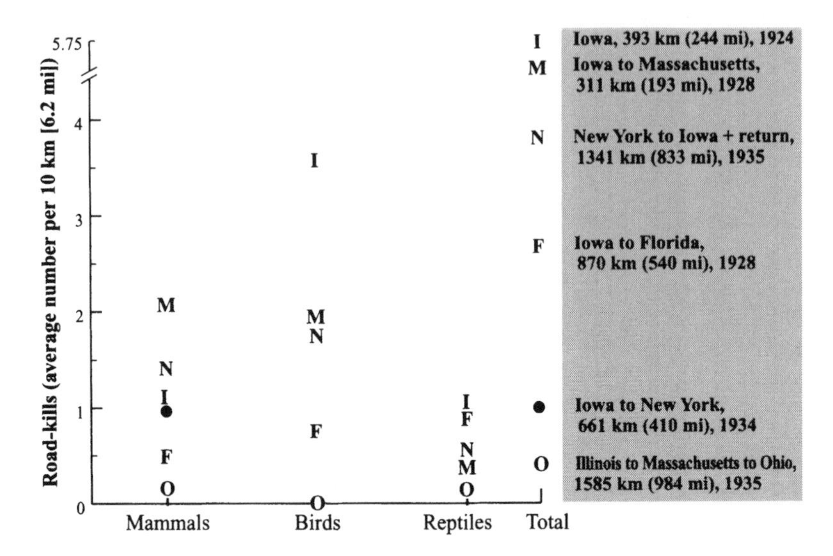

Figure 5.1. Vertebrates road-killed on 13 392 km (8316 mi) of highways in the USA. Total includes small numbers of amphibians and "miscellaneous" recorded. Adapted from Stoner (1936).

Amphibians (frogs and toads) are most often underrepresented in road mortality counts because of their small size.[883] Furthermore, scavenger feeding on smaller vertebrates may account in part for their underrepresentation in the record. Thus, most of the early literature supports the conclusion that birds suffered the greatest mortality, followed by mammals, with usually much smaller numbers of reptiles and amphibians reported.

Economic Costs of Accidents

It should be no surprise that wildlife-vehicle accidents are on the rise in most regions. Increasing human development, sprawling suburbs with their associated increase in daily commuters, increasing road networks, higher traffic flows, and increased traffic pulses in early morning and early evening (when commuting and wildlife activity peak simultaneously) all contribute to these unfortunate encounters that affect both humans and wildlife.

Vehicle collisions with larger-bodied species, such as deer, involve vehicle damage, human casualties, and lost economic opportunities (Figure 5.2). For example, between 1981 and 1991, property damage to Vermont vehicles occurred in 94% (n = 23 285) of all deer-vehicle collisions (Vermont Department of Transportation, unpublished data). The average property damage per accident in the USA has been estimated as US$1577.[182] Using the conservative

Figure 5.2. Deer crossing a lightly traveled earthen road with forest on both sides. Mule deer *(Odocoileus hemionus)*, ponderosa pine *(Pinus ponderosa)*, and junipers *(Juniperus)*. D. H. Lawrence Ranch, Taos, New Mexico. Photo by R. T. T. Forman.

estimate of 720 000 animal-vehicle accidents per year (involving mule deer and white-tailed deer only), approximately 29 000 human injuries occur annually. This includes 211 human fatalities, an incidence rate of 0.029% fatalities.[182] For perspective, there were 44 381 highway-related human fatalities from all causes in 1997, suggesting that fatalities resulting from animal-vehicle collisions constituted approximately one-half of 1% of total annual highway fatalities. A Newfoundland study reported that 5422 moose-vehicle accidents during 1988–94 resulted in 14 human fatalities, or approximately 0.25% deaths per accident.[466] On the other hand, it has been suggested that deer provide more benefits to human society than any other animal species in North America, with an estimated net annual monetary value of more than US$12 billion.[182]

Current Studies of Mortality

Wildlife mortality on roads is in addition to natural causes of mortality, such as predation and disease. The significance of this road mortality for the wildlife population thus depends on its magnitude relative to other mortality sources. Unfortunately, few comparative studies of road-related mortality versus other mortality sources exist. However, estimates of absolute numbers of animals killed on roads in relation to wildlife population sizes indicate that the mortality levels are high for at least some species.

Systematic record keeping of wildlife road mortality on U.S. roads is nonexistent for any species, although national figures for larger species, in particular deer (Figure 5.2), have been pieced together. As of 1994, only 14 studies that described seasonal distributions of deer road-kills by age and sex class were known.[784] Perhaps more telling, only four states (California, Michigan, Pennsylvania, and Wyoming) were represented. Three additional studies focused on population characteristics of road-kills. Although precise numbers are scarce, estimates of the number of deer killed annually on U.S. roads ranges from 720 000[785] to 1.5 million.[181] Approximately 92% and 89% of vehicle collisions resulted in the death of deer and moose, respectively.[13,466] Estimates of the proportion of deer that are hit on roadways and go undocumented, and hence unreported, range from 50% (D. Reed, personal communication, cited in Romin and Bissonette 1996b) to more than six times the reported number.[211] This suggests that actual deer mortalities may be considerably higher than counts indicate.

A total of 205 painted turtles *(Chrysemys picta)* were killed during a four-month summer period on a 7.2-km (4.5-mi) segment of Route 93 adjacent to the Ninepipe National Wildlife Refuge in Mission Valley, Montana.[326] Discounting animals that may have been injured but escaped the road, this is equivalent to a monthly rate of 7 turtles killed per kilometer of road. This two-lane highway bisects a series of rounded marshy wetlands where turtles and other aquatic wildlife thrive. However, roads need not bisect aquatic habitat to have an impact on primarily aquatic species. The seasonal movement patterns from aquatic to terrestrial habitats represent a chronic source of road-related mortality for chicken turtles *(Deirochelys reticularia)*.[124] Roads separating wetland from forested areas can result in high mortality of dragonflies, which need to move between their two critical habitats[778] (S. K. Riffell, personal communication).

Numerous records of salamander and toad seasonal migrations and resulting mass mortalities on roads worldwide exist.[522] Frogs and toads, as well as many species of snakes, are susceptible to road-kill. The probability of amphibians being killed by traffic on a busy highway is reported as 89% to 98%, and 34% to 61% on a road with 3200 veh/d.[412] During a two-year period, more than 32 000 vertebrates (mostly amphibians) were found as road-kills along a 3.6-km (2.2-mi) causeway near Lake Erie, Canada.[30] About 5.5 million reptiles and frogs are reported killed in Australia each year by road traffic.[239]

Three significant trends relating to road-kill of frogs and toads have been identified based on an Ontario study.[255] The number of dead and live amphibians per kilometer on two-lane roads decreased with increasing traffic volume. Moreover, the proportion of frogs and toads that were dead increased with greater traffic. However, the density of frogs and toads in the roadside and adjacent habitat decreased with more traffic. The authors concluded that road mortality had a significant effect on local densities of these amphibians.[255] In Denmark, about 10% of the adult population of *Pelobates fuscus* frogs and brown frogs *(Rana temporaria* and *R. arvalis)* were killed annually by traffic.[412]

Similarly, road traffic causes significant impacts on the snake fauna in the Pa-hay-okee wetlands of Everglades National Park, Florida.[79] Of 1172 individual snakes observed along roads over a two-year period, 73% were either injured or dead on the road. In Arizona, based on over 15 000 km (9315 mi) of road observation over a four-year period, an average of 22.5 snakes per kilometer per year were killed in Organ Pipe Cactus National Monument.[793] Significantly, two species of special interest—the Mexican rosy boa *(Lichanura trivirgata)* and the Organ Pipe shovelnose snake *(Chionactis palarostris)*—were seriously affected by road mortality. Road impacts are more serious when the species are threatened or endangered or otherwise have high conservation interest.

Populations of wide-ranging carnivores are particularly vulnerable to road traffic accidents. Collision with motor vehicles accounted for 49% of mortality of the endangered Florida panther *(Puma concolor coryi)*[585] before mitigation underpasses and fencing greatly reduced this percentage. Road-kill is a significant cause of mortality for the endangered ocelot *(Leopardus pardalis)*.[414] In Southern California, the single most important cause of cougar *(P. concolor)* mortality was motor vehicles.[61] Road mortality is one of the largest mortality factors for Iberian lynx *(Lynx pardalis)* in southwestern Spain[284] and for wolves *(Canis lupus)* in Minnesota.[340]

To date, there are relatively few studies of the *population effects* of animal road mortality. The existing studies suggest that the impacts can be significant. The presence or absence and density of local amphibian populations can be affected by road traffic.[255, 982, 143] The number of roads and road density was the strongest determinant in the decline of badgers *(Meles meles)* in The Netherlands,[958] and mortality of Florida scrub jays *(Aphelocoma coerulescens)* was almost twice as high in populations bordering a two-lane highway compared to populations living away from the road.[660] Furthermore, significant effects of road density have been reported for species richness of reptiles/amphibians, birds, and plants within 2 km (1.25 mi) of wetlands.[255, 288] These studies also highlight the point that other important factors may interact with road mortality to affect a population. For example, in the amphibian and wetland studies, traffic noise and disturbance may be reducing populations in habitat around the roads.

A study of painted turtles at varying distance from roads found both lower density and higher mortality near the roads.[326] Interestingly, higher percentages of adult turtles were found farther from roads. A typical consequence of mortality in a population is a shifting of the age structure to younger age classes. Indirect evidence for population-level effects of roads is provided by studies that quantify the relative importance of different mortality factors for specific animal populations. For example, vehicle collisions have been reported as the largest source of mortality in Florida for panthers, black bears *(U. americanus)*, key deer *(Odocoileus virginianus clavium)*, American crocodiles *(Crocodylus acutus)*, and bald eagles *(Haliaeetus leucocephalus)*.[399] Road-kill is also the largest mortality source for moose in the Kenai National Wildlife Refuge in Alaska[42] and for barn owls *(Tyto alba)* in the United Kingdom.[682]

Factors Affecting Mortality

The many factors affecting road mortality of wildlife may be separated into two useful categories: (1) traffic, road, and landscape influences, and (2) species behavior and ecology.

Traffic, Road, and Landscape Influences

Both vehicle speed and traffic volume influence wildlife collisions. An early study pointed out that vehicles traveling at speeds greater than 40 mph (64 km/h) appeared to have a greater impact on songbirds and rabbits than did cars traveling more slowly.[219] A long-term (1967–85) study of road-kill of raccoons *(Procyon lotor)* in Indiana found almost 14 000 raccoon road-kills during 19.8 million km (12.3 million mi) of road travel.[783] The study implicated vehicle speed as a major cause of mortality. Another study reported that the number of road-killed wildlife was not significantly correlated with average daily traffic on either a monthly or annual basis, but that road mortality was significantly correlated with vehicle speed.[147] In contrast, the armadillo *(Dasypus novemcinctus)* kill rate on Florida roads was highly correlated with traffic volume.[440] In Newfoundland, moose-vehicle accidents occurred where traffic volumes were greatest.[466]

Studies have also looked at the effect of *landscape structure near a road* on deer-vehicle collisions. The probability of deer-vehicle accidents was reported to decrease with increases in the number of residences and buildings nearby.[52] In Pennsylvania, accidents decreased when there was a lower speed limit, a greater distance to wooded areas, and a greater minimum visibility distance. Another study reported a higher tendency for deer-vehicle accidents where there were more bridges (representing wildlife travel corridors) and more traffic lanes.[430] In short, these studies suggest that in addition to population density, two major groups of factors mainly affect road-kill rates: (1) traffic volume and speed, and (2) proximity of habitat cover and wildlife movement corridors.

Species Behavior and Ecology

Some species populations are more vulnerable to road-related mortality than others (Figure 5.3). Species that occur in low densities and have low reproductive rates and long generation times are typically most susceptible to additional mortality.[519, 1051, 325] Some large carnivores are perfect examples of species with those characteristics.

Species with higher intrinsic mobility are generally more vulnerable to road mortality (Figure 5.3). Theoretical studies indicate that dispersal rates decrease with greater mortality during dispersal.[180, 546, 493] Therefore, species with intrinsically high movement rates should be most susceptible to an increase in dispersal mortality, such as by road mortality. This suggestion is supported by two recent studies. In the first study, snake species using frequent long-distance movements to forage for food were found to experience higher mortality than more sedentary foragers.[102] In the second, the effects of roads on the population densities of two frog species of differing vagility or *mobility*—leopard frogs (*Rana pipiens,* more mobile) and green frogs (*Rana clamitans,* less mobile)—were studied. Traffic volume was found to have a significant negative effect on population density of the more mobile species but no effect on density of the less mobile species.[143, 144]

Habitat generalists are more susceptible to road mortality than to the other effects of roads (Chapter 4). For example, individuals of two small mammal species were captured in woodlots and then released in other woodlots, sep-

Characteristics making a species vulnerable to road effects	Road mortality	Habitat loss	Reduced connectivity
Attraction to road habitat	*		
High intrinsic mobility	*		
Habitat generalist	*		
Multiple-resource needs	*		*
Low density/large area requirement	*	*	*
Low reproductive rate	*	*	*
Forest interior species		*	*
Behavioral avoidance of roads			*

Main effects of roads

Figure 5.3. Characteristics making a species vulnerable to three major effects of roads.

arated by varying numbers of roads with different traffic volumes (S. Derrane et al., personal communication). One species, the white-footed mouse *(Peromyscus leucopus),* is a habitat generalist, and the other, the eastern chipmunk *(Tamias striatus),* is a habitat specialist. The ability of the generalist species to return to its home site decreased with increasing traffic volume, whereas traffic level had no effect on the return rate of the specialist species. A possible interpretation is that generalist animals are more likely to venture onto unfamiliar habitats, such as roads, making them more vulnerable to road mortality. Nevertheless, specialists often have lower population sizes, lower reproductive rates, and fewer suitable habitats, and thus may be at risk.

Species that show behavioral *avoidance* (or *road avoidance*) of open habitats and noise are less vulnerable to road mortality than those that will readily attempt to cross roads (Figure 5.3). For example, small forest mammals were found to be reluctant to venture onto road surfaces where the distance between forest margins was more than 20 m (66 ft).[700] An experiment conducted in the Mojave Desert found that even though rodents may travel a great distance, they hesitate to cross roads.[341] In Germany, two species of forest-dwelling mice rarely or never crossed two-lane highways.[583] In Ontario, the probability of a small mammal crossing a lightly traveled road 6–15 m (20–49 ft) wide was less than 10% of movements within the adjacent habitats.[630] Several large animals, including black bears, grizzly bears *(U. arctos),* elk, mule deer *(O. hemionus),* and wolves also appear to avoid roads.[797, 374, 615, 115, 913]

Road mortality of carnivores is often linked to movement patterns. For example, between 1982 and 1986, 17 male long-tailed weasels *(Mustela frenata)* were killed on highways and secondary roads in eastern Washington State.[120] All mortalities were during the summer breeding season, when male weasels increase mobility.

Finally, cases in which animals are attracted to roads, thus increasing their risk of mortality, have been documented. Some animals, such as reptiles and some insects, are attracted to roads and roadsides as basking or nesting sites, making them vulnerable to both traffic mortality and predators[406] (L. Fahrig, personal observation). Reptiles that come to the road surface to bask during the day or to thermoregulate at night can quickly become road-kills and ultimately carrion for scavengers.[793, 492] Animals that perceive a food source on the road, such as road-kills, basking animals, or spilled grain, are also vulnerable to road mortality.[334, 179, 887] On roads requiring snow removal and the application of salt-based de-icing agents, problems may arise even long after winter, when mineral-deficient ungulates come to glean salt from the edge of the road and from cracks in the pavement.[334]

Changes in Amount and Quality of Habitat

Certain other forces that affect animal populations in a roaded landscape may be more important than road-kills as threats to the maintenance of viable species populations. Habitat is a key factor, so we start with three components of habitat change relative to roads and vehicles: (1) habitat loss, (2) reduced habitat quality, and (3) improved habitat quality.

Habitat Loss

Road construction results in *habitat loss* by converting preexisting habitats to pavement and roadsides (verges or rights-of-way). Road construction also can result in the indirect destruction of surrounding habitat through the siltation of streams and drying of wetlands due to interrupted water flow.[463] With 6.25 million km (4 million mi) of public road in the USA, the loss of habitat to roads is far from inconsequential.

Some animal species are more vulnerable to habitat loss than others. The most vulnerable species are large, long-lived species with large area requirements, low densities, and low reproductive rates, such as many large carnivores (Figure 5.3).[342, 325] Based on existing evidence, sustainable populations of large carnivores may be likely only in landscapes containing road densities below about 0.6 km/km^2 (1.0 mi/mi^2).[320, 310] For comparison, the average road density in the United States is 1.2 km/km^2 (1.9 mi/mi^2),[310] whereas average road density in many European countries is over 2 km/km^2 (3.2 mi/mi^2) (see Figure 2.6).

Species whose habitat is limited to forest interior are particularly vulnerable to declining forest patch sizes, because smaller patches have a larger proportion of *edge habitat* (or *forest edge),* which is generally avoided by *forest interior species.*[66] Removal of forest typically involves an increase in the proportional amount of edge habitat in a landscape. For species that cannot live in forest edge, this results in a disproportional decrease in amount of habitat.[254] Because of the long, thin shape of roads, removal of forest for roads creates large amounts of forest edge, resulting in additional loss of habitat for these forest interior species.[752, 760] For strict forest-interior species, the habitat loss due to roads is therefore several times the actual forest removal (Figure 5.4). In other words, the virtual footprint of the road is large.

How mobile a species is can also affect its vulnerability to habitat loss due to roads. Metapopulation theory suggests that more mobile species are better able to cope with habitat loss.[387] However, theory (of metapopulations) does not incorporate mortality of individuals in the "nonhabitat," or surrounding *matrix,* areas of the landscape.[302] Simulation studies[82] suggest that when the

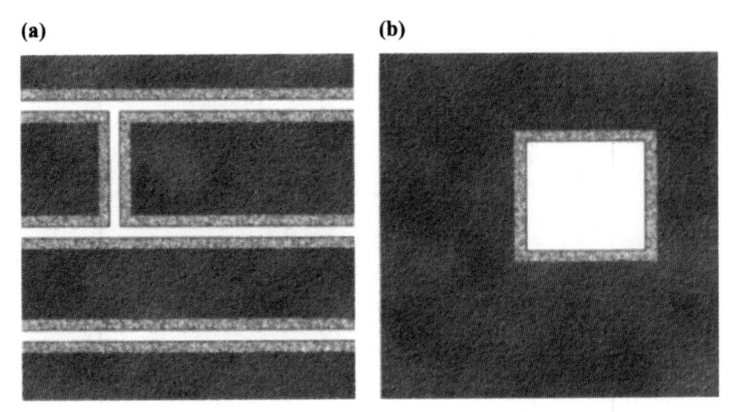

Figure 5.4. Loss of forest-interior habitat due to road and "block" development. Dark shading = forest interior; light shading = forest edge area; white = development. The area of white is the same in both diagrams. (a) Roads in forest where 66% of the original forest-interior habitat remains, subdivided into five separate pieces; (b) a square block of development in forest where 84% of the original forest interior remains in one large connected piece.

mortality rate in the matrix is high (such as when roads are present), highly mobile species are actually more vulnerable to habitat loss. This prediction is supported in a study of effects of habitat loss on amphibians, in which the most mobile species were shown to be the most vulnerable to habitat loss.[349]

Reduced Habitat Quality

In addition to direct removal of habitats for road construction, roads also alter the quality of habitat near roads. Road corridors and the disturbances associated with them may lead to reduced habitat quality as expressed by (1) a *numerical response,* such as a decrease in abundance or density of breeding individuals, or (2) a *behavioral response,* such as road avoidance. In an extreme case, where no usable habitat is left and/or avoidance is 100%, the reduction in habitat quality amounts to further loss of habitat.

While studying the influence of unpaved forest roads on ovenbird *(Seiurus aurocapillus)* density in an extensively forested region of Vermont, researchers found that the bird's territory size decreased with distance from roads.[699] They concluded that habitat quality for ovenbirds may be lower within 150 m (492 ft) of unpaved roads in a forested landscape, thus requiring a larger area for foraging, decreasing overall territory density, and possibly reducing reproductive success. In a similar study, the proximity of Florida scrub jay territories to a paved road with perhaps 500 to 750 vehicles per day had no effect on nesting success.[660] However, roadside territories may have acted as population sinks where immigration was needed to maintain populations.

Breeding birds appear to be heavily affected by *traffic disturbance,* in particular *traffic noise.* A series of three seminal papers, R. Reijnen et al. (1995), Reijnen and Foppen (1994), and Foppen and Reijnen (1994), found that of 43 species of woodland breeding birds, 26 species (60%) showed reduced densities near highways. The distance effects ranged from 1500 m to 2800 m (4920 to 9184 ft) for roads with 10 000 to 60 000 veh/d, respectively. There was evidence, for example, that the presence of a highway reduced willow warbler *(Phylloscopus trochilus)* populations significantly. A regression model showed that road noise was the independent variable that gave the best results in explaining the lowered bird density near the road. Even when car visibility was controlled for, density reductions were much greater on plots with higher noise levels.

In a companion study, disturbance distances and hence grassland bird densities were related to traffic volume and, by extension (and quantified in the earlier papers), traffic noise.[770] At 5000 veh/d, disturbance distances varied from 20 to 1700 m (66 to 5576 ft) from the road; at 50 000 veh/d, from 65 to 3530 m (213 to 11 578 ft) from the road. The number of bird species also was reduced significantly near roads. A more recent study in an outer suburban landscape in Massachusetts found significant effects on grassland birds near roads with 8000 to ≥30 000 veh/d, but not near local collector streets with 3000 to 8000 veh/d[322] (Chapter 10 and Figure 10.7).

Earlier Dutch studies in open grassland found the disturbance effect ranged from 500 to 600 m (1640 to 1968 ft) for rural roads and 1600 to 1800 m (5248 to 5904 ft) for busy highways.[967,956] The total population loss over this distance was approximately 60%. In Illinois (USA) farmland, horned lark *(Eremophila alpestris)* densities in Illinois increased with distance from both interstate highways and county roads during summer.[162] The surrounding habitat was corn and soybean *(Zea, Glycine)* farmland. Similar patterns of bird density in relation to roads were reported from Finland.[514] Flocks of pink-footed geese *(Anser brachyrhynchus)* and greylag geese *(A. anser)* were found to avoid roads when feeding, and flocks were not found within 100 m (328 ft) of the nearest road. Interestingly, fields with centers closer than 100 m from roads were never visited by geese.[477]

Physiological measures of stress can be a valuable tool for evaluating impacts of disturbance, including roads. Recent research shows that northern spotted owls *(Strix occidentalis caurina)* living close to forest roads experienced higher levels of a stress-induced hormone than owls living in areas without roads.[1004]

Contrary to the above examples, researchers in northern Maine found that the total number of breeding birds near a four-lane interstate highway was not significantly different from the number at a greater distance.[285] The roadside verge and median, however, supported half as many breeding birds as an equal area of forest habitat. Densities of breeding birds on the verge along a two-lane highway were 79% of those in forest habitat.

For large, wide-ranging mammals, behavioral responses to roads are more readily detected than numerical responses. Radio-telemetry studies during the last 30 years have allowed researchers to document animal behavior in relation to roads and traffic. Studies of ungulates and large carnivores all concur that buffer areas around roads are generally avoided. The width of the avoidance zone seems to depend on the traffic volume. Furthermore, heavily traveled roads are little used as a route by wildlife, in comparison with lightly traveled roads.[797, 615, 913, 579, 350, 302]

Road density has explained the juxtaposition of home ranges and distribution of bobcats *(Felis rufus)* in northwestern Wisconsin, grizzly bears in the Swan Mountains of Montana, and amphibians near intensively used roads.[565, 579, 982] Bears react to increases in road densities by shifting the locations of their home ranges to areas of lower road densities.[115] If major roads only affect home range and space use boundaries without also causing edge effects, then the substantive influence of roads apparent within home ranges might be less.[25]

For large animals, roads can indirectly cause elevated mortality in habitats near roads, thus reducing the quality of those habitats. Roads increase human contact with wild populations, therefore indirectly contributing to animal mortality through increased intentional or accidental killing, particularly for large predators. In the Swan Valley (Montana), eight grizzly bears were killed by humans in a six-year period, and all deaths were directly influenced by road access.[579] In the Southern Appalachians, an increase in road density heightens the potential for hunters to encounter black bears.[115]

For prey species, greater road density translates to greater access for predators and people. In an industrialized area within caribou *(Rangifer tarandus)* range, individuals that were close to linear corridors were at a higher risk of predation by wolves.[451] Industrial road networks therefore were believed to have a significant effect on caribou population dynamics by increasing predation. This suggests that for some species, areas of high road density may be *biological sinks,*[742] areas of low-quality habitat in which animal populations are not sustainable on their own without constant input.[644]

Improved Habitat Quality

Although most effects of roads on habitat quality are negative, some species respond positively to habitats near roads. Road corridors may provide additional habitat for some wildlife. For example, at certain times, red-winged blackbirds *(Agelaius phoeniceus)* can be more abundant closer to roads.[162] An early paradigm of wildlife management[540] recommended the creation of forest edge habitat (e.g., along roads) for enhanced population growth of primarily game species. Lynx *(Lynx canadensis)* home ranges occur adjacent to a major motorway in Alberta.[25] Rodents appear to be either unaffected or positively influenced by the presence of roads.

The creation of new microhabitats along roads favors certain species, such as amphibians in ditches, plants on concrete rubble, and bats on bridges. One survey reported 4 250 000 bats of 24 species living in 211 highway structures (about 1% of those surveyed) in 25 U.S. states.[472] Most species used concrete crevices sealed at the top, at least 15–30 cm deep, 1.3–3.2 cm wide, at least 3 m above ground, and not over busy roadways.

Animals can be attracted to transportation corridors or roads themselves for a variety of reasons, but most are related to habitat (nesting, living space), facility of movement, or food resources. Habitat enhancements may take the form of added perching sites for birds feeding in the road corridor. In one study, the higher use by raptors of roadsides compared to adjacent habitat was explained not by prey abundance but rather by the greater availability of perch sites and wide roadsides.[633] Ravens *(Corvus corax)* were more numerous along paved highways than in areas away from roads for the same reason.[496]

Road construction and fencing can create high-quality habitat and greatly enhance existing habitat for some species. This is particularly true when the verge and median-strip habitats are left unmowed and allowed to grow naturally.[7] *Exclusion fencing* designed to exclude deer from roads can result in tremendous amounts of forage and cover for small mammals on the highway side of the fence. These seemingly simple changes to a road corridor can positively affect the structure and density of the small mammal community. In an extensive study carried out along interstate highways in the USA, more small mammal species were present and in higher densities on the verge habitat than on adjacent habitat.[8] The study results also indicated that roadside habitat and its adjoining edge habitat were attractive not only to grassland species but also to many less-habitat-specific species. At six interchanges along Highway 417 in Ottawa, the density of woodchucks *(Marmota monax)* per hectare exceeded any density previously reported for this species in any habitat.[1055]

Heavily used motorways also tend to have higher levels of nitrogen than normal in the surrounding environment[23] (Chapter 8). The high nitrogen content of roadside vegetation along a busy motorway in the United Kingdom was believed responsible for the rate of increase and the outbreaks in insect populations.[737] Others have found the densities of grasshoppers *(Chorthippus brunneus)* to be higher on motorway sites compared to sites away from the motorway.[737]

Roads and their construction can create high-quality habitat where food resources are more abundant compared to adjacent areas. Lush forage along medians and verges created by exclusion fencing is attractive to herbivores ranging from elk to voles. Locally abundant small mammal populations found here become targets for predators, such as hawks and coyotes *(C. latrans),* seeking easy and accessible prey (Figure 5.5). With deer, elk, and predators on small mammals foraging near the roadway, collisions with vehicles are inevitable. This results in attractive carrion for avian scavengers (eagles and corvids) and terrestrial scavengers (bears, wolverines *[Gulo gulo],* coyotes). In extensively forested areas, the roadside may be one of the few open habitats present and the first place that greens up after winter, thus attracting deer and related species. The roadside is ideal habitat for dandelions *(Taraxacum officinale)* and fruit-bearing shrubs, which appeal to many wildlife, including bears. Road corridors also may be more attractive than adjacent areas to some birds, including skylarks *(Alauda arvensis),* which forage and nest nearly exclusively on road verges.[528]

Figure 5.5. Tracks used by both off-road and on-road vehicles, along which large predators move at night. In this landscape, coyote *(Canis latrans)* footprints commonly continue for tens to hundreds of meters along tracks, whereas footprints of the much more abundant mule deer *(Odocoileus hemionus)* and elk *(Cervis elaphus)* commonly cross, but infrequently extend more than 10 m (33 ft) along, the tracks. Bobcat *(Lynx rufus)* and cougar *(Puma concolor)* footprints going along tracks have been infrequently observed in this area of grass, sagebrush *(Artemisia tridentata),* and pinon-juniper (*Pinus edulis, Juniperus* spp.) vegetation. Coyote scats tend to be concentrated within 30 m (98 ft) of major intersections of tracks. Tres Piedras, New Mexico. Photo and unpublished data by R. T. T. Forman.

Finally, humans feeding wildlife from vehicles, foods tossed from passing vehicles, and garbage containers along roads can be easy, semipredictable sources of non-natural food for wildlife, including coyotes, bears, crows/ravens, jays, squirrels, and raccoons. In 1997, more than 80 *bear jams* (traffic snarls caused by motorists stopping to look at bears) were reported along roads in Banff National Park, Alberta.[351] This exposure to people and their food results in bears becoming food conditioned and habituated to humans.

Although roadsides provide some benefits to some species, those benefits must be balanced against the negative ecological effects of roadsides. For example, in Banff National Park, the increased habitat quality for bears along roads must be weighed against the increased probabilities of bears being road-killed or (as threats to visitors) removed from the park.[351]

Effects on Landscape Connectivity

Landscape connectivity is the degree to which the landscape facilitates animal movement and other ecological flows.[494, 905, 302, 400, 73] High levels of landscape connectivity occur when the matrix areas of the landscape comprise relatively benign habitat types without barriers, thus allowing organisms to move freely.[919, 920] High landscape connectivity is important for two main reasons. First, many *multihabitat* organisms regularly move through the landscape to different habitat types to obtain their daily and lifetime needs.[232] For example, damselflies (calopterygids) regularly move between streamside habitats for mating and forest habitats for feeding.[464] Some amphibians such as leopard frogs shift seasonally among habitat types, from breeding ponds in the spring to meadows during the summer and to overwintering sites in the fall.[736] Barriers that impede these movements, such as roads, result in higher mortality, lower reproduction and, ultimately, smaller populations and lower population viability.[144]

Second, high landscape connectivity allows for movements to repopulate areas that have suffered local population declines and extinctions, and to minimize the negative effects of inbreeding.[1035, 520, 73] Essentially, high landscape connectivity allows a regional population to take full advantage of the habitats available to it. Reduced movement through the landscape results in habitats that are "empty" or that contain smaller populations than they might otherwise support[970, 144] (Figure 5.6). The consequent lower regional population size increases the probability of loss of the species in that area. When roads act as barriers to movement, they reduce landscape connectivity and the ability of the regional population to inhabit all suitable areas.[450] The result is a lower regional population and lower long-term population persistence.

Effects of roads on landscape connectivity are most straightforward for

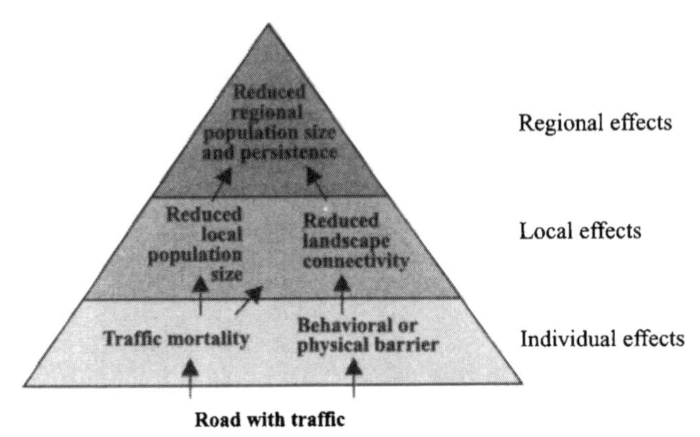

Figure 5.6. Road system effects on individual animals and the wildlife population.

species that avoid roads (Figure 5.3). These are often forest interior species, whose habitats become functionally separated by the presence of roads. For species that do not avoid roads, roads can still represent a barrier to movement when mortality rates are high. However, it is not possible to separate the effects of increased mortality from the effects of decreased landscape connectivity on the regional population without experimentation. In short, the species most vulnerable to the *barrier* effect of roads are those that avoid roads and have multiple resource needs and/or need large amounts of resources, which requires them to travel over large areas. High landscape connectivity is important for an animal foraging within its home range, for dispersal to establish a new home range, and for migration between locations.

Barrier Effects

Roads pose a major threat to the recovery of carnivore populations not only because of vehicle collisions but also because of the indirect effects of barriers[686] (Figures 5.3 and 5.6). Although direct impacts such as road-kills are easy to document, indirect effects are much more difficult to demonstrate.

Little is known about the long-lasting ecological effects of roads on animal populations in terms of reduced mobility, increased isolation, and/or the splitting of gene pools.[39] Recent concern has focused on the influence of human-created barriers, such as interstate highways, on normal distribution patterns of mammals, and perhaps ultimately on *speciation* (the formation of new species).[39] Although few in number, some studies have demonstrated that roads may act as barriers to small mammal movements[700, 48, 583] and large mammal movements.[78,350] These are effectively partial barriers or *filter barriers* that block some but not all movements across.

Isolation caused by physical barriers to movement, such as roads, may reduce gene flow, thus causing genetic effects.[31,1009] Reducing the movement of genes among populations results in more negative effects of inbreeding or *inbreeding depression* that results in weak or sterile offspring (as well as genetic drift). A recent study in Germany showed for the first time the genetic effect of different roadway barriers.[345] The authors found a significant genetic subdivision of bank vole *(Clethrionomys glareolus)* populations separated by a highway, but not within populations separated by a country road or railway. A similar study investigated the effect of land use and roads on the genetic structure of the common frog *(Rana temporaria)* and found that separation by highways reduced the average amount of *heterozygosity* (genetic variation) in the population.[764,765] In addition, the genetic *polymorphism* (diversity of forms) of the local frog population decreased.

Road crossings are as much about *individual behavior* as they are about habitat requirements (Figure 5.6). Little is known about the movement behavior of animals crossing roads; however, the few studies carried out so far provide interesting insights. In a study of how roads affect grizzly bear movements in the highly developed Banff National Park in Alberta, traffic volume was found to be the key variable determining the *corridor permeability* (susceptibility to being crossed) for roads of all types.[350] The four-lane, divided Trans-Canada Highway served as a complete barrier against the movement of adult female grizzly bears and as a partial filter-barrier for adult males. Black bears living on a densely roaded military reserve in coastal North Carolina preferred to cross roads during low traffic volumes.[107] In contrast, the grizzly bears in Alberta crossed primary roads during the day and night regardless of traffic volume.[350] The location of crossings was not random in either study, as black bears preferred to cross at major drainages and areas of dense vegetation, while grizzly bears selected crossing sites close to major drainages, in high-quality habitat, and in rugged terrain.[350]

Elsewhere, studies showed that among the roads that black bears negotiated, low-traffic-volume roads were crossed relatively more frequently than high-traffic-volume roads.[115,78,107] Also, the frequency of road crossing was not affected by age, sex, or season. Similarly, bobcats in Wisconsin and resident lynx in Alberta crossed paved roads less than expected, suggesting that highways influenced the lynx movements.[565,25]

Conversely, several studies have shown that secondary and unpaved roads have little effect on animal movements and are more permeable to wildlife than are primary roads.[115,78,107,350] Bobcats crossed secondary highways and unpaved roads in proportion to their occurrence,[565] and lynx regularly crossed secondary and low-volume roads.[614,656]

The above examples are for large mammals; roads as barriers affect smaller fauna differently. For smaller animals, the width of a road may be more important than the number of vehicles traveling on it. Ultimately, road widths may explain how permeable a transportation corridor or road system network may be. For example, road crossing by small ground mammals was inversely related to road width in Australia,[48] and small road clearances less than 3 m (10 ft) wide have been shown to affect small mammals such as voles and rats.[896] Woodchucks living in highway interchange areas crossed the single-lane on-and-off ramps frequently and successfully, whereas they crossed the wider roads infrequently.[1055]

Roads have been found to severely restrict or stop movements of some Australian small mammals, even when the road was unimproved and partly overgrown with vegetation.[48] An extensive study concluded that a four-lane, divided highway is just as effective a barrier to the dispersal of small forest mammals as is a body of water twice as wide.[700]

The barrier effect of a road will affect species differently depending on an animal's behavior, dispersal ability, and population density (Figure 5.6). Highways do not completely inhibit dispersal of certain rodent species.[700,505,1038] In an Ontario study, medium-sized mammals were essentially the only ground animals crossing the widest roads.[700] Forest roads did not restrict or limit the movement of yellow-necked mice *(Apodemus flavicollis)* but did restrict the movement of bank voles.[40] Small mammals in Kansas did not appear to cross roads readily but would return to their original roadside habitat if displaced across the road.[505] Apparently, that species has established home ranges to which it showed fidelity.

It may not be uncommon for a landscape structure such as a road to be the boundary for home ranges of some species. Highways in southwestern Texas apparently do not completely inhibit the dispersal of certain rodent species.[505,1038] The tendency of rodents to cross a road depends at least in part on the similarity of habitats on opposite sides of the road. When habitats were different, rodents tended to avoid crossing, but they would cross when habitats were similar.

Linear features such as roads may block and hamper the dispersal of even tiny animals and reduce the immigration rate to isolated patches of natural or seminatural habitats.[584] In central Sweden, investigators found that snail *(Arianta arbustorum)* movements were largely confined to roadsides.[56] Only one of their *marked-recaptured* snails crossed a paved road and two crossed an unpaved road, though overgrown paths and earthen roads did not influence the snail movements. Unlike terrestrial invertebrates, wide busy roads were no barrier to movement of butterflies (and burnet moths) in open populations (breeding can occur between populations), though busy roads slightly impeded those in closed populations (breeding within a population).[661] Mark-recapture results showed that 10% to 30% of adults of three species from closed populations crossed the road. Roads were not considered a barrier to gene flow in any of the species studied. The rate of movement of ground-dwelling arthropods along road verges was low compared to that in open areas, but under certain conditions dispersal along verges aids in maintaining metapopulations of ground-dwelling arthropods.[973]

The relative permeability of roadside and adjacent edge habitat is an additional factor influencing the barrier effect of roads. How severe this factor may be depends on the *hardness* (abruptness) of the edge and the species' habitat needs. An open road corridor with grass-covered roadsides can be a formidable barrier to a forest-specialist small mammal or salamander, regardless of mortality risks due to vehicles on the roadway. In the past decade, much work has been carried out assessing barrier permeability and dispersal success of forest animals in landscapes dominated by agriculture.[871, 625, 552] Despite roads being a predominant and permanent feature on our landscapes, surprisingly few studies have specifically addressed the permeability of road systems and their role in habitat fragmentation.[349]

The combined or *cumulative effect* of different phenomena such as linear roads and powerline corridors may affect movement and road-network permeability for certain species. Caribou were found to cross a road and a control area with similar frequency, but when a pipeline ran parallel to a road, crossing frequencies dropped significantly.[194] Wolves frequently crossed multilane highways in Spain, but less so when a smaller frontage road was adjacent to a highway (J. C. Blanco, personal communication). In numerous places, roads run side by side with electric powerlines, railway lines, and aqueducts. The cumulative effect of parallel linear developments on the overall permeability of roads is virtually unknown and certainly merits investigation.[494, 302]

Finally, instead of blocking movements across roads, in some instances roads may channel animals along their length. For example, upon encountering a road, some migratory animals move a short distance parallel to the road before crossing it.[741, 52] Topography, particularly road alignment with major drainages, strongly influenced the movement of deer and elk toward roadways and across them in some areas.[65, 142, 592, 282, 761] Lastly, paved and gravel roads stimulated longitudinal movements and reduced the rate of crossings by arthropods (carabid beetles and lycosid spiders).[584, 973]

Increased Movement

Although roads act as barriers for many species, it is thought that they may aid the dispersal of some other native and non-native species. Construction of the 69 000-km (43 000-mi) U.S. interstate highway system, bordered in most regions by dense grass, has created an extensive array of potential avenues for dispersal by grassland species throughout the country.[347] Although ample evidence of *along-roadside movement* for foraging exists for a limited number of species, there is scant evidence for movement along roadside verges as conduits used by wildlife for dispersal or migration.[863]

In Illinois, a meadow vole *(Microtus pennsylvanicus)* population was able to extend its range after continuous avenues of dense vegetation were established along an interstate highway.[347] Similarly, pocket gophers *(Thomomys bottae)* advanced along a paved desert highway a distance of 60 km (37 mi) in 12 years due to the increased moisture content of the roadside soil compared to adjacent areas.[431] In this case, the pavement served as a microwatershed irrigating the roadside, while precipitation that fell in open country evaporated more rapidly before penetrating the soil. Moreover, dispersal along the highway was facilitated by the underground (fossorial) habits of gophers. Gophers made few above-ground movements and were rarely killed on the road. Presumably, the increased soil moisture facilitated the burrowing and movement.

Predators and their prey can be affected by linear corridors. Wolves are often reported to be less abundant in areas with many roads.[908,456,618,644] However, roads and other corridors that receive little human use may attract wolves as easy travel routes and provide greater access to prey.[913,451] In Nova Scotia, lynx followed road edges and forest trails for considerable distances,[705] and similar observations were made during winter for roads less than 15 m (49 ft) wide in Washington State.[500]

Cane toads *(Bufo marinus)*, a species introduced to Australia, were denser on roads and vehicle tracks than in many surrounding types of vegetation.[833] Toads used roads as activity and dispersal corridors, especially in forested habitats and other habitats with dense ground and tree canopy vegetation. Roads facilitated the toad invasion and extension of range into previously inaccessible areas. However, the extent to which roads influence the distribution and abundance of non-native species, such as foxes, cats, and dingoes, in Australia and the consequences for the native fauna are poorly understood.[604]

Cumulative Effects and Road Density

Most ecological systems display a *time lag* (sometimes called an extinction debt) between when habitat degradation takes place and when its full ecological effects are detectable.[917,562,43,185] Road effects show such a lagged response because the different effects of roads on wildlife populations and assemblages—for example, habitat loss, reduced habitat quality, mortality, and reduced connectivity—typically occur at different rates (Figure 5.7). Habitat loss has the most immediate effect. As habitat is lost to road conversion,

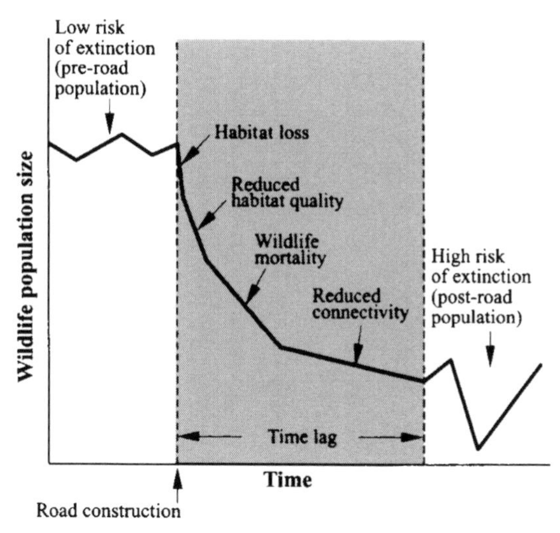

Figure 5.7. Four ecological effects of roads on an animal population and the time lag for the cumulative effect. The four sequential but overlapping road effects are indicated in the shaded area. After the time lag, population size is normally smaller with greater relative fluctuations over time.

roughly equivalent losses in population numbers can be expected to occur. However, reduced habitat quality such as edge effects exacerbate the impact on forest interior species. Furthermore, broad-scale effects may result in nonlinear reductions or loss of populations and species as road density reaches a threshold.

The increase in population mortality due to wildlife-vehicle collisions takes a little longer to be observable in the population, as road traffic increases and wildlife deaths accumulate. If mortality levels are high, their effect on the population should be observable after the road has been in place for about one or two generations of the animal species. Finally, the effect of the road as a barrier, reducing landscape connectivity, will likely take several generations to be observed and will depend on the typical time interval between local extinctions in the animal's regional population. In general, impacts on population density should be observable earlier than impacts on species diversity.

Lags in the loss of wetland biodiversity in response to road construction have been demonstrated.[287] The investigators found that the current levels of species richness in wetlands were more accurately described by road density estimates from three or four decades ago than by the current road density. If this is the case and road density is a primary factor, the estimated time lag for road effects here is about 30 to 40 years. This finding suggests a considerable time lag may occur for the full response of *species assemblages* (natural communities) to road impacts. In some areas, roads and traffic have increased rapidly over the past few years. This implies that mitigation efforts aimed at currently observable effects may address only effects caused by road conditions decades ago. If so, the effective mitigation against ecological impacts of roads currently in place will require preemptive efforts that address effects not presently observable but that may occur in the next few decades.

For species of conservation concern, roads generally reduce population sizes and increase the risk of population extinction. However, most species populations can persist in the presence of at least some roads. Therefore, in the context of road impacts on wildlife, one question is especially important. What is the critical road density in an area, above which a population cannot persist?

Because of the spatial and temporal complexities of road impacts, the question is not easy to answer. As road density increases, wildlife habitat is lost, altered, and fragmented.[450] With increasing road density, what were once larger and more contiguous expanses of habitat have become smaller and more isolated patches, often of remnant habitat. Even though so-called edge species (e.g., white-tailed deer [*O. virginianus*], most grouse species, and rabbits) benefit from increased habitat heterogeneity, interior core-sensitive species are disadvantaged. Further, with the creation of smaller and smaller habitat fragments, we are beginning to understand that species have certain spatial requirements, regardless of whether they are "edge" or "core sensitive" species. Species' response to whatever habitats they inherit is mediated through their movement patterns.[626, 627] As species populations become increasingly fragmented, the probability of extinction of local populations increases, especially if dispersal between populations is hindered. Adding the effects of a dense road network to a patchy landscape threatens the persistence of key populations.[401, 390, 388, 386, 1032]

A review of landscape ecology theory has suggested that when landscapes change over time, under certain conditions the *rate of change* is often far more important than the spatial pattern itself in affecting species longevity and survival.[402] Disturbance regimes therefore play a pivotal role in species persistence.[253, 413, 402] The key idea here is that the rate of habitat change can outstrip the genetic ability of species to successfully adapt. The rate of disturbance is only one of the keys to understanding why species respond as they do. Another key is the spatial pattern of the landscape, which will affect species proportionally more in the presence of high disturbance rates.

Numerical responses of large mammals to roads are generally interpreted as a road density threshold necessary for sustainable populations and coexistence. Several models have been developed to predict wolf pack occurrence or survival in relation to road density in Minnesota and Wisconsin.[908, 618, 643, 644] A road density threshold of 0.45 km/km^2 (0.72 mi/mi^2) best classified pack and nonpack areas for wolves.[643] Similar road density thresholds were reported for cougars *(P. concolor)* and brown bears.[960, 171] However, these studies only scratch the surface of the research problem of estimating critical road density.

Conclusion

The length of roads in the USA is about 24 km (15 mi) per 1000 persons, with almost 6.5 million km (4 million mi) of roads crisscrossing the surface (Chapter 2). This means that every square kilometer of land area is matched by 1.1 km of road. Think about that virtual ecological footprint introduced at the beginning of this chapter. What are the ecological implications? Road density varies heterogeneously across the country, and roads have come to represent major barriers to wildlife movement in developed landscapes.[398] Barriers hinder normal animal movement, effectively creating smaller patches of habitat and restricting movement across those habitats. These primary impacts show cascading effects throughout animal populations, affecting daily activities and inevitably leading to higher death rates and decreased birth and survivorship rates. The end result of a highly connected road system is a decrease in both the number and the abundance of the species that once inhabited the landscape.

But to society and the transportation community, roads are the lifelines that connect people to people, communities to communities, and cities to cities. Without these lifelines, the quality of life would dramatically decrease. How does one resolve this wildlife and societal paradox?

An introspective look at the nature of the problem by all people who have the knowledge and responsibility to solve it is needed. The broad viewpoint of the transportation community is different from that of environmentalists and land managers, as it should be. Think about a hologram: its hidden pattern emerges only if one gains just the right perspective. Some are unable to see the pattern even after significant attempts to do so. But with persistence, nearly everyone can discover the pattern. When this occurs, something significant has happened. A neural connection has been made revealing something new.

The hidden pattern in the landscape is the virtual ecological footprint. Recognizing that it is sensible to lessen the impact of that footprint is the wisdom that comes with seeing the pattern. It then becomes clearer that to a large degree the ecological landscape is the mirror image of the road network. The more connected and impermeable the road system, the less connected is the landscape matrix in which it is embedded. Increasing road density fragments the landscape more and more, creating semipermeable or impermeable barriers (e.g., solid Jersey barriers) to animal movement. The result is fragmented wildlife populations whose longevity is challenged and whose survival is increasingly affected by random events.

"The flip side of a [wildlife and landscape] corridor strategy is to diminish barriers," as ecologist Reed Noss has written.[685] The key is to be able to read the hologram, to turn the coin over and see the flip side, the virtual footprint. Specifically, the key involves making sensible choices that maintain the lifelines upon which people and communities depend while at the same time achieving a much more permeable landscape that maintains the lifelines upon which ecological processes and animal movements depend.

The choices made should both lessen direct mortality of wildlife and reduce the barrier effect. Reducing mortality will involve the installation and maintenance of wildlife exclusion fences, underpasses, and overpasses (hereafter often referred to as wildlife crossing structures or passages) as well as other techniques. Installing fences and structures that allow animals to escape the right-of-way is a necessary adjunct to installing crossing structures. As these kinds of sensible mitigation choices are made and implemented, the ideal of environmentally friendly or "green" road networks comes closer to being a reality.

A handful of principles or guidelines emerges from the preceding concepts:

- A larger virtual ecological footprint is associated with the physical footprint of roads. Road planners/builders and environmentalists need to be concerned with the broad landscape rather than the one-dimensional road corridor.
- Animal mortality on roads is largely determined by the interactions among driver behavior, animal movement behavior, structure of the road, and structure of the nearby landscape
- For viable animal populations and their persistence over time, animals need to be able to move through the landscape, move among populations, disperse freely, and recolonize lost areas.
- The effect of road mortality on wildlife populations increases one or two wildlife generations after the road has been in place, whereas the effects of a road as a barrier will likely take several wildlife generations to be observed.

CHAPTER 6

Mitigation for Wildlife

A sedan driven by a forty-year-old woman approached. She saw the turtle and swung to the right, off the highway, the wheels screamed and a cloud of dust boiled up. The turtle had jerked into its shell, but now it hurried on, for the highway was burning hot.

And now a light truck approached, and as it came near, the driver saw the turtle and swerved to hit it. His front wheel struck the edge of the shell, flipped the turtle like a tiddly-wink, spun it like a coin, and rolled it off the highway. . . . Lying on its back, the turtle was tight in its shell for a long time. But at last its legs waved in the air, reaching for something to pull it over. Its front foot caught a piece of quartz and little by little the shell pulled over and flopped upright. . . . The turtle . . . jerked itself along, drawing a wavy shallow trench in the dust with its shell. The old humorous eyes looked ahead, and the horny beak opened a little.

—John Steinbeck, *The Grapes of Wrath,* 1939

In front of a motel room in Ottumwa I finger-scrape the dry, stiff carcasses of bumblebees, wasps, and butterflies from the grille and headlight mountings, and I scrub with a wet cloth to soften and wipe away the nap of crumbles, the insects, the aerial plankton of spiders and mites. I am uneasy carrying so many of the dead. The carnage is so obvious.

—Barry Lopez, *Roadkills,* 1998

The preceding chapter unravels numerous ways that roads affect wildlife, not only as individuals but as populations. The impacts depend on both the characteristics of the road system and the wildlife populations interacting with it. Some effects are subtle, not readily perceptible to the human eye, yet others, such as road-killed animals, are glaringly visible as one motors along.

Although some rarely used roads are occasionally removed,[35, 376] roads are continually being built and traffic is growing at a much greater rate.[44] This process produces a densely woven fabric of transportation on which economies heavily depend and with which wildlife has immense trouble coping. The new realism in transportation planning recognizes the impact of roads on local ecology and ecosystems[799,674,725,675] and the key role of transport planners in promoting sustainable development.

Mitigation measures designed to lessen the impact of roads on wildlife are necessary components of a sustainable transportation strategy.[654,611] Techniques for reducing wildlife-vehicle collisions and providing safe passage across roads have been used for decades. Yet, for mitigation measures to contribute usefully to a sustainable or "environmentally friendly" transportation system, they must be effective themselves. Poorly designed measures do little to *mitigate,* or minimize, road impacts and are largely a waste of taxpayers' money. Furthermore, they have the potential to disrupt naturally occurring processes, such as movement patterns across the landscape, which may lead to severe overgrazing, increased erosion, or a decline in population size.

We begin this chapter with sections on both mitigation for mortality and mitigation for habitat loss and reduced habitat quality. Then we explore wildlife underpasses and overpasses, including their design and ability to work successfully. Finally, we introduce some distinctive case studies.

Mitigation for Mortality, Habitat Loss, and Reduced Habitat Quality

This section begins with approaches to minimize road-kill and concludes with approaches that address habitat loss and degradation.

Mitigation for Mortality

Several mitigation techniques have been used over recent decades to reduce mortality of wildlife on roads (Figure 6.1). Most, but not all, have been directed at reducing *large animal–vehicle collisions.* Eleven mitigation techniques employed by 42 states in the USA to reduce deer-vehicle collisions are particularly important: wildlife fencing, underpasses or overpasses, warning signs, lower speed limits, mirrors, reflectors, highway lighting, ultrasonic warning

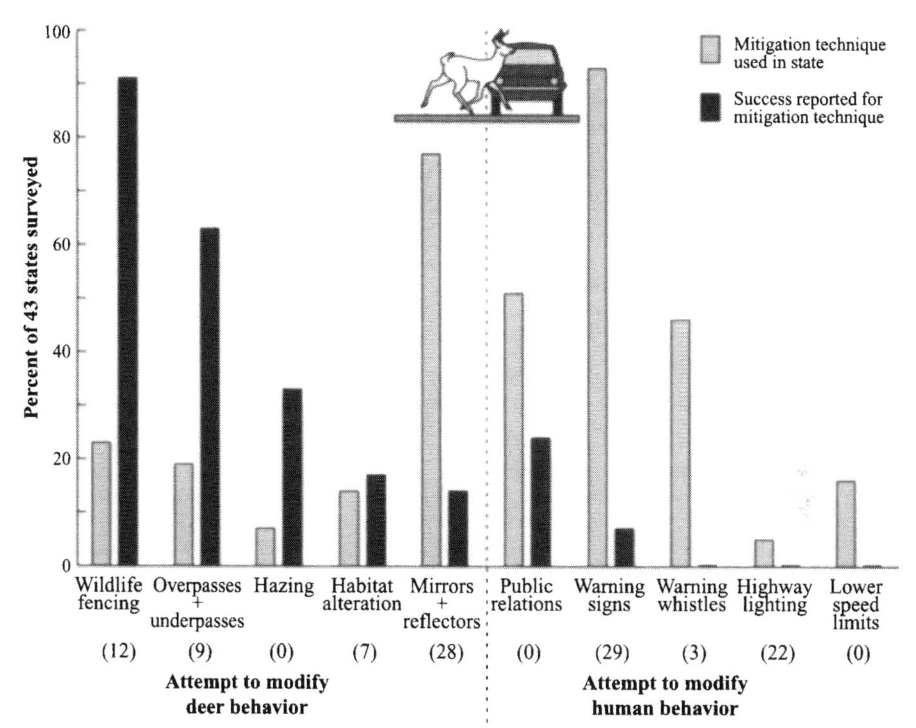

Figure 6.1. Mitigation techniques used in states to reduce deer-vehicle highway collisions. Data are based on responses to a 1992 survey by natural resource agencies in 43 states. Numbers in parentheses are the number of states where use of the mitigation technique is based on one or more research studies. Adapted from Romin and Bissonette (1996a).

whistles, habitat alteration, hazing animals from the road, and public awareness programs.[785]

Curiously, the measures most used by states do not correlate with the perceived success of the measures (Figure 6.1). In general, the most promising measures are among the least used. Moreover, in many cases, evaluations of success are based mainly on opinion rather than research. Of all the techniques shown in Figure 6.1, fencing and wildlife crossing structures were believed to be the most effective (91% and 63%, respectively), yet few states had based that conclusion on research results. There is an urgent need for rigorous evaluations of these measures, and of new ones that may appear, in an adaptive mitigation process. Such a process will allow new research results to be incorporated into future mitigation schemes and be further tested and evaluated. Incrementally and over time, deficiencies will be corrected and the effectiveness of mitigation improved.

Most of the measures used are quite simple. Wildlife fences in North America typically consist of 2.0–2.4 m (7–8 ft) high wire-mesh fence material, often with one-way gates or ramps that allow deer and related ungulates to get off a highway right-of-way.[757, 570, 170] It is a continual struggle to maintain fences and one-way gates so that wildlife do not slip through and reach a roadside. Earth slumping or gullying on hillslopes, inadequate installation techniques resulting in gaps between the earth and the fence bottom, and breaches of the fence by the public (e.g., by hunters and snowmobilers) allow animals to gain entry to the roadside. Fence maintenance is usually neglected shortly after construction and is a major concern because priorities and budgets change over time. Meanwhile, fence damage and gaps are a continual problem.

Properly constructed and maintained fences used in conjunction with overpasses and underpasses, as described later in this chapter, can effectively deter deer trying to access roadside vegetation[256, 570, 282] (Figure 6.2). However,

Figure 6.2. A particularly effective underpass for crossing a multilane highway by both large predators and herbivores. Note the relatively wide view in the upper portion and the vegetation visible on the far side. This underpass is 11 m (39 ft) wide, 3 m (10 ft) high, and 40 m (131 ft) long (from far edge of one highway span to far edge of the other) on the divided, four-lane Trans-Canada Highway in Banff National Park, Alberta. Wolves, grizzly bears, black bears, and cougars commonly cross the highway in this underpass; see Figure 6.6 for wildlife crossing data. Photo by A. P. Clevenger.

deer and other ungulates avoid situations where they become vulnerable and thus are reluctant to use confining structures. These structures may be threatening to deer[757,997] in part because they place limitations on escape. Similarly, long and narrow underpasses may pose formidable threats to ungulate passage.

For these and other reasons, roadway planning and design can be effective only with a strong understanding of animal behavior. *One-way gates,* which require deer, elk, and other large animals to squeeze through a ladder of interdigitated steel rods, are not the best design to ensure the ability of ungulates to exit from a roadside. Earthen *one-way ramps* have been used in Wyoming, Utah, The Netherlands, and elsewhere.[63, 792] Positioned against the roadside fence, these ramps are designed to allow animals to jump over the fence and escape from the roadside. One-way ramps are 10 to 12 times more effective than one-way gates.[92]

The effectiveness of some of the other mitigation techniques (Figure 6.1) has also been tested. Increased highway lighting[755] and ultrasonic whistles[787] have proven ineffective in reducing deer-vehicle collisions. A popular accident-reduction technique, "Swareflex" reflectors are red reflectors mounted on posts at the average height of vehicle headlights. The posts are placed at regular intervals along both sides of a highway where deer-vehicle collisions are to be reduced. The system is designed to reflect vehicle headlight beams, creating a low-intensity red beam that acts as a moving "lighted fence." However, the underlying assumption of deer avoiding reflected red light has been seriously questioned.[1066, 999] Tests of reflector effectiveness have produced mixed results, presumably because animals quickly get used to the reflected beams.[819, 762, 938]

Little evaluation has been conducted of warning signs for drivers, habitat alteration, *hazing* (where animals are harassed away from the road surface by highway personnel), or public awareness efforts (Figure 6.1).[538, 632] However, it is apparent that warning signs will be effective only if they result in heightened driver caution and slower speeds in the areas of high mortality. Familiar signs, such as a yellow leaping-deer sign in some areas, are normally unsuccessful (even if antlers are attached backward, few motorists notice). However, novel signs, and perhaps changing or relocating signs to capture attention, combined with increased law enforcement may slow drivers down in key wildlife crossing locations.

Habitat alteration, such as planting unpalatable species or vegetation removal along roadways, especially in the more arid regions of the continent, may help reduce the attractiveness of roadsides to deer and other herbivores.[744] This is an opportunity for public awareness education. Brochures, as well as local workshops or information sessions, represent a little-tested way to get the message across. Such public education efforts could be sponsored or supported by transportation agencies and other interested parties.

At-grade crossings (crossings where the road and surrounding area are at the same level) with wildlife fencing, diagonal boulder strips on roadsides, and cross-lines on road surfaces were studied in northern Utah. They were found to reduce deer mortality by 36% to 40%.[539] However, because the specially designed crossings channel deer across the road surface, a large number of deer-vehicle mortalities continued to occur in areas of high traffic and on divided highways. The use of at-grade crossings offers some promise in reducing deer-vehicle collisions on two-lane, low-traffic-volume highways.

Earlier studies have shown that vehicle speed, traffic volume, and timing of traffic pulses are among the primary factors that affect animal-vehicle collisions. These are variables over which transportation agencies have some control through construction alternatives, signage, and so forth. Highway speeds of 90 km/h (55 mph) or higher sharply increase the chances of vehicle-animal collisions. Driver reaction time and the distance at which cryptically colored animals are visible to the speeding motorist contribute to the high animal-vehicle collision rate. If these primary variables could be widely mitigated, the gain would be substantial. In particular, if traffic speed were significantly reduced in areas of high road-kill, this single action would result in a major reduction in animal-vehicle collisions.

The options for mitigating high road-kill rates fall into two categories: (1) changes that affect *motorist behavior* and (2) changes that affect *animal behavior* around roads. Survey results indicate that current mitigation efforts tend to focus on attempting to modify human behavior (Figure 6.1). Less effort overall attempts to modify animal behavior which, being perceived as more successful, represents a tangible opportunity for progress. Motorist response can be influenced by the following:

- Changing motorist behavior (education, signage, movement-sensing roadside signal lights, rumble strips [also see Noise and Animals section in Chapter 10], highly visible speed indicators, law enforcement)
- Improving the motorist's field-of-view to better see roadside animals (wider open roadsides in key locations, movement-sensitive lights and indicators along roads)
- Managing the traffic on roads (restricting use of roads during critical breeding or migration times or when accident risk is highest)
- Implementing *traffic calming* techniques (adding structures that reduce speeds, redesigning roads for lower speeds, channeling traffic onto specific routes)

Animal behavior can be influenced by the following:

- Structures (creating wide roadsides in key locations, wildlife crossing structures, partial or full fencing, one-way ramps, reflectors, highway lighting)
- Habitat modification (unpalatable vegetation along roadsides, channeling wildlife to designated crossing locations)

Mitigation for Habitat Loss and Reduced Habitat Quality

The first step in the avoidance of ecological impacts from roads and vehicles on sensitive habitats is to make road alignment adjustments to prevent conflicts (Table 6.1). If this cannot be done, the second step focuses on minimizing or eliminating the effects on habitat loss or quality by using ecological mitigation measures at the problematic location. If neither avoidance nor mitigation can

Table 6.1. Three steps in ecologically sensitive road planning. Examples listed for each.[192, 193, 191]

Avoidance: prevent or avoid ecological impact.

- Do not build the road.
- Change the route.
- Put the road underground.
- Close the road to motorized vehicles.
- Remove the road.

Mitigation: minimize the ecological impact.

- Perforate the road corridor with underpasses and/or overpasses for animal movement across.
- Depress roads and/or add soil berms to reduce traffic-disturbance/noise effects.
- Make quieter road surfaces, tires, motors, and vehicle aerodynamics.
- Use stormwater best-management practices for reducing dispersion of pollutants.
- Use cleaner fuels and engine modifications to reduce air pollutants and greenhouse gases.
- Reduce vehicle distance traveled (VKT or VMT), especially on secondary roads.

Compensation: provide an equivalent amount of ecological improvement in the local region to balance the ecological impact that cannot be avoided or mitigated.

- The following guidelines address habitat loss, habitat degradation, and habitat isolation:
  - An area larger than the impacted area is better than providing improvement for the same amount of space.
  - Closer to the impact location (but generally not in the zone of road effects or where future road effects may occur) is better than farther.
  - Replacing the same type of ecological conditions is better than improving a different type.
  - Improvement that exceeds the quality of ecological conditions before an impact was added is better than providing the same quality.
- Create a greater protected area of quality bird habitat than the area degraded by traffic noise.
- Enlarge a nearby, large natural-vegetation area.
- Restore nearby altered streams or wetlands to their natural conditions.
- Establish appropriate corridors and stepping stones to enhance wildlife movement.
- Enhance habitat conditions for rare species and special sites of high biodiversity.

(a) Jersey barrier with "wildlife scupper"

(b) Stump line for small and midsized animal movement

(c) Wildlife culvert

(d) Canopy connection for arboreal animals

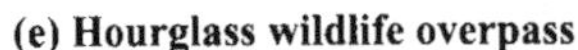

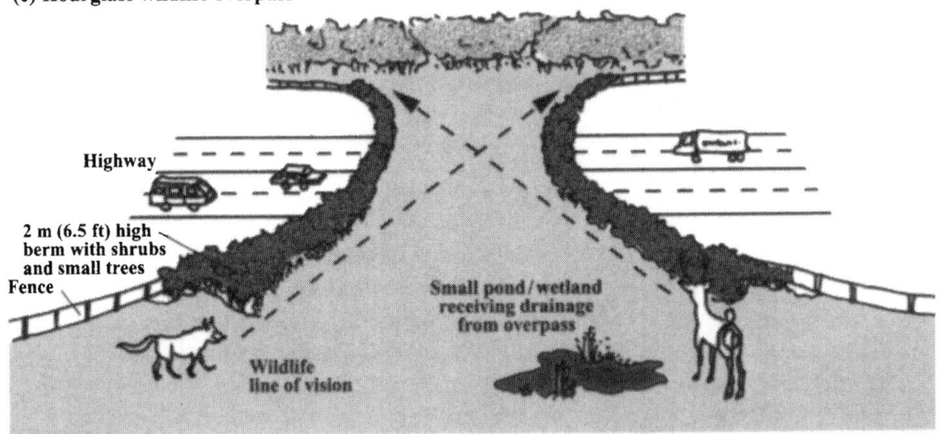

Figure 6.3. Diverse structures used to enhance highway crossing by wildlife. (a) Concrete Jersey barrier between traffic lanes designed for crossing of wildlife and water; California and Trans-Canada Highway, Alberta. (b) Line or wall of stumps, logs, and branches that greatly facilitates animal movement; The Netherlands. (c) Culvert with two 40 cm (16 in) wide wildlife paths that are only briefly covered by water after heavy rainstorms; The Netherlands. (d) Canopy structure providing connectivity for arboreal mammals and other animals across a (road) barrier; rainforest in Queensland, Australia. (e) Hourglass-shaped overpass with grassland (some are savanna-like or wooded), in which subsurface drainage pipes carry water to a tiny pond that attracts wildlife; The Netherlands and France (where diagonal lead-in sides are straight). Note the wide angle of view for approaching wildlife.

be effectively accomplished, the third step involves implementing compensation measures, usually at another site. The ecological *compensation principle* implies that, for new road developments, there is a goal of "no-net-loss" of natural processes and biodiversity. This principle has been law in The Netherlands since 1993 and is beginning to be applied elsewhere.[191,192,193] Results from six Dutch projects reveal that (1) compensation costs are often marginal compared to total project costs, and (2) the compensation principle may play a valuable role in the development of alternative route alignments.[192]

Other means of mitigating habitat loss or diminished habitat quality focus on making the area around roads suitable for wildlife. The construction of wetlands and ponds may enhance the habitat for amphibians (Figure 6.3). Small ponds have been constructed on some wildlife overpasses that enhance the value of the area for amphibians in the surroundings.[63] The construction of vegetated earth berms along roadsides bordering wetlands and above wildlife underpasses can help reduce traffic noise levels and disturbance to wildlife. Furthermore, when landscape connectivity and road permeability (Chapter 5) are reestablished, habitat loss is partially mitigated.

Types of Wildlife Underpasses and Overpasses

Attempts to mitigate barrier effects can be found in road construction and upgrading projects in the USA, Canada, and other nations. Apparently the first (documented) passage for wildlife was constructed in Florida in the 1950s.[247] Today, passages are used for road mitigation in many parts of the world. Yet relatively few performance evaluations have been reported. Effective designs for mitigation structures need to be included in highway planning for more effective mitigation. Transportation initiatives such as the TEA-21 legislation in the USA and program COST-341 in the European Community have recently heightened the concern for sustainable transport systems and the regular incorporation of mitigation passages in transportation planning schemes.[131,946,792,725,279]

The general function of a *wildlife crossing structure,* or *passage,* is to get animals safely across a roadway, thereby providing for natural movements and usually reducing road-kills.[722] When this is achieved, individual animals and their populations as a whole will benefit. Essential biological requirements—such as finding food, cover, and mates, as well as facilitating the dispersal of young and aiding in the recolonization of vacant areas—can be met. The latter is important if populations are to remain viable within fragmented and human-modified landscapes. A variety of mitigation passage designs is currently in use. Choosing a design is usually based on the target species (e.g., cougars) or species groups (e.g., carnivores) expected to use the passage.

Numerous factors influence animal movement across a passage, including the design features of the passage, surrounding landscape features, and levels of human activity occurring in close proximity.[1060, 722, 168] Monitoring and performance evaluations show whether passages are meeting criteria for success and whether retrofitting or landscape enhancements are needed. More importantly, they reveal what design features may be critical for passage construction in the future. Still, much can be learned from the more than two previous decades of constructing numerous types of wildlife passages.[785]

Small Passages

Small passages (less than 1.5 m [5 ft] diameter or height) include amphibian tunnels and assorted "ecopipes" and "ecoculverts." *Amphibian tunnels* are widely installed and used in Europe where annual migrations to and from breeding sites are blocked by roads.[522, 445, 339, 792] However, as yet, few are used in North America. Diameters range from 30 to 100 cm (1 to 3.3 ft) for tunnels less than 20 m (66 ft) long. Diameters are at least 150 cm (5 ft) for tunnels more than 50 m (164 ft) long. Concrete tunnels are preferable to metal or plastic, and are more effective if the tunnel floor has an earthen substrate, because most amphibians require moist conditions. Many different amphibian tunnel designs are used across Europe.

Wildlife pipes, or *ecopipes,* are small, dry tunnels (30 to 40 cm [1 to 1.3 ft] diameter) primarily designed for the passage of small and medium-sized mammals.[63, 339, 64] They are placed so that water normally does not flow through them. Pipes up to 90 cm (3 ft) diameter have been installed as otter *(Lutra lutra)* crossings in the United Kingdom.[832] Originally referred to as "badger tunnels" in The Netherlands, more than 300 wildlife pipes have been installed along Dutch motorways and have aided in the dramatic recovery of the species.[64]

Wildlife culverts, or *ecoculverts,* are similar to the preceding, but are located over waterways. These structures are up to 120 cm (4 ft) wide, and have raised dry ledges on each side of the central water channel to facilitate animal movement[968] (Figure 6.3c). Monitoring has shown that wildlife culverts with ledges greater than 40 cm (1.3 ft) wide are most effective, though many wildlife culvert designs are used.[683, 792] Also, in Australia, *talus tunnels,* locally called "tunnels of love" since they permit males to reach areas with females, were successfully designed for pygmy-possums *(Burramys parvus),* an endangered species entirely dependent on rock talus slopes for survival.[591]

These passages were specifically designed for animal crossing. Animals also cross roads through a myriad of existing drainage pipes, culverts, and other conduits not expressly designed for wildlife movement. Limited monitoring of these "non-wildlife-engineered" passages has shown that they can also be important linkages for local wildlife.[1060, 781, 791, 167, 846] Small passages are often combined with fencing to prevent animals from reaching the road and to direct them to the passage entrances. *Drift fencing* (barrier used to channel the movement of animals), earth berms, and vegetation are effective ways of leading animals to a tunnel entrance.[522, 63, 432]

In studies conducted in Spain, small mammals, rabbits, and small carnivores preferred short culverts (<15 m [49 ft]) as opposed to long culverts.[1060, 792] Vegetation cover near underpass openings was important for small mammals, including carnivores. Small mammals used small-diameter culverts (≤1 m [3 ft]) more than large ones.[433, 781, 167] Reptiles crossed more frequently through circular than rectangular structures; amphibians, rabbits, and domestic animals preferred rectangular structures; and no difference was evident for small mammals and carnivores.[791]

Large Passages

Large passages (>1.5 m [5 ft] diameter or height), which include wildlife underpasses and overpasses, are designed primarily for large mammals or a wide variety of animal types.[318, 310, 63, 446] Although rare and scattered throughout North America and Europe, large passages continue to be built across Europe and in Florida and are being planned or under consideration in many regions.

Wildlife underpasses, including narrow tunnels, are below-road passages that range from 2 m (6.5 ft) diameter metal or concrete culverts to extended bridge or viaduct-type structures more than 100 m (328 ft) wide.[324, 168] The height of most underpasses is about 2 m and can reach up to 4 to 5 m (13 to 16 ft). There are many variations of underpass design, but in North America today most built specifically for wildlife can be categorized as one of three types: (1) metal multiplate culvert (circular or elliptical), (2) prefabricated concrete box, and (3) open-span bridge (over land or waterways) (Figure 6.2). Viaducts designed to minimize noise are the most effective underpasses.

Wildlife overpasses, or *ecoducts,* are also designed for large mammals. Most are 30 to 50 m (98 to 164 ft) wide (for instance, as viewed by an animal crossing over a highway), but they can be as wide as 200 m (656 ft) or more (Figure 6.3e). Unpaved roads may be situated on top of the wider structures, though the primary objective is the passage of wildlife. There are close to 50 wildlife overpasses worldwide but only six in North America. Two in New Jersey are about 30 m (100 ft) wide, one in Utah is 8m (26 ft) wide, one in British Columbia is 5 m (16.4 ft) wide, and two in Alberta are 52 m (171 ft) wide. However, as highways widen and traffic grows, interest in wildlife overpasses as habitat linkages across transportation corridors is increasing.

The term *green bridge* is sometimes used to refer to a structure, such as a wildlife overpass, with a relatively wide continuous strip of natural (including woody) vegetation crossing over a road.[63, 475] These structures are usually designed to enhance large mammal crossing, but many have the additional

Figure 6.4. Two green bridges or landscape-connector overpasses on the edge of a major city. These overpasses, with continuous vegetation and typically a lightly used earthen road, provide connectivity for local residents as well as animals and plants. In addition, there are five green bridges over two-lane highways within the 50 km (30 mi) radius semicircle around the city. Carretera de los Tuneles, Barcelona, Spain. Photo by R. T. T. Forman.

major role of enhancing the crossing of local people or hikers or both (Figure 6.4; also see Figure 1.5 in Chapter 1). Underpasses and green bridges are combined with fencing to keep animals off the road and lead them to the entrances. Landscaping around and on the structures creates an attractive, reassuring environment for approaching wildlife (J. Weader, personal communication).

The term *landscape connector* is sometimes considered synonymous with green bridge. Alternatively, a landscape connector is sometimes considered to be an especially wide overpass that effectively maintains the connectivity of many horizontal ecological flows across the landscape, including wildlife movement.[320] Where a highway is placed in a relatively long tunnel with vegetation above, as for example in Germany and northern Minnesota, it should be the most effective passage, functioning as a landscape connector.

Passages Originally Designed for Other Purposes

The types of small and large passages described above are used only where animal crossing was purposely considered in the structure design. However, there are literally millions of below-road passages expressly designed for draining road runoff or moving livestock, water, or people (Chapters 2 and 9). Most presumably are also used by wildlife at least in a limited manner (see Figure 9.8 in Chapter 9). A few studies monitoring these *non-wildlife-engineered passages,* or *multipurpose passages,* have shown that the structures can be important linkages for local wildlife.[135,167,846,792,169]

Of the huge number of below-road passages, many could be used more effectively by wildlife. Road impacts on local wildlife could be lessened if efforts to examine habitat needs and movements around the structures were implemented. Both transportation agencies and natural resource agencies have too often overlooked these existing passages and their potential for improvement, even though substantial gains for little investment could be realized. In an ecological planning program, the Florida Department of Transportation has identified "regional ecological zones" (or "bottlenecks") in the state road system (see Figure 14.6 in Chapter 14).[851] Areas were prioritized by ecological significance (such as riparian and/or conservation areas). Then geographic information system (GIS) models were used to identify ecological corridors or animal movement linkages across roads. For several road projects, suitable existing passage structures were identified, and structural and land-planning recommendations were made to improve wildlife movement at unsuitable passage sites. Likewise, the periodic reconstruction of highway bridges that span waterways is a valuable opportunity to improve wildlife passage along riparian corridors by widening bridge spans and enhancing habitat.

Factors Affecting the Use of Wildlife Passages

Knowing how the different crossing-structure types perform, particularly from a cost-benefit perspective, is of inherent interest. Because the type of surrounding habitat, the animal populations present, and the dimensions of a structure may vary widely, it is difficult to make direct comparisons. Having structures side by side would minimize confounding variables, but that rarely occurs and usually is not economically feasible.

However, a close approximation to a side-by-side design is found in Banff National Park, Alberta. An underpass is within 200 m (656 ft) of each of two overpasses. After four years of study, some interesting species-specific patterns have emerged. Grizzly bears, wolves, and all ungulates (deer, elk, moose, bighorn sheep) tend to prefer wildlife overpasses (Figure 6.5). Cougars (mountain lions) prefer underpasses. Black bears do not appear to have a preference (Figure 6.5). These results are supported by similar findings from subsequent analyses of crossing structures in the same area[359] (A. P. Clevenger and N. Waltho, unpublished data).

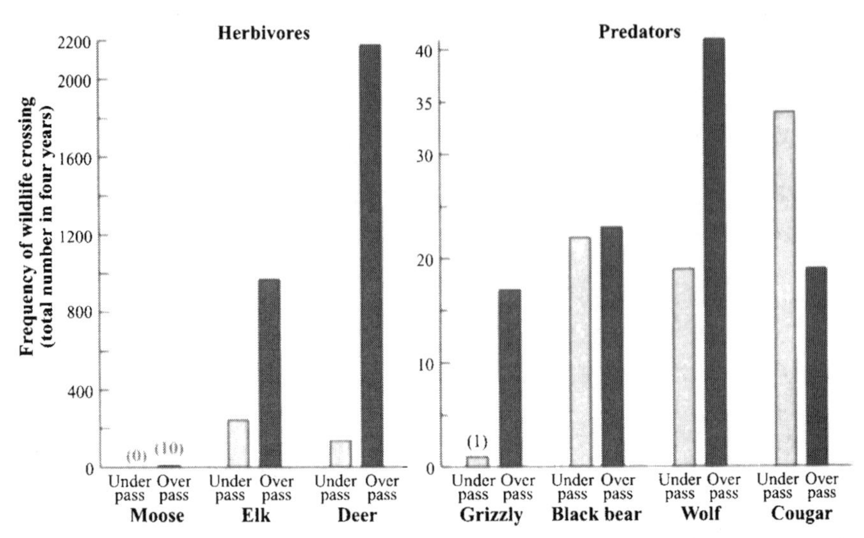

Figure 6.5. Comparison of wildlife usage of overpasses and underpasses for crossing a multilane highway. Two 52 m (171 ft) wide overpasses and the nearest (<200 m [656 ft] distant) underpass (3 m and 7 m wide) for each were sampled for four years (1997–2001). Trans-Canada Highway, Banff National Park, Alberta. Based on Clevenger (2001).

Some underpass design recommendations have been made based on observations of how often animals use the designs. For instance, deer and other ungulates tended to prefer underpasses that are at least 7 m (23 ft) wide and 2.4 m (8 ft) high, and that have vegetation cover nearby.[758, 145, 63, 324, 611] Although it will be hard to determine the optimal passage type, we can learn what factors are important in facilitating wildlife passage.[1060, 781, 168] This is a key step in learning about structures in order to create still more effective designs.

A simple index that incorporates structural width, height, and length dimensions and characterizes the amount of light or openness has been proposed (width × height/length).[756, 792] The authors found that underpasses with index numbers less than a certain level were not used by ungulates, and thus used the number as a recommended minimum for better wildlife-crossing results in an underpass. The relatively few tests of the index available seem to have produced mixed results. For example, the index number was not an important factor in a study of 11 old underpasses in Alberta, but was one of several significant correlates in a study of new underpasses in the same landscape[168] (A. P. Clevenger and N. Waltho, unpublished data).

Design requirements for prey species such as deer and elk will not necessarily be the same for large predatory beasts. Florida panthers *(Puma concolor coryi)* used underpasses 2.1 m (7 ft) high. However, the amount of and proximity to human activity consistently ranked as a significant factor influencing carnivore and ungulate use of 11 underpasses in Banff National Park.[168] Underpass dimensions may have had little effect on use because animals might have become behaviorally adapted to the 12-year-old underpasses. Individuals require time to learn to adjust to crossing structures.[758,1005,125,696] Once this adjustment has occurred, the dynamics of human activity and landscape heterogeneity features may be more decisive in determining whether the structures will be optimally used than the size dimensions themselves. Results indicated that the best-designed and best-landscaped underpasses may be ineffective if human activity is not controlled.

A recent sequel to the Banff study examined a completely new set of underpasses and overpasses, which animals would have had little time to become familiar with (A. P. Clevenger and N. Waltho, unpublished data). Contrary to the earlier findings, crossing structure dimensions best explained use by both predator and prey species (Figure 6.6), whereas landscape and human-related factors were of little importance. Grizzly bears, wolves, elk, and deer preferred large open passages, while black bears and cougars preferred small, more constricted passages. These patterns conform to the evolved species behaviors and life history traits, with some species preferring open areas and others needing cover. The new findings emphasize that animals apparently need time to adjust their behavior and that different species prefer different passages based on their behaviors and life history needs.

With relatively few overpasses existent, the information about these structures is still somewhat limited; most data come from Europe. One study compared several overpass designs in Europe.[722] For large mammals, the location and width were more important than the substrate or the type of vegetation cover. Overpasses less than 20 m (66 ft) wide were used significantly less by wildlife than were wider structures. Surveillance by infrared video camera indicated that animals were more relaxed on wide structures than on narrow ones. A key point was that successful overpasses provided habitat connectivity not just to adjacent habitats but also for a much broader landscape scale. The authors concluded that the standard width for overpasses depends on the purpose of the structure and the requirements of the target species. Nevertheless, widths of 50 to 60 m (164 to 197 ft) appeared to fulfill the requirements of all wildlife in their study.

Hourglass-shaped (or parabolic-shaped)overpasses have sometimes been built in Europe[63,792] (Figure 6.3). In France, these were not considered effective for crossing by red deer *(Cervus elaphus)*, a relative of North American elk.[964,965] Yet wild boar *(Sus scrofa)* regularly used the overpasses. In The Netherlands, the first wildlife overpass built was hourglass-shaped, 15 m (49 ft) wide in the center and 30 m (98 ft) wide at the ends, and although target animals generally crossed at low rates, many tended to "spook" when approaching the center.

Crossing structure type	Predators					Herbivores				Total	Human use
	Grizzly bear	Black bear	Wolf	Coyote	Cougar	Elk	Deer	Moose	Sheep		
(a) "Old structures" (built 1986–88)											
Creek bridge with path	2	37	131	50	73	425	158	0	0	876	83
Open-span underpass	*13	*148	*285	57	*272	1804	915	0	9	3503	26
Open-span underpass	*5	89	*823	*81	167	2185	578	0	0	3928	54
Open-span underpass	2	17	118	*79	116	1504	*1229	0	*175	3240	*2368
Open-span underpass	2	6	185	64	*219	*3236	448	0	*727	*4887	554
Open-span underpass	0	1	237	13	197	*3989	215	0	0	*4652	*1553
Open-span underpass	0	31	140	69	159	1425	*1937	0	1	3762	19
Open-span underpass	2	33	252	41	86	1713	553	0	2	2682	933
4 m diam. round culvert	0	*99	146	61	68	490	949	0	0	1813	42
Total	**26**	**461**	**2317**	**515**	**1357**	**16 771**	**6982**	**0**	**914**	**29 343**	**5632**
(b) "New structures" (built 1997)											
52-m Wildlife overpass	*15	14	*29	18	57	206	*1025	*6	0	*1370	23
52-m Wildlife overpass	2	9	*19	1	73	*795	*1327	*8	0	*2234	18
Creek bridge with path	2	4	14	17	50	139	177	0	0	403	*152
Creek bridge with path	1	4	7	*29	95	218	41	0	0	395	18
7×4 m elliptical culvert	2	*19	17	0	51	* 942	281	0	0	1312	*115
7×4 m elliptical culvert	0	5	9	17	*113	234	236	1	0	390	5
7×4 m elliptical culvert	1	8	1	7	73	232	201	0	0	523	9
7×4 m elliptical culvert	0	6	7	*22	36	135	111	0	0	317	13
3×2.5 m box culvert	1	18	13	14	88	142	36	0	0	312	29
3×2.5 m box culvert	1	*22	3	8	62	107	41	0	0	244	19
3×2.5 m box culvert	0	17	10	17	*176	27	18	0	0	265	7
3×2.5 m box culvert	0	2	3	2	38	83	27	0	0	155	23
1.8-m diam. round culvert	0	14	0	18	73	8	3	0	0	116	5
Total	**25**	**142**	**132**	**170**	**985**	**3268**	**3524**	**15**	**0**	**8036**	**436**
Grand total	**51**	**603**	**2449**	**685**	**2342**	**20 039**	**10 506**	**15**	**914**	**37 379**	**6068**

*=One of the two most frequently used crossing structures by each species in each study.

Figure 6.6. Frequency of use of various types of wildlife crossing structures by large predators and herbivores to cross a multilane highway in Banff National Park, Alberta. "Old structures" were monitored for five years and three months beginning November 1996; "new structures," for four years and three months beginning November 1997. Figure 6.2 illustrates the first open-span underpass in the list. Creek bridge with path indicates a 11–25 m (36–49 ft) wide bridge over a creek with a 2–3 m (6.5–10 ft) wide path along one side. Based on Clevenger and Waltho (2000) and Clevenger (2001).

The most recent overpass completed, 30 m (98 ft) wide in the center and 80 m (262 ft) wide at the ends, is considered by the Dutch Ministry of Transport to be the optimum design for the large mammals and other species present.

The information to date also seems to indicate that placement is critical in determining the functioning of structures.[62, 324, 722] An unobstructed view of the habitat on the far side of a wildlife passage is particularly important for carnivores[62] (Figures 6.2 and 6.7). In Florida, an unobstructed view by the crossing animals was considered to be probably as important as the exact width or height of an underpass.[324]

Because species respond differently to wildlife-crossing-structure designs and adjacent landscape features, mitigation planning for a multiple-species ecosystem will not be simple. Further, the structures will only be as effective as the land and resource management strategies around them. For these measures to fulfill their function, mitigation strategies need to be contemplated at two levels. *Local-level impacts* from development and human activity near crossing structures will decrease habitat quality and likely disturb animal

Figure 6.7. Box culvert underpass used by Florida panther *(Puma concolor coryi)*, black bear *(Ursus americanus)*, and many other animals. Precast structure with 8 × 25 ft (2.4 × 8 m) opening is dropped into place on a two-lane highway lined nearby with fencing. Panther tracks and an infrared-sensor flash camera were present in this culvert when photographed. Route 29, South Florida. Photo by R. T. T. Forman.

movements, particularly of large predators.[851, 168] Similarly, *broad-scale impacts* or habitat disturbance could impede or obstruct movements across the landscape toward the structures, thus rendering them ineffective. Mitigating a highway network for wildlife is a long-term process that will last for decades and affect individuals and populations alike.[696] Thus highway mitigation strategies developed around land-use planning do not terminate with the construction process but rather are proactive at both spatial scales to ensure that crossing structures remain functional over time.

When planning and designing mitigation passages, it is important to remember that every mitigation scheme has unique characteristics. What seems to work in Florida may have less than desirable results in Québec and vice versa. Each mitigation scheme has its own set of faunal components, connectivity concerns, and land management priorities. Nevertheless, general principles and guidelines are emerging from the existing passages and studies which will increasingly be the foundations for mitigation ahead.

Wildlife Passages: Summary and Effectiveness

We now summarize the status of the information available and then examine the criteria and design characteristics for successful wildlife crossing structures.

State-of-Our-Knowledge

To assess the state of the research and identify knowledge gaps, Table 6.2 summarizes 17 published studies evaluating wildlife passages for animal use. Although not exhaustive, the list nonetheless captures the majority of accessible reports in international journals and conference proceedings. Features examined include how studies were designed, how and what data were collected, and how they were analyzed. Studies have been carried out on three continents: Europe, Australia, and North America. Of the 17 studies, 15 had stated objectives and 3 stated hypotheses to test. On average, passages were evaluated for 15 months (range = 1–56), using sand transects (n = 12), infrared-operated 35mm cameras (n = 4), and direct observations of animals (n = 1) to detect animal passage. All but 3 studies assessed the passage performance of individual species, 4 looked at groups of species (e.g., small mammals, amphibians), and 1 examined responses at the community level (large mammals). Of the individual species studies, 75% evaluated two or more species.

All but one study focused on mammal passage, 6 reported on reptile passage, 2 on birds, and 1 on amphibians (Table 6.2). Of the mammal studies, 11 reported on small- and medium-sized mammals and ungulates, while 5 assessed large carnivore use. Only 3 studies had predefined criteria for measuring passage effectiveness. All but one study measured how effective passages were based on the frequency of animal use. Unfortunately, this does not take into account what the expected passage rate of the animals may be, based on the population size or how abundant they are in the vicinity of the passage (see below). Of the 17 studies, only 4 looked at how human activity might affect animal passage at the structures.

The state of the research and knowledge gaps can be highlighted in nine points:

1. The results support the general sense that few rigorous studies have been carried out to date.[785] Hypothesis testing has not been a part of most studies, and rarely have investigators predefined criteria for what constitutes a successful passage. To accurately judge passage performance, predefined criteria have to be developed, at whatever scale the research is focused, in order to measure how well a passage meets those criteria. Results can then be used to design future structures.
2. Several studies limited their analysis to a single species, whereas others took a broader *multispecies community approach*. Species do not function in isolation but are linked to other species on multiple spatial and temporal scales.[690, 286] Therefore, by focusing only at the species level, passage requirements for other species and ecological processes may not be accommodated.
3. A large research void exists in addressing factors that affect how large carnivores use passages (Figures 6.2, 6.3, and 6.7). A similar research need exists for reptiles and amphibians relative to passages. Major transportation corridors bisect and potentially fragment most of the major ecosystems that still support wide-ranging large carnivores.[803, 152] Increased concerns expressed by transportation and natural resource agencies regarding mitigation planning for large carnivores highlight the need for more information and research in this area.
4. Surprisingly, few studies have contemplated how human activity may affect passage by wildlife. Humans are an integral part of most landscapes and systems, and managing human activities around important highway passages may lead to greater road permeability for wildlife.
5. Most evaluations used frequency of observed passage as the measure of success. Measures of effectiveness can hardly be based on total counts of animal passage alone, however, but must be in the context of how abundant the animal is and its distribution around a structure. For example, 300 deer crossings at one passage versus 10 passes at another may simply reflect the rarity of deer in the vicinity of the second passage. Also, 300 deer passes cannot be directly compared to 2 grizzly bear passes. Expected passage based on a probability of occurrence measure at each passage will

Table 6.2. Performance evaluation studies of mitigation passages for wildlife.

	Design		Data Collection				Analysis			
Source and Location	*Hypothesis stated?*	*Objectives stated?*	*N structures*	*Method*[a]	*Duration (months)*	*Monitoring frequency*	*Level*[b]	*Species*[c]	*Criteria for success?*	*Observed/ Expected*[d]
Reed et al. 1975 Wyoming, USA	No	Yes Do deer use underpass and what is the extent of behavioral reluctance associated with it?	1	Counters Transects	48	Weekly	S (S)	Mammal (u)	No	Obs
Ballon 1985 Upper Rhine, France	No	No	4	Transects	9	Weekly	S (M)	Mammal (u)	No	Obs
Hunt et al. 1987 NSW, Australia	No	Yes Do tunnels facilitate movement?	5	Traps Transects	2	1 per 8 days	S (M)	Mammal (s, m)	No	Obs
Jackson and Tyning 1989 Massachusetts, USA	No	Yes Are tunnels effective?	2	Observation	<1	Daily	S (S)	Amphibian	Yes	Obs
Woods 1990 Alberta, Canada	No	Yes Are underpasses used, and do they allow access to seasonal ranges?	8	Transects Telemetry	36	1 per 3 days	S (M)	Mammal (u)	Yes	Obs
Foster and Humphrey 1995 Florida, USA	No	Yes Do underpasses effectively allow panthers to cross highway?	4	35mm camera	2–16	Continuous	S (M)	Mammal (m, lc, u), Bird, Reptile Human	No	Obs

Yanes et al. 1995 Central Spain	No	Yes Are passages used, and what are the features influencing use?	17	Transects	12	16 days per year	G (M)	Mammal (s, m), Reptile	No	Obs
Land and Lotz 1996 Florida, USA	No	Yes Are underpasses effective?	4	35mm camera Transects	24	nr[e]	S (M)	Mammal (m, lc, u), Reptile	No	Obs
Rodríguez et al. 1996 South-central Spain	Yes	Yes Which taxa use underpasses, how frequently, and what are the features affecting use?	17	Transects	11	1 per 3 days	G (M)	Mammal (s, m, u), Reptile, Amphibian, Human	No	Obs
Roof and Wooding 1996 Florida, USA	No	No	1	Transects 35mm camera Telemetry	12	1 per 3 days	S (M)	Mammal (s, m, lc)	No	Obs
AMBS Consulting 1997 NSW, Australia	No	Yes Inventory species in the area; assess the level of use at underpass.	3	35mm camera	9	Continuous	S (M)	Mammal (s, m)	No	Obs
Pfister et al. 1997 Switzerland, Germany, France, Netherlands	No	Yes Which species use overpasses? How often? Are they effective?	16	Video camera	24	nr[e]	S (M)	Mammal (s, m, u) Bird, Reptile, Amphibian, Invertebrate	Yes	Obs
Rodríguez et al. 1997 South-central Spain	Yes	Yes Which taxa use underpasses, how frequently, and what are the features affecting use?	17	Transects	10	1 per 3 days	S (M)	Mammal (m), Human	No	Obs

(continues)

Table 6.2. Continued

	Design		Data Collection				Analysis			
Source and Location	*Hypothesis stated?*	*Objectives stated?*	*N structures*	*Method*[a]	*Duration (months)*	*Monitoring frequency*	*Level*[b]	*Species*[c]	*Criteria for success?*	*Observed/ Expected*[d]
Rosell et al. 1997 Catalonia, Spain	No	Yes Identify variables that explain underpass use.	56	Transects	11	16 days per year	G (M)	Mammal (s, m, u) Reptile, Amphibian	No	Obs
Clevenger 1998 Alberta, Canada	No	Yes Determine relative importance of variables.	11	Transects	12	1 per 3 days	S (M)	Mammal (lc, u) Human	No	Obs
Veenbaas and Brandjes 1999 Netherlands	No	Yes Are underpasses used? What factors influence use?	31	Transects	5	nr[e]	S (M)	Mammal (s, m, u)	No	Obs
Clevenger and Waltho 2000 Alberta, Canada	Yes	Yes Do species respond to underpass features similarly? What features facilitate passage?	11	Transects	35	1 per 3 days	S (M), G, C	Mammal (lc, u), Human	No	Exp

[a]Method: Transects = sand traps (see Bider 1968; Clevenger and Waltho 2000); Traps = live-trapping; Observation = direct observation; 35mm camera = remote camera monitoring (see Kucera and Barrett 1993); Telemetry = radio-telemetry; Counters = motion-sensitive game/trail counters (see Kucera and Barrett 1993); Video camera = remote-operated video camera monitoring.
[b]Level = level of analysis: individual species (S) [single-species, S (S), or multiple species, S (M)]; species groups or guilds (G); or community level (C).
[c]Species types: s = small mammals; m = medium-sized mammals; lc = large carnivore; u = ungulate; Human = human impact on passage analyzed.
[d]Observed/Expected: Obs = observed passage frequency counts; Exp = expected passage frequency based on probability of occurrence in vicinity of passage.
[e]nr = not recorded in the publication or report.

likely be more accurate and reflect the true potential for passage and structure preference. Crossing probabilities along a highway can be estimated using relative animal abundance data from surveys, radio-telemetry, or habitat suitability maps.[948] In addition, changes in home-range distribution[964, 965] and animal movement[12] may be useful evaluation measures for mitigation success. Reduction of animal mortality after establishment of passages can thus be put into sharper context.

6. The studies seemed to generally agree that mitigation passages were functional and effective if "animals used them to cross the road." However, the frequency of crossing is more to the point. Does one passage a day, one a month, one a year, or one in an animal's life span indicate success? The last frequency may be satisfactory to maintain genetic variation in a population, but one of the greater frequencies is more appropriate for maintaining population size and minimizing local extinction.
7. Evidence that predators wait by passages and consume crossing herbivores is scant, largely anecdotal, and indicative of infrequent opportunism rather than a pattern of recurring predation.[557] Most studies record no evidence of predation. Habitat near busy roads may simply be inhospitable. Also, predators may tend to use different passages than do their prey.[222,169]
8. Greatly lacking in the literature were studies comparing the efficacy of passage types, particularly underpass versus overpass designs. A key variable in mitigation planning is cost and where to best invest mitigation funds. Should we build a few expensive wildlife passages or many smaller inexpensive ones? Which can we predict will be more effective for particular types of species and locations?
9. Finally, knowledge for determining where to place passages and spaced at what interval remains inadequate. Existing information plus general principles provides some insight. Thus, in risk spreading, two is much better than one, three is better than two, but six adds little benefit over five.

Effectiveness Criteria and Design Considerations for Wildlife Passages

What are some ways to determine whether a wildlife passage is effective? Knowing the intended purpose of the passage, the focal species or group of concern, and the amount of postconstruction funding available for monitoring and research will be important. The overall objective of wildlife passages is to increase the *permeability of a road corridor.* Success *reduces barrier effects* and usually *reduces road-kills* (Figure 6.8). One or more subcriteria can be measured to assess how well the two main criteria have been met. Meeting the first three

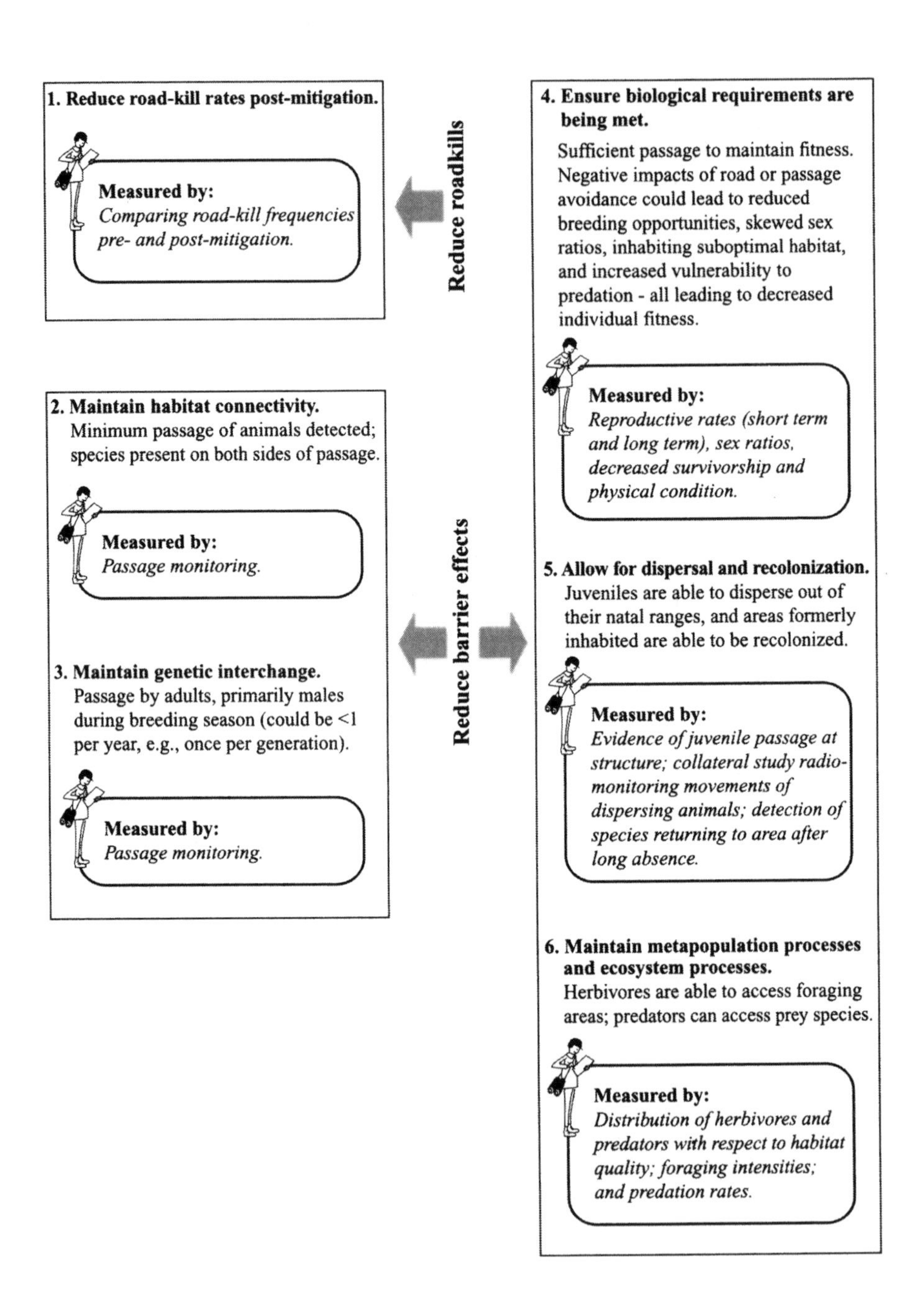

Figure 6.8. Goals, and how to measure them, for reducing road-kills and reducing the barrier effects of roads.

subcriteria in Figure 6.8 implies that a passage is functional. Measurements to evaluate success of these three are relatively simple to do. In contrast, achieving the last three subcriteria in Figure 6.8 explicitly defines passage performance and function. Measurements to demonstrate that the latter three are met require a greater time commitment and financial support. Long-term monitoring will be required in addition to collateral studies of wildlife populations in the transportation corridor. Another means of measuring passage effectiveness might include comparing treatment (mitigated) and control (unmitigated) areas, though randomization and replication of experimental units are often impossible. A range of new modeling and statistical approaches is available and especially useful when controlled experiments are not possible.[935, 441,130]

A comparison of use of wildlife crossing structures by large mammal species in Banff National Park (Figure 6.6) emphasizes that the frequency of passage differs considerably between species at a given structure as well as for a given species between different structure types. One of the more relevant questions is whether it is necessary to build costly wildlife overpasses or whether more inexpensive underpass designs could suffice (Table 6.3). If inexpensive measures are equally effective as more costly measures, then more of the former can be used to provide greater permeability across roads.

Results from Banff monitoring indicate that the number of large mammal crossings was highest at wildlife overpasses (average = 1105 crossings) (Figure 6.6). This was followed in turn by metal culverts 7 m (23 ft) wide and 4 m (13 ft) high (average = 245), dry pathways under creek bridges (average = 183), and lowest at 2- by 2-m (6.5- by 6.5-ft) concrete box culverts (average = 111). Crossing rates for large carnivores differed from those for large herbivores, although structures are suitable for both groups. Average crossing rates by carnivores were similar among wildlife overpasses, creek bridge pathways, and box culverts (averages = 32, 30, and 30 crossings, respectively) (Figure 6.6). Carnivore crossings were far less frequent at 4- by 7-m (13- by 23-ft) metal culverts (average = 19).

Information on structural attributes that facilitate passage by large mammals suggests that placement is critical in determining functional success.[62, 324, 167] Inexpensive structures placed in the right location (optimal habitat for crossings) will likely be more effective than costly structures placed in suboptimal habitat. Preconstruction study in areas of planned transportation projects is critical for determining mitigation placement.

Wildlife passage location has often been derived from the spatial distribution of wildlife road-kills, primarily where road-kill densities are highest.[758,247] Other methods include radio-monitoring of animal movements or tracking surveys along roads. While road-kill and tracking data are useful, animals can learn and adjust their movement patterns. Thus a passage placed where road-kills are abundant may be less effective than one that overcomes a road barrier

Table 6.3. Specifications and costs of mitigation passages for wildlife.

Mitigation type	*Dimensions (width × height)*	*Materials[a] Cost/m US$*	*Unit cost US$*	*Comments*
Large passages				
Concrete box	3 × 2.5 m (9.8 × 8.2 ft)	Concrete: 1880	120 000	Less cover/fill required compared to metal culverts (= less cost)
Metal culvert—elliptical	7 m × 4 m (23 × 13.1 ft)	Corrugated metal: 3630	150 000–170 000	Greater time to install due to bolting together pieces
Open-span bridge—over land	~12 m × ~5 m (39 × 16 ft)	Concrete: 33 600–40 300	470 000–670 000	Based on 12 m (39.3 ft) wide structure. Greater than 15-m (49.2-ft) width requires center pier and expansion joints (= greater cost)
Open-span bridge—over waterway	Variable: ≥50 m × ≥5 m (≥164 × ≤16.4 ft)	Concrete: 375 (× 2)	Minimal added cost	Bridge construction is included in highway plan. Adaptation for wildlife passage is minimal added cost.
Overpass	52 m (170.5 ft) wide	Concrete: 22 600	1.15 million	Prefabricated concrete arches; one to two days to install arches. Benefit = ease of rerouting traffic.
Elevated roadway[b]	10 m (33 ft) above the ground	Concrete spans: 42 000	8.5 million	——
Tunnel[c]	27 m × 5 m (88.5 × 16.4 ft)	80 000	16 million	——
Fencing				
Wood post—no apron	2.4 m (7.9 ft) high	Page wire	23	——
Wood post—with apron	2.4 m (7.9 ft) high	Page wire	35	——
Steel post—with apron	2.4 m (7.9 ft) high	Page wire	60	——

[a]Costs based on engineering costs during 1995–97 Trans-Canada Highway upgrade project, Banff National Park, Alberta, Canada.
[b]Calculated for four-lane highway, center-to-center width of 31 m (102 ft) and overall (edge-to-edge) width of 56 m (184 ft).
[c]Calculated for 200-m (656-ft) section of highway.

between two high-quality habitats or that reestablishes a major wildlife route. Digital habitat suitability and land-use data used with GIS tools and applications can help in making passage location decisions.[928, 172]

There is a need for innovative engineering design and techniques in wildlife passage construction, particularly of wildlife overpasses. Although relatively few overpasses currently exist worldwide, they continue to be built, and interest in the structures as landscape connectors and as habitat linkages across transportation corridors is growing. However, a few basic overpass designs have been replicated, thus resulting in little variation in design. Furthermore, little investigation of novel overpass or underpass designs that could result in significant cost reductions has yet been done. For example, shallow, light, sandy soil would reduce the weight of soil and water on a structure and support drought-resistant woody plants as wildlife cover. Spots or lines could be mounded with loam to support other shrubs and trees. Engineering or natural designs could prevent vehicle access and facilitate animal crossing on a structure. Designs should be possible that permit people (and perhaps even dogs) to cross without compromising the effectiveness of wildlife crossing and landscape connectivity.

In addition, creative, inexpensive passage design can benefit wildlife big and small. Simple and cheap landscaping may include the creation of small ponds and brush piles on overpasses. Or a row of stumps and branches as a *stump line* or *stump wall* can be used in underpasses (Figure 6.3) to enhance movements and connectivity for ground-dwelling insects, spiders, small mammals, and amphibians.[63, 475] In The Netherlands, lines of stumps and branches have greatly increased the movement of these animals under bridges on floodplains.

Mitigation Case Studies

Several case studies of effective mitigation passages warrant mention. Perhaps the most cited example is how the habitat of the pygmy-possum was restored by talus tunnels.[591] The habitat of this rare Australian marsupial was fragmented by roads and other developments, causing skewed sex ratios and low survival rates. By constructing tunnels filled with rocks, imitating the tiny animal's natural scree habitat, natural movements were restored and population structure and survival rates were returned to normal levels.

On Christmas Island in the Indian Ocean, millions of large red crabs annually migrate from the land to the coast to breed. A large fraction of the crabs have to cross main roads, and hundreds of thousands of crabs may be smashed by vehicles each year (P. S. Lake and P. Green, personal communications; also see *Ranger Rick,* January 1988). Three management techniques used attempt to reduce road-kills and the road barrier effect during the pulse of migration. Roads receiving the heaviest flows of crabs are temporarily closed with gates and signage. Several portable, blue plastic drift fences extend diagonally out from certain roads to channel movement to specific crossing locations (at times, it helps to have a person with a broom keep the crabs moving). Third, concrete gutters are built along a road to help channel crabs to special concrete culverts, where the crabs cross the road. The crab culverts, topped with cattle-guard rails on the road surface, are highly successful in providing connectivity for this distinctive migrating wildlife.

In the 1980s, road mortality accounted for 20% to 25% of the annual mortality of badgers *(Meles meles)* in The Netherlands.[958,64] Ecopipes, or "badger tunnels," were installed throughout the Dutch road network; as a result, the badger population has increased from 1500 to 2500, with annual road-kills of about 10%.

Similarly, road-related mortality and habitat fragmentation threatened the isolated Florida panther population in South Florida. Collisions with vehicles accounted for 49% of documented deaths.[585] Prior to upgrading a two-lane highway to a four-lane divided interstate highway (I-75), 5 panthers of a total population of roughly 50 were killed by vehicles annually. Ten years after the construction of underpasses and bridge replacements designed to allow panthers to cross the I-75 highway, panther road-kills were reduced sharply and successful movements across the highway increased[324,518] (G. Evink, personal communication). Wildlife underpasses (2.4 m or 8 ft high, and 7.6 m or 25 ft wide) have been added on a nearby two-lane highway (Route 29) and used by the animals (Figure 6.7). The Florida panther population is slowly growing.

In southern France, the construction of a major motorway threatened an already vulnerable population of Hermann's tortoise *(Testudo hermanni)*.[380] One measure adopted to reduce construction impacts was to build fences and tunnels to allow movements between highway-fragmented habitats. Four years after construction, only five tortoises had been killed on the highway, the annual survival rate of reintroduced tortoises was 78%, and mark-recapture studies indicated that the tunnels were frequently used and that the adult population had stabilized. At least in the short term, this transportation project was considered highly successful.

Animals capable of flight are generally not considered to be dependent on mitigation passages, though traffic disturbance and habitat fragmented by roads can disrupt movements and breeding.[285,771,514,699] Nonetheless, study of bird crossings along a multilane highway in Switzerland showed that woodland bird species used overpass structures for crossing significantly more than they did open areas without an overpass.[476] Subsequent work on overpasses supported this finding and further suggested that overpasses can have a guiding-line function, encouraging birds and even butterflies to cross roads.[722]

Conclusion

Eight principles or guidelines emerge from the preceding evidence from existing transportation mitigation projects:

1. A major discrepancy exists between mitigation measures currently employed and their apparent success in reducing wildlife-vehicle collisions. More systematic monitoring and evaluation of these measures, and of how the road system interacts with wildlife populations, would offer transportation planners and land managers a more informed means of predicting problems and devising preventive measures.
2. The most effective mitigation measures need not be the most expensive nor the most difficult to achieve.
3. Preconstruction road planning for highway expansion or upgrade projects is a more economical means of mitigating highways for wildlife than retrofitting measures on an existing highway. Also, ongoing bridge reconstruction and highway expansion projects are cost-effective opportunities to mitigate roads for wildlife passage.
4. The cumulative effects and lag-time effects of roads on populations and biodiversity are important considerations when planning and evaluating mitigation schemes.
5. Mitigation that uses an integrated system of wildlife crossing structures and fencing (with animal-escape mechanisms) is more effective than using one or the other alone for maintaining habitat connectivity and reducing road-kills.
6. No one-size-fits-all solution exists for wildlife crossing structures. Inevitably, species vary in their comfort level with and adjustment to different structure types and designs.
7. The location of a wildlife passage is a key decision for achieving objectives. Rather than selecting locations based mainly on road-kill and tracking data, crossing structures are likely to be most effective long term if located to fit with overall patterns of the landscape and to coincide with or create effective zones of landscape connectivity.
8. Wildlife crossing structures are built to last. Accommodating changes in species, their demographics, animal behavior, habitat conditions, and nearby human activities is important for sustained effectiveness.

Using these overlying principles of making roads more permeable for crossing accomplishes the dual objective of reducing the barrier fragmentation effect of roads and reducing road-kills. As a consequence, this practice maintains landscape connectivity for the fauna and the future.

PART III

Water, Chemicals, and Atmosphere

CHAPTER 7

Water and Sediment Flows

Every path has its puddle.

—English proverb

Major Roman roads . . . were bordered on both sides by carefully constructed longitudinal drains. Where the roads served a military function, the roadside was cleared of vegetation for 60 m on either side to prevent attack by hidden archers.

The next step was the construction of the *agger,* an earthen structure raised up to a meter above ground level and . . . typically about 15 m wide, made up of two 5 m wide edgestrips on either side of a 5 m pavement. In this respect the cross section is very similar to a modern road design.

—M. G. Lay, *Ways of the World,* 1992

Anyone driving through a pouring rainstorm can appreciate the problem of too much water on the road. If water does not run off of paved roads efficiently, even a small depth of water on the road can lead to treacherous driving conditions. Perhaps the two most common, and dangerous, of these conditions are hydroplaning of vehicles and water spray onto windshields from passing vehicles. *Hydroplaning* leads to loss of vehicle control, and water spray can completely blind a driver, sometimes for too long. Too much water on unpaved roads can cause slipping and sliding, even getting stuck in the mud. These troubles led to the paving of roads in the first place.

Roads also affect the movement of water. How often have we noticed, especially in flat terrain, that a lake or wetland adjoining a road on one side looks a little different from the one on the other side? Most likely, these two water bodies were one before the road was built. Even with the installation of culverts to connect the separated water bodies, roads commonly disrupt the natural flow and circulation of water. Similarly, where roads cross streams in hilly or mountainous landscapes, the *riparian areas* (floodplains or portions thereof near streams and rivers) seem a little different upstream compared to those downstream of a bridge or culvert. The difference probably is caused by the road, which may have affected the natural flow and possibly the chemistry of the stream.

The presence and movement of water affect roads, and roads affect the movement of water and the material transported by water. Water transports chemicals that are dissolved in it (the *dissolved load*). But water also transports the solid material that is suspended in it (the *suspended load*) or moved along a stream bottom (the *bed load*). This solid material can include all types and sizes of materials, from clay to boulders and from leaves to trees. Roads can affect the movement of water and these materials by blocking and/or rerouting their flows. Conversely, water can affect roads by (1) flooding, (2) destroying bridges and culverts, (3) eroding unpaved roads and the shoulders of paved roads, (4) inducing landslides onto roads or sliding of the road itself, (5) deteriorating road surfaces through the freeze-thaw cycle in some climates, and (6) discharging groundwater, which can saturate roadbeds, making them unstable.

This chapter focuses on the movement of water and the material it transports, because this process is an integral part of terrestrial and aquatic ecosystems. The quantities and rates of movement of water, sediment, and chemicals in a "natural" landscape may be critical to how natural ecosystems function. Increases or decreases in the natural flow and character of the material may substantially alter ecosystems, but they are natural processes of ecosystem disturbance and renewal. For example, landslides, floods, floating debris, and deposition of sediment may kill some organisms, but they may also create habitat that favors reestablishment of the affected species as well as establishment of other species. Thus floods, landslides, and flows of chemicals, sediment, or wood can be viewed as natural processes of terrestrial and aquatic ecosystems in many landscapes. On the other hand, substantial human-induced increases or decreases in these processes in such landscapes, caused by the building of roads, for example, can have potentially significant ecological consequences.

We begin with erosion and sediment control. The remainder of the chapter then addresses the many important dimensions related to the reciprocal effect of road systems and water.

Erosion and Sediment Control

Roads are a key contributor to erosion processes because of the abundance of exposed soil in roadsides and on unpaved road surfaces. Erosion and sedimenta-

Figure 7.1. Road erosion and sediment transport processes related to animal and vehicle usage plus wind and water flow. Central Nicaragua. Photo by R. T. T. Forman.

tion are natural processes that can be accelerated by human activities.[184,516,267] (Figure 7.1). *Erosion* occurs at the surface and can be described as the detachment of soil or rock particles by water or wind. Once detached, *sediment transport* describes the movement of particles by three forces. Rainfall erosion is predominant in most areas. In dry climates, wind erosion often predominates[110] (Chapter 10). Finally, channel erosion occurs along rivers, streams, and intermittent channels. The end product of transport is *sedimentation,* the process of deposition of particles.

Rainfall Erosion Processes

Rainfall erosion involves four processes that may occur separately or in a linear sequence[231,184,766,516,267]: raindrop splash, sheet erosion, rilling, and gullying. All are important in roadsides and unpaved roads. *Raindrop splash* initiates erosion, as raindrops falling with tremendous energy dislodge or detach particles. Falling at 6 to 9 m (20 to 30 ft) per second, raindrops can send soil particles 60 cm (24 in) high and 1.5 m (5 ft) laterally. On bare ground, a heavy storm can splash as much as 205 000 kg/ha (90 tons/acre) of particles into the air. Rainfall on bare earth such as roadsides also increases the compaction of particles, which may slow the rate of plant establishment. *Stabilization* of the earth surface with vegetation or other appropriate material minimizes the number of raindrops directly impacting on surface particles.

Sheet erosion is the removal of earth or soil in thin layers or sheets from sloping land.[231, 186] The common transport mechanism for soil loosened by raindrop splash (Chapter 1), sheet erosion is especially important in road construction due to the frequency of cutbanks and fillslopes created upslope and downslope of many roads. The potential for sheet erosion depends on soil type as well as on depth and velocity of water flow on the slope. For especially erodible soil, the potential also increases with length and steepness of slope as well as with upslope drainage area.[516] Stabilization, as for raindrop splash, and diversion away of as much water flow as possible are the control methods for sheet erosion.

Rill erosion, or *rilling,* occurs where *sheetflow* (water moving in a continuous layer across a surface) becomes concentrated in small defined channels typically a few centimeters deep.[231, 186, 516] Rills are common along steep fillslopes by roads. In total, rilling may carry more soil than the other rainfall erosion processes. Prevention requires stabilization and often water diversion, while repair may be done with mechanical *disking* or *tilling* (harrowing or plowing). Stabilization and repair are important to prevent gullying.

Gully erosion, or *gullying,* occurs from concentrated water flows much greater than in rills and when the resulting gullies cannot be covered by tilling or disking. Gullies are often created on fillslopes due to concentrated water flows downslope of road culvert outlets. Gully erosion, often the result of unrepaired rills, is costly to repair.

Factors Affecting Rainfall Erosion

These rainfall erosion processes are a function of four major factors linked in a "universal" *soil loss equation*[231, 455, 516, 267]: climate, soil, topography, and vegetation cover. First, *climate* involves the frequency of intense rainfalls as well as the presence of conditions moist enough to support a relatively complete vegetation cover. Seasonality is also important, for instance, so that construction work is done when stream ecosystems are least damaged and so that vegetation establishment avoids dry or cold seasons.

Second, soil texture, organic matter content, structure, and permeability influence erodibility.[455] *Soil texture* refers to the proportions of sand, silt, and clay particles. *Soil organic matter,* composed mainly of decaying plant leaves, stems, and roots, tends to cement soil particles together and improve soil fertility. *Soil structure,* the arrangement of soil particles into aggregates, decreases surface water runoff and may increase the capacity of soil to hold water. *Soil permeability* enhances water penetration, thus reducing surface water runoff that may cause soil erosion along roads.

Third, for topography, *slope length* and *steepness* are keys to erosion potential.[231, 186, 516] Flatter slopes tend to maintain sheet erosion, whereas steeper slopes tend to have rill and gully erosion. For example, moderate or steep slopes longer than 30 m (100 ft) tend to be difficult to stabilize.[267] Exposure direction (relative to the sun) affects heat and moisture conditions that largely determine vegetation establishment. In addition, the size, shape, and slope of the drainage area upslope affect erosion potential.[647]

Finally, *vegetation cover* is the most critical factor influencing erosion.[726, 516] Vegetation provides six major benefits.[267]

1. Reduces raindrop impact
2. Reduces runoff velocity
3. Provides, via the root system, structural integrity to the soil
4. Filters chemical pollutants and sediments from runoff
5. Increases water infiltration into the soil
6. Increases evapo-transpiration, the vertical movement of water to the air

These benefits are easy to visualize. Raindrop impact is reduced by the cover of foliage and leaf litter on the ground. Runoff velocity and sediment transport decrease with stem density and litter cover. Structural integrity of the soil is enhanced by a mixture of species, which provide both shallow and deep dense root networks.[455] *Pollutant filtering* increases with more blackish organic matter, assuming that water infiltrates into the soil. Infiltration increases with lower runoff velocity and an abundance of pores produced by soil animals, such as earthworms.[455] Evapo-transpiration pumps out more water vertically as plant density and vegetation cover increase. Diverse human-related activities tend to accelerate erosion (Figure 7.1), primarily by altering or removing the vegetation cover.

Numerous soil types exist that can be readily grouped and ranked according to water runoff potential.[860, 861, 421] *Low-runoff soils* have high infiltration rates even when drenched, and are deep and sandy or gravelly. Water readily passes through these well- to excessively drained soils. For example, think about how fast beach sand dries out after a rain or emptying a bucket of water onto it. At the other extreme are very *high-runoff soils,* which have an extremely slow infiltration rate when saturated. These soils are mainly clay (with a high swelling potential), have a permanent high watertable, or appear as a shallow layer over clay or bedrock. Water moves slowly through such soils. Puddles and the mud that sticks to boots are common on the surface.

Sediment Control

Preventing erosion is the core principle for controlling sediment transport. This is much more effective than filtering and trapping sediment or other control measures. Stabilizing slopes and waterways with vegetation, such as seeding fillslopes and cutbanks with fast-growing plants, is the essential step in preventing erosion.[726, 516] A high-percent plant cover, a shallow-and-deep dense root network, and a thick layer of blackish soil organic matter are the overriding goals. Mulching and erosion-control mats and blankets made of various fibrous materials that decompose are temporary actions used to reach these goals more quickly.

Three other principles supplement the core principle of preventing erosion.[267, 1068, 578] First, *limiting exposure* involves reducing the area as well as the duration of exposed soil. Second, *retaining sediment on site* is achieved with devices that filter sediment from low-flow water runoff, such as silt fences, straw bales, or brush barriers. Third, *managing runoff* means preventing off-site water runoff from entering the site, so that only on-site runoff and sediment have to be managed. Off-site water is diverted with berms or channels, and streams and swales are maintained in their natural state so that clean water passes by or through the site. These principles may be applied in different combinations for each phase of road construction: the initial clearing, the middle grading, and the final site-stabilizing phase.

Where erosion potential is low, roadside slopes may be stabilized by natural plant colonization and succession. In areas of higher erosion potential, *temporary seeding* may be used to stabilize denuded areas that will remain disturbed for relatively short periods, such as slopes not ready for permanent stabilization, areas that will be redisturbed, and soil stockpiles.[267] *Permanent seeding* is appropriate for bare areas that will be left undisturbed for longer periods.

Seedbed preparation on roadsides may involve surface loosening for compacted earth, fertilizing nutrient-poor soils, liming or acidifying to adjust the pH, adding topsoil on gradual slopes with poor soil conditions, or mulching to reduce erosion and enhance seed germination. However, some of these techniques are often inappropriate for preventing pollution or degradation of a stream, a lake, or groundwater, or for facilitating the establishment of native species and natural plant communities. For example, loosening of compacted earth may cause an elevated erosion rate and muddy streams, and fertilizing a nutrient-poor soil tends to cause aggressive grasses or non-native species to eliminate native species.

Currently, seeding of roadsides is often done with either annuals or perennials, though research may identify mixtures that provide the benefits of both plant types.[516, 267, 123] Annual grasses or legumes, such as winter rye, oats, and annual lespedeza *(Secale, Avena, Lespedeza)*, are normally quick to germinate and provide extensive plant cover. Annuals provide temporary cover, since the plants die in a year, but also serve as nurse plants for slower-growing perennials. Perennial grasses and legumes, such as tall fescue, perennial ryegrass, and crown vetch (*Festuca, Elymus, Vicia*), generally require more moisture and compete poorly with annual grasses. Legumes that add nitrogen to the soil are especially planted on infertile soil.

Many *structural control measures* to reduce erosion or sediment transport are used in special situations along roads (see Chapter 8). These include check dams, diversions, temporary slope drains, energy dissipators, silt fences, straw bales, brush barriers, inlet protectors, and outlet protectors.[267] In high-runoff locations, temporary, relatively small sediment traps may be used; in still-higher-flow sites, whole sediment basins may be appropriate.

In short, erosion and sediment transport are major concerns, both during and after road construction. Vegetation stabilization of slopes and ditches and supplementary technological solutions are used to control or reduce the threats. Such action is important to minimize chemical pollution of groundwater and to minimize both chemicals and sediment from entering surface water in roadsides and adjoining land. Control measures are also critical for the rapid establishment of natural plant communities with a richness of species, a deep dense root network, and ample blackish organic matter in the soil.

Reciprocal Effect of Road Systems and Water

The two components of the water and road system feedback are quite distinctive and thus are presented separately. We begin with the effects of water on roads and then the effects of roads on water and sediment.

Effects of Water on Roads and Traffic

The flow of water over road surfaces can result in deterioration of road surfaces, especially if the road is unpaved.[1068] On gravel roads, vehicle tracks may become incised, leading to the creation of gullies on the road surface. In other instances, overland flow from a sloping road surface may lead to gully erosion of the fillslope. These conditions may be common on rural roads, such as logging roads.[1018] Road surface deterioration also may result from precipitation, freezing, and thawing. Copious precipitation can flood roads or result in a sufficiently deep layer of water to cause hydroplaning of vehicles. Water also seeps into cracks in the pavement, causing deterioration of the road surface, which is especially serious in climates where the water goes through freeze-thaw cycles. If enough water seeps into roadbeds, it may erode some of the roadbed material. Roads placed in areas of substantial groundwater discharge can be unstable if the roadbed is not engineered to remove the continual flow of the upwelling groundwater. Such conditions are common where regional groundwater flow systems discharge near the edges of major river valleys.

Roads placed on hillsides can disrupt the flow of surface runoff and cause erosion and road failure, unless they are properly engineered to control and

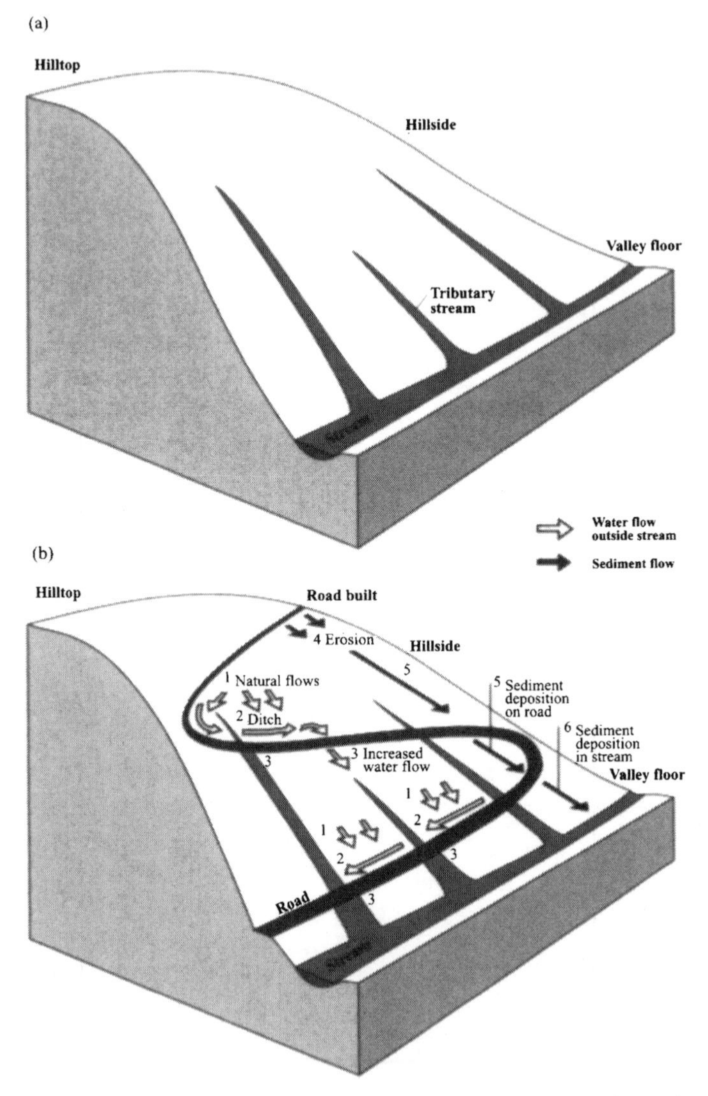

Figure 7.2. Water and sediment flows on a hillside before (a) and after (b) road construction. Numbers refer to processes described in chapter text. Adapted from Jones et al. (2000).

route the flow beneath the roadbed (Figure 7.2). This may occur when the volume of water reaching the road is greater than the ditch system was designed to handle, when a culvert becomes plugged, or where culverts are too small to accommodate the water flow. Such flow can cause gully incision and saturation of the fillslope, which may initiate subsequent sliding of the fillslope.[1018] Where groundwater discharges at the lower parts of hillslopes, it is difficult to build roads, because of the persistently saturated soils. Roads constructed on flat land, such as valley bottoms, terraces, or floodplains, may be undermined by flooding and bank erosion.

The substantial flow of some floodwaters also can remove a bridge or causeway, if the structure has not been engineered to withstand large flood flows. Removal of road segments by landslides on hillslopes may also occur during storms.[932] Such events can have major effects on the larger transportation network when the road segment, bridge, or causeway is the only connector in the road network.

Effects of Roads on Water and Water-Transported Material

Roads commonly affect how water and its various loads move through watersheds. A road segment relative to water has four major functions characteristic of essentially all corridors in a landscape[302]:

1. A *source* of water when water runs off of a road, especially a paved road (in some cases, impermeable road surfaces can produce high rates of runoff that lead to flooding in downstream areas)
2. A *sink* for water when water accumulates on a road, although the volume of water that accumulates on a road generally is not large
3. A *barrier* to the flow of water down a hillside
4. A *conduit,* or corridor, for the flow of water when water runs down ruts and depressions in the road surface[463, 1068]

In these ways, roads can disrupt natural flows of surface water and groundwater, create new routes for the flow of water, or serve as sources of chemicals and sediment that are introduced into surface water and groundwater. Such functions help describe the ecology of roads but also loom large in the ecology of landscapes.

Roads may intercept surface and subsurface flow and reroute it along roadside ditches.[578] The water intercepted by roadside ditches may be routed to streams, thus effectively increasing the density of stream channels in a watershed. This rerouting of water could potentially accelerate flows in the entire stream drainage network during rainstorms, thus increasing flooding.[1017, 517] Roads may divert stream flow from one small drainage basin to another, thus increasing discharge in a receiving channel and increasing the potential for localized erosion. In places, ditches may intercept groundwater flow, adding more water to a ditch network designed only to remove excess surface water.

Roads have often been built across small wetlands and shallow lakes, but in some places roads cross large lakes, extensive wetlands, and estuaries. The presence of such roads can disrupt water circulation patterns, and in some cases the movement of organisms, so much that the separated water bodies exhibit quite different ecological characteristics.[893] These differences commonly include differences in (1) water color due to algal content or suspended sediment, (2) *aquatic macrophytes* (rooted plants in water), or (3) the types of trees in the case of a forested wetland or swamp.

Runoff from roads commonly delivers chemicals to surface water and groundwater (Chapter 8). The chemicals originate from (1) materials from which roadbeds and road surfaces are constructed; (2) emissions, fluids, vehicle parts, and tire deposits that vehicles leave on roads; and (3) application of chemicals to roads to make them safe, such as de-icers, or to control dust on gravel roads, such as liquid calcium chloride. The effects of road-related chemicals on aquatic environments are discussed in Chapter 8.

This section has introduced the many ways that roads and water interact. To place these varied factors into a unifying framework, so that road and water interactions can be understood and managed consistently, some underlying concepts will be presented. Furthermore, a framework is provided that addresses road placement in the context of different types of landscapes.

A Conceptual Framework for Roads and Water Flow

To build a framework for understanding the mutual interaction of roads and water, including the various materials transported by water, it is convenient to use a schematic diagram of a hillside (Figure 7.2).[463,1018] In the hillside's natural condition, three tributary streams flow down the hillside and join the larger stream flowing along the valley floor (Figure 7.2a). A road constructed on this hillside can substantially alter the flow of water down the hillside (Figure 7.2b). The altered stream flow could lead to damage to roads, driving hazards, and degradation of stream habitat. The road can intercept the natural flows (number 1 in Figure 7.2b) and reroute the water by way of roadside ditches from one stream to another (number 2 in Figure 7.2b). This extends the channel network and changes the flow and channel characteristics in different segments of the tributary streams (number 3 in Figure 7.2b). Roads may also alter movement of sediment. Road construction can initiate erosion of materials in unstable road fills (number 4 in Figure 7.2b) and deposit these materials on roads (number 5 in Figure 7.2b) or on the valley floor (number 6 in Figure 7.2b). Overall, roads may divert water and sediment from the paths they follow in the natural landscape, as well as initiate multiple new flow paths.

We generally consider the effect of roads on the movement of water and the material it transports, as depicted in Figure 7.2, only in terms of surface flows. Water that infiltrates the land surface and moves downhill through the subsurface of hillsides also can affect roads, and in turn roads can affect the movement of subsurface water.[463] A cross-sectional view provides a generalized picture of the movement of both surface and subsurface water down a hillside and the interaction of the water with roads (Figure 7.3).[231, 463]

Natural flow paths of water on and through the hillside include *infiltration* (number 1 in Figure 7.3a), *overland flow* (number 2), shallow saturated *subsurface flow* (number 3), saturated overland flow (number 4), and *groundwater flow* (number 5). In Figure 7.3b, a road is present on the top, side, and bottom of

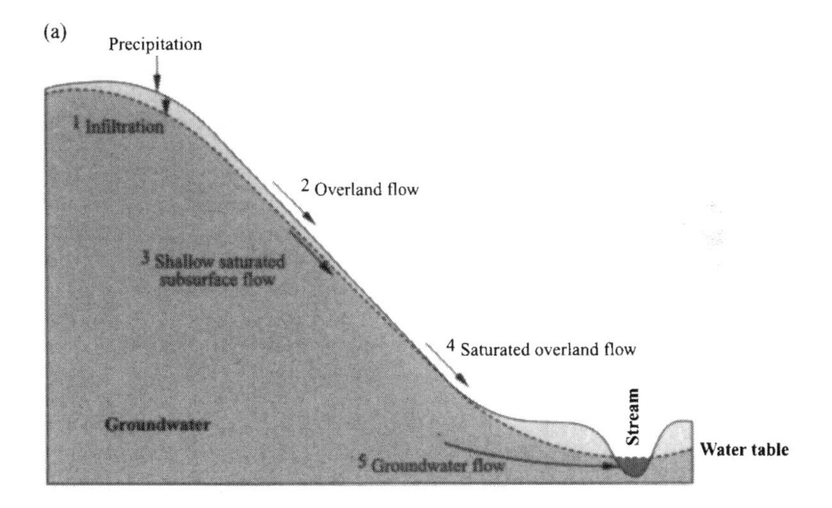

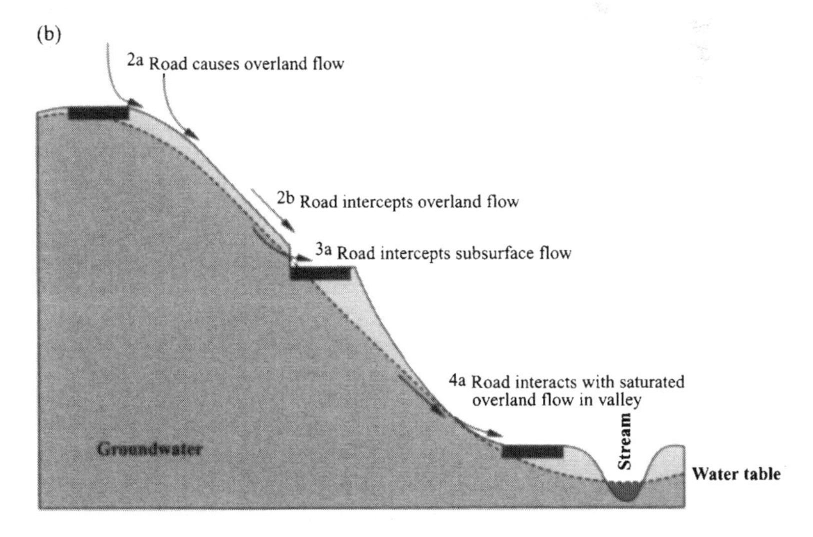

Figure 7.3. Road location affecting interactions between water and road. Black structure = road; slope exaggerated for illustration. (a) Natural water flows; (b) natural flows altered by roads. Adapted from Dunne and Leopold 1978, Jones et al. 2000, and Wemple et al. 2001.

the hill. Roads may generate overland flow (2a in Figure 7.3b), intercept overland flow (2b) and subsurface flow (3a), or interact with saturated overland flow in valley bottoms (4a). In many cases, shallow saturated subsurface flow results in landslides onto the road, such as might happen at location 3a in Figure 7.3b.

Upwelling groundwater, such as at location 4a in Figure 7.3b, may cause continuing problems with roadbed stability. In cold climates, the upwelling groundwater commonly freezes when it reaches land surface, resulting in a buildup of ice near, and sometimes on, roads. Such conditions require special engineering of the roadbed to drain the groundwater from beneath the roadbed.

The flow of water and transported material on hillsides does not cover the full range of road and water interactions. At the other extreme are the many parts of the earth that are relatively flat—that is, areas especially sought out for the placement of roads. Water affects roads and vice versa in flat landscapes as well as on hillsides. Perhaps the greatest effect water has on roads in flat landscapes is deterioration caused by flooding. Repeated saturation of a roadbed tends to cause slumping with associated cracking and destabilizing of the road surface. This is especially true for roads placed in flat lowlands. A major effect of roads on water in flatlands is related to the building of roads across wetlands and shallow lakes. This commonly results in the separated water bodies developing distinctly different aquatic ecosystems, even if culverts connect them.

The earth's surface has a rich variety of shapes, from steeply sloping mountainsides to vast areas of flat land. These shapes are controlled to some extent by the rock formations that make up the earth's surface, but they are also related to geologic processes, such as uplift and weathering, and the effects of climate. Although rocks and landforms have a great deal of variability from place to place, they also have some features in common. If understanding of the mutual interactions of roads and water is to be transferable, a conceptual framework for the movement of water in all landscapes would be useful. The next step in building such a framework is to place roads in the context of how water moves over and through the subsurface of the generalized landscapes that make up the earth's surface.[885, 1046]

Most landscapes consist of uplands adjacent to lowlands separated by steeper slopes. The uplands and lowlands can vary from being narrow to wide, and they can have a variety of slopes. In addition, smaller uplands and lowlands may be nested within larger ones. Six landscape types are widespread on the earth (Figure 7.4):

- *Mountainous terrain.* Consists of narrow lowlands and uplands separated by high and steep valley sides (Figure 7.4a).
- *Basin and range terrain.* Consists of very wide lowlands separated from much

narrower uplands by steep valley sides, as well as basins of interior drainage that commonly contain playas (Figure 7.4b).

- *Plateau and high plains.* Consists of narrow lowlands separated from very broad uplands by valley sides of various slopes and heights (Figure 7.4c).
- *Riverine valleys.* Consists of one or more small uplands and lowlands nested within a larger river-valley lowland (Figure 7.4d).
- *Coastal terrain.* Consists of one or more small uplands and lowlands nested within a larger coastal-plain lowland (Figure 7.4e).
- *Hummocky glacial and dune terrain.* Consists of numerous small uplands and lowlands superimposed on larger uplands and lowlands (Figure 7.4f).

Because the focus is on water flow as a key process tied to the structure of the terrain, these six basic types are usefully referred to as *hydrologic landscapes*.[1046] The slope, roughness, and permeability of their surfaces control movement of water over the surfaces.[885, 231] For example, for a given surface roughness, surface runoff will flow faster over landscape surfaces that have steeper slopes than over landscape surfaces with gentler slopes. In addition, the quantity and rate of runoff versus infiltration is dependent on the permeability of the (surficial) geologic material.[556, 336] Topographic relief and the permeability of the geologic material control movement of groundwater through the subsurface of these hydrologic landscapes. If the watertable is at a higher altitude in the upland, groundwater will flow from the upland to the lowland. This is evident for many scales of uplands and lowlands in the six examples of generalized hydrologic landscapes (Figure 7.4). In addition to the effects of land-surface form and geology on surface runoff and groundwater flows, the exchange of water with the atmosphere (precipitation and evapo-transpiration) further affects the movement of water.

By describing the earth in terms of generalized hydrologic landscapes, it is possible to discuss the interaction of roads and water relative to features common among different landscape types. For instance, hillsides are present in all of the landscapes (Figure 7.4), and roads are constructed on hillsides in all of them. Therefore, the general interactions of roads with the hydrologic system on hillsides, discussed above, apply to the hillsides of all landscapes. Similarly, most hydrologic landscapes have flat areas, and the general interactions of roads with the hydrologic system in flat areas, discussed above, apply to the flat areas of all landscapes. Therefore, the next two sections of this chapter present more detailed information and examples of how roads and water interact on hillsides and how they interact in flat lands. All examples are placed in the context of generalized landscapes, so the transfer value of the information should be readily apparent.

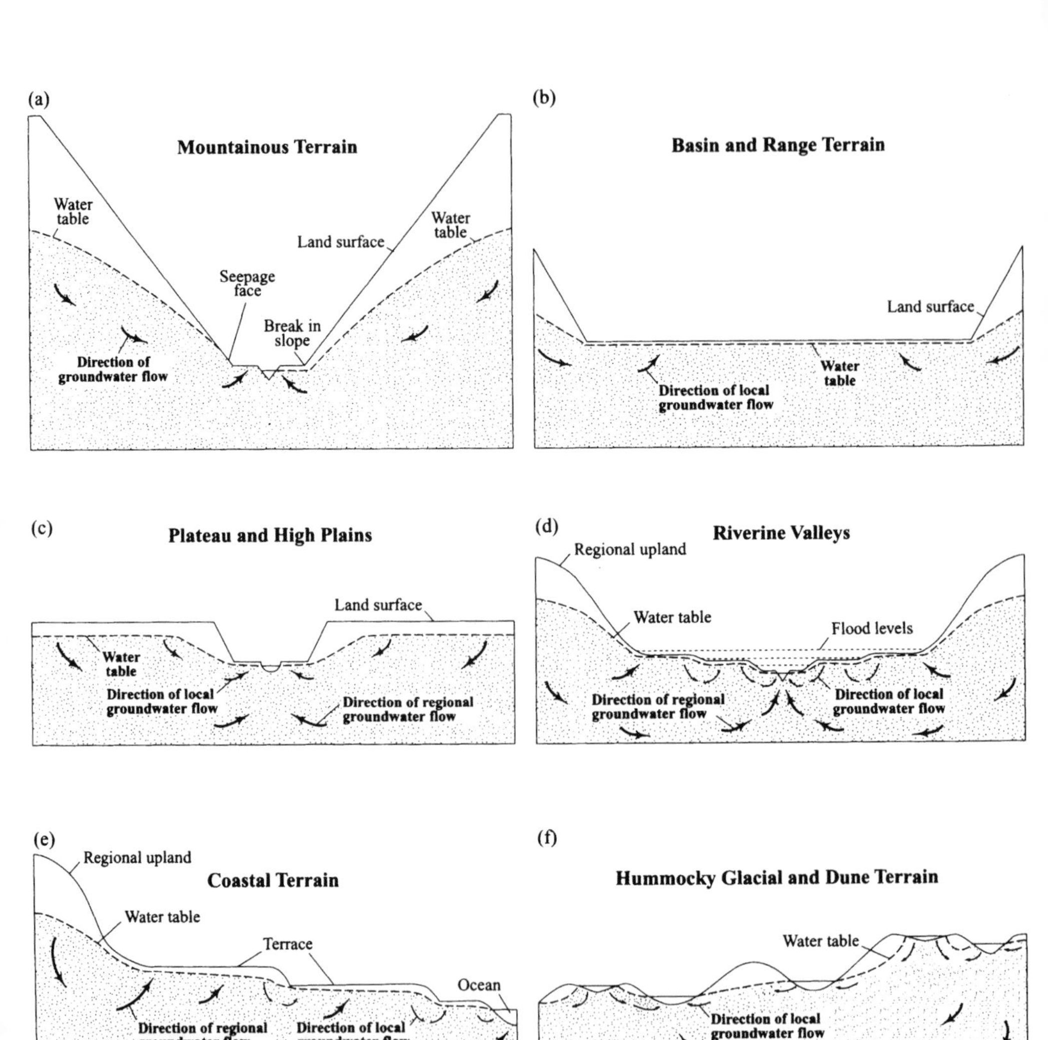

Figure 7.4. Six major hydrologic landscapes based on combining water flows and terrain. Roads placed in different locations across a landscape are affected by, and affect, different hydrologic conditions. Adapted from Winter (2001).

Roads and Water on Hillsides and Flat Lands

In view of these major diverse hydrologic landscapes, the reciprocal effects of road systems and water are now related to topography or slope. Thus, we first consider roads and water on hillsides and then roads and water in flat lands.

Hillsides

Roads built on hillsides in all landscapes affect natural *surface-water flow paths* (Figures 7.2 and 7.3).[463] These roads need to be carefully engineered to route stream flows from the uphill side to the downhill side of the road. However, even if the roads are carefully engineered, the process of road building and the modifications to the hillside may result in the movement of substantial quantities of sediment into streams by surface erosion or landslides. Much of this sediment is mobilized by surface erosion following the removal of soil and rock on the uphill side of a road and deposition of that material on the downhill side.

Roads that traverse the base of slopes where groundwater seeps onto the land surface can disrupt the natural groundwater discharge (Figure 7.3). The engineering necessary to stabilize the roadbed in such areas can reduce or cut off the groundwater source for wetland communities that may be present at the base of slopes. In other cases, the water intercepted by roadside ditches may be routed to streams, which may affect flood levels in a stream drainage network.[1017] Roads may divert stream flow from one small drainage basin to another, thus increasing discharge in a headwater channel and increasing the potential for gully formation or small rapid landslides down that channel.[647,1018]

Road construction on mountainsides can also lead to landslide movement on either side of the road.[647,932] *Landslides* are especially common where the water table is at or close to the land surface. Groundwater-saturated soils on the uphill side of a road can result in landslides that cover roads. Saturation on the downhill side can cause the roadbed to fail and move downslope. Sediment and wood transported by water may affect roads when landslides on hillslopes (or debris flows along stream channels) deposit sediment and wood onto roads.

Many types of landslides, ranging from a few cubic meters to millions of cubic meters in volume and moving very slowly to very rapidly, affect roads.[188] Avalanches (snow slides) also deliver substantial amounts of snow and other debris onto roads and are common in many steep mountain areas. Although most avalanches do not significantly damage roads, they frequently disrupt traffic flow.

The effects of roads on sediment production, transport, and storage are well documented by a study of logging roads in the Cascade Range of Oregon.[1018] This study indicated that the position of road segments on hillslopes strongly affects the types and frequency of erosion and deposition. Road segments near ridges function primarily as sources of sediment to areas down the hillslope or down stream channels. Road segments located low on hillslopes may trap sediment transported from upslope or upstream areas. Roads can change the processes of sediment transport, such as where a landslide plugs a culvert, directing stream flow over a road and producing a gully. In this sequence, the road forces a transition from the landslide process of sediment transport to erosion by excessive water runoff.

The effect of roads on *sediment production* has been considered in a number of studies, especially on unpaved roads in steep forest landscapes.[85,766,621,1068] In the U.S. Pacific Northwest, gravel-surface forest roads are associated with elevated sediment production from erosion of the road surface and from landslides, both during construction and over many decades after road construction. In a four-year study after forest road construction in the Idaho mountains, erosion rates during the first winter were about five times greater compared to those in the subsequent three years (pre-road rates were not estimated).[621] In California and Washington, surface erosion on gravel roads that had been in place for many years led to increased sediment production relative to conditions without roads.[85,766]

Surface erosion from roads is more closely related to traffic than to time since construction.[766,578,1068] On unpaved roads, tires of passing vehicles tend to break down soil aggregates and promote the detachment of fine particles. Some particles become airborne and transported as dust (Chapter 10) and some may be carried in water runoff. More traffic and more heavy vehicles passing can be expected to produce more surface erosion from roads.

During *flood-generating storms,* two- to five-decade-old forest roads on steep slopes in Oregon were sources of elevated sediment production in the form of frequent landslides.[891,892] Of over 3224 landslides inventoried in the Pacific Northwest, more than half were associated with roads.[647] In an Oregon study of forest roads and a large flood, failures of road fill at stream crossings on middle hillslope positions caused the greatest sediment movement (Figure 7.3).[1018]

Also during flood-generating storms, gravel roads on steep forested slopes are involved in complex process interactions called disturbance cascades.[668] Here water and sediment from hillslopes and stream channels are intercepted by roads, diverted through road drainage structures, and erode larger channels downslope with greater flows (Figure 7.2).[1018] The eroded material, combined with small landslides generated by the flows, is deposited on lower-slope roads, which blocks traffic.

Roads commonly increase the incidence of landslides in steep, landslide-prone terrain, which usually is underlain by poorly consolidated soil and rocks.[932] In such terrain, construction of road fills on steep slopes creates the potential for landslides that can course down hillslopes and into streams or other water bodies below roads. This can even happen along well-maintained major highways, such as California Highway 1 along the Big Sur coast, which traverses steep mountain slopes consisting of poorly consolidated rocks. In some areas, for example, forest roads are the main sources of landslides in landscapes where they occupy only a small percentage of the total watershed area.[891, 843] Poorly engineered roads in unstable positions or made too wide may be especially prone to landslides. Roads can increase the potential for landslide initiation by (1) placing soil and rock material on steep, marginally stable hillslopes; (2) containing fill material that is not compacted or may contain decomposing organic matter, which may be less stable than the native soil on the slope; (3) accumulating surface and subsurface waters in soil and rock materials that are subject to landslides; and (4) having sediment and organic debris plug road drainage structures, such as culverts, diverting flow into areas with unstable substrate.

Roads built on valley bottoms in mountainous terrain commonly are not much higher than streams (Figure 7.3). Because of this, they can be flooded by high stream stages during times of snowmelt and heavy precipitation. If the floods are large enough, the water may erode or even remove segments of the road. For example, a major flood in the Big Thompson Canyon in the Rocky Mountains in Colorado washed out large segments of U.S. Highway 34 after a deluge in the headwaters of the Big Thompson River in 1976.[841] This flash flood was caused by rainfall levels as great as 0.3 m (1 ft) in some parts of the canyon. The resulting flood, which had flow velocities as high as 7 m (23 ft) per second, far exceeded the probability of occurring once during a 100-year period.

Most of the information presented in this section on the interactions of roads and water on hillsides has concentrated on steep slopes in mountainous terrain. Interactions are most pronounced in this type of terrain, and thus a fairly large and recent literature base is available. However, the same interactions between roads and water occur on hillsides in all landscapes, including riverine, coastal, glacial, dune, plains, and plateau terrain (Figure 7.4). In terrain having less relief, the effects are generally more localized. Nevertheless, the interactions are important and can have significant local effects. For example, in northwestern Wisconsin, where clay deposits are dissected by river valleys with less than 30 m (100 ft) of relief, parts of roads constructed on the clay-containing valley sides commonly slide downslope, frequently disrupting the road network. Landslides onto roads and sliding of parts of roads downslope are a common problem for road construction and maintenance wherever roads are located on water-saturated clays, shales, and other poorly consolidated rocks.

Flat Lands

Even though flat lands have low slopes, erosion and sediment movement can still be a major problem. In addition, ditching and rerouting of water are common engineering practices in flat lands. Some of the interactions between roads and water that occur on hillsides also take place on flat lands. For instance, water leads to the deterioration of road surfaces (especially gravel roads) and can undermine roadbeds. Roads affect water by disrupting natural surface water and groundwater flow paths and creating new flow paths. Roads also serve as sources of sediment to surface water and of chemicals to surface water and groundwater. In flat terrain, such as in plains, riverine, coastal, and glacial terrain (Figure 7.4), extensive ditching and routing of surface water are particularly common.

A major difference in the movement of water and the material it transports between steep and flat landscapes, however, is related to differences in the available energy expended to transport materials. For the same flow depth and channel or surface roughness, the ability of a flow to mobilize and transport sediment decreases as the water-surface slope decreases. For this reason, the erosive power and the ability to move coarse sediment and other large materials are less in flat lands than in steeply sloping lands. This is not to say that the erosive power is nonexistent in flat lands, because even there, floodwaters, especially for large rivers, can cause extensive damage to bridge piers and levees. On the other hand, the low gradient of flat land results in flooding of roads as a major water issue.

Effects of Water on Roads

In flat lands, water commonly floods roads or saturates roadbeds, making them unstable. In a saturated roadbed, water filling the pore space between grains acts as a lubricant, reducing the structural integrity of the roadbed, especially if clay is present. Water standing on road surfaces (especially gravel roads) for long periods of time due to little slope leads to deterioration of the road surface if the road has heavy traffic. Floods commonly cover roads in river valleys, but after the floodwaters recede, the roads, usually after some repair, become usable again.

However, roads in flat coastal areas that are subject to storm surges from hurricanes and other large storms can sustain substantial damage and need to be rebuilt. For example, in 1999, Hurricane Dennis removed part of a coastal road on the Outer Banks of North Carolina. River flooding associated with copious rainfall also can seriously affect roads in coastal areas. Thus flooding of the Pearl River following massive storms in 1980 and 1983 caused considerable damage to U.S. Highway 90 near Slidell, Louisiana.[355]

Floods in river valleys can cover and wash out segments of roads, but one of the most pervasive effects of floodwaters on roads is the *scouring* (sediment removal by water flow) of riverbeds and banks near bridge piers and abutments.[657] Scour problems are exacerbated where gravel has been excavated from rivers for uses that include road construction. The economic costs of bridge scour are substantial. Between 1980 and 1990, the U.S. government spent an average of US$20 million annually to fund bridge restoration projects. As a result of extensive flooding in the central USA in 1993, more than 2500 bridges were damaged or destroyed, requiring US$173 million in repair costs. Bridge scour is such a widespread problem that the U.S. Geological Survey (USGS) has a program with the Federal Highway Administration (FHWA) and state highway departments across the country to obtain data on peak stream flows and, in some cases, on riverbed scour. These data are then used by highway engineers to design bridges for larger stream crossings and to use appropriately sized culverts for smaller stream crossings.

Although the problem of bridge scour has been recognized since bridges were first built, the problem gained critical attention in the USA when the New York State Thruway (Interstate Highway 90) Bridge over Schoharie Creek near Fort Hunter, New York, failed in 1987[1068] (Figure 7.5). This incident resulted in 10 deaths and led to the establishment of the USGS/FHWA program mentioned above.

In riverine settings, roads are generally flooded for only days or weeks, as

Figure 7.5. Bridge failure due to scour of the riverbed during flood flows. Multilane interstate highway 90 at Schoharie Creek, Fort Hunter, New York, 1987. Photo courtesy of D. S. Mueller, Sidney Brown, and U.S. Geological Survey.

water drains back to the river after the river level falls. Yet it is quite a different story if a closed-basin lake or wetland rises and floods nearby roads. With no drainage outlet, roads in these settings may be flooded for months or even years.[1031] This has been a serious and costly problem in south-central Canada and the adjacent north-central USA in recent years. This region consists of glaciated terrain characterized by thousands of small depressions occupied by lakes and wetlands called *prairie potholes.* Water remains in these depressions because an integrated drainage network has not developed. When the region receives abundant precipitation, the water levels in the depressions simply rise. In most of the region, the geologic substrate is poorly permeable, so ponded water does not readily infiltrate. Prairie potholes are closely spaced and shallow, so it is easier and cheaper to build roads across them than to wind around them. Roads that do wind around the depressions commonly are not much higher than the lakes and wetlands.

Under normal climate conditions, roads in this landscape are sometimes affected locally by rising water levels. However, a deluge that fell in the central USA in 1993 has been followed by highly unusual wet conditions that continued for eight years, resulting in the wettest conditions in more than 150 years.[1030] Many road segments throughout the prairie pothole region have become inundated by water. As a result, the road network in this region has become greatly disjointed. The blocking of roads by flood waters has nearly isolated some farms, decreased road density, and substantially altered traffic routes through the region. The common solution to the problem has been to raise the roads by adding fill and new road surfaces. Since 1993, 844 projects to raise roads in eastern North Dakota alone have cost more than US$175 million (R. Hartl, personal communication).

Effects of Roads on Water

Roads can be barriers to (1) the efficient drainage of flood water, (2) the encroachment of water during high water levels, and (3) the natural flow of water. They also commonly dissect shallow surface water into separate water bodies.

Flooding in the Red River of the North drainage basin in North Dakota provides an excellent example of how roads can be a barrier to removal of flood water.[580] The Red River of the North flows in an ancient glacial lake bed that is extremely flat, with a slope of as little as 0.02% in places. The area is heavily farmed and has an extensive network of roads spaced about 1.6 km (1 mi) apart, generally forming a square-mile grid. The roads are built a few feet higher than the surrounding fields. During major floods, the water spreads over large areas, including over the roads. However, when the river level falls following floods, the water on the fields is dammed by the roads, producing square and rectangular temporary lakes (Figure 7.6). Water can return to the

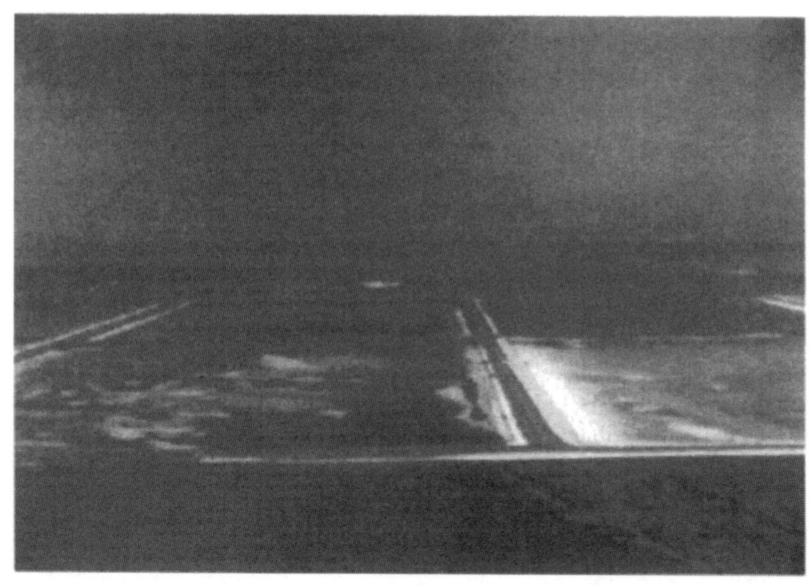

Figure 7.6. Temporary lakes formed by raised roadbeds as barriers to water drainage after the peak of a flood. Flooded rectangular fields of midwestern grid near Red River of the North in North Dakota. Photo courtesy of U.S. Geological Survey.

river only by way of roadside ditches and culverts. Because the culverts are not designed to quickly allow large quantities of water to pass through them, they commonly become blocked or plugged. Consequently, fields can remain flooded for weeks, delaying or sometimes preventing the planting of crops.

Just as roads can be a barrier to water drainage, they also can be a barrier to water encroachment.[1031] Again, the extremely wet period in the north-central part of the USA mentioned above provides an example. In northeastern North Dakota, the Devils Lake drainage basin covers 8600 km^2 (3320 mi^2). Similar to prairie potholes, the basin does not have a surface drainage outlet. Therefore, substantial precipitation within the basin raises the level of Devils Lake. Since 1992, the lake level has risen more than 7.5 m (25 ft), inundating more than 12 150 ha (30 000 acres) of land, including roads. The flooded area is mostly farmland, but some of the town of Devils Lake has also been flooded. In anticipation of further rises in the lake level, roads have been raised around the perimeter of the lake so they will not need to be hurriedly raised when flooding is imminent. Yet, as the lake has continued to rise, some roads have essentially become dams, preventing further expansion of the lake in the vicinity of those roads. Roadbeds are typically composed of porous material to enhance water drainage, whereas dams are designed and constructed to block water passage. Because the roads were not designed to be dams, the safety of the roads for transportation and for protection of property across the roads from the lake has come into question. The U.S. Army Corps of Engineers has estimated that it would cost as much as US$50 million to construct protection levees for these roads (V. F. Schimmoller, personal communication).

In the coastal areas of some states that border the Gulf of Mexico, such as Louisiana and Florida, shallow water moves as sheetflow through marsh grasses from inland areas toward the coast (see Figure 13.1 in Chapter 13). In Florida, much of the original water supply to the Everglades was a result of this type of surface-water flow. Roads that cross such areas commonly obstruct the flow, affecting the ecosystems on both sides of the road. The effect on ecosystems on the upslope (upgradient) side of the road is more localized near the road because the road essentially acts as a dam and pools water. The effect on ecosystems on the downslope (downgradient) side of the road is more extensive because the road cuts off the water supply to that ecosystem. Even if scattered culverts and bridges are built into the road system, water is routed past the road through these localized features, thereby substantially altering natural flow paths and connectivity of the ecosystem. In this situation, large numbers of culverts would tend to maintain relatively natural conditions in ecosystems on both sides of the road (Figure 13.1).

Where a road transects a shallow lake, wetland, or estuary, commonly the water bodies on opposite sides of the road appear distinctly different. For example, the color of the water may be different or the water bodies may have different plant communities (Figure 9.8 in Chapter 9). Curiously, data and studies that have documented the changed ecological characteristics of the dissected water bodies are scarce. A few examples here illustrate striking ecological effects.

The Great Salt Lake in Utah receives most of its fresh water from streams flowing into the southern part of the lake.[567] In 1903, a railroad trestle on posts was constructed in an east-west direction across the entire lake, but in 1959 an earthen causeway was constructed to replace the trestle. The trestle permitted water to circulate freely throughout the huge lake, but the causeway was built with only two 4.5-m (15-ft) culverts and an 88-m (290-ft) bridge opening. This design severely restricted water movement between the northern and southern parts of the lake, and the two parts quickly took on different characteristics.

Before the causeway was constructed, the concentrations of dissolved salt (largely sodium chloride) in the northern and southern parts of the lake were about equal. Within only a few years, a net flow of salt water from the southern to the northern part was reported,[10] resulting in a thick layer of salt being deposited in the northern lakebed. Because nearly all inflowing streams are on the southern side, the causeway caused the southern part to have a higher level, or head, forcing flow to the north and preventing flow back to the south.

A recent study indicated that by 1972 salt concentration in the northern part was about 200 g/L (grams per liter) greater than in the southern part, and by 1998 it was 250 g/L greater.[567] These differences in concentration result in water density differences between the two parts of the lake. This in turn causes a striking two-way flow through the culverts. Denser, more-saline water from the northern side flows southward beneath the fresher water from the southern side that flows northward.

On the western side of Florida, Tampa Bay is a large, shallow Y-shaped estuary. One arm of the bay, Old Tampa Bay, is transected by three major rock fill–based *causeways,* each containing a major bridge (Figure 7.7). The other arm, Hillsborough Bay, has small causeways near its upper end. A fourth major causeway is present near the mouth of Tampa Bay. These engineering works, together with dredge and fill operations and shoreline landfills, have had a substantial effect on water circulation patterns in the bay. A study to evaluate the extent of these engineering works on water circulation patterns was conducted by the USGS in cooperation with the Tampa Port Authority and the Corps of Engineers.[368] One of the study's goals was to develop a computer model of present water circulation patterns, then run the model again with all of the engineering works removed, representing predevelopment conditions about 1880. As might be expected, the largest changes in the circulation patterns for a typical floodtide between 1880 and 1972 are in the vicinity of the causeways and bridges (Figure 7.7).

The effect of the changes has essentially redirected the natural flow paths of water from the head of Tampa Bay to the mouth, thereby increasing the rate that water, and the material transported by water, moves through the bay. Thus any contaminants that enter the head of the bay are transported to the Gulf of Mexico more rapidly than they would have been prior to human involvement. In addition, although not studied extensively, the aquatic life in the areas of changed circulation patterns are likely to be different following construction of the causeways (Figure 7.7). This is probably particularly true for bottom-dwelling organisms, because sediment types and transport of sediments are likely to be different in parts of the bay where causeways prevent free circulation of water throughout the estuary.

In addition to disrupting surface-water flow patterns, roads also can affect surface-water bodies dependent on groundwater for their ecological characteristics.[893] For example, near Crystal Springs in North Dakota, a railroad, a county highway, a U.S. highway, and an interstate highway cut across two small prairie lakes (Figure 7.8). In their natural condition, these two lakes received water and dissolved chemicals from groundwater, which were then circulated freely throughout the lakes. Disruption of the circulation pattern of the lakes by the roadbeds resulted in each of the individual pools having substantially

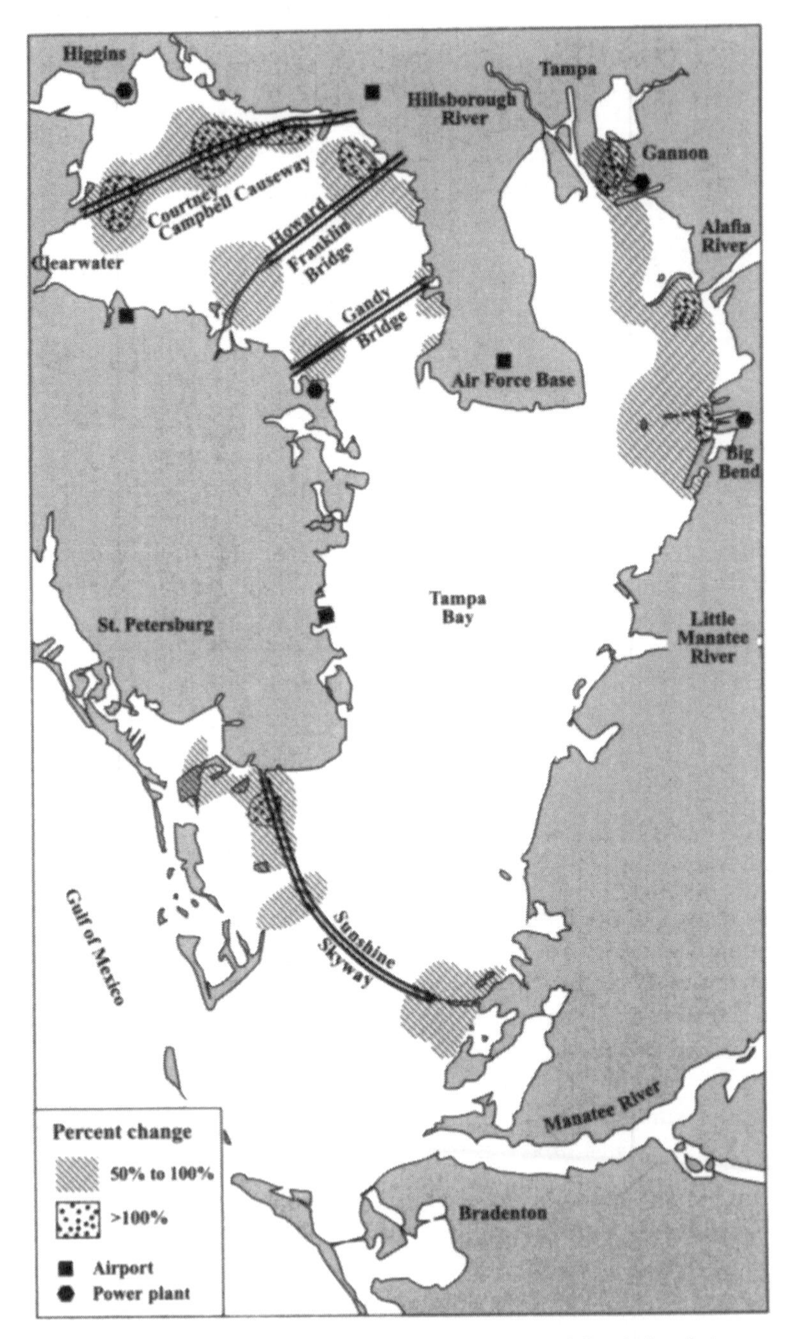

Figure 7.7. Changes in water circulation patterns for a typical floodtide relative to constructed-fill causeways in a large bay. Changes based on simulation modeling of Tampa Bay, Florida, from 1880 to 1972. Adapted from Goodwin (1987).

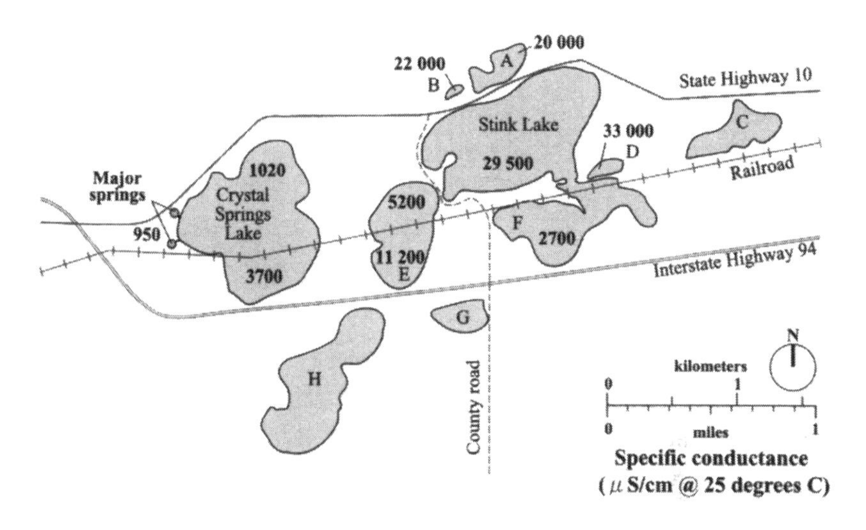

Figure 7.8. Water quality altered by transportation corridors in an area where shallow lakes depend on groundwater inputs. High specific conductance reflects the accumulation of dissolved chemicals, mainly resulting here from roads disrupting the circulation of lake water, such that the water gained from groundwater on the upgradient (upslope) side of the lake could no longer circulate freely throughout the lake. Near Crystal Springs, North Dakota. Adapted from Swanson et al. (1988).

different chemical and biological characteristics because some no longer received the constituents brought in by groundwater.[893]

The pools that continued to receive groundwater input following road construction, Crystal Springs Lake north of the railroad and pool F of Stink Lake, remained the freshest (Figure 7.8). Crystal Springs Lake continued to support game fish. Because the fresh groundwater input to pool F of Stink Lake could no longer circulate throughout the lake after road construction, chemical conditions in the main body of Stink Lake and in pools A and B of Stink Lake changed rather dramatically (reaching conductances of 20 000 microSiemens/cm or greater). Saline conditions in Stink Lake itself increased enough to provide habitat for brine shrimp.

Road Types, Land Use, and Water

Roads are built for a wide variety of reasons in all types of terrain. In some cases, roads are built to develop a resource, such as minerals or timber, and in other cases they may be built to connect two points by the shortest distance at the lowest cost. The purpose of a road has a direct bearing on the type of road and where it is placed within landscapes. General relationships of roads, land use, and water can be summarized by examining four types of roads that differ by the amount of engineering used to construct them: (1) field roads and tracks, (2) constructed gravel roads, (3) secondary paved roads, and (4) primary paved roads. These different types of roads are all affected by water and also affect the movement of water and water-transported material, but generally to different degrees.

Tracks, Field Roads, and Constructed Gravel Roads

Field or *dirt roads* and *tracks* are earthen roads that have little engineering design or construction, such as preparation of roadbeds or use of bridges and culverts. Most roads of this type are created simply by continually driving over the same route. Field roads generally are present in agricultural areas and therefore are found mostly in flat landscapes. Tracks generally are used to drive only occasionally to a location, such as a favorite fishing hole or a remote part of a ranch. Tracks are present in undeveloped areas and can be found in all types of landscapes, from coastal areas to mountains. Water can temporarily flood field roads in flat landscapes and make them muddy and difficult to use. The roads themselves generally do not substantially affect the movement of water. However, if field roads have even a small slope, wheel ruts can erode into gullies and cause some diversion of water. In areas of steep slopes, such as sides of mountains, plateaus, and river valleys, rivulets flowing in gullies caused by the erosion of tracks can become very large, carry substantial amounts of sediment, and cause major diversions of water.

Field roads provide some important ecological functions. For example, large animals, including coyotes, wolves, cougars, lions, and dingoes, may use field roads with little traffic as major foraging routes.[302] During warm seasons, insects, including butterflies, are often attracted to the mud puddles common in field roads.

Gravel roads are constructed for a variety of reasons. Most are probably constructed to provide general transportation routes in rural areas (Figure 7.1), but some are used to extract resources. Gravel roads for transportation in rural areas generally are intended to be permanent. Therefore, the roadbed is usually raised (unlike field roads), and the roads, bridges, and culverts are designed and constructed to last a reasonably long time. On the other hand, most gravel roads are not expected to carry large volumes of traffic.[3] The routing of water via ditches and culverts along rural roads is expected to have a permanent effect on the flow of water. Roads built to extract nonrenewable resources commonly are temporary, in that they are not used extensively after the resource has been extracted. In addition, the cost of such roads needs to be in accord with the benefit obtained from the resource. The quality of these roads generally indicates that they are not designed to last indefinitely. Although these roads are not permanent, they need to be designed and constructed well enough to repeatedly carry heavy equipment. In contrast, gravel roads constructed for extraction of renewable resources generally are well designed and constructed because they are intended to be in place for a long time.

When gravel roads in flat terrain become flooded, the water does not drain away quickly and the water-sodden road can become unstable, resulting in extensive damage to the structure of the road. These types of roads generally follow the land-surface topography. Therefore, cutslopes and fillslopes tend to be few and small, and the erosion and transport of sediment from these slopes usually are not serious problems. However, traffic volume can affect the amount of sediment released from road surfaces because tires detach and pulverize particles. Some particles become airborne and transported as dust (Chapter 10). Runoff from the surface of gravel roads and erosion of roadsides contribute sediment to flowing water, but only the finer-grained sediment is transported very far, because of the low gradient of ditches and streams. In contrast, gravel roads built in steeply sloping terrain can result in substantial modifications to water flow paths down the hillsides. As a result, sediment is contributed not only from the road surface but also from the extensive movement of earth needed to construct the roads. This is especially a problem when roads are built for logging in mountainous areas, because the road network can be quite dense (Chapters 11 and 12).

Secondary and Primary Paved Roads

Secondary paved roads, such as county roads and some state highways in the USA, traverse all types of landscapes and land uses. They are intended to be used for extended periods of time and to move all types of vehicles at greater speeds than on gravel roads. These roads are better engineered than gravel roads, including better roadbeds and water-routing features, such as ditches, culverts, and bridges.[2,3] Such roads do not follow the land-surface topography as closely as gravel roads, and consequently they require much more cutting and filling during construction. As a result, a considerable amount of sediment can be generated from road construction. In addition, even though the surface is paved, the shoulders of paved roads commonly erode, resulting in a continuing source of sediment to receiving water bodies. This problem is especially common in areas of steep slopes, such as mountains and the sides of plateaus and river valleys. The large flow velocity of water running over the hard road surface before the water reaches the unpaved road shoulders has considerable erosive power.

Primary paved roads include freeways and other major federal and state arterial highways. These roads are designed to carry large volumes of traffic through all types of landscapes and land uses.[2] Furthermore, they are designed to carry large numbers of heavy vehicles that travel at high speeds. Major highways are perhaps the best engineered for dealing with water. Generally, great care and extensive technical and financial resources go into bridge and culvert design, and roadside ditches are designed to carry large quantities of water.

As a result, construction of primary paved roads usually involves extensive cutting and filling. In mountainous areas, it is not unusual for whole mountainsides to be removed to accommodate the routing of major highways. Even in flat terrain, construction of these types of roads usually results in moving large amounts of earth material in grading the land where the roadbed will be placed and the generally large ditches constructed. In addition, massive amounts of roadbed material are usually transported from gravel pits or quarries. Both of these activities can result in considerable movement of sediment into receiving waters, even though typically the most care is taken in these major construction projects to alleviate the problem. In some cold climates, sand and other aggregates used on icy highways are an additional source of sediment.

Conclusion

The interactions of road systems and water are numerous and complex. Much of the damage to roadways and much of the environmental impact of roads are related to water. Flowing, freezing, and seeping water can severely damage roads. Roads alter the movement of water, sediment, and contaminants in landscapes. Some roads and ditches may become a part of stream networks.

The structure and function of road systems are relatively easy to understand as human-engineered works designed and built for specific purposes. The structure and function of natural hydrologic systems, such as the movement of water through the atmosphere, surface water, and groundwater systems, and their interactions, are much more difficult to understand. Yet understanding the interactions of roads and water needs to be grounded in an understanding of both. The hydrologic system includes streams, lakes, wetlands, overland flow, soil water, and groundwater. Since the amount of water and loads transported by water vary along flow paths, interactions of road and water systems depend strongly on where a road is positioned in a landscape. Some general principles are evident with respect to the interactions of roads and water in different parts of generalized landscapes.

Roads on uplands, whether along mountain ridges, plateaus, uplands adjacent to river valleys, or high river terraces, generally are not affected much by water, even during flooding. However, chemical contaminants and sediment from roads in these landscape positions have the potential to affect the headwaters of streams and to contaminate recharge areas of regional groundwater flow systems. Therefore, great care needs to be taken in designing and constructing roads in these upland settings to contain contaminants and prevent them from entering surface-water and groundwater systems.

Roads on hillsides can be affected by water because of the sometimes swiftly flowing water in such settings. Water can erode cutslopes and fillslopes, wash out culverts and sections of roads, erode ditches into gullies, and trigger landslides that cover roads or remove road segments. Roads on hillsides also affect the movement of water because the roads block natural flow paths. Contaminants from roads on hillsides can readily move into streams and quickly travel long distances. However, contamination of groundwater generally is only local in extent because of the prevalence of surface flow and shallow subsurface flow paths on hillsides. Control of water near roads on hillsides is paramount for the sake of both the roads and the environment. Flow-routing structures and subsurface-water drainage systems can be designed to protect roads from washouts and landslides, both uphill and downhill from the roads.

Roads in low areas are affected by water largely through flooding and occasional washouts. Roads in such settings can be barriers to the removal of flood water or to the encroachment of flood water. Effective routing of surface water and controlling groundwater discharge to the land surface are important for road design in lowlands. In addition, containing pollutants originating from roads is especially important for preventing contamination of groundwater and nearby surface water.

Finally, it bears emphasis that water is the major enemy of roads. Thus a rich array of road design and engineering practices to combat water problems has evolved. Yet ironically, roads are major enemies of water bodies. Thus preventing and disconnecting linkages between them is now an engineering and ecological challenge before us.

CHAPTER 8

Chemicals along Roads

He pulled out in a cloud of dust
Laying rubber and spewing rust
And on any road he'd take
He'd have his foot on the pedal and my heart on the brake

—Mary Chapin Carpenter, "A Road Is Just a Road," 1987

I have an "Ecology Now" sticker on a car that drips oil everywhere it's parked.

—Mark Sagoff, *Earth Ethics,* 1990

A rich assortment of invisible chemicals accumulates along roads. The mixture is served up by vehicles, roadside management, and the roads themselves. In low amounts, the substances are just as benign as the chemical array served up by nature. But in high amounts—that is, levels unwanted by society—the added chemicals become *pollutants* or *contaminants.*

Some chemicals accumulating along roads are transported short distances through the air, but most are carried by water washing off (or seeping through) a road. Chapter 9 focuses on chemical effects in aquatic ecosystems, while Chapter 10 highlights the effects of chemicals transported longer distances through the atmosphere.

The effect of roads on water quality is a concern to water resource managers, highway departments, and many other people concerned with environmental conditions.[243, 268, 227] Several comprehensive reviews or bibliographies on stormwater runoff from roads are particularly useful, including a synthesis of previous documentation and research[1063]; a user's manual on highway water-quality impact assessment and mitigation[499]; an assessment of pollutant concentrations from urban and rural highways[230] (see Figure 3.4 in Chapter 3); and a national-scale assessment of chemical runoff from roads (U.S. Geological Survey website 2002). But the chemicals added along roads affect far more than water. They build up in the soil, in plants, and in animals, with consequent cascading effects through terrestrial ecosystems.

Sources of Chemical Pollutants

Chemical substances, of course, are everywhere on earth, but polluting levels of chemicals cause damage or toxic reactions to human health, to ecological systems, or to both.[675] Here we focus on the chemicals that inhibit natural processes and native species. Major sources of roadside pollutants are vehicles, roads and bridges, and dry and wet (dust and rain) atmospheric deposition. Localized, less-frequent sources include spills of oil, gasoline (petrol), industrial chemicals, and other substances, and losses of materials in accidents involving vehicles and roadside structures. In addition, objects discarded from vehicles accumulate along many roads. Roadway maintenance practices, such as sanding and de-icing road surfaces and applying herbicides to roadsides, usually add pollutants. Also, both the road surface and the tires rolling on it gradually degrade.

One assessment of chemicals found along roads indicates that 19 of the 23 important pollutants (83%) come from vehicles[499, 268] (Figure 8.1). Within vehicles, there are several major sources of pollutants, and a particular type of pollutant can come from a number of sources. Thus one-third (35%) of the types of roadside pollutants come from oil, grease, and hydraulic fluids. Engine and parts wear produces 30% of the pollutant types; metal plating and rust, 22%; tire wear, 22%; fuel and exhaust, 22%; and brake lining wear, 17%. Similarly, nonvehicular sources produce many types of pollutants. Sanding and de-icing agents produce one-fifth (22%) of the pollutant types; roadbed and road surface wear, 17%; and herbicide and pesticide use, 13%. These figures do not include heavy metals and other chemicals that leach from bridges into streams and other water bodies. In short, chemical pollutants along roads originate from diverse sources, and even significantly reducing a single pollutant would normally require control of a number of the sources.

Roads, Roadsides, and Management

Since road networks slice through or surround most ecosystems, and because road surfaces degrade between resurfacings, the composition of road surfaces is of ecological interest. Two billion metric tons of total solid waste are produced annually in the USA.[277] Eighty percent is mineral waste from mining,

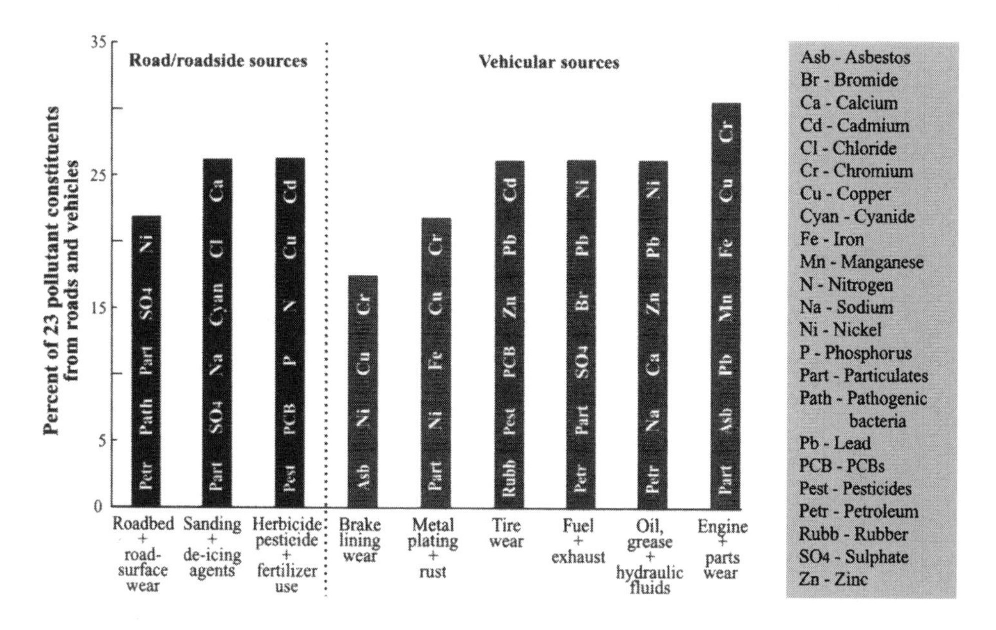

Figure 8.1. Sources of 23 pollutant constituents in stormwater runoff. Adapted from Kobringer (1984) and FHWA (1996a).

10% is municipal solid waste, and 5% (about 100 000 tons) is asphalt road-surface material annually removed from roads. To reduce the amount of waste for disposal and the amount of new asphalt containing petroleum products, 80% of the road-surface material removed is recycled back into new road surfaces. Although few data are readily available on amounts and frequency, in some cases road recycling or resurfacing projects include recycled asphalt pavement, blast furnace slag, coal fly ash, cement kiln dust, or steel slag.[277] Apparently, smaller amounts of nonferrous slags, reclaimed concrete, coal bottom ash, and boiler ash may also be included in recycled road material. In various European countries, many other materials are components of recycled road material, including municipal solid waste, coal mining waste, dredged material (e.g., from harbors), gypsum, phosphorus slag, slightly contaminated soil (and heavily contaminated soil after cleaning), demolition waste, coal fly ash (silico-aluminius), tires, and plastics.

Road surface degradation over several years leads to surface distress, deformation, or cracks.[274] This may permit rainwater or snowmelt to penetrate into the surface and base of a road, leaching chemicals that reach surrounding ecosystems. Resurfacing follows, increasingly using recycled materials in the USA. Water and wind sweep across road surfaces carrying road particles into nearby or downwind ecosystems. Eroded road particles are also lifted into the air by the turbulence of passing vehicles. Unfortunately, little information is available on the chemical composition of different roads or whether these inputs are important ecologically in ecosystems.[277]

Finally, in road construction, *fill material* trucked in to construct a road (or topsoil used on the roadside) may be chemically quite distinct from that of the roadside or adjacent area. Limestone-derived fill material placed over an acidic granite-derived substrate, or vice versa, illustrates this point. In this case, chemicals from the roadbed would change the soil pH and plants of an adjacent zone. Aquatic ecosystems in streams alongside of or downstream from such a roadbed would have an altered water quality.

Vehicles, Tires, and Fuel

Several types of chemical pollutants originate from vehicles[499, 268] (Figure 8.1). Gradual or chronic wear, leaks, and emissions from exhaust of moving vehicles tend to widely distribute the chemicals along a road, where they accumulate.

Vehicles leak or spill several polluting materials:

- *Mineral nutrients.* Road runoff may include nitrogen or phosphorus, which can be nutrient-polluting (eutrophication) sources in aquatic ecosystems.
- *Heavy metals.* Runoff may include metals such as zinc and cadmium from vehicle wear, combustion products, catalytic converters, abrasives in brake linings, and uncombusted fuel additives.
- *Organic compounds.* Most organic compounds found in runoff are from exhaust (uncombusted products), spilled fuels, lubricants, coolants, and hydraulic fluids. Petroleum products are mainly composed of diverse *hydrocarbons,* which primarily consist of carbon and hydrogen. *Polycyclic aromatic hydrocarbons* are usually formed in incomplete combustion and are generally volatile, but they also can be found in runoff sediments. *Mono-aromatic hydrocarbon compounds* are common in crude oil and petroleum products, and find their way into runoff primarily through spills and leaks of gasoline and other petroleum products.[545]

One increasing concern regarding hydrocarbons is the impact of ultraviolet (UV) radiation. The toxicity of hydrocarbons to aquatic organisms seems to be greatly increased by UV radiation from sunlight.[698]

A series of chemical additives to gasoline used to increase combustion efficiency has been a particular pollution problem.[675] *Tetraethyl lead* was introduced as an antiknock compound for gasoline engines in 1922. The resulting lead pollution (and consequent problems with newly introduced catalytic converters on vehicles) caused the USA to shift to unleaded gasoline in 1974. Canada and Europe switched to unleaded fuels 15 to 20 years later.[878, 536]

Although some of the lead from fuel remains in the roadside environment, particularly somewhat deeper in the soil profile,[132,577] lead levels in plants and animals overall have dropped significantly since lead was removed from fuel.

As a substitute for lead, manganese-containing fuel additives were used extensively in the mid-1970s and early 1980s but supposedly were discontinued soon thereafter.[465] *MMT* (methyl cyclopentadienyl manganese tricarbonyl) then replaced tetraethyl lead as the antiknock compound in some gasolines. The combustion of fuels with MMT produces vehicle exhaust containing manganese oxides.[906,575,878]

Although manganese oxides can clearly cause negative health effects in animals and humans in laboratory studies,[473,503] the amount of manganese added to the environment from exhaust is not currently considered a significant ecological risk.[575] Nonetheless, a consensus exists calling for a better understanding of human-caused manganese oxides in the environment and their potential effects on humans and nature.[623,1057] One recent study in Utah reported soil manganese concentrations along interstate freeways with high traffic volume (70 000–148 000 veh/d) up to 100 times higher than historic levels.[577] The investigators also found that roadside aquatic plants were higher in leaf-tissue manganese than were herbs or grasses, and that submerged and emergent aquatic plants were particularly sensitive bio-indicators of manganese contamination.

One of the most ubiquitous indirect examples of motor vehicle–associated pollution in recent years has been the spread of *MTBE* (methyl tertiary-butyl ether), which in a study of 16 cities was found in 7% of the stormwater drains. MTBE improves the combustion of gasoline and is an additive to reduce ozone emissions in some regions. Although MTBE is volatile in the air, the primary pollution concern is the presence of leaking fuel tanks among the thousands of gas stations that use underground storage (regulations require testing and replacement of tanks as appropriate). Most of the hydrocarbons that compose gasoline move relatively slowly through soil, which means that they accumulate and may be amenable to site-specific cleanup. In contrast, MTBE is highly soluble, spreads rapidly through groundwater, and persists a long time. That poses a different and more serious problem. The MTBE additive readily reaches and contaminates aquifers as well as streams and lakes.

Areas adjacent to major highways receive the greatest input of heavy metal particles. On both sides of such highways, a distinct gradient of elevated concentration typically extends outward for up to about 50 to 100 m (165 to 330 ft). Increased concentrations have been found in air, soil, and plants within this zone. Toxic levels of heavy metals may extend outward for meters rather than tens of meters, though this certainly varies by species.[639] It is important to know how these contaminants are distributed and their potential effects, because they may adversely affect all forms of life in an area, including humans.

Rubber from tires also accumulates along roads. Highway travel usually means passing "road-kill mimics," the black chunks of shredded truck tires that at first glance resemble creature forms, from dead dogs to squashed snakes. Unnoticed and more important environmentally are the fine particles of rubber and its various synthetic forms that result from incessant tire wear along the road. Roughly one tire per year per car on average is discarded after distributing its tread in particles along the road network. For example, in the former West Germany, which has a much smaller area but a highway density more than twice that of the USA, it is estimated that 1 mm of road surface material is annually eroded from the highways (1 in 25 years).[204] On the same road system, 100 000 tons of tire dust are annually generated.[338, 121] Wind and the turbulence of passing vehicles raise the tire dust and other particles into the air to be deposited downwind. Furthermore, rain and snowmelt periodically wash the particles into aquatic environments.

Chemical Spills

The road network is a thoroughfare for passenger vehicles and for trucks carrying freight, some of it toxic chemical substances. More than a half million shipments of hazardous materials are carried daily on the U.S. road network.[674] A small fraction is spilled, though this includes some large spills and some next to water supplies and sensitive ecosystems (Figure 8.2).

In the USA, about 2400 accidental chemical spills are reported each year to the federal government. The actual number of spills, plus the illegal dumping of chemicals (and washing of chemical-carrying trucks) while standing or moving along roads, is unknown. In addition, 7 million vehicle accidents, most of which release some chemical pollution onto the road, are annually distributed over 6.2 million km (3.9 million mi) of public roads. This combined process distributes the gasoline, oil, and other leakage pollutants through much of the road network. Roadsides, groundwater, and nearby aquatic ecosystems are major recipients of the concentrated pollutant mixture. Most roads are bordered by ditches that carry water rapidly and directly to streams, lakes, and other water bodies (Figure 8.2). Therefore, chemical spills that pollute aquatic environments are probably a fairly common occurrence, though the frequency, amounts, and ecological effects are unknown.

An overview of the transportation-related pollutants and their primary sources (Figure 8.1) emphasizes that no "magic answer" will be found as a solution. Most pollutants come from multiple sources, which can be *temporary* (pollution due to road construction or maintenance), *chronic* (vehicle exhaust, pavement and tire wear), *seasonal* (de-icing in winter), or *accidental* (spillage). The most persistent and problematic sources are chronic and seasonal. Temporary and accidental pollution sources tend to be localized but can be highly

Figure 8.2. Equipment containing and attempting to clean up a significant chemical spill by wetlands and a drinking-water-supply reservoir. A truck delivering home heating oil tipped over just to the left of the white car and about 75 m (250 ft) from the shallow City of Cambridge, Massachusetts, reservoir. At least 29 vehicles and major pieces of equipment were being used when this photograph was taken seven days after the truck spilled 1700 gallons of oil onto the porous glacial-material substrate soil. Interstate Highway 95. Photo by R. T. T. Forman.

concentrated. All can be toxic to different organisms and can degrade local ecosystems.

In view of the array of chemical pollutants originating from roads and vehicles, a set of best management practices has evolved to reduce environmental impacts, especially near sensitive ecosystems and drinking water supplies. These practices, which reduce source amounts and mitigate effects, are introduced at the end of this chapter.

Chemicals and Their Dispersion

The major contaminants involved in highway runoff are mineral nutrients, metals, petroleum-related organic compounds, sediment washed off the road, and agricultural chemicals used in road and roadside maintenance (Figure 8.1). Petroleum-related compounds may include polycyclic aromatic hydrocarbons, benzene, toluene, thylbenzene, xylene, and MTBE. Organic compounds derived from oil, grease, and tires enter water bodies by way of water and air transport. Roadsalt and other de-icing agents are considered in a separate section below.

Studies of the effects of contaminants, such as organic compounds and *trace metals* (*heavy metals* plus lighter-weight metals), on water resources and ecosystems[590] are not as common as for de-icing salts. Part of the reason for this is the high analytical cost of measuring many of the organic compounds associated with roads and motorized vehicles. Another reason is that selection of an appropriate sampling location for either groundwater or surface water can be a problem, due to natural (microbial and biogeochemical) processes that often occur once the compounds enter soil or subsurface water.

A study of storm runoff from the Bayside Bridge in Tampa Bay, Florida, is particularly informative from this perspective, because the water samples were collected in a container directly below the road surface of the bridge.[880] The Bayside Bridge is 4.3 km (2.7 mi) long, with separate northbound and southbound spans and concrete road surfaces. The bridge opened to traffic on June 2, 1993. The average daily traffic volume in 1993 was about 37 400 vehicles and increased to about 48 800 vehicles in 1994.

Storms flushed the road surface of chemical constituents. As expected, concentrations in the collection container were highest in the early part of storms that followed long periods of no rainfall. The concentrations of selected chemical constituents collected during 24 storms from May 1993 to September 1995 on a northbound section of the bridge are shown in Table 8.1. For some constituents (copper, iron, lead, and mercury), average values exceeded the state-approved standard level. For others (aluminum, nickel, and zinc), average values were less than the standard but their maximum values exceeded the standard.

This flushing of chemicals is especially important in the case of organic chemicals.[1063] Accumulation of oil, rubber, and grease on road surfaces causes slick driving conditions when a road surface first becomes wet, before runoff water removes such chemical accumulation from the road.

Roadways and their associated drainage systems carry organic and inorganic materials (soil particles, eroded pavement material, tire rubber, and so forth) suspended in stormwater into aquatic systems.[716, 1053] Attached to these particles, in turn, are many other chemical compounds discussed above, such as mineral nutrients and metals,[688] which therefore are also carried into water bodies. *Road runoff* or *stormwater runoff* refers to rainwater or snowmelt flowing off road surfaces and containing substances in solution or tied up in *sludge* (slushy sediment).[160] Although sludge is a small part of runoff, it normally contains most of the contaminants. Most metals and almost all polycyclic aromatic hydrocarbons are bound to the sludge that spills off the road.

Table 8.1. Water quality characteristics of stormwater runoff from a Florida bridge. The data are from 24 storms between May 1993 and September 1995, Bayside Bridge, Tampa Bay. Specific conductance is in microsiemens per centimeter at 25 degrees C (77 degrees F); constituents listed from alkalinity to phosphorus are in milligrams per liter; constituents from aluminum to zinc (trace metals) are in micrograms per liter. Each characteristic is based on between 159 and 186 samples. Water quality standards are for Class II surface water.[293] Bridge is by Clearwater (see Figure 7.7 in Chapter 7). Source: Stoker 1996.

Constituent	*Surface water quality standards*	*Minimum*	*Maximum*	*Mean*	*Median*
pH (pH units)	6.5–8.5	6.6	7.8	7.0	7.0
Specific conductance	—	29	730	142	109
Alkalinity	—	11	137	34.2	28.0
Total suspended solids	—	1.0	270	36.8	20.0
Volatile suspended solids	—	<1	250	24.9	16
Total organic carbon	—	0.5	96	11.9	8.0
Nitrite nitrogen	—	0.01	0.42	0.04	0.03
Nitrite nitrogen plus nitrate nitrogen	—	0.02	10.0	0.99	0.64
Ammonia nitrogen	—	0.01	1.4	0.16	0.13
Ammonia nitrogen plus organic nitrogen	—	<0.01	5.8	0.78	0.46
Orthophosphorus as Phosphorus	—	0.01	0.27	0.06	0.05
Phosphorus	—	0.02	0.81	0.14	0.10
Aluminum	≤1500	80	2300	483	370
Arsenic	≤50	<1	4	0.98	0.77
Cadmium	≤9.3	<1	3	0.29	0.14
Chromium	≤50	<5	22	4.3	3.2
Copper	≤2.9	1.0	110	12.4	8.0
Iron	≤300	30	9600	823	530
Lead	≤5.6	1.0	440	21	11
Mercury	≤0.025	<0.1	0.2	0.08	0.07
Nickel	≤8.3	<1	20	4.0	3.0
Zinc	≤86	4.0	470	84	50

Wind dispersion along a road is the process through which substances are dispersed by airflows, including traffic turbulence. Most such substances are deposited right beside the road. The quantities deposited decrease rapidly with increasing distance from the road traffic.

The amount of contaminant dispersion resulting from wind or runoff largely depends on the type of road surface—concrete, asphalt, gravel, or dirt. Research in The Netherlands compared the effects of porous and dense asphalt. The study revealed that wind dispersion was the major cause of pollutant movement on heavily used highways with dense asphalt.[501] Also, the magnitude of material transported along dense asphalt surfaces tended to be 10 to 20 times greater than on porous asphalt. From a range of possible mitigation measures, the construction of a porous-asphalt top layer seems to be an effective measure to reduce pollution levels near roads. While porous road surfaces may need to be resurfaced a bit more frequently, they serve as partial absorbers of chemical pollutants, thus reducing dispersion and facilitating cleanup.[426] The study also raises the point that in ecologically sensitive areas or drinking-water supply areas, this mitigation approach may need to be supplemented with periodic steam-cleaning of road surfaces and the collection and purification of runoff water, using existing equipment and technology.

Pollutant dispersion patterns for heavy metals are quite variable. However, atmospheric metal deposition from vehicles is strongly influenced by local climatic factors (such as prevailing wind direction) and traffic flow characteristics.[745, 67] The particles and pollutants deposited on a road surface through normal operation and wear can be redistributed by *vehicle turbulence* from the road surface to roadsides and adjoining land.[482, 50] Thus traffic volume not only determines the amount of pollutants deposited but also affects the amount of turbulence and dispersion of chemicals into adjoining ecosystems.

Two measures of traffic volume are commonly considered: *average daily traffic* (ADT)—or simply *traffic volume* (number of vehicles per 24-hour commuter day)—and *vehicles during a storm* (VDS).[268] Mixed results have been reported in correlating pollutant concentrations with average daily traffic.[50] However, one extensive study of more than 900 storm events in 31 states suggests that average daily traffic influences the concentrations of several pollutants in highway runoff.[230, 268] Busy urban highways (with ADT >30 000 veh/d) in general had two to five times higher pollutant levels in runoff than from rural highways with ADT less than 30 000 veh/d (Figure 8.3). However, individual sites within each highway type correlated poorly with traffic volume. This suggested that the overall difference may be due more to urban versus rural conditions than to the amount of traffic. Traffic volume during a storm, however, may be a stronger predictor than average daily traffic volume for pollutant concentrations in highway runoff. The concentrations of lead, zinc, chemical

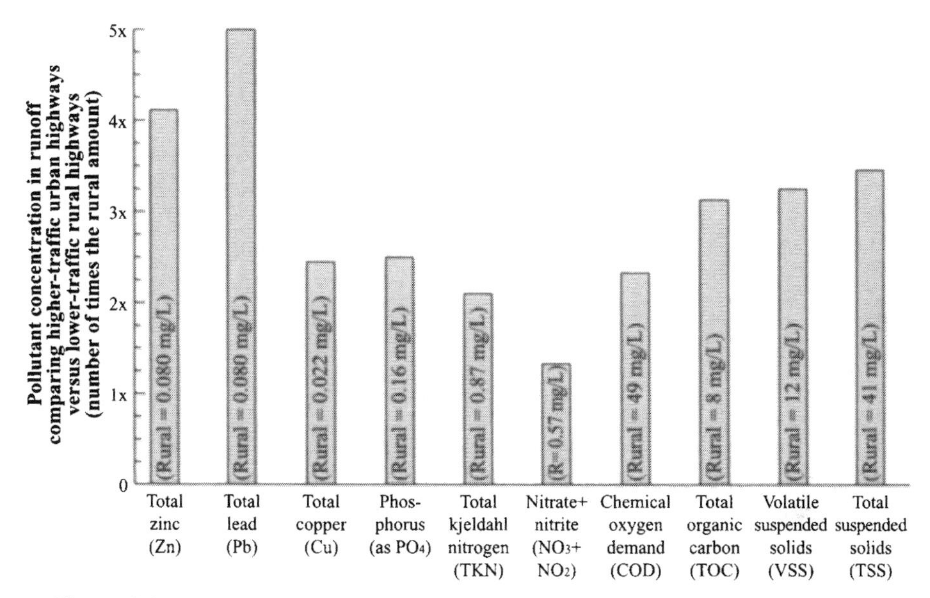

Figure 8.3. Pollutant concentrations in stormwater runoff from urban high-traffic highways versus rural low-traffic highways. Based on average daily traffic volume: urban >30 000 veh/d; rural <30 000 veh/d. The abbreviation mg/L = milligrams per liter. Adapted from Driscoll et al. 1990, Federal Highway Administration 1996a.

oxygen demand, nitrogen (TKN), and filterable residue were found to correlate significantly with traffic volume during a storm.[748, 482]

In urban areas, the pollutants present in highway runoff and in urban stormwater passing through underground pipes are similar.[268] Indeed, the concentrations of the pollutants are also similar. The major exception to these patterns is the presence of, or higher levels of, heavy metals in highway runoff. Heavy metals here largely come from vehicle use, wear, and emissions.

Types of Ecological Effects

The ecological effects of many chemicals added by roads and vehicles are little known. The effects of a few pollutants on aquatic ecosystems will be discussed in the next chapter. Thus, this section briefly introduces several effects of roads and vehicles to illustrate the types of ecological effects to be expected. Terrestrial species and ecosystems, including natural communities in roadsides, are emphasized.

Concern that crops grown close to busy roads might contain dangerous levels of lead prompted numerous studies of lead accumulation in vegetation. These studies showed that the lead levels present in plants are more variable than those in the soil. Also, the concentrations vary not only by distance from road but also by type of vegetation, time of year, and prevailing wind.[1040] Chemical pollution from vehicle exhaust, primarily oxides of nitrogen (NOx), may change plant species composition radically along highways.[23, 1077] The high level of added nitrogen (functioning like a heavy dose of fertilizer) stimulates the growth and dominance of a few species at the expense of many others, thus altering, and in some cases even transforming, certain natural plant communities. The extent of this effect can range up to 200 m (650 ft) from major multilane highways. The effect of this nitrogen enrichment pollution is lower along smaller, less traveled roads with fewer vehicles.

Invertebrates such as insects are essential food for many birds, amphibians, reptiles, and mammals, so it is important to know how lead toxicity might affect insects and the animals that consume them. Early studies suggested that the *abundance* (number) and diversity of invertebrates do not decline with increasing amount of metal pollution in roadside habitats.[1040, 664] Some invertebrate groups appeared to increase with proximity to certain major roads sampled. These studies found that overall lead levels were low in insects but high in earthworms. For earthworms, the highest lead concentrations were in individuals living closest to highways, and especially to highways carrying more than 21 000 veh/d.[366, 593] Furthermore, there was no evidence of lead concentration increasing from one *trophic* (feeding) level to the next (known as *biological magnification*); predatory animals had lower heavy metal concentrations than did their prey.[1040, 986]

Species vary widely in their response to heavy metal concentrations, depending largely upon differences in metabolism, type of diet, amount of food consumed, home range, and life span. Of the heavy metals, lead is the most studied ecologically. Lead concentrations measured in little brown bats, short-tailed shrews, and meadow voles *(Myotis lucifugus, Blarina brevicauda, Microtus pennsylvanicus)* living adjacent to the Baltimore-Washington Parkway (35 000 veh/d) equaled or exceeded levels that have caused mortality or reproductive impairment in domestic animals.[161] Small mammals living adjacent to high-traffic-volume highways (>19 000 veh/d) tended to have greater concentrations of lead than did individuals living near low-volume roads or near sites more than 50 m (164 ft) from roads.[454, 366, 348] Lead in roadsides and median strips of a major highway in Maryland (52 500 veh/d) was not considered a threat to adult ground-foraging songbirds.[375] Lead concentrations in songbird populations in urban Champaign-Urbana, Illinois, were found to be significantly higher than in similar populations in rural areas, but the absolute concentrations were considered to be below levels reported to have toxic effects.[348]

It is unknown how representative these early studies are. Although the results are somewhat at variance with predictions from modern ecology, the results may be robust and they need to be tested. Most lead was finally eliminated from gasoline by the mid-1980s in North America, and lead levels in plants and animals are now relatively low. Unfortunately, the results of the lead studies illustrated give little insight into the ecological effects of other heavy metals. Copper, zinc, cadmium, nickel, mercury, and chromium (Table 8.1) each function differently in living organisms, and each is highly toxic to humans, laboratory animals, and plants at high levels. We know vehicles give off heavy metals. The actual levels along most roads and the levels that cause ecological effects remain poorly understood.

Heavy metals, especially copper, lead, and zinc, are considered to be priority pollutants in highway runoff because of their toxic effects on aquatic organisms.[927] Elevated levels of mineral nutrients, particularly nitrogen and phosphorus, cause *eutrophication* (overenrichment causing algae blooms) and disruption of aquatic food webs. Organic pollutants commonly cause direct toxic effects. They also frequently produce bacterial explosions resulting in loss of oxygen and loss of fish in aquatic systems. Particulate matter tends to hold and transport phosphorus and various metals to water bodies. *Turbidity* (opaqueness due to suspended sediment, as in muddy water) and elevated metal concentrations inhibit fish and other organisms, while phosphorus may eutrophicate the water body. In general, heavy metals tend to be more serious in streams; phosphorus, more serious in lakes.

Roadsalt

The *salting* of roads—using *sodium chloride,* which lowers the freezing point of a water solution, to melt snow and ice—has been common practice throughout the colder regions of the world for many years.[926] Applications of roadsalt mixed with abrasives such as sand began in the 1930s.[157] By the 1970s, 8 million tons were applied annually to roads in the USA. Although sodium chloride is an abundantly available chemical that is commonly extracted from seawater and underground salt domes, the relative purity of the salt varies greatly depending on the source. Besides adding sodium chloride to road runoff, the application of roadsalt can also include doses of heavy metals and essential plant nutrients (such as iron), which eventually reach water bodies.[688]

Forty years ago, it was discovered that lake production may be limited by low levels of chemical elements in *trace* (tiny) amounts.[362] Thus, in some situations, the impurities in roadsalt provide sufficient iron or essential trace metals to relieve nutrient limitation for algae growth, which in turn contributes to eutrophication. Fortunately, aquatic organisms tend to be fairly tolerant of salt content (unless it reaches concentrations at which the osmotic stress is too great). A few species, such as brine shrimp (see Figure 7.8 and associated text in Chapter 7), can exist at a salinity level over twice that of sea water.

In California, sodium chloride has been used extensively on the roads of the Sierra Nevada Mountains to maintain a constant flow of traffic, except during the most severe blizzards.[364] An examination of a variety of roadsalts applied by the state sand- and salt-spreading trucks revealed important differences in their purity. It was also determined that the quality of salt varied greatly among suppliers, and that cyanide was used as a de-caking additive. The study led to a recommendation to purchase salt from the least-contaminated sources and to discontinue the use of cyanide statewide.[688] Pollutants can also be introduced in the application of sand and silt to roads.

Snow and ice control on roads can be mechanical or chemical. Due to its low price, sodium chloride is the de-icing agent used almost exclusively. It is spread mechanically on the roads as dry salt or "moistured salt" or in solution with abrasives such as sand. In open land, salt can be transported and spread by air movement several hundreds of meters from a road.[480] At forested sites, vegetation filters the air, leading to higher salt deposits on the ground under the vegetation.[404] Under some conditions, such salt deposition can damage the vegetation and alter infiltration and percolation to groundwater.[364, 735, 926, 34]

In studies of airborne movement of de-icing salt, up to 90% of the deposited salt was found within 15 to 20 m (49 to 66 ft) of the road.[605] Between 20% and 63% of the de-icing salt applied on moderate-volume roads (5500 to 8000 veh/d) in Sweden was transported by air and deposited on the ground 2 to 40 m (7 to 131 ft) from the road edge.[101] In this study, the deposition pattern was not related to the amount of salt applied on the road. A greater retention of salt near roads is often found when there are greater amounts of snowfall, which in turn leads to greater splash generated from traffic and more snow plowing. Thus less salt is left on the road to be blown greater distances during dry periods. The use of de-icing salts can also increase the retention of heavy metal contaminants in the environment.[926, 57, 536]

Roadsalt and Lake Ecosystems

Compared to that on other contaminants, the literature on contamination of surface water and groundwater from roadsalt is voluminous. Thus this section briefly presents examples involving three types of lakes.

Salt used for de-icing in the Rochester, New York, area caused the chloride concentration in Irondequoit Bay of Lake Ontario to increase by 10 times between 1910 and the late 1960s; the increase was by 5 times during the 1950s and 1960s alone.[118] The dissolved salt reaches the bay largely by urban stormwater drainage. The quantity and salinity of the road runoff during the winter of 1969–70 was enough to prevent complete vertical mixing of the bay water in the spring. Comparison of these conditions with conditions in Irondequoit Bay in 1939 indicated that the period of summer stratification was prolonged a month because of the increased density of the lower bay waters caused by the salt runoff. This was a major seasonal change in lake dynamics and, consequently, for the aquatic ecosystem and fish.

In the glaciated mountains of New Hampshire, Mirror Lake lies at the lower end of the Hubbard Brook valley, a site of extensive ecosystem research since the 1960s.[550, 103] The lake has had increasing concentrations of sodium and chloride since Interstate Highway 93 was constructed through the eastern part of its watershed during 1969–71 (Figure 8.4). The highway cut across the lower end of a stream draining the watershed east of the lake, thereby inter-

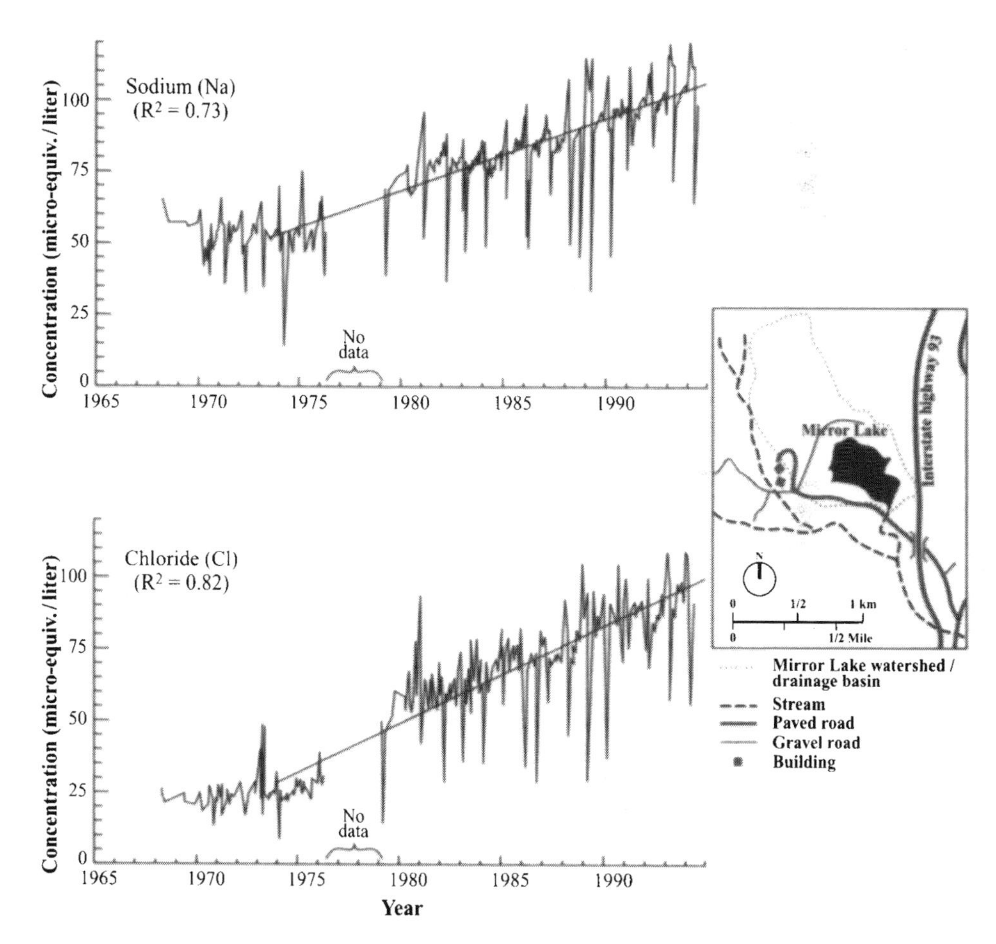

Figure 8.4. Sodium and chloride concentrations in a lake affected by winter roadsalt from a multilane highway 150 m (500 ft) away. Straight lines are regressions for 1974–94; R^2 = coefficient of determination indicating the proportion of the variability explained by this chemical element-to-time relationship. Hubbard Brook ecosystem-study area, West Thornton, New Hampshire. Adapted from Rosenberry et al. (1999).

cepting most of the water that would normally drain to the lake from that side. To prevent road runoff from reaching the lake, a diversion earthen berm was constructed across this eastern inlet stream between the highway and the lake. However, data collected prior to and following road construction indicated that the sodium and chloride concentrations in the lake started to increase a few years after the road was completed and have continued to increase to the present (Figure 8.4). A recent study designed to determine the hydrologic pathway of the roadsalt[795] found that most salt-laden runoff was diverted by the berm, but that some of it was seeping beneath the berm and was then being carried to the lake via the eastern inlet stream. However, the study also revealed that groundwater in the fractured crystalline bedrock beneath the highway was contaminated to a depth of at least 123 m (403 ft).

Finally, shallow roadside lakes appear to be especially sensitive to the additions of roadsalt. This is because salt accumulates and forms a dense layer of salty water on the bottom, which does not easily mix with the overlying fresher water. This layering may create a permanently stratified *(meromictic)* lake.[467,420,490] The salty water that covers the bottom is likely to have no oxygen and will exclude the normal rich production of *benthic* organisms, which dwell on the lake bottom. As an example, following periods of heavy salt application, Putts Lake, near the crest of the Sierra Nevada along Interstate Highway 80 in California, developed these conditions. Thus the three types of lakes illustrated here highlight the special sensitivity to roadsalt of water bodies in confined basins with limited water flowing through them.

Diverse Ecological Effects of Roadsalt

To develop a general appreciation of the effect of road chemicals on water resources, it is convenient to use the generalized hydrologic landscape types (see Figure 7.4) to view contaminant dispersal in the context of surface water and groundwater flow paths.[1046] Chemicals introduced into the environment by the use of roads have different effects depending on where the roads are located within landscapes. Contaminants from roads in uplands have the potential to contaminate streams in their upstream reaches and, therefore, to affect the stream ecosystem for a considerable length. Similarly, contaminants from roads in uplands can contaminate recharge areas of local and regional groundwater flow systems, thereby affecting large portions of the groundwater system.[735,926]

Streams receiving contaminants from roads on hillsides (Figure 7.3) generally flow swiftly if the flow is not too restricted by channel roughness. As a result, contaminants from roads that enter these types of streams can move large distances in a relatively short time. Chemicals from roads on hillsides are less likely to contaminate groundwater, except on a local scale near the base of the slopes. This is because steep slopes favor shallow downslope movement of subsurface water.

From the early 1950s to the early 1970s, sodium transported by the Mohawk River in New York increased by 72 percent, and transport of chloride increased by 145 percent.[721] Although land development in the Mohawk basin during this period led to increases in many chemicals in the river, the increases in sodium and chloride were largely attributed to roadsalt. Because of dilution by running water, especially in late winter and early spring, the ecological effects of roadsalt in most streams and rivers may be minor. Impacts are most likely where a stream flows slowly or flows alongside a salted road.[926] However, roadsalt corrodes bridges, which may cause heavy metals as well as salt to enter a stream.[926]

Contaminants from roads in lowlands commonly affect lakes and wetlands because these types of water bodies are more common in lowlands[1046] (Figure 7.3). Chemicals that enter groundwater in lowlands[735, 926] generally affect smaller portions of groundwater flow systems, because most lowlands are near discharge areas where groundwater enters surface-water bodies such as streams and lakes (Figure 7.4). However, because of the short flow paths, natural attenuation of contaminants is limited and the chemicals are likely to move into and contaminate the surface-water body.

There is relatively little information on the toxicity of roadsalt (sodium- or calcium-based) to amphibians. Animals can maintain osmotic balance in slightly brackish environments, but there is a limit to their ability to do this. The effect of de-icing agents on amphibians living in roadside environments may depend on the amount of salt entering the ponds and on whether other salt-free water sources are entering the pond simultaneously, thereby reducing the salt load over a season. Some long-term data from Michigan suggest that runoff from salted roads has had little effect on local populations of amphibians.[554] However, given the dearth of information on the subject and a large body of literature on osmotic balance in amphibians, field observations and experiments on natural populations should be informative. For example, roadsalt applied to reduce dust on an unpaved forest road inhibited crossing of the road by salamanders.[214] Other effects of roadsalt[214] include the attraction of certain animals such as moose *(Alces)* to the vicinity of a road, where they are subject to higher road-kill rates.[334]

The roadside plant community, on the other hand, can suffer severe losses adjacent to routes where salt is applied.[364, 926] This problem is most serious during drought periods, which allow a buildup of sodium chloride to levels that may cause mortality in the vegetation at some distance from a roadway. Another somewhat subtle impact of roadsalt is that it causes soil compaction, which greatly reduces soil permeability along roadsides. Thus runoff occurs at a higher rate and there is less infiltration of rainwater into the ground to support adjacent plant life. The gradual buildup of salt in the soil along the roadside, together with lowered moisture conditions, can make it more difficult to maintain natural vegetation on roadsides.[910] Also, if plants die, it can be more difficult to revegetate roadsides to reduce erosion.

Other De-icing Agents

CMA, as an alternative to sodium chloride, neither reduces the permeability of soils nor kills roadside vegetation.[364,926] It has been used extensively at airports to de-ice planes and runways. Calcium magnesium acetate (CMA) was developed for areas where sodium chloride accumulation was a threat to either orchards or groundwater supplies,[735] but it is also important where roadside vegetation would be killed or damaged by sodium chloride applications. Some applicators have objected to the vinegar-like odor of acetate, and the high cost of CMA has limited its use.

To determine the possible effects of CMA on aquatic environments, a series of *bioassay* experiments (effects on certain key species are measured) on 10 lakes in Northern California was performed.[365] Of the 10 lakes studied, 8 showed no significant impact of CMA on chlorophyll levels (a measure of algae production in a lake).

Several other icing retardants have been applied to roads. *Calcium chloride* has been the most widely used substitute for sodium chloride.[157] Its low melting point in water (–60 degrees F) makes it more effective than conventional salt at very low temperatures. However, it has the same corrosive properties as sodium chloride and also degrades the roadside environment.[926] *Magnesium chloride,* called "Freeze Guard," is as effective as sodium chloride and is less corrosive.

Sodium chloride has remained the predominant de-icing agent to dissolve snow and ice on roads due to its low cost.[364,926] Several alternatives to sodium chloride have appeared over the years, and although some cost about the same as sodium chloride, those with certain major ecological benefits are significantly more costly. The approximate costs per ton (US$ in 2001) are: sodium chloride, $50; magnesium chloride, $45; "Quik Salt" with corrosion-inhibiting additive, $45; calcium chloride, $150; urea, $200; sodium formate, $200; and calcium magnesium acetate (CMA), $450–$600.

Ironically, the destruction of road surfaces, vehicle corrosion, corrosion of bridges and reinforcing steel, and the loss of erosion-preventing vegetation are almost never factored into the cost of roadsalt. This brings up an interesting economic paradox. It has been estimated that the real cost of applying sodium chloride to highways is about US$1600 per ton, when these deferred (but real) costs are factored into the equation. However, with road safety a major concern and many competing interests for public dollars, application of the relatively cheap and effective sodium chloride is likely to continue to dominate roadway de-icing. The associated corrosive effects on structures and vehicles and detrimental effects on the environment will also continue.

The Mirror Lake, New Hampshire, study mentioned above suggested another sobering result[795] (Figure 8.4). Salt is not "breaking down" to become innocuous in the environment. Rather, it is accumulating. The contamination of wells by salt provides added insight. From 1982 to 1989 the Massachusetts Department of Public Works received complaints of salt contamination of public and private water supplies from about 100 of the 351 cities and towns, representing all areas of the state.[735] Groundwater generally moves slowly, from a few feet to a few hundred feet per year in this glaciated state. Consequently, wells in the vicinity of roads are often contaminated by salt that was spread on the roads several years earlier. Areas along some continuously salted roads may become in the future distinctive saline strips.

Sand and, in some instances, cinders may be applied together with, or instead of, roadsalt. Once ground into small particles by traffic, sand particles are easily transported by runoff or become airborne from wind or vehicle turbulence. Depending on the source, such particles can contain significant levels of heavy metals and phosphorus. The metals can produce toxic effects, and the phosphorus can speed the eutrophication of wetlands, streams, and lakes. In addition, if the sand and cinders are transported into a water body, they may cover the existing bottom with a new substrate. Some aquatic plants and bottom-living animals can adjust to the covering, but others cannot.

Best Management Practices for Pollutant Mitigation and Source Control

Best management practices have been developed over time as guidelines in many applied fields, from soil conservation to forestry, range management, and fisheries. Although the general concept could apply to roadside vegetation management, in the transportation field it mainly refers to stormwater and chemical pollutants. Thus a range of approaches has evolved to minimize the effects of such developments as roads with vehicles on water quality.[243, 268]

The previous sections have highlighted the array of chemical pollutants produced by roads and vehicles along roadways. Thus a best management practice may reduce the rate of buildup of pollutants. However, the overall effort or combination of approaches may be considered successful if the pollutant absorption and removal rate exceeds the rate of pollutant input. Thus, instead of pollutants continuing to increase though at a slower rate, a successful effort causes the pollutant level in the environment to decrease.

There are two basic ways to reduce the levels of pollutants along roads.[268] Indeed, best management practices are divided into two groups accordingly. The first is *pollutant mitigation,* in which the effects of existing and future pollutants are minimized. "Structural" practices, such as building detention ponds, wetlands, and sand filters, are used, many of which are somewhat expensive and have significant maintenance effort and costs over time. The second group is *source control,* in which the amount of future pollutants is minimized using "nonstructural" practices, such as land-use planning, integrated pest management, chemical storage, and bridge maintenance. Many of these are analogous to rigorous housekeeping, and though costs are often modest, sustained commitment to such non–"squeaky wheel" tasks is difficult.

For each of the two categories of best management practices below, a list of the primary current approaches in the USA is given and a few approaches are briefly described.[268]

Pollutant Mitigation with Structural Approaches

Structural approaches operate by trapping and detaining runoff until pollutants settle out or are filtered through the underlying earthen material or soil. The capability for removing pollutants is basically determined by three interrelated factors: (1) the removal mechanisms used, (2) the fraction of annual runoff volume that is treated, and (3) the nature of the pollutant being removed. Design and management can affect the first two.

Ten major or conventional structural approaches are recognized[268]: (1) extended detention ponds, (2) wet ponds, (3) infiltration trenches, (4) infiltration basins, (5) sand filters, (6) water quality inlets, (7) grassed swales, (8) filter strips, (9) constructed wetlands, and (10) porous pavement. Each requires a reasonable amount of space near a road. Where little space is available, a series of "space-limited" approaches is recommended, including sand filter alternatives, storage tanks, bioretention areas, manhole filter systems, stream channel retrofits, and pollution removal.

In selecting which approach or approaches will be used for a given runoff situation, six broad criteria are considered:

- Site considerations, including soil, slope, water availability, and so forth
- Area served
- Site constraints, including depth to watertable, proximity of buildings and wells, and so on
- Soil types and associated infiltration rates, including sand, loam, clay loam, and so forth
- Removal efficiency of targeted constituents focusing on sediments, total phosphorus, bacteria, and the like
- Implementation requirements, such as capital costs and maintenance

Two examples of structural approaches for removing pollutants from road runoff are briefly described. A third example is included to emphasize that, in addition to the major engineering approaches, in the future more natural approaches and hybrid natural-and-engineering approaches can be developed jointly by engineers and ecologists.

Infiltration/detention ponds are the first example. Road runoff as an important source of pollutants may be responsible for serious environmental impacts, especially in the long term. Runoff may be discharged directly into the natural environment or first treated as *effluent*. The types and concentrations of chemicals depend on road characteristics and on rainfall pattern. Infiltration ponds are used to hold runoff water for a period of time sufficient to allow the sedimentation of pollutants or their infiltration and absorption by soil.[437, 268] As an illustration, the effectiveness of infiltration ponds in removing heavy metals from runoff was investigated in northern Portugal.[45] During rainfall, 60% to 70% of the total zinc, copper, and lead and total suspended solids was transported into the detention pond in the first half of runoff volume. If ponds are also to break down heavy metal compounds, they must have the right soil characteristics. Soils with a high absorption capacity and the ability to retain pollutants at low pH are considered ideal for infiltration ponds. Like any filter, the soil may become saturated or clogged at some point, and thereafter no longer be effective in trapping a pollutant.

Constructed wetlands are the second example (see Chapter 9). Road runoff also can be treated in wetlands that are constructed for the purpose.[138, 842] Important design criteria include traffic volume, road drainage area, land availability, cost, and the size and type of receiving water body. Several features are considered important to include in the wetland design, including an oil separator and silt trap, spillage containment, a settlement pond, vegetated wetland, and a final settlement tank.[842] Unfortunately, the history of constructed wetlands for mitigation projects has been uneven. Two primary messages emerge from this history: (1) get the hydrology right, and (2) plan for periods of extreme conditions. Establishing the appropriate flow rates into and out of a wetland is generally more important than planting wetland plants and seeing wetland animals. Wetland species tend to colonize naturally and rapidly. Providing for floods and severe droughts and for rapid repair mechanisms (such as the immediate reconstruction of dams by beaver) means the wetland will persevere over time.

Shelterbelts, the third example, have long been used in the United Kingdom and other countries to reduce windspeed and visual intrusion. However, far less attention has been paid to their role as a barrier or filter for pollutants, particularly along the edge of highways. Along a major motorway in Britain, shelterbelts were found to effectively entrap heavy metals and, indeed, to override the effect of prevailing winds on airborne metal deposition.[407] In northern France, hedgerows running across a farmland slope typically have a ditch and a raised soil bank.[129] During dry periods, the bank catches any water runoff so that it soaks into the soil, and during wet periods the ditch drains excess water off the slope. Properly designed shelterbelts or hedgerows near roads could provide stormwater mitigation as well as other significant benefits.

Source Control with Nonstructural Approaches

Nonstructural approaches for pollutant removal are source control systems designed to reduce initial concentrations and minimize accumulations. While developed for other development activities, these approaches, many in the arena of commonsense housekeeping, apply nicely to transportation. Nine groups of approaches are recognized[268]: (1) land-use and comprehensive site planning, (2) landscaping and vegetative practices, (3) pesticide and fertilizer management, (4) litter and debris controls, (5) illicit discharge controls, (6) bridge cleaning, maintenance, and deck drainage, (7) bridge painting, (8) chemical storage, and (9) maintenance of stormwater facilities.

The removal, transport, and deposition of accumulated pollutants to ecologically appropriate locations are especially important. Other nonstructural approaches, such as simple street cleaning, could be added to the list above. Just as for structural approaches, many source-control best management practices also lend themselves to natural solutions. Thus the use of natural communities and combined engineering practices with nature should offer benefits in the future.

Normally both nonstructural and structural approaches are required to attain the successful result of pollutant absorption exceeding input.[268] The more intensively nonstructural approaches are used, the better. They lower the number of structures built, the maintenance effort and cost, and the pollutant accumulation to be transported and deposited at a suitable spot. Some approaches provide additional environmental or aesthetic benefits.

Conclusion

Although pollutants arise everywhere in society, those along roads are directly tied to surface transportation. Since the pollutants disperse into surrounding areas and since roads slice through most ecosystems and other land uses, chemicals cause widespread ecological effects. Worse still, the chemicals come in an array of types, they accumulate to different levels in different places, and diverse species have varying sensitivities to them. Consequently, transportation and society are left in a quandary. Somewhere between "We barely understand anything" and "The obvious degradation is everywhere" lies the proverbial truth.

Certainly, we know the types of pollutants that come from vehicles and roads. We have some information on their concentrations in parts of the environment, including roadsides and aquatic systems. We know how wind works, how air turbulence comes from vehicles moving along a road, and how rainwater flushes pollutants from roads into ditches. We know less about how much is transported, what route it takes, how fast it moves, and where it is deposited. We know of some species that get hit hard by certain pollutants, but what happens to most species in the face of most pollutants and at what levels has yet to be explored. Indeed, the landscape ecology effects of concentrated chemicals lining the road network is an unopened frontier.

This all means that in painting the big picture, some tiny spots are crystal clear, many areas are shrouded, and others are still blank. Can we decipher, or even guess, what the picture will look like? Probably, although with humility. Surprises, both good and bad, doubtless lie ahead.

In the face of this uncertainty, what action should be taken, if any? Source-control practices make sense no matter what. Mitigation practices normally do too, and a jump in research certainly does. Dragging ecologists into the transportation field to work hand in hand with transportation engineers and other experts certainly does. Indeed, that could be the catalyst. Source control, mitigation, and design and accomplishment of research with this joint synergy could literally clean up the future environment along roads.

CHAPTER 9

Aquatic Ecosystems

. . . . [P]robably at 171 B.C. . . . prior to the building of the Via Cassia [Cassian Way near Rome] such water as entered Lago di Monterosi was mainly precipitation on its surface. . . . After clearing on either side of the road . . . run-off over the surface of the terrain and consequent erosion must have increased . . . increase in . . . calcium and . . . phosphorus . . . a great development of blue-green algae. . . . The lake at this time, at least in the summer, would have appeared as a typical eutrophic [green] body of water.

—G. Evelyn Hutchinson and U. Cowgill, *Transactions of the American Philosophical Society,* 1970

. . . . [T]he plowman's road . . . surface was kept relatively horizontal and raised up to 2 m above the surrounding countryside, in the manner of a Roman road. . . . Complaints were rife when this form of raised construction and maintenance occurred in the towns. Houses were flooded, ground floors became underground rooms, and dampness was widespread. The alternative of sinking roads to lower their levels by a meter or so became commonplace but brought with it a new set of drainage problems. . . . Both travelers and road managers were clearly floundering.

—M. G. Lay, *Ways of the World,* 1992

A drive along the shore of a placid lake can be one of the joys of life, rivaling even the surprises along the bends of a winding river. The preceding two chapters have explored the flows of water and sediment and chemicals in the vicinity of roads. These flows lead somewhere. Most enter lakes, rivers, streams, wetlands, vernal pools, estuaries, and other surface water bodies. Now we will examine the water bodies themselves, especially the aquatic ecosystems they hold.

This chapter explores the ecological effects of roads on aquatic ecosystems. We begin with habitat structure within the water body and how the water body is connected to the surrounding land. Then we sequentially examine lakes, vernal pools, wetlands, streams and rivers (including bridges and culverts), and estuaries.

Habitat Structure, Connectivity, and Roads

An aquatic ecosystem, such as in a lake or stream, reveals an intriguing structure on which countless types of aquatic organisms depend. However, the aquatic ecosystem does not stand alone but is tightly linked to the surrounding land in important ways. Linear roads with vehicles running along those roads affect the linkages and aquatic habitat structure all the time. Thus we begin with habitat structure, types of connectivity, and the effects of roads on them.

Physical Habitat Structure

Looking into the water reveals aquatic habitats at several scales.[423,1022] These are best teased apart by considering the organisms and processes in the ecosystem. At a fine scale, for example, a streambed and even individual pebbles function as habitats for specific organisms, such as trout and caddis flies (Trichoptera). Even over a few meters, the streambed often varies markedly in composition and texture. The size distribution of sediment particles and the pore spaces among them determine the suitability of a streambed as habitat for many species of vertebrates and invertebrates. Some use it for feeding, some for spawning, and some to seek refuge during floods. Also, the water column above is a habitat for swimming and floating organisms, such as fish and *phytoplankton* (floating algae). Not surprisingly, the physical, chemical, and optical properties of the water column affect how it functions as a habitat. Current velocity and habitat structure are influenced by the presence of logs and branches, which strongly influence the numbers, species, and sizes of fish in the individual pools of a stream.

At a broad scale, the form of water bodies, such as coves and points on a shoreline or pools and riffles in a river, also molds aquatic habitat structure.[423]

Some species are adapted to life in one habitat; some, in several habitat types. The form of habitat units, such as pools or rock faces, may affect the numbers, species, and sizes of algae or minnows. For example, deeper pools may support larger and more types of fish. More structurally complex pools may reduce competition among species, permitting more-complex natural communities to occupy a given habitat volume. Some species may rely on the juxtaposition of such habitats, as in the case of fish that reside in a pool and feed on drifting invertebrates from upstream riffles.

A subtle aspect of aquatic habitat structure is water in the shallow subsurface environment (the *hyporheic zone*) adjacent to lakes and stream channels.[872, 354, 1047] Exchanges of surface water and shallow subsurface water within floodplains can help regulate water quality through filtration. Some stream invertebrates depend on these subterranean environments during phases of their life histories.

The interactions of streams and lakes with adjacent *riparian vegetation* are so numerous and ecologically significant that the status of this vegetation is widely considered to be an indicator of the well-being of small or narrow aquatic systems, in terms of both habitat and water quality.[91, 373, 302] Streamside forests, for example, shade streams, thereby regulating the light available to support growth of aquatic plants and partially regulating temperature. Riparian vegetation of any size can also supply dead leaves and branches to water bodies. This dead organic matter is the predominant base of the food chain in low-nutrient lakes and small forest streams. Large pieces of wood from trees in the riparian zone create important habitat complexity in river channels. Consequently, modifying adjacent vegetation can profoundly affect the habitat structure in streams and lakes. Overall, aquatic organisms and processes are highly dependent on the richness of habitats underwater.

Connectivity

Connectivity is just as critical in aquatic ecosystems as in transportation systems. As drivers on roads, we expect complete connectivity to get from starting point to destination. Even when connectivity is interrupted by construction zones, detours are provided though a delay may occur. Connectivity is a central principle in both transportation theory and landscape ecology.[8,98,568,302,73,441] In aquatic systems, *connectivity* is essential to sustain flows of organisms, genes, water, nutrients, and energy, plus materials that build and sustain habitat.

Many vernal pools owe their existence to connections with groundwater in the surroundings.[438] Amphibians migrating across the land are sometimes said to view wetlands as aquatic islands in a barren terrestrial sea. Leaves and wood falling into small, nutrient-scarce lakes from the surrounding shore may be a major base of a lake's food web.[328]

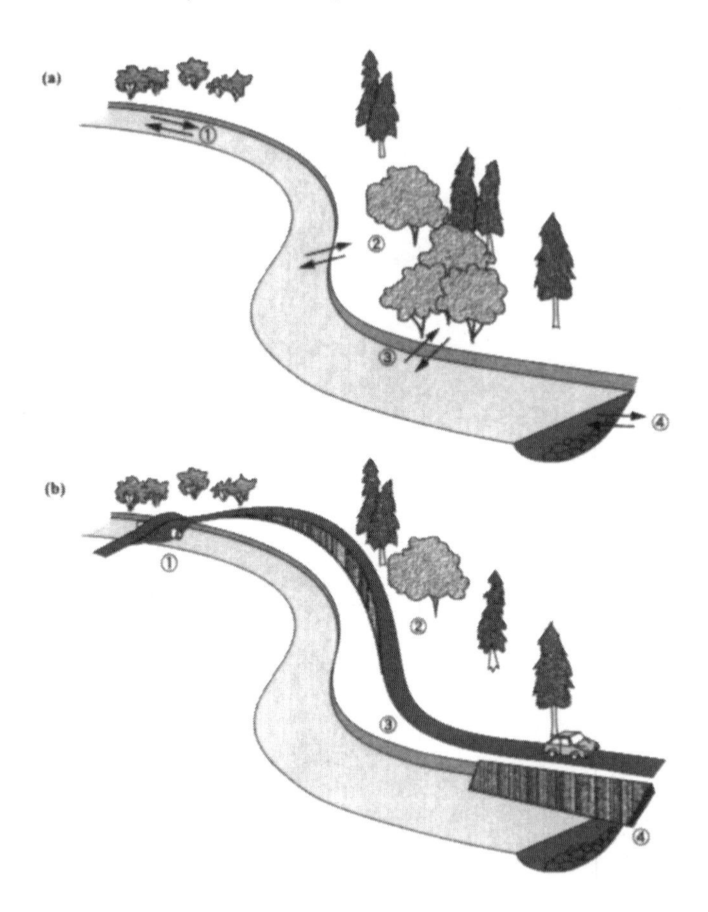

Figure 9.1. Road affecting four aspects of stream connectivity. (a) Upstream-downstream (1), floodplain-stream (2), forest-stream (3), and surface-subsurface water (4) connections. (b) The connections severed or disrupted by a road in the floodplain.

Connectivity with stream (river) systems has at least four major dimensions—*upstream-downstream, floodplain-stream, forest-stream* and *surface-subsurface* connections[998] (Figure 9.1). This connectivity is a critical property of stream ecosystems, in part because of the high mobility and long travel distances of water, sediment, plant material, and many species in streams and rivers.

The principal theory of river ecosystems, the *river continuum concept,*[963,190,856] holds that communities of aquatic organisms and natural processes change in an orderly manner from upstream to downstream. Upstream areas supply critical nutrients and other resources to downstream areas. Upstream-downstream connections in river systems are essential for migratory species such as *anadromous* (migratory) salmon. Rivers are tightly connected with adjacent forest through a great variety of processes, including shade, fallen branches and logs, control of water temperature and oxygen, control of riverbank erosion, and cover for predators and people fishing.[373] Interactions between the channel and the adjoining floodplain are considered critical to river functions (as described in the *flood pulse concept*).[468, 58] Large, floodplain-bordered rivers are believed to depend substantially on inundation of floodplain areas for supporting river productivity and maintaining the biological diversity of river ecosystems. Finally, the interactions between surface and subsurface waters are now known to be important in diverse ways in rivers, lakes, and wetlands.[354, 1047]

Road Effects on Habitat and Connectivity

Road systems and associated land use can have dramatic effects on aquatic ecosystems. These effects may be directly evident or may be indirect as a consequence of altered connectivity. *Direct effects* result from the local immediate alteration of the natural aquatic conditions. For example, the relocation of stream channels by construction of roads on valley floors has both immediate and protracted consequences. Any straightening of meandering stream channels alters aquatic habitat, as shorter, steeper channel segments lose some natural pool-riffle structure. Alternating pools and riffles are replaced by more-uniform channel habitat structure in the streambed. Stream crossings on bridges and *culverts* (drainage structures crossing under a road) locally alter aquatic habitat by changing channel form and the hydraulics of the site.

Indirect effects occur where roads intentionally or inadvertently disrupt connectivity within aquatic systems. Thus roads may block or accentuate the natural flows of organisms, materials, and energy, sometimes with unintended consequences far removed from the site of disturbance. Altered connectivity can result through changes in water quality, physical habitat, and interactions with adjacent terrestrial ecosystems (see Figure 7.8 in Chapter 7). Upstream-downstream connections may be severed where culverts limit upstream migration of fish due to water drops or excessive current velocity. Floodplain-stream interactions are impeded where roadbeds on valley floors interrupt the natural flow of water over floodplain areas and through secondary channels. Connections between surface and subsurface waters are altered where floodplain drainage is modified and where water bodies are lined with impermeable material, such as in a concrete channel.

Roads can also enhance connectivity within river systems. For instance, stormwater flows along roads or ditches often effectively create new segments connected to the natural stream network.[1017] This increases the ability of a watershed to produce flood runoff, resulting in higher flood levels. Extensive areas of paving also greatly increase runoff.[733] The resulting change in the natural flow regime can negatively affect downstream rivers, lakes, and wetlands.

In essence, aquatic habitat structure within water bodies and connectivity of the aquatic ecosystems with surrounding environments are two keys to understanding the interactions of road systems and aquatic ecosystems. The third key is the organisms themselves. These range from floating phytoplankton to *zooplankton* (tiny floating animals), aquatic insects, bottom-dwelling (benthic) species, minnows, mighty fish, and more. These three keys are now examined more carefully for—in sequence—lakes, wetlands, streams and rivers, bridges and culverts, and estuarine salt marshes.

Lakes

Other than chemical pollutants from roads (Chapter 8), probably the most important concern for aquatic ecosystems is erosion along paved and unpaved roads (Figure 9.2). Among the greatest sources of sediment yield are unpaved

Figure 9.2. The direct linkage between vehicles/roads and water quality/clarity in a lake. During peak tourist season, more than 50 000 veh/d use this lakeshore road. Note the abundance and depth of exposed tree roots and deep gullies, indicating severe erosion by water runoff from the road area (as well as people walking on the bank). Nutrients and pollutants, washed by rainwater from the surrounding road system into stormwater drains, pour onto the bank and into the lake through an outlet drain with large rocks (center). The resulting opaque water here contrasts with the reputation of Lake Tahoe as one of the world's clearest lakes. State Route 50, South Lake Tahoe, California. Photo courtesy of U.S. FHWA.

timber-harvesting roads. While the roads are in active use, vehicles dislodge road-surface particles that then may be transported by wind or water to lakes.[578, 1067] Left unattended, these roads can become deeply gullied and contribute significantly to the sediment and nutrient input to streams and lakes.

Furthermore, as sediment washes across the hard surface of paved roads, it is subject to vehicle traffic, which grinds the sediment into finer particles. These particles then easily become airborne with subsequent passing vehicles or with wind. Water transport deposits sediment especially into the borders of lakes, either directly from the lakeshore or via streams. On the other hand, wind tends to spread the sediment more widely over a lake.

Road Dust and the Lake Ecosystem

Watch a vehicle pass close to the edge of a road on a dry day. A little whirl of *road dust* comes up from the displaced air as the vehicle speeds past. It may be a surprise that those fine particles typically stay airborne in suspension for hours. After a winter application of salt, sand, and cinders, vehicles grind this mixture into extremely small particles, which can also remain suspended in the air for hours.

Once airborne, these fine particles have the potential to form yet another link between roads and aquatic environments. The particles, and hence the elements they consist of, plus the chemicals *adsorbed* (attached) to their surface may have come from almost anywhere. If small enough, they may be carried thousands of kilometers, whereas larger particles may redeposit within a few centimeters. Small, airborne soil particles from the Saharan desert have been blown to the Amazon basin,[894] and particles from the drying shores of the ever-shrinking Sea of Aral in southern Russia have been found beyond the Arctic Circle. The atmospheric deposition of particles has been shown to be a significant source of elements and chemicals, including lead, iron, and PCBs (polychlorinated biphenyls).[213]

Airborne particles are particularly important to aquatic environments because they can transport nutrients or chemicals that are toxic to aquatic life. *Dry deposition* is the process of particles settling from the air onto a surface such as a lake or field.[423,1022] In contrast, precipitation actually "scrubs" the particles out of the air and deposits them as *wet deposition. Bulk deposition,* a combination of wet and dry deposition, was collected from July to October 2000 along the north shore of Lake Tahoe, a large lake in California (Figure 9.2). Since there is little precipitation during this time of year, most of the deposition collected was dry deposition. Particles deposited during the dry season tend to be larger because smaller particles often need precipitation to be removed from the atmosphere. The larger particles tend to originate locally because their settling velocities are too rapid for the particles to travel far.[663] The average diameter of the mass of particles collected at Lake Tahoe was approximately 1.5 micrometers. This relatively small size means that a significant number of particles were transported across the lake before being deposited.[561]

Atmospheric deposition at Lake Tahoe is of particular concern because phosphorus has a strong affinity to soil particles, especially clays (which are fine particles possessing an adhesive negative charge). In other words, adding phosphorus in this way is like fertilizing the huge lake. It stimulates algae to grow, which tends to give the crystal-clear lake a greenish tinge.

These very small deposited particles also have a slow settling rate in the water. Consequently, they remain suspended in the water column for long periods. Together with the stimulation of algae growth by phosphorus, these particles add to the lake's declining water clarity.

As evening approaches, air temperature drops. A typical pattern develops of colder air moving downslope from the surrounding mountains toward the lake. This downslope flow of air transports airborne dust in suspension out over the lake, where it is then deposited. Based on atmospheric deposition collectors placed beside roads, halfway between the road and lake, and at the lakeshore, an estimated 11 000 kg of phosphorus are deposited on Lake Tahoe each year in this manner[452] (M. Liu, unpublished data). Since stream flows to the lake are greatly curtailed during the typical dry summer, this daily application of airborne dust to the surface of the lake can be an important contributor to the eutrophication of this unique water body. In all probability, this contribution is especially important in the summertime, when the stratified surface waters of the lake become low in phosphorus and contributions from stream flow are at their minimum levels.

Once airborne particles enter the lake, they join a diverse assemblage of organic particles and fine inorganic sediment. Particles in the lake range from submicrometer-sized bacteria, viruses, and colloidal particles through phytoplankton cells tens of micrometers in diameter. The majority of the particles in the lake are smaller than 2 micrometers in diameter, and the number of particles increases rapidly with decreasing size (a hyperbolic relationship: $N = 2210\ (\text{diameter})^{-3.1}$; $R^2 = 0.95$).[645, 176]

However, the concentration of particles in the lake varies considerably with depth and time, from approximately 2000 particles/ml to 130 000 particles/ml (in one observation) (Figure 9.3). Even though tiny particles are predominant, optical microscope studies find phytoplankton cells much larger than 20 micrometers in size. Particle volume increases with the cube of the diameter. Thus, by volume, the relatively large but less numerous phytoplankton cells seem to equal or exceed the total volume of small particles.

After the particles are deposited into the lake, the chemicals in the particles and those attached to particles can dissolve into the water. In the case of phosphorus at Lake Tahoe during the 2000 sampling period, nearly 50% of all

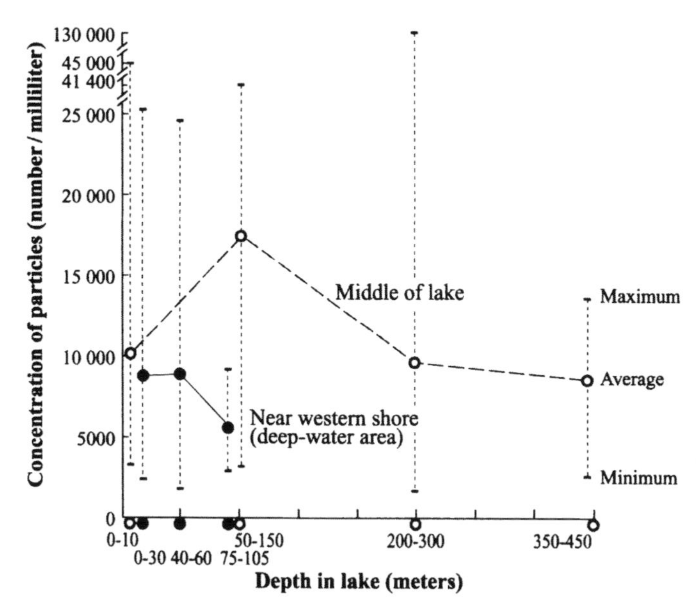

Figure 9.3. Particulate matter at different locations and depths in a large deep lake. Lake Tahoe, California (500 km^2 area; 505 m deep). Based on Coker (2000) and Swift (2001).

phosphorus originally deposited on particles dissolved to form orthophosphate. This is the form of phosphorus most biologically available to phytoplankton and to algae attached to plants *periphyton* near the shore. Indeed, it was the luxurious growth of this periphyton in the *littoral* (near-shore) zone that initially alerted the largely shorebound public to the fact that Lake Tahoe was changing.[363]

In comparison, only about 13% of the phosphorus from streams is immediately available biologically.[453] Nevertheless, phosphorus from streams and water runoff can also significantly affect lake ecosystems. A watershed nutrient budget for Lake Chocorua in New Hampshire revealed that the largest source of phosphorus entering the lake was runoff from a multilane highway that passed near the eastern shoreline.[821] These results highlight the importance of including atmospheric deposition as well as runoff from roadways and streams when developing comprehensive lake management strategies.

Although plankton species in general are fairly tolerant of roadsalt, they may experience reduction in growth rate from changes in their salt balance or may suffer from toxic pollutants. Conversely, as noted above, planktonic algae and periphyton may benefit from stimulating contaminants, such as phosphorus, iron, and a variety of trace metals. When as many as 100 different species of algae may be present in a lake, subtle changes (such as slightly altering the monovalent-to-divalent ratio by introducing sodium chloride) may greatly alter the success of one species in comparison to others. Exactly how this translates into food web changes and eutrophication has yet to be investigated.

Aquatic food webs are among the most complex of all aquatic systems. Since sodium chloride and its associated contaminants can subtly alter the algae species (primary producers) and their population sizes in the food chain, runoff of roadsalt and its various accompanying chemicals could be an important factor in altering the behavior of higher levels in a food chain. The quality of phytoplankton as food for zooplankton consumers varies greatly from species to species. Phytoplankton with high levels of *highly unsaturated fatty acids* (HUFA) are a much better food than those with lower levels. Although phosphorus can be an important component, recent studies indicate that the quality of food for zooplankton can be better predicted by the level of HUFAs.[112,658] If runoff from roadways alters the species composition of algae, or if toxic substances accumulate through the food chain, the impact on animals at higher levels in the food chain is likely to be important.

While we understand the importance of atmospheric deposition of nutrients to aquatic ecosystems, especially clear-water lakes, it is still difficult to accurately quantify how much of the deposited materials originate from unpaved road shoulders and from road surfaces. The relative importance of roads to atmospheric deposition will vary greatly from area to area depending on such factors as the prevalence of unpaved shoulders and the traffic volume. Also, sufficient vegetation between the road and the water body can serve as a filter to catch some of the road dust.

A promising evaluation technique would compare the concentration of unique elements contained in potential sources, such as road surfaces, roadbeds, and vehicle deposits, to the concentrations deposited on water bodies. Also, particle size, shape, and spatial distribution can provide important clues to better quantify the contribution of atmospherically deposited materials from roads, traffic, and other sources.

Lakes as Integrators of Erosion Processes in a Drainage Basin

Direct runoff from roads and associated development is also a significant source of pollution for lakes, streams, and coastal areas (Figure 9.2). Precipitation directly striking an impermeable or *hard surface,* such as a road, roof, or parking lot, has no possibility of interception by overhead vegetation or infiltration into soil (Chapter 12). This means that the nutrients and pollutants contained in rain also have little chance to be absorbed by vegetation in the watershed or to infiltrate into the groundwater. Where roadsides are not adequately vegetated or are too steep to absorb water, these hard surfaces speed water runoff at sufficient velocity to cause roadside erosion. Also, the transport of sediment particles and dissolved substances is accelerated downslope, where they often eventually enter streams, lakes, and reservoirs. Road surfaces accumulate various pollutants from automobile tires, brakes, and exhaust, as well as dust from the application of roadsalt, cinders, and sand. These chemicals are washed from road surfaces during storms and are added to the eroded soil and chemicals in roadside ditches. Recent urbanization of the Lake Tahoe basin is a good example of accelerated runoff from a highly disturbed, road-laced watershed. Thus a major impact of roads on large lakes results from both wind erosion and runoff from road and roadside surfaces (Figure 9.2).

In coastal areas, accelerated road runoff can cause problems for tide pools and other near-shore life. Sediment plumes resulting from wave action that resuspends silt from nearby roads may inhibit coastal species. This situation is paralleled in lakes, where large culverts often direct water from roadside ditches with high sediment loads directly to near-shore areas. Here the sediment plume blocks light penetration for periphyton and smothers bottom-dwelling organisms with sediment. At Lake Tahoe, large culverts deliver sediment-bearing road runoff at high velocity to the littoral zone (Figure 9.2). Efforts are being made to intercept this runoff in settling basins.

Spilled toxic chemicals may quickly reach lakes through road ditches, streams, and groundwater. Chemicals that accumulate on road surfaces tend to wash off into nearby lakes or coastal areas with each major rainstorm. Providing cleaner roads by vacuuming, steaming, or washing road surfaces (and then removing the sediments and chemicals) near sensitive ecosystems may emerge as an important future management strategy. Road-cleaning equipment has been assigned to the Lake Tahoe basin, and a stormwater management program has been established.

In view of the air and water transport of particles and chemicals to a lake, many mitigation approaches to minimize the amount deposited could be useful. These may range from covering road shoulders with wood chips and planting roadside vegetation as dust filters to controlling off-road-vehicle use and closing forestry roads no longer used for logging. In view of the extensive evidence of contaminants from road runoff, slow mitigation progress is being made at Lake Tahoe. Meanwhile, though, the water clarity of this flagship lake continues to degrade.

Lakes, once envisioned as self-contained systems, are increasingly considered as integrated components in a functioning landscape.[548] During the normal process of lake development, silt, soil, and other sediments are transported from watersheds to accumulate in lake basins. In this respect, lakes act as natural receptacles that integrate watershed processes of erosion. However, elevated erosion rates resulting from road construction and subsequent deposition in receiving lakes produce a wide range of harmful effects on aquatic life.[183]

Most lake studies have been based on the discharge of suspended and dissolved solids entering from streams, normally over the period of a few years. Yet annual sediment yields from a river can fluctuate by as much as five times from one year to the next.[224] *Lake sediment cores,* on the other hand, allow us to extend that time scale to study changes over decades, centuries, or more.[94] By coring and analyzing lake-bottom sediments, it becomes possible to reconstruct the landscape's erosional history.[203, 210, 478] One of the most visible consequences of clearcut logging in a watershed is a substantial increase in the production of sediment,[712] most of which typically originates from the construction and maintenance of logging roads.[81, 766, 84, 328]

Three lake-sediment studies illustrate how road erosion in a watershed has been recorded. First, the increased sedimentation rates in Lago di Monterosi in Italy were related to building the Via Cassia in 171 B.C. (the famous Cassian Way, which enters Rome; see the epigraph at the beginning of this chapter).[436] Second, three mountain lakes in Montana showed strong evidence of a link between the construction of new logging roads and increased lake sediment (Figure 9.4). Sedimentation in the lakes was two to three times the preharvest

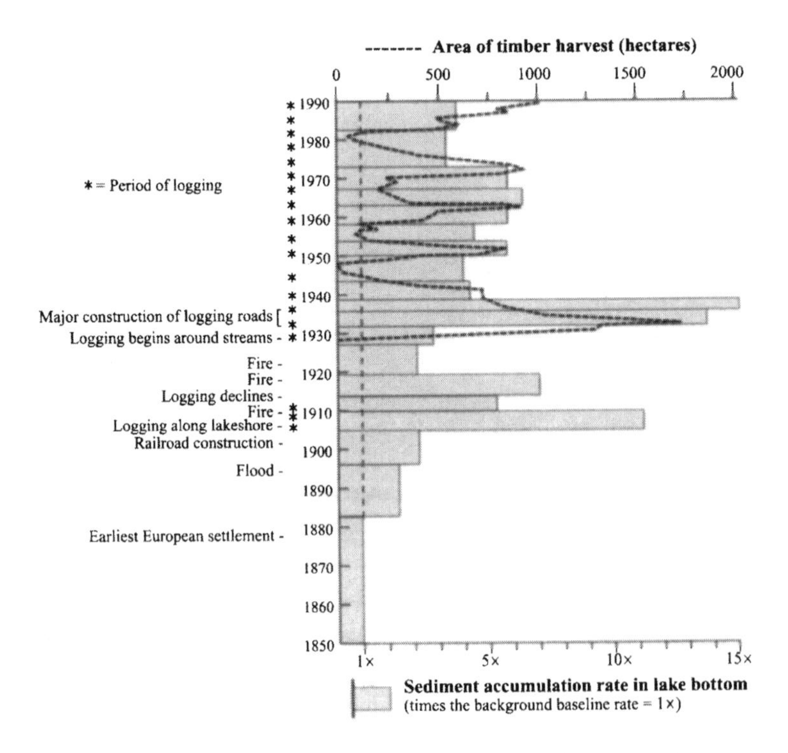

Figure 9.4. Lake sedimentation related to timber harvest, logging road construction, and other factors. Timber harvest is a five-year cumulative average; baseline sedimentation rate (before arrival of eastern USA settlers) was calculated to be 13 mg/cm^2 per year. Whitefish Lake, Montana. Adapted from Spencer and Schelske (1998).

1930s background level. Interestingly, a later period of timber harvesting during the 1970s and 1980s, which essentially reused the preexisting roads, did not generate elevated sediment accumulation in the nearby lakes. In contrast, a third study of nine headwater lake basins in northwestern Ontario found no discernable increases in sediment accumulation nor changes in phytoplankton community composition after logging road construction and clearcutting.[94, 710, 711] The authors attributed the paucity of eroded sediment to a prolonged regional drought. Furthermore, it appeared that the larger the area of a watershed relative to that of a lake within it, the more likely sediment from road erosion will be funneled into the lake.[94]

The plastic silt fences and hay bales periodically seen on roadsides across North America are testament to the progress made in reducing soil erosion, sediment flows, and sedimentation in water bodies. Yet the frequency with which these barriers have been breached emphasizes that much remains to be done. Another sign of progress is the increasing number of stormwater management plans being developed. While the mitigations resulting from these are visible only to the perceptive traveler, such plans tend to provide a welcome overview linking the road to processes well beyond the road. The preceding pages emphasize the importance of including both wind erosion and stormwater in road management plans as well as the importance of the plans themselves.

Vernal Pools

Organisms such as amphibians that depend on *vernal pools* (temporary pools) and on wetlands can be particularly susceptible to changes in hydroperiod, caused by road construction.[1034] If *hydroperiod* (duration of water level at or above the substrate surface) is shortened, amphibian larvae may be killed before metamorphosis from tadpole to adult can occur, causing local extinction of the population. Also, lengthening of the hydroperiod can be problematic for amphibians. The persistence of amphibian populations depends on pond drying following metamorphosis, to eliminate populations of predatory fish. If the hydroperiod is lengthened by nearby road construction that alters hydrologic flows, amphibian populations will disappear through increased fish predation.[836] Also, since amphibians can tolerate only slightly brackish conditions, the effect of de-icing agents depends on both the amount of salt and the amount of fresh water entering a pool or pond (Chapter 8).

On the other hand, roadside ditches can act as additional vernal pool habitats. For example, several threatened aquatic invertebrates are found in roadside ditches in the Central Valley of California. These species have also been found in roadside ditches and, interestingly, measures have been proposed barring further development that might eliminate any of the ditches where these species are found,[502] reminiscent of the protected roadsides in Britain (Chapter 4).[944]

The value of roadside ditches as vernal pools will, however, depend on the degree to which they are contaminated by highway runoff. Thus, in comparing the embryonic survivorship of a salamander species in temporary woodland pools versus roadside pools, survivorship was much lower in roadside pools.[936] The roadside pools were heavily contaminated by de-icing salts, which the author suggested was responsible for the lower salamander survivorship.

Wetlands

Wetlands occupy over 8 million km^2 (3.1 million mi^2) worldwide and provide at least four valuable roles for society.[641, 330]

1. *Hydrology modifiers.* Wetlands are generally flat areas that slow down stormwater runoff from hard surfaces, such as roads, and then gradually release the water over a prolonged period. In so doing, they reduce peak flows or flood levels downstream. Furthermore, by reducing the velocity of floodwaters, sediments are deposited and downstream erosion is reduced.
2. *Contaminant sinks.* Wetlands operate somewhat like kidneys in that they remove, retain, and break down contaminants through a host of physical, chemical, and biological processes.
3. *Wildlife centers.* Wetlands are the most botanically productive habitats on the earth and, as a result, support an exceptionally high abundance and diversity of animals per surface area. For example, in North America, about half of all waterfowl species nest in wetlands, and two-thirds of the catch from commercial shellfish and sport-fish fisheries derive benefits from wetlands. Although wetlands constitute only 5% of the total land surface area in the USA, over a third of all rare and endangered animal species reside there.
4. *Human amenities.* To some people, wetlands are among the most beautiful of all landscapes due to their great diversity in location, surroundings, size, shape, and life-form composition. Sometimes situated between built areas, wetlands provide a wild buffer to ameliorate the stresses of increasingly urbanized lifestyles.

Road Effects on Wetlands

All that is the good news. The bad news is that there are few environments that have suffered the same extent of abuse as wetlands.[977] Wetlands include *marshes,* inundated for long periods and dominated by grasses or other herbaceous plants, as well as *swamps,* inundated for shorter periods and dominated by trees or shrubs.[641, 138] Today, only about half of the original 80 million ha (200 million acres) of wetlands still exist in the USA, and this amount may be as low as 10% in some states. Although the vast majority of these losses historically resulted from drainage and clearing for agriculture, more recently road construction has also played an important role in affecting the health of the surviving wetlands. In fact, some of the first manuals on wetland restoration practices were produced by state and federal highway agencies.

Whereas lakes receive plumes of sediment erosion and river channels are realigned by bridges, entire wetlands may disappear if they are situated in the path of an intended road. However, with various wetland laws and regulations in place in the USA, today modification rather than disappearance is more common.[811] Swamps and marshes are modified in many ways, including by filling, draining, excavating, clearing, water diversion, flooding, sedimentation, and impeding water circulation.

Fill-and-culvert crossing structures have commonly been used for building roads across swamps and marshes. A culvert is often placed near the center of the structure (on an enclosed wetland) with ditching on both sides to facilitate water flow and minimize flooding on the upslope side. Studies of two swamps in the North Carolina Coastal Plain[773,486] suggest that a few key readily measured or estimated patterns will usefully assess changes in wetland functions resulting from a road crossing structure: (1) water surface elevation (also indicating changes in wetland area) and water depth, (2) tree density, composition, and mortality, (3) herbaceous layer, (4) soil/water invertebrates, and (5) sedimentation and soil phosphorus and nitrogen. In the wetlands studied, certain patterns differed on upslope and downslope sides, while some patterns varied with distance from road (typically up to 10 to 60 m). Presumably increasing the number of culverts in the fill-and-culvert crossing structures would reduce the number and magnitude of ecological effects.

However, road networks can affect wetlands on a broader landscape scale by altering the hydrology and nutrient flows of a watershed or by altering movement patterns of wildlife. Indeed, a curious imbalance often exists between the scale at which the significance of wetland losses is felt by society (the drainage basin or watershed) and the scale at which the protection of wetlands is regulated (the individual body of water).[330] Often, the most important role that wetlands play is unrelated to any single wetland but, rather, derives from the cumulative function of many wetlands. In other words, they operate together as *wetland complexes* in a functionally interdependent manner, so that the loss of one affects the others as well as the whole.[515] For example, a strong positive relationship exists between the percentage of upstream wetlands that have been lost and the percentage increase in watershed flooding. The extent of wetland loss also correlates with the export of contaminants and nutrients from a watershed to receiving waters downstream.

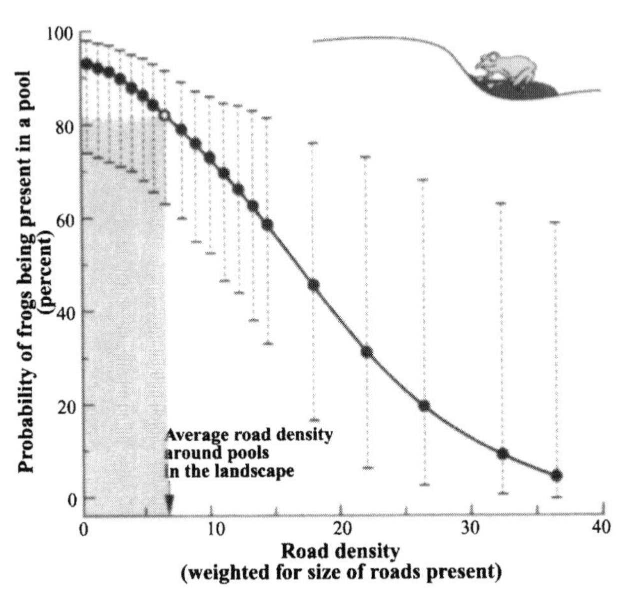

Figure 9.5. The probability of frogs in a pool as related to road density in the area surrounding the pool. Road density is within 750 m (2460 ft) of a pool. For the average road density, there was an 82% chance of moor frogs *(Rana arvalis)* being present, but only a 5% chance where the highest road density was present. The Netherlands. Adapted from Vos (1997).

Furthermore, in the sense that wetlands exist as habitat islands in a terrestrial sea, their loss or fragmentation into smaller isolated patches can have profound implications on mobile species that naturally require several nearby wetlands to sustain their life cycles. For example, the total road length within 0.75 km (0.5 mi) from 109 Dutch moorland pools significantly explained the probability as to whether or not moor frogs were present in a pool[981] (Figure 9.5). The author suggested that this finding could be due to either direct mortality effects from crossing roads or to the barrier effect that reduces movements.

One particularly informative study illustrates the landscape-level effects of roads on wetland biodiversity.[288] A strong correlation was observed to exist between wetland area and species richness. Surprisingly, though, the species richness of birds, amphibians/reptiles, and plants was found to be negatively correlated to the density of paved roads around a wetland (Figure 9.6). In the case of birds, road density within 500 m (1640 ft) was the best predictor of species number in a wetland. For plants, the effect extended outward 1 km (0.6 mi), and for amphibians/reptiles, road density within 2 km (1.2 mi) was the best predictor (Figure 9.6). Species richness was also correlated with the percent forest cover on lands within 2 km of wetlands. Indeed, the removal of forest cover

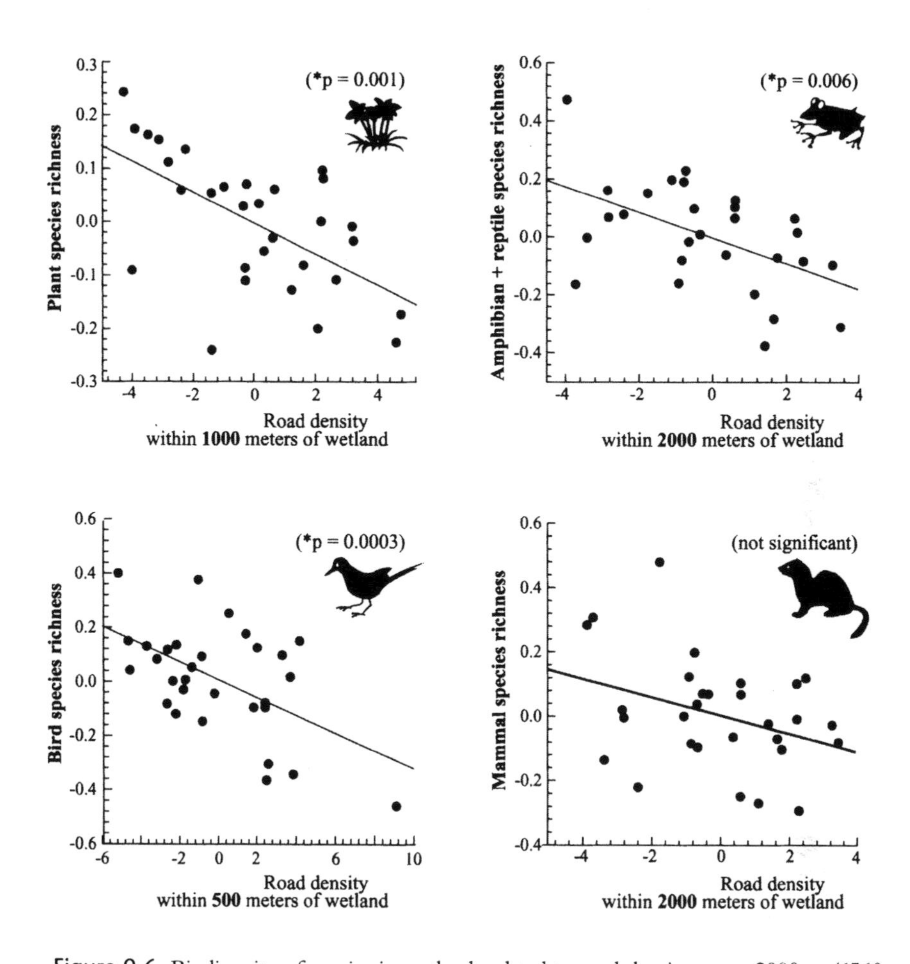

Figure 9.6. Biodiversity of species in wetlands related to road density up to 2000 m (6560 ft) from wetlands. The number of species for each group was compared with road density within 500 m (1640 ft), 1000 m (3280 ft), and 2000 m of a wetland; only the best correlation (*p = a significant effect of road density [probability of Type I error; $p < 0.05$]) for each group is plotted. The graphs show the standardized residuals of the species-richness-versus-area regression plotted against standardized paved-road density. Ontario. Adapted from Findlay and Houlahan (1997).

and the increase in paved road density were predicted to have the same impact on wetland species richness as would the physical loss of a portion of the wetland area itself. A subsequent study found that the present levels of species richness in the wetlands were more accurately described by road density estimates from 30 or 40 years ago than by the current road density.[287]

Thus land-use patterns around wetlands and the manner in which roads are situated within landscapes may be just as important as the actual size of wetlands. These results indicate that road density affects biodiversity in wetlands and that a road need not be next to a wetland to have an impact. Furthermore, in mitigation and compensation projects, it seems prudent to consider addressing likely time lags for ecological effects to show up (Chapter 5).

Wetland Mitigation and Compensation

Fortunately, once a wetland has been destroyed the story need not be over. The process of restoring degraded wetlands or actually creating new wetlands from scratch, both in relation to road-related degradation, is a rapidly expanding field. *Mitigation* is normally considered to be the minimization of impact at a site, and *compensation* the providing of equivalent benefit elsewhere.[191, 192, 193] However, sometimes the term wetland mitigation has been used to include the compensation principle, at least in the sense of creating a new wetland elsewhere to compensate for the loss of a wetland from seemingly unavoidable development.[811] This procedure can take the form of direct "in kind" or "one-for-one" functional creation of new wetlands, or the complete restoration or functional enhancement of existing wetlands.

Depending on regulations, replacement wetlands are constructed as near as feasible to the site of the original lost wetland (though the existing road-effect zone [Chapter 11] and the potential for road widening are key considerations). Alternatively, new wetlands are located well off site, occasionally aggregated in a large *mitigation bank* as a cluster of wetlands, wetland complex or portion of a large wetland. One advantage of on-site mitigation is that it may be easier to successfully replicate many of the environmental conditions suitable for sustaining a wetland. However, proponents of mitigation banking stress that the creation of numerous small, isolated wetlands into one large hydrologically and ecologically favorable site is easier to both monitor and manage.[601]

Unfortunately, the success rate at creating mitigation wetlands has been low.[811,747] Many approaches in the past have been based on matching the size of the new wetland to that of the lost one. But size is only one of the attributes to be considered in evaluating the ecological condition and importance of wetlands destroyed or damaged by roads. Functional attributes, such as hydrology, stability of the shoreline, nutrient supply, sediment and contaminant retention, and wildlife diversity and productivity, also are key considerations.[51] Consequently, mitigation or compensation projects work best when wetlands are reintegrated into the rest of the landscape, especially with hydrologic flows.[330]

As roads bisect a landscape, they often pass along the edge of a wetland, interfering with hydrologic flows, which leaves the wetland severed or uncoupled from its watershed. One approach to recovering wetlands lost in this way is to attempt to relocate them elsewhere.[811] For example, a 29-km (18-mi) multilane highway constructed outside Chicago covered 30 ha (74 acres) of wetlands, 26 of which, from three separate sites, were extremely rare and irreplaceable wet prairie and saturated meadows. These three sites contained over 300 plant species, including two that were endangered in the state. The wetland mitigation plan called for the creation of 48.6 ha at five nearby sites as well as relocating 1.2 ha (3 acres) of one particularly high quality wetland (both plants and topsoil) to a new, specially prepared location. "Donor wetlands"—those to be lost due to highway construction—served as the source material for creation of the new mitigation wetlands. Within three years, all new wetlands were completely covered by vegetation, though provision for sustaining the more important hydrologic flows was less clear. Some people considered the wetland mitigation project a success.

Another example illustrates the complexities and controversies that may occur after wetland mitigation, in this case relative to the type of wetland replaced. In south-central Pennsylvania, 119 small groundwater seepage wetlands (55% classified as forested, 25% scrub/shrub, 20% emergent herbaceous vegetation, and about 1% open water) totaling 15 ha (37 acres) in a river floodplain were disturbed by construction of a new highway.[1075] An elaborate and exemplary process involving a hierarchical ranking of groundwater and water budget analysis, soil characterization, land-use availability, archaeological considerations, wildlife evaluation, and excavation costs was used to identify potential replacement sites for compensation or mitigation wetlands. Five wetland sites totaling 22 ha (54 acres) were finally selected to replace the functions and values of the lost wetlands, and another 60 ha (148 acres) of buffer zone areas were purchased to provide additional habitat for wildlife. Fifty *habitat-enhancement structures,* such as birdhouses and standing dead-wood snags, were installed to further attract wildlife. Based on the wide variety of wildlife occupying the replacement wetlands, which became popular for school groups and birdwatchers, the project was considered by some to be a success.[1075]

For some wetland scientists, however, this award-winning replacement scheme proved unsatisfactory.[177] Because the new replacement wetlands were not located in a floodplain and not established with seepage from groundwater but, instead, were established with flows of surface water, the new wetlands, strictly speaking, did not recapture the functionality of the wetlands lost. Moreover, the new wetlands contained perennially open water rather than periodically drying out as the original natural wetlands did, thus favoring different plants and animals. Indeed, the abundant wildlife, while photogenic, was not typical of that inhabiting natural floodplain wetlands. Finally, the critics concluded that because it was apparently impossible to adequately duplicate the natural groundwater-seepage wetlands, impacts such as the highway construction should have been avoided.

An additional broad issue arose. The original 119 natural wetlands were small and scattered over the landscape. The five replacement wetlands represented a mitigation banking project where they were aggregated into a few large *wetland complexes*. Both the scattered, small wetlands that dry out and the large wetland complexes are ecologically valuable. But they provide different values, have different functions, and support different species. Priority goes to maintaining the original natural wetland values through avoidance, perhaps on-site mitigation and, in a pinch, off-site compensation.[191, 192, 193] If these three options are deemed impossible by society, compensation to provide different ecological values, either on site or off site, is the remaining ecological option.

An important shift in focus from reaction to proaction would get ahead of and help streamline both the road planning and road construction processes. For this, sites are identified and mapped in advance that would provide the greatest ecological gain to the watershed after wetland restoration or creation due to road construction.[601, 330]

Finally, it bears reminding that the self-designing capacity of nature remains the premier player in ecosystem development, especially in undertaking wetland replication due to road construction. A simple motto for wetland mitigation in land-use planning might be "Design less; understand more." Thus the best strategies are those that plan for self-design, self-regulation, and self-maintenance.

Streams, Rivers, Bridges, and Culverts

Roads and vehicles interact with stream and river ecosystems in diverse ways.[587, 470, 159] The most direct effects are seen around bridges and culverts, where the two basic types of corridors cross. Therefore, in this section, we begin with the broad types of interactions between road systems and streams and rivers, followed by a closer look at bridge and culvert effects.

Roads, Streams, and Rivers

As flowing-water systems, streams and rivers are highly vulnerable to alteration by roads[423,1022] (Figure 9.1). Road networks and stream networks both occupy small percentages of the total land area but are widely distributed because of their transport roles. Consequently, roads and streams have many intersections and opportunities for interaction. Furthermore, the engineering structures involved in stream crossings, such as culverts and bridges, can profoundly affect stream processes and aquatic ecosystems. Effects may be local or, by altering upstream-downstream, stream-floodplain, and stream-forest connectivity (Figure 9.1), may extend a considerable distance.

Roads can impede the downstream movement of materials. Thus constricted flow occurs where road fills, bridge approaches, and undersized culverts effectively dam streams, causing a backup of water and consequent deposition of sediment and sometimes wood. Similarly, upstream movements can be impeded where a culvert outlet is too high for fish, salamanders, and other species to pass through the culvert.

Riparian forests exert an enormous control over small stream channels.[87, 587, 1022, 470, 667] The trees provide shade (which regulates water temperature and light available for aquatic primary production), a source of fine organic matter (a food source used by many aquatic organisms), and a source of large wood (which creates complex habitat structure beneficial to many aquatic species).[373] Roads, however, are often built on riparian areas, thereby disrupting many of the benefits of riparian vegetation.

Exchanges between surface and subsurface waters especially occur where streams change a bit in slope, such as the transitions between pools and riffles. These exchange flows have a variety of effects on stream systems, including the delivery of cool, relatively high quality water to certain areas of streambed habitat. Such areas may be selected as spawning grounds by fish, including salmon. Because of *stream migration* (the natural tendency of channels to move laterally across a floodplain), sometimes streambanks are "armored" with rocks or concrete to protect roads from washout. Such armored streambanks disrupt the surface-subsurface exchanges and degrade streams in many other ways.

The connectivity between large river channels and floodplains can also be severed by roads on floodplains. During important flooding events, the roads, often on raised earthen causeways, impede water flows. That results in a buildup of organic matter and sediment on the floodplain in some places and less material entering the river to support aquatic food chains in other places. Where roads block flows to secondary channels on a floodplain, fish cannot use the channels as refuges during floods.

Bridge and Culvert Effects

As roads snake their way through most landscapes, it is almost inevitable that they will meet with a stream or river. Occasionally, a road can be rerouted alongside a river. But for the water and sediment reasons just outlined, and because wildlife frequently use riparian zones as movement corridors, it often makes ecological sense to bridge over the flowing water in as unobtrusive a manner as possible. In many situations, however, inadequate attention has been paid to how roads intersecting rivers lead to deleterious repercussions for the aquatic fauna. In essence, road crossings can affect stream ecology in two major (and not independent) ways: (1) by *altering flow regimes,* and (2) by *scouring sediments* and *increasing sedimentation.*

Waterways must, of course, remain unblocked to enable the free passage of water, which provides passage for fish. Culverts, however, by their very nature,

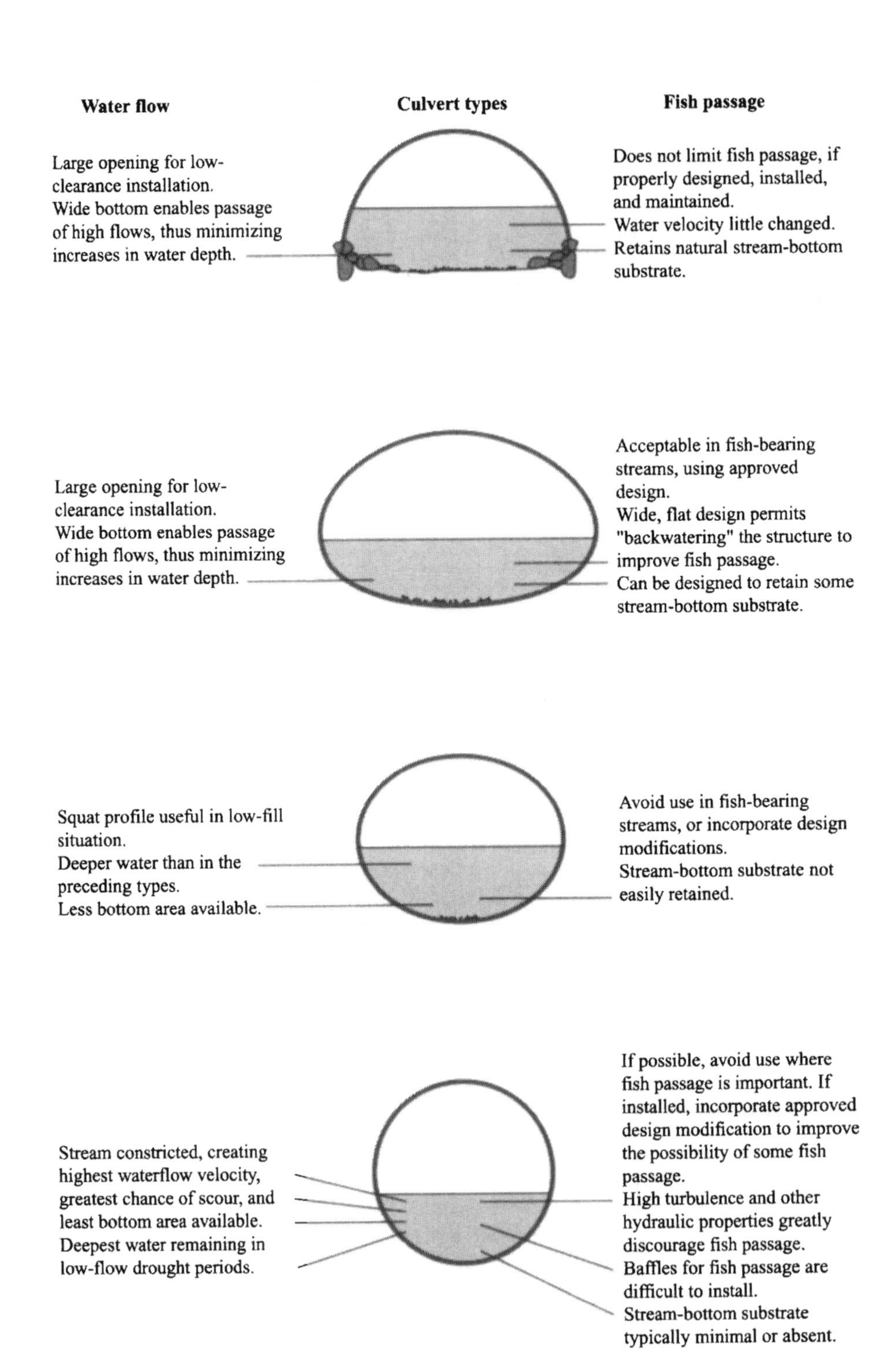

Figure 9.7. Culvert types affecting water flow and fish passage. Adapted from Saremba and Mattison (1984).

restrict flows and therefore require special attention in order to maintain both water quality and fish passage[263] (Figure 9.7). The potential for road crossings to act as barriers to fish migration often results from altering current velocity. For example, in one study, the movement of fish through crossings was inversely proportional to the water velocity.[1002] The fish barrier effect depended on how much a culvert accelerated flows through the restricted opening. As a result, fish movement was about 10 times lower through culverts having the most restricted openings compared with passage through bridges or natural stream conditions.

Road-stream crossings are also a major source of erosion and consequent sedimentation, which can suffocate fish eggs and impede the emergence of baby fish.[19] Moreover, inadequate opening sizes of culverts can lead to erosional washouts during floods, which not only damage roads but cause massive sedimentation for a stream.

Crossings can also exert more subtle effects on stream ecology by influencing gradual sedimentation patterns. One study quantifying the effects on sediment deposition by stream crossings of a dozen forestry roads found, on average, more fine clay and silt downstream compared to upstream of the crossings.[868] Reduced current velocities on the downstream side may have led to the accumulated deposition, though erosion followed by sedimentation of roadbed material may also have been a factor. The filling in of holes and the creation of a smoother stream bottom by silt downstream of a bridge or culvert may help explain why fishermen often park and head upstream. Sediments have reduced fish habitat downstream, especially for large fish.

The locations of road-water crossings are best selected at the earliest possible planning stage. Reducing the number of crossings minimizes site disturbance and aquatic ecosystem effects. Aligning a road crossing at a right angle rather than obliquely to a stream disrupts less of the stream shoreline. However, in the case of an obliquely oriented road, if its direction has to be changed to cross the stream at a right angle, more total area is disrupted (though less riparian area might be affected). The construction of bridges causes more site disturbance than it does for culverts. But, over time, bridges cause much less disruption of both water flow and fish passage. Still, various approaches exist to minimize the influence of culverts on aquatic systems.[291,694,825]

For streams used by migrating or spawning fish, *arch culverts,* which retain the natural stream bottom and slope, are to be preferred over *pipe culverts*[263] (Figure 9.7). These open-bottom culverts retain most of the natural stream characteristics and enhance water passage during droughty, low-flow conditions.

Culvert openings are appropriately sized to ensure that upstream water levels will be acceptable, and that flow velocities will not be too high to inhibit fish movement through the structures. For instance, culverts sized so that normal water levels rise no higher than half the diameter of the pipe generally make good ecological sense. A continuous flow of water in sufficient volume to attract and pass fish is important during periods of fish migration. An average velocity of 0.5 m (1.6 ft) per second for warm-water fish (e.g., bass or pike) and of 0.9 m (3 ft) per second for cold-water fish (e.g., trout) have been considered appropriate.[291] If such low current velocities within the culvert cannot be attained, energy dissipators such as *baffles* (devices that deflect flowing water and produce a range of water velocities) may be useful.

Multiple culverts may be an option where water flows are expected to be quite variable, because a single culvert can create excessive velocities during high flows but not provide adequate depth during times of low flow. Indeed, the bottoms of paired culverts may be constructed at different levels to allow for some flow during dry conditions. Overall, multiple smaller openings will generate a lower flow velocity than will a single large opening and thus may be preferred by some migrating fish.

Amphibians and many other terrestrial wildlife species also move through culverts (Chapter 6), and diverse culvert designs are in use. Many wildlife species are highly sensitive to the design, and especially characteristics of the bottom, which may enhance or inhibit passage.

Many concerns addressed in Chapter 8 involving the chemical pollutant effects of runoff from road surfaces to receiving waters are exacerbated where roads pass directly over water bodies. Perhaps the aspect of water quality most threatened by water crossings in northern temperate climates results from materials used to prevent ice formation over paved bridges. Because bridges are open to cold air underneath, ice will form on the bridges long before it forms on roads underlain by insulating ground. The application of salt or sand to prevent ice buildup on bridges results in the runoff of some of this material to streams below.

Today, it is not uncommon to see a sign declaring a "Salt-Free Zone" as one approaches a bridge over a particularly sensitive water body, such as a drinking water supply. Use of sand rather than salt as a de-icing agent is preferable where salt accumulation may damage the aquatic ecosystem or drinking water supply.[926] However, sanding is appropriately avoided around a bridge located immediately upstream of an important fish-spawning area, which would be damaged by additional sediment.

We presently know little about the effects of bridges or culverts on the subtleties of aquatic food web dynamics. One unusual anecdote, however, underscores just how unpredictable and surprising such effects can be. Along the jagged coastline of Vancouver Island, British Columbia, highway bridges cross over the many rivers that tumble down from the mountains to the sea. Salmon return up those rivers of origin to breed, and the resulting young fish then make their oceanbound journey back down the same rivers. Harbor seals

(Phoca), keyed to follow the salmon, have learned how to orient themselves beneath several of the illuminated bridges that span such rivers (P. Olesiuk, personal communication). Here the fat and "happy" seals line up in groups and roll onto their backs. They merely lift their heads and mouths to gorge on the oceanbound young salmon, as the latter become momentarily silhouetted against the bright bridge lights overhead. The overall effect that such a harvest might have on returning salmon stocks is being monitored.

This story highlights another important ecological role of bridges and culverts. They serve as habitat, not just for fish and other aquatic species but for distinctive plants and terrestrial animals. Indeed, bats roost in certain bridges during the day.[472]

Estuarine Salt Marshes

We are becoming a "marginalized species" in the sense that more and more of us are seeking to live along the world's coastlines. In Massachusetts, for example, several coastal townships experienced 50% to 80% increases in the number of houses during the 1980s and 1990s. Certain coastal regions of the mid-Atlantic states have experienced population growth rates in excess of 300% over this time period. It is no surprise, therefore, that a strong correlation exists between human population density and the rate of coastal wetland loss. Whereas the major source of freshwater wetland loss is agriculture, for coastal wetlands the greatest threat is building development, including road construction.[811] Given that estuarine salt marshes are the most productive ecosystems on earth, and play key roles in sustaining biodiversity and fisheries, the legacy of their loss is profound.[874,977]

Today, around the world, vast areas of salt marsh are being impacted by road crossings that restrict tidal flows to inland (upstream) areas of marsh, due simply to inadequate culverts (or hinged "flapper tidegates" that allow only one-way water flows).[743] For example, in one small region in northern Massachusetts, of 125 culverts on raised roads crossing salt marsh streams, half were found to be so small that they restricted tidal flows.[706] Similarly, over 100 restrictions of tidal flow by culverts were identified on the scattered New Hampshire and Maine salt marshes.[731] For one site—deemed the worst in Massachusetts—the tidal crossing was so restrictive that it held back more than 13 cm (5 in) of tide during the ebb-and-flow cycle.

Vegetation-transect analyses upstream and downstream from this road crossing demonstrated the dramatic effects of severe tidally restricted flows on plant dominance and composition (Figure 9.8). Upstream vegetation was dominated by high tide bush and salt hay grass. Yet downstream, only a few meters away, the vegetation dominated by smooth cordgrass and spike grass was completely different. Particularly bothersome in many locations is the

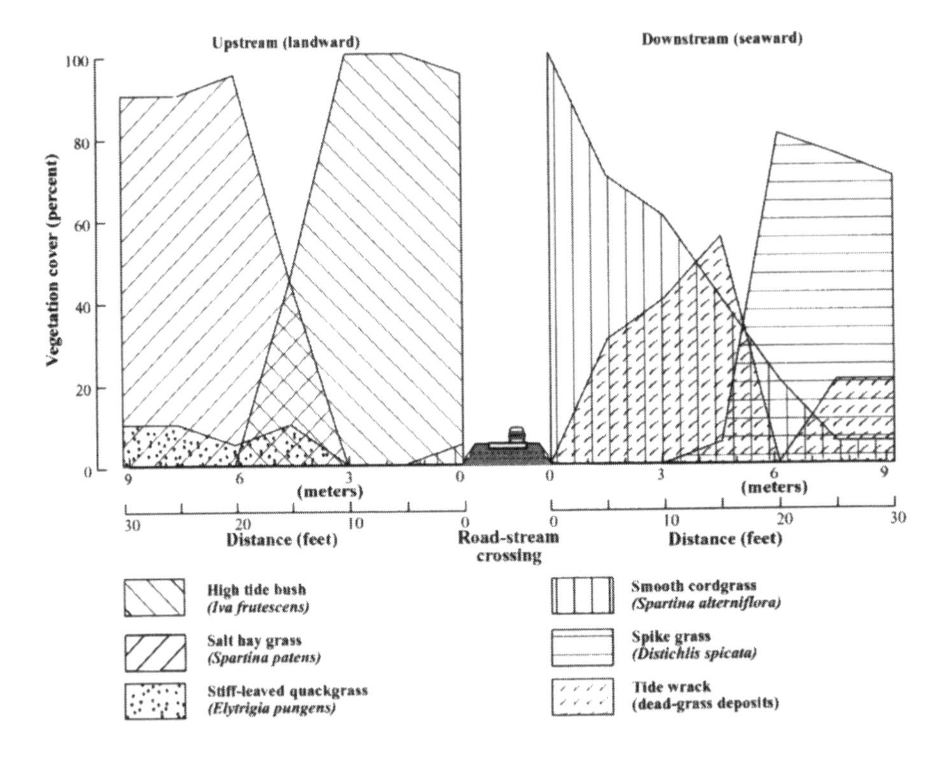

Figure 9.8. Vegetation upstream and downstream of a culvert on a road causeway that crosses a salt marsh. Tidal flows are severely constricted by the culvert, resulting in sharp hydrologic differences on opposite sides of the road. Route 1A, Rowley, Massachusetts. Adapted from Parker River Clean Water Association (1996).

invasion and colonization of the freshwater common reed *(Phragmites communis)* as a result of low salinity on the inland side of tidally restricted gates underneath coastal roads.[600]

Numerous adverse impacts develop in tidally restricted salt marshes.[772] These include (1) decreased tidal flow and flushing, (2) increased bank erosion due to tidal flow constriction and elevated flow velocity, (3) decreased salinity and nutrient exchange, (4) loss of shellfish, (5) loss of migratory (anadromous) fish, (6) lowered wetland water levels, (7) loss of native salt marsh vegetation (Figure 9.8), (8) invasion of the common reed, *Phragmites,* (9) low wildlife use and high fire hazard of dense *Phragmites* monocultures, (10) invasion of upland vegetation, (11) water quality degradation (nonpoint source pollution resulting from oxidation of marsh soils, including acid sulfate soil development, sulfur mobilization, stream acidification, aluminum mobilization, and summertime oxygen depletion), (12) loss of detritus export to estuaries and bays, and (13) mosquito proliferation.

The positive benefits of restoring tidal flows are also manifold. Indeed, many of the benefits are the reciprocal of the damaging effects,[772] such as increased tidal flow and flushing, decreased bank erosion due to less fluctuation in tide levels, and so forth. Mitigation measures to correct restricted tidal flows are being implemented in the Massachusetts area and elsewhere.[706, 743] The hydrologic techniques used vary widely and are tailored to the idiosyncrasies of the sites being restored. The approaches include (1) removing existing standard flapper tidegates with self-regulating sluice gates, (2) widening existing tidegates or increasing the diameter of culverts, (3) notching and adding culverts to road berms to increase water flow, (4) installing flow dissipators to reduce erosion scouring, and (5) dredging the inland side to increase water flow.

Often, these engineering approaches must be complemented by active management in the degraded salt marsh to improve conditions necessary for restoration. Open marsh approaches include (1) ditching to facilitate saltwater intrusion from the ocean (rather than marsh drainage ditches to reduce mosquitoes), (2) creating *salt pans* (pools) to retain water at low tides and thus provide habitat for mosquito-eating fish, and (3) harvesting *Phragmites* and correspondingly planting native salt marsh grass.

Many effective restoration projects have been implemented with consequent improvements in the condition of marshes and fish communities that support large numbers of birds.[221, 854] Mitigation and restoration occur after errors are made by society. The next stage moves from retrofitting tidal crossing gates to proactively designing better coastal roads in the first place.

Conclusion

Several perspectives are critical for understanding, anticipating, and ameliorating the effects of roads on aquatic ecosystems. These ecosystems may be sensitive to the effects of roads even when separated by a considerable distance, in part because of connections provided by flowing surface water and groundwater. Connectivity is important at several spatial scales, both within aquatic ecosystems and between them and the surroundings. Roads readily interrupt this connectivity with quite undesirable consequences. On the other hand, substantial improvements in the design of road-stream interactions have reduced the types of impacts common from earlier design approaches.

Road effects on aquatic ecosystems result from the transport of materials by wind, by streams, and by direct runoff where roads and water bodies are adjacent. Soil particles have an impact by being in suspension in the water and by producing sedimentation on the bottom. Chemicals both composing the particles and attached to them produce major effects on aquatic ecosystems. Nutrients tend to cause eutrophication, while toxic pollutants inhibit or kill aquatic organisms. Vernal pools are especially sensitive to roads because of their dependence on groundwater flows and chemical effects in a small water body. Similarly, wetlands, including salt marshes, are sensitive to roads because of conspicuous disruptions of hydrologic flows. Bridges and culverts also represent an especially sensitive interaction between roads and aquatic ecosystems. The replacement or upgrading of each bridge and culvert is a unique opportunity to create a significant visible improvement in major ecological processes and patterns.

Although mitigation and compensation for impacts overall have had a mixed history of failures and successes, numerous approaches have been developed. Therefore, in a new ecological era for transportation, a huge improvement in aquatic ecosystems near roads could become visible to the public rather quickly.

CHAPTER 10

Wind and Atmospheric Effects

> The air is so polluted that you wake up in the morning listening to the birds cough.
>
> —Anonymous, quoted in a letter to *Le Monde,* 6 August 1982

> In addition to the appearance and the smell and the noise and the fright, the most obviously annoying characteristics of the car at the turn of the century were that it was dangerous and it raised dust, with the latter looming larger in the eyes of the public. Once car speeds exceeded 30 km/h, dust covered everything within 20 m of a highway. Adjacent property values dropped by about 30 percent.
>
> —M. G. Lay, *Ways of the World,* 1992

Imagine being the self-appointed captain of the great ship *Titanic* reconstructed for tourists on a mountainside surrounded by nature. The rusting ship's massive diesel engines run every day, and you become accustomed to pouring in fuel and oil and breathing large amounts of exhaust. Pairs of animals keep slipping into the ship, and the inevitable feeding frenzy on the huge ark leaves carcasses strewn along the deck and aromas emanating from below. Your only assistance is a small army of sanders and painters trying to keep up with the rust and eroding particles that spread over your nature reserve. As captain, you accomplish the herculean task of keeping the ship operative, and you simply assume that nature around you continues to function reasonably well.

Road systems seem more manageable, though some aspects of the great-ship analogy are disconcerting. The airborne materials from the ship—both known and unknown—include rust, sand particles, smoke, air pollutants from the engines, paint fumes, aromas, discarded wastes, and noise. Road systems produce a different mixture of materials but with somewhat similar ecological effects: particles and chemicals from maintenance activities, discarded solid waste, liquids that leak from vehicles, and various gases emitted from engines. Heat, noise, and light energy also emanate from roads and vehicles. Some materials and energy are limited to the local area, others are carried by wind across the region or to other regions, and some reach the upper atmosphere to encircle the globe. Ecological effects are significant at all three spatial scales: local, regional, and global.

This chapter uses a landscape ecology perspective to link land and air. It begins by discussing the microclimate around roads, followed by a focus on wind and windbreaks, an exploration of the effects of dust and snow, and a discussion of traffic disturbance and noise effects. Finally, the ecological impacts of atmospheric additions at local, regional, and global scales are briefly introduced.

Microclimate, Wind, and Windbreaks

Consider a pleasant stroll across a wide-open pasture during which a small road appears. As you begin walking on the road, the air passing your face somehow feels different. If you had been walking through a forest and then began walking along a road without an overhead tree canopy, the difference in sensation would be much greater. In both situations, sun and wind tend to determine the main differences you feel. However, temperature and moisture, which are largely controlled by sun and wind, also affect your face.

Microclimate of a Road

Little seems to have been published on the local climate around a road. However, the basic principles of microclimate are well known,[343, 794, 11, 650] so few surprises can be expected when considering the microclimate of roads and their vicinity. Detailed studies of roads through open pastureland and through forest in Germany illustrate patterns that are probably widespread.[241]

The major microclimate differences around a road in open land are close to the ground surface and relate to *surface microtopography*. This pattern was demonstrated by a study of air temperatures along a north-south highway passing through open pastureland.[241] Microclimatic measurements were made within the ground, just above the ground surface, and 50 cm (1.6 ft) above the ground. Soil temperature was highest on the west-facing road shoulder, that is, facing the afternoon sun, whereas air temperature changed little with distance from the road surface. Evapo-transpiration rate was also highest on the road shoulder. No significant differences in temperature, relative humidity, or evapo-transpiration were evident beyond 8 m (26 ft) from the road surface in this open land. Indeed, ground temperature differed outward only about 4 m (13 ft) from the road surface. At 50 cm above the ground, microclimatic conditions differed outward only about half that distance.

Some of these microclimatic patterns were similar by a four-lane highway that ran through forest and had open roadsides 20–25 m (66–82 ft) wide on each side.[241] Highest temperatures were recorded just above the ground surface on the road shoulder and in the outermost edge of the roadside at the base of a west-facing forest boundary. The highest evapo-transpiration and the lowest relative humidity occurred on the road shoulder. The highest relative humidity was recorded in the outer roadside. Not much change in these microclimatic variables extended into the forest beyond the roadside-forest boundary. Also, little change beyond the forest border was evident for several key soil and plant characteristics measured, including species richness, number of disturbance-related species, percent of salt-resistant plants, soil acidity, and soil nitrogen.[241] Roadside microclimate was similar on the eastern and western sides of the highway, but differences became more pronounced closer to the adjacent forest borders.

Both the open-land and forest-road examples indicate that a distinctive road-related microclimate is present only in a narrow zone. Heat absorption and radiation by the road surface itself are probably central factors in determining this microclimate. In general, the changed conditions do not extend much beyond the roadside, and indeed, some of the microclimate differences do not extend much beyond the road shoulder. The greatest variability tracks surface microtopography, at or just above the ground surface, and also occurs near adjacent forest.

A study of a two-lane (6 m [20 ft] wide) forest road also in Germany, with 5 m (16 ft) wide roadsides and no tree canopy, reemphasizes some of the same patterns and highlights others.[754] Sunlight, subsurface ground temperature, and evapo-transpiration rate were recorded at intervals outward to 50 m (164 ft) from the road. The narrowest significant road effect was recorded for soil temperature, which increased outward only approximately 2–3 m (7–10 ft) (mainly on the road shoulder). Light levels were elevated outward about 3–6 m (10–20 ft) from the road surface. Elevated evapo-transpiration rate, however, extended outward 15 m (49 ft) or more. The *forest edge* (or border zone or edge effect) averaged about 15 m wide (based on vegetation differences compared with the forest interior). Thus much of the forest edge had an elevated evapo-transpiration rate (relative to the forest interior) due to the presence of the road. For all three microclimatic variables measured, the middle of the two-lane road had the highest values, with values rapidly decreasing in the road shoulder and ditch areas and decreasing more slowly across the outer roadsides.

These microclimatic patterns result from study of specific places and times but illustrate basic microclimate principles[343,794,11,650] that should provide general predictions of the microclimatic effects of roads. Differences in effect should be evident in a series of contrasts: (1) day versus night, (2) summer versus winter, (3) tropics versus arctic, (4) roadsides on the north versus the south side of a road, (5) high- versus low-watertable roadsides, (6) forest road with a complete overhead tree canopy versus no canopy, (7) presence or absence of a vegetated median strip, and (8) black asphalt versus white concrete. Also, the effects of road-surface width, roadside width, and roadside objects, such as guard rails, signs, and concrete structures, should be generally understandable with microclimate principles.

Forest edge widths next to fields and clearings, based on microclimatic and vegetation measurements, are typically considerably wider (10 to 200 m)[397,523,154,1062,155,302,806] than those suggested by the road studies just described. Thus, empirical studies of forest edge widths next to roads of different type, width, and direction should be quite informative, especially to understand the role of roads in habitat loss and fragmentation and as barriers to wildlife movement.

Wind and Windbreaks

Wind is the giant of road microclimate because its effects reach so far (Figure 10.1). Air movement or wind carries energy, such as heat and noise. It transports lightweight particles, including: dust and snow; *aerosols* (tiny liquid particles suspended in air), such as nitrates and smog; and gases, including CO_2 and NOx. These transports of materials tightly link road systems to adjoining local areas as well as more broadly to regions and the globe.

Wind erodes the surfaces of many dirt roads and roadsides, especially in dry periods and dry climates. It deposits particles, including sand and snow, on road surfaces, usually as a barely visible covering but sometimes in massive piles that block road traffic. Wind can cause vehicle accidents, especially where it funnels up valleys perpendicular to traffic flow (C in Figure 10.1). Wind drops trees and wires onto roads and vehicles. It clears the air of vehicular pollutants. It carries scents that attract or repel animals from crossing a road, and it carries seeds along roads. Wind blows insects into the paths of onrushing windshields.[778] Some of these airflows are along and some across the road, while others connect the road to the broader landscape with inputs and outputs.

Moving vehicles, particularly trucks (lorries), cause local winds that lift particles from the road surface, a familiar sight when passing or, worse, following a vehicle on a dusty road. Some particles lifted into the air include heavy metals, tire fragments, and body rust. Vehicle-caused wind may transport seeds along roads[916] and blow dust or snow that blinds a driver.

Rows of trees or shrubs, either planted or natural, are especially common

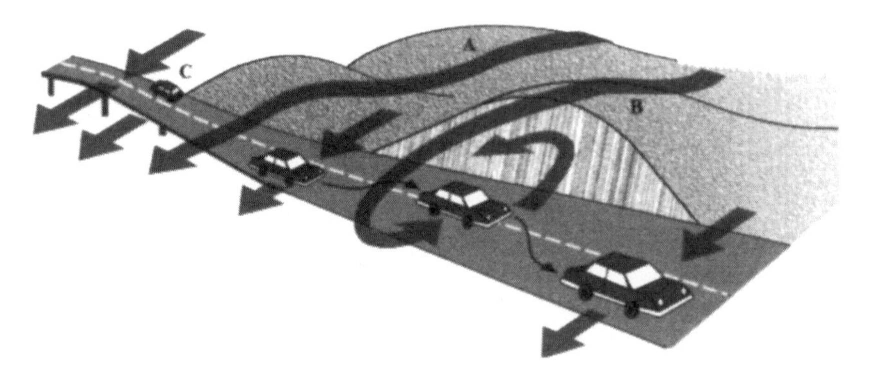

Figure 10.1. Vehicles encountering streamlined and turbulent airflows along a road. (A) Streamlined airflow on smooth hilly terrain; (B) turbulence downwind of a steep cutbank; (C) accelerated airflow in valley and past bridge. Adapted from Ahrens (1991).

near roads. They effectively function as windbreaks, which typically reduce windspeed. However, depending on arrangement or design, windbreaks may also accelerate windspeed and produce turbulence or vortices.[734,110,129] Indeed, wind direction can be changed, even reversed.[302] Electric and telephone poles and wires, plus road signs and advertising signs, doubtless complicate airflows around roads. This section presents basic principles of wind and windbreaks.

Wind or air movement takes three basic forms.[110,411] *Streamlines* are normally parallel horizontal layers of moving air (A in Figure 10.1). *Turbulence* is irregular air movement typically characterized by up and down currents forming *eddies* (small circular flows within turbulence) (B in Figure 10.1). A *vortex* is a helical or spiral flow, often with a central vertical axis.

A *streamlined object,* such as a boomerang placed in horizontal wind, splits the streamlines and produces little or no turbulence, because the streamlines come back together just downwind of the object. On the other hand, a *bluff object,* characterized by a blunt upwind or downwind side, splits the streamlines and produces turbulence (B in Figure 10.1). If the blunt side is upwind, turbulence forms in front of the object and usually also on the downwind side. But if the upwind side is streamlined and the downwind side blunt, turbulence essentially forms only downwind. If an object is porous, *bleed flow* (airflow through the object) reduces turbulence on both the upwind and downwind sides by decreasing air pressure differences.

Windbreaks, such as shelterbelts, hedgerows, and snowfences, are often long, thin barriers on an open surface. Assuming airflow across an extensive smooth surface encounters a windbreak perpendicularly, windspeed can be calculated with some confidence at varying distances both upwind and downwind of the windbreak and at different heights above the ground.[961,110,411,302] Windspeed is highly sensitive to *windbreak height* (*h*), measured as the average maximum height along a length of windbreak. *Porosity* is also an important determinant of windspeed and can be estimated optically with visual or photographic measurements. For instance, a porosity of 0.15 means that 15% of a vertical plane is unobscured by the windbreak. Optical measurements are less useful for wide windbreaks with three-dimensional spaces or in differentiating pore size or sharp-versus-smooth-edged pores; therefore, aeronautical engineers calculate a resistance coefficient to estimate porosity.[33,411]

Windbreaks thus alter horizontal windspeed, turbulence, and vortex airflows. Height and porosity of windbreaks are the two major controls on airflows, and both are amenable to design and management.

Upwind of a windbreak, measurable reductions in windspeed normally extend to a distance of a few times the windbreak height (*h*), resulting in a partially sheltered area called the *upwind zone*.[33,110,302] Downwind of a windbreak, reduced windspeed typically extends to a distance of a few tens of *h*. Two relatively distinct microclimatic zones—the quiet zone and the wake zone—are distinguished downwind. The quiet zone extends downwind at ground level to a distance of roughly 8*h*.[616] The location of this zone appears to be independent of porosity and is characterized by lower horizontal windspeed and higher turbulence than in airflow in the open (far upwind). The wake zone extends from about 8*h* to 24–30*h* or more. Relative to the preceding quiet zone, the first part of the wake zone is characterized by higher windspeed, greater turbulence, and larger eddies. Beyond about 12*h*, turbulence drops and wind patterns in the wake zone gradually change downwind until they are indistinguishable from those in the open. Windspeed reduction of about 20% commonly extends some 25*h* downwind of vegetation windbreaks. However, if the surface upwind of the windbreak is rough, such that somewhat turbulent airflow reaches the windbreak, windspeed reduction will extend only a short distance both up- and downwind of a windbreak.

Windbreak height largely controls the distance that a windbreak effect extends, whereas porosity primarily affects windspeed over that distance.[961,110,411] A highly porous windbreak only slightly decreases windspeed but has the advantage of minimizing turbulence. Also, a porous windbreak provides a relatively long distance of reduced airflow downwind, although windspeed is only slightly lower than airflow in the open. On the other hand, a low-porosity windbreak, such as a wall or thick dense evergreens, produces a short downwind zone of much lower windspeed. However, in this case, high levels of turbulence are created both upwind and downwind. In many situations, a medium-porous windbreak is optimal. Windspeed reduction in such cases is nearly as great as for the dense barrier, but the shelter effect extends over a longer distance downwind. A medium-porous windbreak also generates much less turbulence.

Rows of trees or shrubs can be designed and managed in many ways to affect airflows on and near roads. For example, the roadside manager can vary several properties of linear roadside vegetation[110]: (1) single or many species present, (2) deciduous or evergreen species, (3) shrubs or trees, (4) number of rows and total width, (5) distance between rows, (6) distance between woody plants, (7) degree of porosity, (8) height and trimming of lower branches, and (9) shape of the upper windbreak surface. These characteristics, in turn, help determine the: strength and location of crosswinds on a road; windspeed along the road; sand, snow, and ice accumulation; and locations of cold, warm, wet, and dry sites along a roadside.

In considering the degree of porosity to wind, a forest almost mimics a wall of boards, because so little wind can move through it. Thus, downwind of a forest, high turbulence and reduced horizontal windspeed occur for only a relatively short distance.[343, 302] A rough forest-canopy surface, such as occurs in tropical rainforest or old-growth forest, causes air turbulence and hence further decreases the downwind distance of windspeed reduction. Also, airflow reaching a windbreak at an oblique angle tends to be forced over the plants and then bends toward the windbreak, resulting in a short downwind area of reduced windspeed.[961, 110] Short windbreaks are relatively ineffective at reducing wind, because of the bending of wind around their ends.

Gaps, or breaks, in windbreaks (where gap width is less than *h*) often cause increases in windspeed up to 20% or more (due to the Venturi effect) with consequent effects on passing automobiles or cyclists[270] (Figure 10.2). A vehicle traveling in the quiet zone downwind of a windbreak is hit by accelerated airflow through a windbreak gap.

In snow-prone areas, trees along a road may also be important in shading portions of the road. This reduces the rate of snowmelt and leads to ice formation on the road surface, a significant safety hazard for vehicles.

In short, the arrangement of lines of woody plants near roads is a major controller of wind and other microclimatic conditions affecting vehicles, roads, and roadsides. Weather reports on television barely suggest the richness of microclimatic variables. Of these variables, windspeed is modified the farthest downwind of a windbreak, commonly some 24–30*h*.[734, 794, 616, 411, 302] Also, evapo-transpiration is affected a considerable distance, often to about 16*h*. Other temperature, light, and moisture variables typically are affected only to about 8–10*h*. Nevertheless, because road corridors are generally narrow, even shrub windbreaks in open land near roads can affect most microclimatic variables over the width of most road corridors.

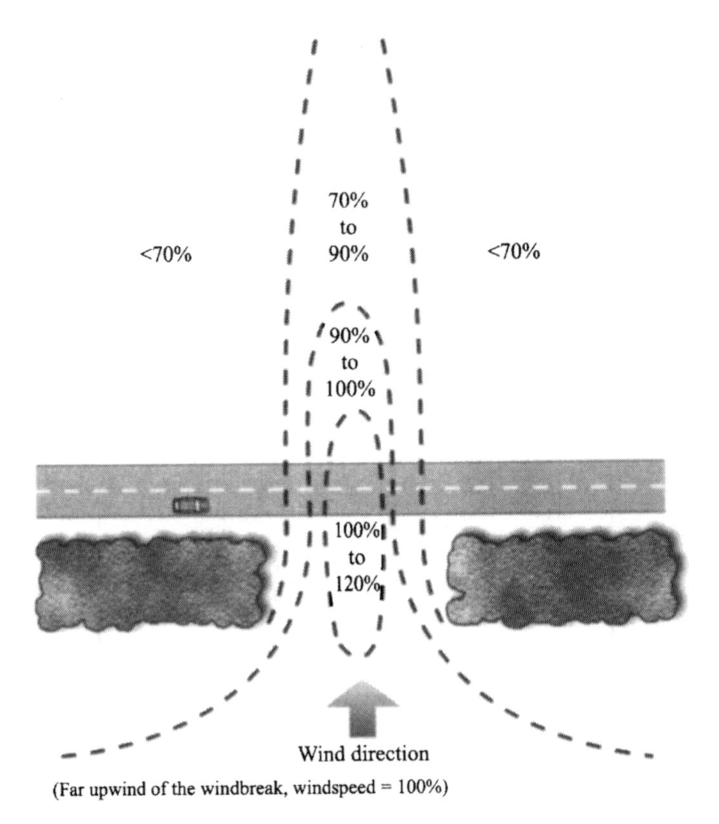

Figure 10.2. Airflow through a gap in a windbreak, showing areas of accelerated and reduced windspeed. Based on FHWA (1996b) and other sources.

Dust and Erosion

Anyone who has followed another vehicle along a dry, dusty road knows to shut the windows, stay far behind, and watch out for airborne debris following trucks going in the opposite direction. The ritual of car washing often follows such a trip. Indeed, car washing is also common in snowy areas, where roadsalt covers and "eats away" at cars.

Wind Erosion

The *Dust Bowl* is a France-sized area in Colorado, New Mexico, Texas, Oklahoma, and Kansas where massive amounts of topsoil blew away in the 1930s and were deposited, often far downwind (Figure 10.3). Wind erosion, the removal of particles from a surface by airflow, consists of a sequence of overlapping processes of particle initiation, transport, abrasion, sorting, and

Figure 10.3. Wind erosion from local fields separating and distributing particles by size. Lightweight clay and silt are suspended as dust in air and carried considerable distances; heavier silt and sand are deposited in nearby ditches and fencelines. Twelve cars were involved in accidents in this strong wind. Sully County, South Dakota. Photo courtesy of USDA Soil Conservation Service.

deposition.[36,156,574] *Particle initiation* involves oscillation of the particle before it is dislodged, either indirectly by another moving particle or directly by wind.

Most eroded soil is transported short distances.[1065] Coarse sand- and gravel-sized particles or aggregates generally move along a surface by *soil creep,* as they are rolled, pushed, or knocked by other particles. Midsized particles tend to move by *saltation* or hopping, rising a few decimeters (a foot or two) and ending up in nearby roadsides, ditches, windbreaks, and woodland edges. Fine particles, such as clay, are suspended in the air and can move meters to thousands of kilometers before sedimentation or deposition returns them to the earth surface.

Two effects of wind erosion are particularly noteworthy.[110] First, modest wind erosion selectively removes fine nutrient-rich particles, leaving a less-fertile soil. This loss of soil fertility can be arrested or reversed by windbreaks, which themselves can contribute to soil fertility. Second, more severe erosion accelerates *sorting,* the separation of soil into deposits of similar particle size. Thus a loam soil of mixed particle sizes in one location may be converted into sand, silt, and clay accumulations deposited in separate downwind locations. Dust deposits on vegetation near roads during dry periods or in dry climates tend to be silt (often mixed with fine sand and clay) particles that were wind-eroded from a bare roadside or unpaved road.

Five principles for reducing soil erosion by wind are commonly recognized[914, 574]:

1. Reduce field sizes in the predominant wind direction (thus reducing *fetch* or *run*, the distance of air movement along a continuous surface) to reduce windspeed.
2. Maintain vegetation or plant residues to protect the soil surface.
3. Maintain soil aggregates or clods large enough to resist the wind force.
4. Roughen the surface to reduce windspeed and trap moving particles.
5. Cover (e.g., with vegetation) hilltops and other spots susceptible to accelerated streamline airflows, turbulence, or vortices.

A combination of practices is always recommended to minimize wind erosion. Windbreaks may contribute in all five cases, but their primary role addresses the first principle, reducing the length of erosive surface. A threshold windspeed between 19 and 24 km/hr (12 and 15 mph) is required to start particle movement on most erosive soils.[914, 110] Thus a key goal for reducing the impact of wind is to maintain windspeed over the entire ground surface below this threshold, even during high winds. Roadsides and unpaved roads oriented in the direction of preponderant winds are particularly susceptible to wind erosion and appropriate for management.

Fortunately, windbreaks reduce wind erosion forces much more than they reduce windspeed.[847, 914] This difference in effect occurs because the force causing soil movement varies with the cube of wind velocity. In other words, for a particular percent reduction in windspeed, the same percent reduction in wind-erosion force extends much farther downwind. Consequently, windbreaks are especially effective in reducing soil erosion by wind over large distances.

Dust

Perhaps the most physically extensive impact of roads is on *airsheds,* the areas susceptible to downwind deposition of materials originating from roads and roadsides. Considerable research on the effects of wind erosion and silt deposition in arid regions and on beaches has shown that the downwind deposition of *aeolian* (windblown) material may take place over wide areas, depending on particle size, windspeed, and terrain features.[36] Considering the large number of unpaved roads with loose surface material (Chapter 2), surprisingly little is known about these rough surfaces as sources of airborne particles or about the ecological effects of the particles later deposited.[902, 444, 990, 875]

Fugitive dust consists of lightweight particles suspended in the air that originate from ground surfaces, such as soil and roads[361] (Figure 10.4). Most fugitive dust particles are microscopic silt and clay particles. The material is

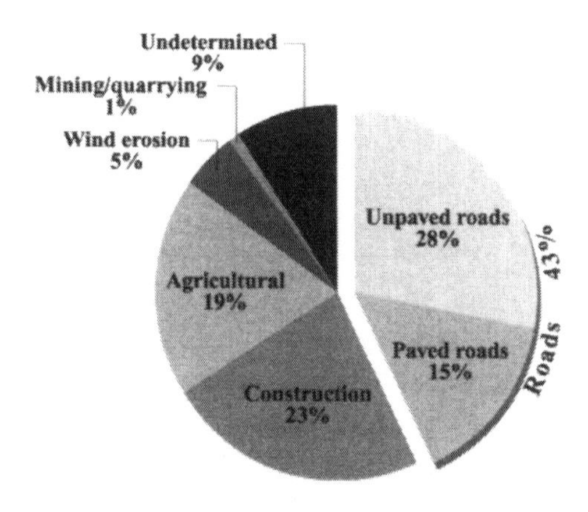

Figure 10.4. Major sources and amounts of fugitive dust in the USA. Adapted from Goff (1999).

generated by two basic physical processes: (1) mechanical abrasion or pulverization of surfaces by wind, and (2) lifting of existing particles by wind scouring.

Each car traveling 1.6 km (1 mi) over a dirt road daily for a year has been estimated to generate on average about 1 ton (900 kg) of dust.[74] Furthermore, soil particles or dirt tracked onto paved roads from unpaved surfaces is highly transportable, being quickly picked up even by small winds and made airborne.[74] It is no surprise, therefore, that roads represent the single largest source, generating 43% of fugitive dust globally (Figure 10.4). Unpaved road surfaces are estimated to contribute 28% and paved roads 15% of the total amount of fugitive dust.

Wind-transported dust carries a host of detrimental elements, though some other elements may be benign. Human health has been the overwhelming concern with airborne dust, especially the finer particles.[537, 674, 675, 750] Because of many cases of respiratory afflictions, the U.S. Environmental Protection Agency has issued a series of exposure warnings.[74] However, the focus here is on the ecological effects of fugitive dust.

The benign effects of dust can be illustrated by elevated concentrations of calcium, potassium, sodium, and phosphorus detected in dust collected at distances up to 20 m (66 ft) from a well-traveled gravel road in Sweden.[902] If the added amounts of these important growth-limiting nutrients are sufficient, such deposits stimulate the growth of certain plant species and, therefore, change the plant communities alongside gravel roads.[258]

Surface dust from roads also carries contaminants, which may accumulate where particles are deposited.[922] For example, concentrations of metals were studied in wetland moss samples near a gravel road in New Brunswick (Canada) to determine how far outward the contaminants spread in road dust.[814] Aluminum, chromium, iron, and vanadium concentrations decreased exponentially with distance from the road. Most of the material was deposited within 50 m (164 ft) of the road, though elevated levels of aluminum extended outward 10 times that distance. A study of lead content in the air near a California highway (during the era of leaded gasoline) found elevated lead concentrations extending outward about 200 m (656 ft) downwind, but barely at all upwind.[241] The accumulation of road dust in gutters along streets may represent a concentration of contaminants, which then is quickly washed by rain into groundwater, streams, and other water bodies. Chemical contaminants from roads may be toxic to certain plants or animals[258] or may *biomagnify* (accumulate through the food chain) in the tissues of organisms, with greatest concentration and potential toxicity in top predators.

A distinct gradient is evident in the deposition of airborne sediment originating from road surfaces. In a Swedish study, coarse and medium sand (0.2–2 mm diameter particles) were rarely deposited more than 5–10 m (16–33 ft) from roads, with most fine and very fine sand (0.02–0.2 mm) being deposited at distances up to 20 m (66 ft).[902] Silt particles were mainly deposited within 40 m (131 ft) of the road source.

Microscopic dust particles, the major component of fugitive dust, can be transported regionally or globally. However, a major portion of the dust falls out near its source of origin. For example, an Ontario study examined dust around logging (forestry) roads and the off-road ruts of log-removal (skidder) equipment within a forest.[875] The authors concluded that 75% of the dust was deposited within 10 m (33 ft) of a logging road, and 93% within 30 m (98 ft) of a road.

As discussed above for wind erosion, the generation of road dust depends on several factors.[455, 574, 631] Moisture on the road surface plays a major role in the susceptibility of particles to being dislodged by wind. Although high windspeed is needed to lift sand-sized particles, even a gentle breeze can lift small particles. Particles on paved roads come from road-surface degradation, tire degradation, vehicle degradation and emissions, and soil deposited by wind or vehicles on a road. Both air movement across the land and wind caused by a passing vehicle can lift particles. Abundant traffic acts as an ongoing wind machine, creating fugitive dust. The wind machine goes faster and slower depending on the types, speeds, and numbers of vehicles streaming along the road.

The extent of nearby vegetation and its influence on wind patterns play a critical role in the amount and transport distance of airborne particles and thus the potential deposition into nearby water bodies.[875] In northwestern Ontario, forested buffer strips situated between logging roads and surface waters were found to substantially decrease the generation of airborne particles from logging roads (presumably by decreasing fetch) and their subsequent deposition into lakes. Also, many "nonstructural best management practices" (Chapter 8), such as reducing vehicular traffic volume and speed, would reduce dust generation.[361]

Other approaches to decreasing the generation of dust focus on stabilizing road surfaces by applying *dust-suppression* or *dust-control agents*.[631] Watering is the most common method but provides only temporary dust control and adds to surface erosion. In the past, many petroleum-based waste products were used as dust suppressors, but dispersing these materials is now prohibited in the USA. A particularly poignant story reputedly involves a family who had to move from their home at Love Canal, a New York community contaminated with industrial toxic waste. The family selected a remote midwestern farm community far from any industrial sites but then was forced to move a second time after a contractor sprayed oil laden with toxic PCBs (polychlorinated biphenols) on the local roads.

Today, many products designed to bind dust and make it immobile are promoted as environmentally sensitive.[631, 361] These materials include recycled paper products with organic binding agents, calcium chloride flakes or brine, lignin sulfonates, vegetable oils, and acrylic copolymers. One study compared a suite of surface treatment techniques for forestry roads and found calcium chloride to be the most economical choice.[14] With a high surface tension and an ability to hold moisture at low humidity and high temperature, calcium chloride helps to bind aggregate particles together and, therefore, to stabilize an unpaved road surface.

In short, road dust from both unpaved and paved surfaces continues to be a serious environmental concern. The direct effects are illustrated by damage to roadside vegetation as well as by sediment deposition that affects fish spawning. The indirect effects involve dust as a vector carrying chemical contaminants, which, in addition to contributing to human health problems, inhibit plants and animals near roads, alter lake ecosystems, and may accumulate through a food web to a high concentration.

Snow and Snowbreaks

Snow causes additional problems for travelers and the environment. Blowing snow sharply reduces drivers' visibility. Furthermore, accumulating snow on roadways reduces the capacity of vehicles to travel and increases snow-removal maintenance costs. However, roadside vegetation and road structures have major effects on both blowing snow and snow accumulation patterns. *Snowbreaks*—wind-reduction structures that affect snow movement, accumulation, and disappearance—may be made of vegetation or construction material (the latter usually referred to as a *snowfence* or *artificial snowbreak*). Although both types are considered below, the focus is on vegetation snowbreaks, which are widespread and are particularly important for biodiversity and wildlife movement patterns.

Live snowbreaks and snowfences decrease windspeed, causing airborne snow particles to settle out and be deposited nearby in a specific pattern. This process also decreases snow deposition further downwind. A well-designed windbreak can keep a highway nearly snow-free, whereas a poorly designed snowbreak may concentrate mammoth snowdrifts directly on the highway.

The end portions (several times *h* from the end) of snowbreaks are relatively ineffective at snow accumulation, because wind accelerates and bends around the ends.[900, 302, 270] If snow arrives obliquely to a windbreak, a huge mound of snow accumulates at a well-defined spot near one end of the windbreak, where wet soil persists in summer. Snow blowing parallel to a windbreak tends to form deep, narrow parallel drifts,[961] which may cause travel problems where a road with parallel hedgerows is oriented in the preponderant wind direction.

Like filling a swimming pool with water, a windbreak can accumulate a finite amount of snow before reaching equilibrium, where output equals input. This *windbreak capacity* for accumulating particulate matter is often important in designing a snowbreak sufficient to accommodate the largest snowstorms. Windbreak capacity depends primarily on height and porosity, though width is a useful secondary predictive factor.[838, 888, 270] Other things being equal, a doubling in snowbreak height increases snow storage capacity roughly four times. Medium-porous windbreaks, which produce smooth, gently sloping, reasonably deep (relative to *h*) snowdrifts, apparently have the highest windbreak capacity.

Several windbreak types are normally avoided due to low snow capacities. Short windbreaks, oblique windbreaks, highly porous windbreaks, and streamlined windbreaks all usually trap relatively little snow.[110, 270] A low-porosity windbreak—such as dense evergreens or, in the extreme, a concrete wall—produces considerable downwind turbulence, usually resulting in two parallel snowdrifts. Minimizing turbulence and vortices is normally desirable in snow management.

Artificial snowfences with 50% to 60% porosity have been recommended as optimum for accumulating snow upwind of a road surface.[838, 888] A gap at the bottom that is 10% to 15% of the snowfence height increases the downwind length of the snowdrift and, apparently, increases total windbreak capacity. Although the gap at the bottom reduces upwind snow accumulation, it normally prevents snow buildup, which could bury the snowfence. A buried snowfence results in a streamlined snowdrift, which sharply reduces the ability of the windbreak to catch snow.

Snow accumulation in sheltered zones results in more infiltration of subsequent snowmelt water into the soil.[110] This increased percolation can help maintain the watertable, support wetlands with waterfowl and other wildlife, and maintain soil moisture that reduces wind erosion. Deep snowdrifts in some spots and no snow in others create a wide range of soil moisture conditions, which can typically support a variety of plant communities.

Snow piles have several other consequences. Large snow piles along suburban roads alter wind and microclimate and also help reduce traffic noise, resulting in a noticeably quieter neighborhood. Snow piles that remain for a period may be rich in roadsalt, sand, and chemical contaminants. As the snow eventually melts in place, these materials reach local water bodies through either roadside ditches or stormwater drainage pipes. If snow piles are removed and dumped elsewhere, snowmelt water then carries the material into the nearby ecosystems.[831]

In contrast to artificial snowfences, living vegetation snowbreaks are preferable for enhancing wildlife habitat, aesthetics, and long-term economic benefits.[838] The effective service life of living snowbreaks is considered to be about 50 years. This longevity compares with 5 to 7 years for the standard 1.2 m (4.5 ft) high vertical-slat fence (Canadian structural fence) and with 25 years for the up to 3.8 m (12 ft) high "Wyoming structural barrier."[888, 270] Disadvantages of vegetation snowbreaks include the difficulty of establishment on some sites, length of time to reach effective height, initial cost and maintenance, and amount of land required. Several key arrangement or design features determine the effectiveness of vegetation snowbreaks of any sort[141, 110, 270]: (1) distance from the road, (2) length, (3) species, (4) number of rows, (5) spacing of plants, and (6) habitat quality for wildlife and biodiversity.

The species of trees and/or shrubs present affect both leading factors determining snow accumulation patterns: height *(h)* and porosity.[141] Taller plants have a greater effect than shorter plants. Plants with highly porous canopies trap less snow than those with dense canopies but distribute the drifts over longer downwind and upwind distances. Dense canopies also produce more turbulence. The number of rows and the spacing of plants are important determinants of porosity.

The species of woody plants as well as the number of rows and spacing of plants are important in determining wildlife habitat and biodiversity.[141] A diversity of native woody plants and variability in the spacing of plants can be expected to maximize biodiversity. However, to provide long-term effective snow management along the entire snowbreak, a natural-vegetation windbreak normally will need to be relatively wide in order to compensate for the different heights, porosities, and lifetimes of different plant species.

Standard planted snowbreaks ("snowbreak forests") in Hokkaido, in northern Japan, are 10, 22, and 32 m (33, 72, and 105 ft) wide, based on systematic experience with managing snow accumulation and movement[270] (Figure 10.5). The 10-m-wide snowbreak produces a high snowdrift just downwind, whereas the 32-m and 22-m-wide tree plantings have the maximum drift upwind and at the upwind edge, respectively. In the two wider cases, little snow accumulates downwind, so these windbreaks could be relatively close to the road. In this region, a 7.5-m (24-ft) distance between trees and road

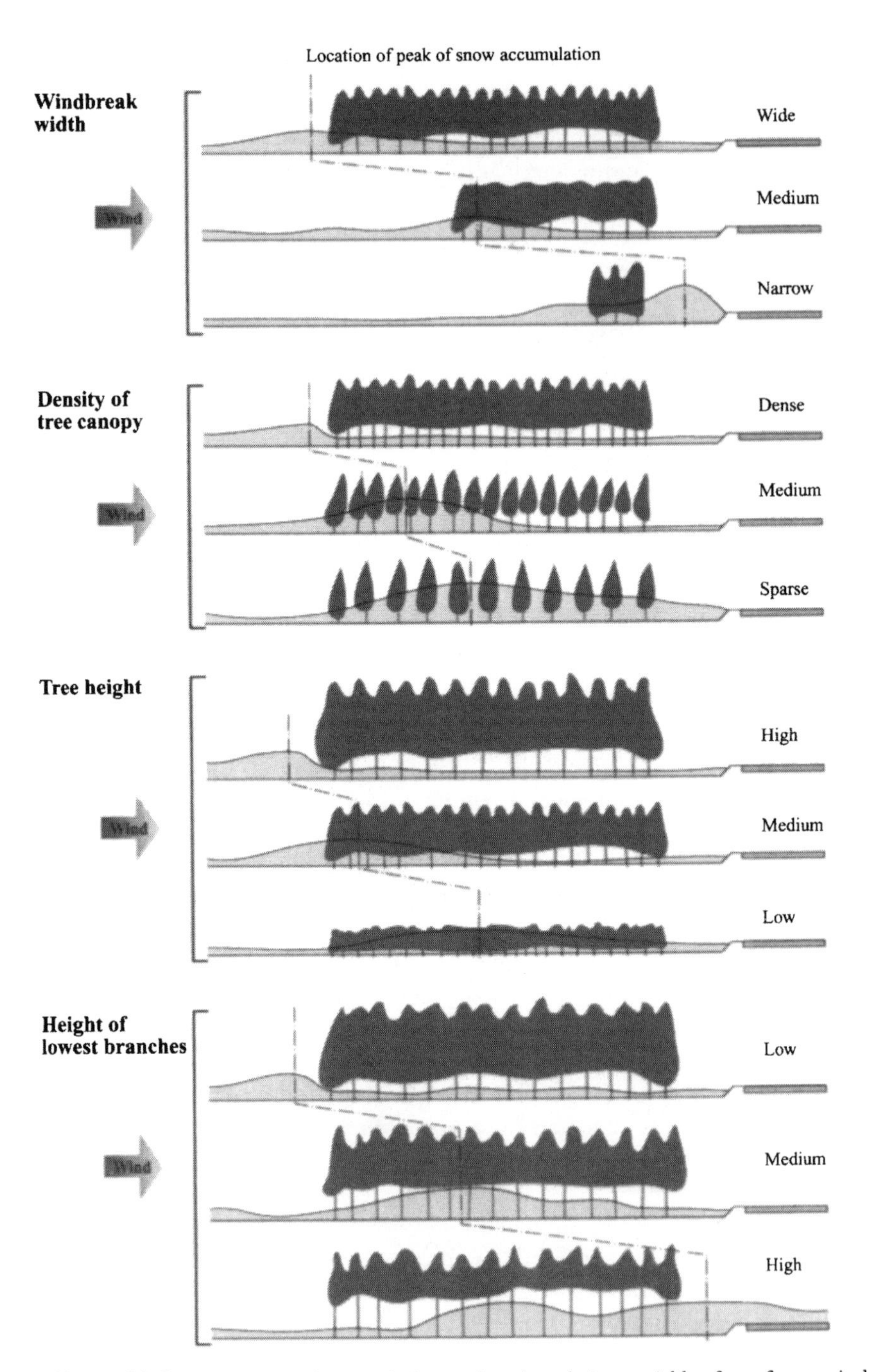

Figure 10.5. Snow accumulation relative to four key design variables for a forest windbreak. Shading indicates snow, with wind coming from the left and the road on the right; vertical dashed line indicates location of the snowdrift's peak. Standard windbreak widths are 32 m (105 ft), 22 m (72 ft), and 10 m (33 ft); windbreaks are 7.5 m (24.5 ft) from the edge of the road shoulder. Based on measurements, modeling, and experience in reducing snow accumulation and blowing-snow visibility problems in Hokkaido, Japan. Adapted from FHWA (1996b).

shoulder is recommended for many situations and also provides space for deposits from snow removal equipment on the road.

Tree height, canopy porosity (or density), and height of branches off the ground are also major factors in determining snow accumulation patterns[270] (Figure 10.5). Dense canopies, tall trees, and low branches trap more snow upwind and leave small accumulations downwind by the road. In contrast, three snowbreak designs result in major snowdrifts downwind: (1) sparse, highly porous canopy; (2) narrow, 10-m-wide snowbreak; and (3) bottom tree branches high off the ground. The last design is the worst for keeping snowdrifts off a road, but in all three cases, the snowbreak would be more effective if located farther upwind, away from the road surface.

Snowfences can be combined with vegetation snowbreaks. For example, in Hokkaido, a tall, solid ("windstopping") fence placed 1*h* distance upwind of a tree windbreak considerably increased snow accumulation.[270] A porous ("wind-reducing") fence placed 5*h* or more upwind of the trees also considerably increased total accumulation as well as accumulation for some distance upwind. Even a tall fence placed on the road shoulder, with horizontal slats angled to force the wind down onto the road surface, significantly increased snow accumulation, which piled up in the roadside upwind. Such a snowfence used on a narrow road with relatively constant wind direction presumably would be removed during non-snow seasons, since it funnels wind onto the road and blocks the roadside view.

Increasing the height of the roadbed or road surface above the surrounding land tends to reduce snow accumulation on the road.[270] For instance, many rural roads in the Canadian Great Plains with raised roadbeds and no bordering windbreaks do not require snowplowing, because snow simply blows off the road surface. However, if a windbreak is present and the difference between roadbed height and windbreak height is small, the effectiveness of the snowbreak is reduced. A smooth soil berm, used instead of a vegetation or artificial windbreak, would function as a streamlined object, analogous to a multirow dense-canopy windbreak with trees in the center and shrubs on the outside.[110]

Snowbreaks are designed not only to minimize snow accumulation on roads but also to improve driver visibility by reducing *blowing snow*.[888, 270] Reducing blowing snow is important in Hokkaido, where winter road closures result from snow accumulation, yet about half of the winter accidents are related to reduced visibility, especially due to blowing snow. Trees several times the height of an artificial snowfence and located upwind relatively close to roads are usually effective in reducing blowing snow across roadways. Tall, solid fences upwind of a vegetation windbreak also help improve visibility over the road. One winter road study comparing areas with and without snowbreaks found that the amount of time when driver visibility was slightly limited

(<1 km [0.62 mi]) decreased from 5% to 2% of the time when driving next to snowbreaks.[270] However, during worse driving conditions, driver visibility was below 300 m (984 ft) 96% of the time without snowbreaks but only 58% of the time with adjacent snowbreaks. Thus the snowbreaks significantly improved visibility by reducing blowing snow, which causes the most dangerous driving conditions.

Vehicle Disturbance and Traffic Noise

Which is worse, a passing truck that covers us with soot and dust or trucks passing in the night, producing a continuous racket in the neighborhood? In the first case, we rub our eyes and dust off our clothes, but in the second case, we lose sleep night after night. The first is an *infrequent disturbance,* a sporadic event causing a usually negative response.[724, 302, 547] The second is a *chronic* or *repeated disturbance.* A rare disturbance takes us by surprise. If it is somewhat frequent, we adjust our behavior and move away from the road when a truck approaches. If the disturbance is repeated frequently or continuously, we may move away from the noisy highway in order to sleep soundly. Alternatively, we may become *habituated* (accustomed) to the noise so that it no longer disturbs us. Then we can sleep soundly next to the highway, maybe even dreaming or snoring as if we could drown out the trucks passing in the night.

Vertebrate wildlife species—mammals, birds, reptiles, and amphibians—are in many ways like us. So, not surprisingly, many wildlife responses parallel our responses to disturbance.[105] Animals learn, adjust behavior, avoid places, and become habituated, and over generations they adapt to disturbance.

Energy is transmitted through air, along with materials in particle, aerosol, and gaseous forms. Three types of energy, heat, light, and sound, are sometimes minor and sometimes major disturbances. *Heat energy* generated by vehicles may have negligible ecological effects, though this area has received little scrutiny. However, heat from road surfaces certainly has numerous effects on animals as well as plants on and around roads. At cool times, reptiles and other animals may be attracted to warm road surfaces. Unfortunately, a road marked with squished snakes and turtles commonly results. Heat from the road also tends to desiccate plants on the road shoulder and may blow into adjoining woods, altering seedling survival and plant communities.

Animals also react to the dazzling lights of passing vehicles. *Light energy* transmission is evident in seeing a vehicle's headlights or taillights at night or an onrushing stop sign readable by reflected daylight. Light energy moves virtually instantaneously through air, irrespective of wind speed. Burdening the air with particles or with aerosols such as snow or fog reduces light transmission and, hence, visibility. Although roadside lighting may affect nocturnal frogs,[119] little is known about the ecological effects of vehicle headlights.

However, visual disturbance, in the sense of responding simply to the sight of a vehicle or vehicles, is likely to be important to wildlife.[547]

Sound and noise, which are particularly important to wildlife, are introduced in the next section. That leads to a section on wildlife responses to busy roads and to lightly traveled roads. Finally, some key factors affecting noise, and hence important for noise reduction efforts, are highlighted.

Noise and Animals

Noise in road ecology results primarily from the movement of vehicles. The subject is complex due to the numerous types, intensities, and durations of sound, the ways of measuring and modeling it, and the many types of responses by animals to sound.[767,533,105,534,405] However, some key concepts can simplify the subject. *Noise* is usually described as unwanted, or annoying, sound and therefore is a subjective human-response concept rather than simply a loudness concept. A birder may consider a distant train to be making noise but not a louder group of crows chasing a hawk. A train buff may consider the raucous birds to be noisy while the passing train makes a pleasing sound. It is hard for humans to guess what an animal considers to be unwanted sound—that is, what noise is to the animal. Therefore, scientists study ecological patterns or responses that correlate with measurable levels of sound. If a negative (unwanted) ecological pattern (e.g., a degraded animal community or the existence of a threat to a population) correlates with some attribute of sound, we may call the sound "noise."

Many animals detect and depend on sound to communicate, navigate, avoid dangers, and find food.[105] Human-made noise, however, alters the behavior of animals or interferes with their normal functioning. This noise can harm the health of animals as well as alter reproduction, survivorship, habitat use, distribution, abundance, or genetic composition.[105] Noise in the sense of a disturbance can also *harass* an animal, threatening it or causing it discomfort.

Sounds differ in *pitch* (frequency), duration, and loudness or level. Mammals as a whole hear in the (logarithmic) frequency bandwidth from below 10 hertz (Hz) to over 150 000 Hz.[259,105] Birds hear well in a somewhat narrower range, from 100 Hz to about 10 000 Hz, though some species detect very low frequency noise.[259,506] Reptiles hear only between about 50 and 2000 Hz, and snakes and turtles hear quite poorly. Overall, amphibians detect a still narrower bandwidth, between 100 and 2000 Hz. Naturally, most individual species have narrower hearing ranges than does the group as a whole, and some species extend above or below the general ranges.

Vibrations induced in an animal or substrate by low-frequency noise are also readily detectable by some animals, especially birds and reptiles.[839,105] Sensitivity to vibration is an important way to detect approaching prey or predators. This form of detection is probably especially important for reptiles in view of their poor hearing. Little is known about traffic-caused vibrations and wildlife. One study found that vibrations associated with traffic in Britain, perhaps directly transmitted from vehicles through the soil, may increase the emergence of earthworms from soil.[901] An abundance of predators such as crows *(Corvus)* was observed feeding on the earthworms. A paucity of snakes near roads[801] might be related to traffic-caused vibrations. Traffic also shakes plants, which affects their physiology.[547]

Sound level, or *loudness,* measured in decibels (dB[A]) usually as "continuous sound level"[533] includes the following key levels[105] (Figure 10.6):

- –20 dB(A) is the sound just audible to a nocturnal carnivore.
- 0 dB(A) is the sound just audible to humans.
- 20 dB(A) characterizes a quiet desert.
- 40 dB(A) describes nighttime at home.
- 60 dB(A) characterizes normal human speech.
- 80 dB(A) is the safe human limit for continuous noise.
- 100 dB(A) is where the "startle reflex" stops habituating.
- 120 dB(A) is the threshold of auditory pain in humans.

For comparison, bird population decreases in a Dutch study occurred between about 70 dB(A) next to highways and about 40 dB(A) hundreds of meters

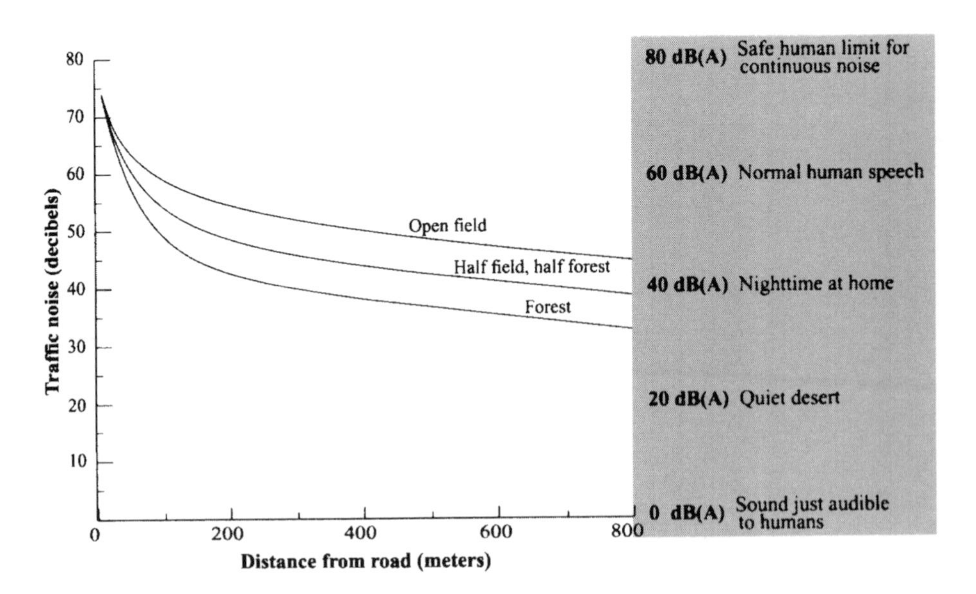

Figure 10.6. Traffic noise through forest and across open field. Based on a busy Dutch highway with 50 000 veh/d and traffic speed of 120 km/h (75 mph) (R. Reijnen et al. 1995). Key decibel levels from Bowles (1997).

from a highway, beyond which significant avian responses were not detected.[770,771]

Attraction, tolerance, and aversion are three general responses to human-caused noise.[105] The first two normally lead to an animal being in close proximity to roads and vehicles, inherently a dangerous milieu for most wild animals. The third response, aversion, is of primary interest here. Many of the aversion responses for wild animals are the same as for humans, though differences exist. The list of possible responses is long: (1) annoyance, (2) hearing loss, (3) "speech" and sleep interference, (4) possible stress-related illness, (5) humans perceived as predators, (6) many population changes, (7) genetic change over generations, and (8) numerous other behavioral responses. *Habituation,* the process of getting used to common repeated noise and activity, is of special importance relative to roads, traffic, and wildlife.

The response of an individual animal, as well as of a whole population or community, may correlate with chronic noise, such as busy traffic on a road. Or it may respond to a single noise, such as a brief blast of a siren or the approach of an off-road vehicle. Thus the following section focuses on the ecological responses of wildlife near busy roads and low-traffic roads, respectively.

Wildlife, Busy Roads, and Low-Traffic Roads

Many native wildlife species are less common or absent near roads, which effectively results in a *road-avoidance zone.*[310] Ignoring cases where habitat conditions, including food and predators, are less suitable due to soil and microclimatic conditions, the potential causes of an avoidance zone include visual disturbance, vehicular pollution, road-kills, and traffic noise.[767,310]

It is possible to sort out likely causes. Researchers studying songbirds near busy highways found no effect of differences in visual disturbance level on bird distribution when traffic noise level was constant.[771] But when visual disturbance was kept constant in the study, bird distribution patterns correlated with traffic noise. The researchers noted that visual disturbance and vehicular pollutants extended outward only a short distance from the highway, whereas both traffic noise and reduced bird densities extended outward much farther. Road-kills of birds, as well as predators and scavengers moving along roads feeding on roadside animals and road-kills, may affect overall avian density negligibly or for only a short distance from a road, and hence probably rarely cause an avoidance zone.[660]

Studies of the ecological effects of highways on bird communities in The Netherlands point to an important pattern. In both woodlands and grasslands (pastureland) adjacent to roads, about 60% of the bird species present had lower densities near highways.[770,771] In the affected zone, both total bird density and species richness were approximately one-third lower, and species progressively disappeared with proximity to the road. *Effect-distance,* the distance from a road at which a significant effect is detected (in this case, a population-density decrease), was greatest for birds in grasslands. The effect-distance was intermediate for birds in deciduous woods and least for birds in evergreen coniferous woods.

Many songbird populations appear to be inhibited by remarkably low noise levels.[767, 770, 771] The noise level at which the total population density of all woodland birds began to decline averaged 42 dB(A), compared with an average of 48 dB(A) for grassland birds. The most sensitive woodland species (cuckoo) showed a decline in density at 35 dB(A), and the most sensitive grassland bird (black-tailed godwit, *Limosa limosa*) responded at 43 dB(A). These noise levels affecting songbirds are similar to those in a typical library reading room or during nighttime in one's home.

Effect-distances in the Dutch studies (with an average traffic speed of 120 km/h, or 75 mph) were also sensitive to traffic volume.[771, 767] The effect-distances were 305 m (0.2 mi) in woodland by roads with a traffic volume of 10 000 veh/d, and 810 m (0.5 mi) in woodland by highways with 50 000 veh/d. In grassland, effect-distances were even greater: 365 m (0.2 mi) by roads with 10 000 veh/d, and 930 m (0.6 mi) by highways with 50 000 veh/d. In addition, most grassland bird species showed population decreases near smaller roads with 5000 veh/d or less. The effect-distances for both woodland and grassland birds were estimated to increase steadily with traffic volume from 3000 to 140 000 veh/d and with traffic speed up to 120 km/h.[767] These road effects were more severe in years when overall bird population sizes were low.[769]

A Massachusetts (USA) study on grassland birds in open patches near roads in a suburban landscape supported and extended these overall results.[315, 322] For a traffic volume of 30 000 veh/d or more (a busy multilane highway, with average traffic speed of about 80–85 km/h [50–55 mph]), both bird presence and regular breeding were significantly reduced up to 1200 m (0.75 mi) from a road (Figure 10.7). Within this avoidance zone, no site had regular breeding birds, and the presence of the grassland birds during breeding season was rare. For the relatively heavy traffic of 15 000 to 30 000 veh/d (two-lane highway), both bird presence and regular breeding were reduced outward for 700 m (0.4 mi). For moderate traffic of 8000 to 15 000 veh/d (a through street), there was no effect on bird presence, though regular breeding was reduced for 400 m (0.25 mi) from a road. Finally, a relatively light traffic volume of 3000 to 8000 veh/d (local collector street) had no significant effect on grassland bird distribution.

Deer tend to have lower densities near roads with traffic in the Rocky Mountains, suggesting an avoidance zone of 100–300 m (328–984 ft) from

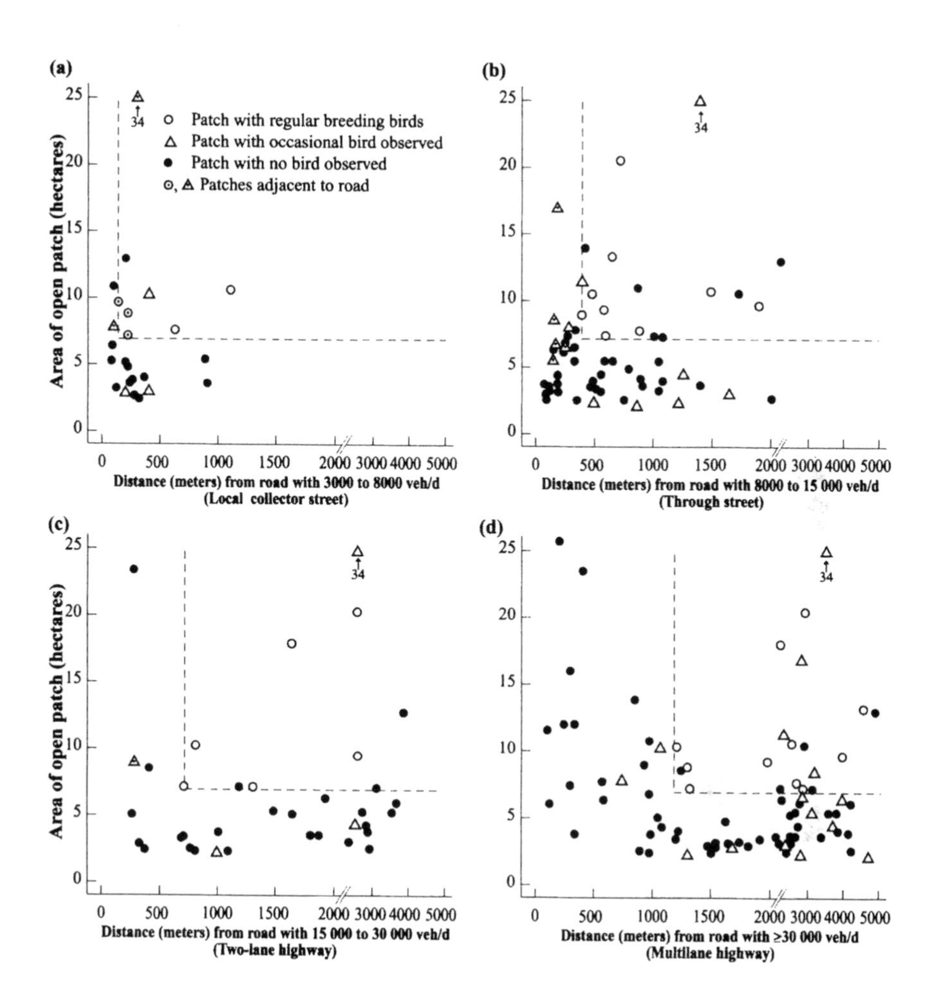

Figure 10.7. Birds relative to distance from roads with different traffic volumes. Regular breeding and the presence of grassland birds based on five years of data for all 84 potentially suitable open patches in a 240-km² (93-mi²) outer-suburban landscape of forest and built area west of Boston. Distance measured to center of opening. All regular breeding occurred in patches above and to the right of the dashed lines. From Forman et al. (2002).

roads.[223, 797, 844] Elk *(Cervus elaphus)* may have a road-avoidance zone several hundred meters wide,[797, 238, 547] depending on the number of vehicles passing per day. Female caribou *(Rangifer tarandus)* in Alaska apparently tend to avoid remote roads.[1027] Snakes in East Texas were found to be lower in density for at least 650 m (0.4 mi) from forestry roads, for which soil vibration and road-kill may be important reasons.[105, 801] Further evidence of a road-avoidance zone exists for coyotes *(Canis latrans),*[346] small mammals,[347] birds,[285,371] and amphibians.[255] Such road-avoidance zones, extending outward tens or hundreds of meters from a road, generally exhibit lower population densities compared with control sites. Considering the density of roads, plus the total area of avoidance zones, the ecological impact of road avoidance probably exceeds the impact of either road-kills or habitat loss (Chapter 5) in road corridors.

Despite the paucity of studies on the ecological effects of traffic volume, some preliminary trends are evident from the existing evidence. Effect-distance apparently increases with traffic volume for all species or groups studied. Grassland birds show the greatest effect-distances yet recorded for birds (see Figure 11.6 in Chapter 11). Yet, at relatively low traffic volumes, grassland birds seem to be unaffected, while many other species studied are affected. This highlights the importance of analyzing wildlife near lightly traveled roads.

In a few cases, reduced population densities have been found near roads with low traffic volume. A grassland bird (horned lark, *Eremophila alpestris*) had lower densities in farm fields within 200 m (656 ft) of country roads (or "county highways" with perhaps 300 to 3000 veh/d) in Illinois, USA.[162] On a two-lane paved road with perhaps 500 to 750 veh/d, the mortality of Florida scrub-jay fledglings *(Aphelocoma coerulescens)* from territories that encompassed or were adjacent to a road was significantly higher than fledgling mortality from territories not adjacent to the road.[660] The affected scrub-jay territories extended outward on average about 250 m from the road. Elk are reported to leave a 500- to 1000-m (0.3- to 0.6-mi) buffer zone around logging roads with a few vehicles (especially trucks) passing per day.[238,547] Spotted owls *(Strix occidentalis caurina)* in the U.S. Pacific Northwest with nests within about 400 m (0.25 mi) of a logging road with perhaps 5 to 50 veh/d had higher levels of stress hormones than did owls farther from the road.[1004]

At some low level of traffic, animals may be mainly responding to the noise of an individual vehicle approaching or passing, rather than to the relatively continuous noise of a stream of passing vehicles. This raises the complex question of what an individual animal does in response to a single disturbance event.[547,105] For example, a deer hearing a snowmobile approaching 2.5 km (1.5 mi) distant may move away from a trail, whereas the deer hearing a hiker approaching may not move away from the trail until the hiker is some 250 m (820 ft) away.[105] In both cases, after the disturbance has passed, the deer may move back toward its original location near the trail. Alternatively, faced with an approaching or passing vehicle, an animal may remain in place due to habituation, being "petrified" with fright, or responding to a greater "need," such as protecting eggs or young. The response of an individual animal to the movement of different types of vehicles remains an important research frontier.

Some Key Factors Affecting Noise

Unlike materials in the air, noise carries rapidly through still air. Also, whereas windborne materials commonly move a mile in several minutes, noise moves a mile through air in only 5 seconds. Noise even carries upwind, though more energy is carried downwind than upwind. The trajectory of noise tends to be straight, yet because wind can bend, a noise trajectory can also bend. Noise also changes direction by reflecting off surfaces, such as concrete walls, roadbanks, and dense evergreens. Noise reflection is greater if the object encountered has low porosity and if its surface is smooth. Porous objects and rough convoluted surfaces absorb more energy and reflect less. Noise, like materials, also travels through porous objects. Indeed, traffic noise carries a relatively long distance through vegetation, meaning that wide vegetation bands are required to greatly decrease noise. An echo reminds us that vertical surfaces, such as rock faces and noise walls, may reflect much of the noise back to the source location. Angled roadbanks and soil berms reflect much of the noise upward, where it is largely dissipated in the atmosphere. However, especially under low clouds or conditions of high relative humidity, the atmosphere may reflect considerable noise back downward. This is why areas beyond roadside soil berms and noise walls receive traffic noise at times.

The greatest noise increase related to traffic volume occurs when traffic increases up to 1000 veh/d and noise rises to over 50 dB(A)[241] (Figure 10.8). However, an increase in traffic volume up to about 5000 veh/d continues to cause a steep increase in noise level. At 10 000 veh/d, the noise level is nearly 70 dB(A). Above about 10 000 veh/d, noise level increases gradually and almost linearly. For example, a tripling of traffic from 12 000 to 36 000 veh/d increases noise only 2 to 3 dB(A), almost equal to the 1 to 2 dB(A) minimal difference that people can detect.[533] A tenfold traffic increase from 5000 veh/d (typically a through street or lightly used two-lane highway) to 50 000 veh/d (a relatively busy multilane highway) increases noise level about 10 dB(A).

Noise level decreases curvilinearly with distance from a road. Thus, next to a multilane highway with 50 000 veh/d, noise level drops sharply in the first 100 m (328 ft) outward[771] (Figure 10.6). At greater distance, noise decreases less rapidly and almost linearly. In open land, noise drops from about 75 to 45 dB(A) at 800 m (0.5 mi) from the highway, whereas in woodland, noise drops from 75 to 35 dB(A) at 800 m. A landscape with a mixture of forest and field has an intermediate decrease in noise. In the first 100 m from the highway, the difference in noise level between field and forest is relatively small. Beyond about 200 m (656 ft) out, the difference is larger but remains relatively constant with increasing distance from the highway.

Traffic volume and distance from the road are the primary factors affecting traffic noise levels and hence are at the core of noise models for evaluating the

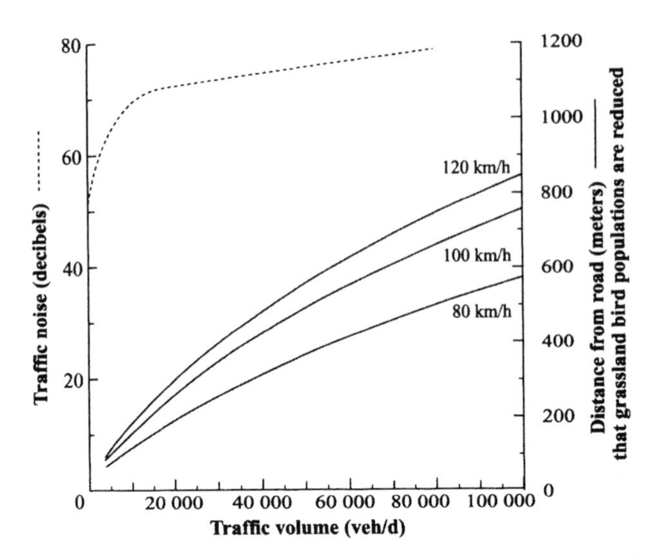

Figure 10.8. Noise and effect-distance for birds as related to traffic volume and speed. Traffic-noise pattern from Ellenberg et al. (1981). Effect-distance on grassland birds at various traffic speeds based on Dutch studies (M. Reijnen et al. 1995).

potential effects of highways on communities, schools, hospitals, and so forth.[264, 533, 534] Nevertheless, many other factors affect traffic noise. Microclimate was mentioned above, and standard traffic-noise measurements include windspeed and direction, temperature, and relative humidity.[533] Higher traffic speed increases noise levels, though the rate of increase moderates above about 100 km/h (62 mph)[767] (Figure 10.8). For example, by a highway with 50 000 veh/d and traffic flows averaging 80 km/h (50 mph), grassland birds in open areas are significantly affected outward to approximately 300 m (0.2 mi). However, the effect on birds is predicted to extend outward to 500 m (0.3 mi) when traffic moves faster at 120 km/h (75 mph).

Medium trucks (six or more tires on two axles) and *heavy trucks* (three or more axles) in traffic flows also significantly affect noise. A heavy truck passing produces approximately 10 dB(A) more noise than a passing automobile.[533] In the USA, when the percentage of medium and heavy trucks reaches perhaps 5% to 10% on a busy suburban highway, noise levels in the surrounding neighborhoods apparently are high enough to bother people, sometimes resulting in pressures to decrease noise propagation with soil berms, noise walls, or other approaches. Noise at these levels also may harass wildlife.

Road surface texture, as well as roughness and porosity, has a major effect on noise levels.[512] The orientation, spacing, and depth of *tining* (essentially a micro-raking) of a road surface can be especially important for noise generation.

Road surfaces are relatively easy to control and adjust. Shortly after the Dutch studies on birds near highways were completed,[294,770,771] the Dutch Ministry of Transport began laying down a quieter road surface in its program of resurfacing highways with recycled pavement material (P. Opdam, personal communication).

Other engineering and design factors significantly affect noise levels. Tire (tyre) surfaces vary from quiet to noisy. The type and design of motors in vehicles affect noise. Mufflers may muffle almost all sound or emit the proverbial racket. Vehicle aerodynamics strongly affects the sound level of a passing vehicle. These factors are all amenable to engineering improvements to reduce noise.[264]

The road-avoidance zone is evident from correlations of wildlife density with distance from roads and is interpreted as mainly due to traffic noise. However, experiments testing this relationship apparently have not been done. Many approaches are possible, including decreasing or increasing traffic volume, altering traffic speed, reducing noise levels near busy highways with berms and other barriers, and making the engineering changes just mentioned. In addition, noise experiments in remote areas could be created. A large portion of the land containing enormous numbers of people and wildlife is presently subject to elevated traffic noise levels[322] and would benefit from a less noisy environment.

Atmospheric Effects at Local, Regional, and Global Scales

Vehicles emit a rich assortment of gases, aerosols, and particles. Some, such as carbon monoxide, decompose in a matter of seconds, while others, including carbon dioxide, persist for centuries. Some are pollutants, directly harming humans, animals, or plants, while others alter climatic patterns, causing global impacts to natural ecosystems.

Roads and roadsides also are the source for many airborne pollutants. Most are in particulate form, such as sand and clay, though some appear in aerosol form, such as herbicides. Most road-related pollutants produce local or regional ecological effects.

The increasing accumulation of these diverse human additions to the atmosphere has major implications from global climate change to human health and agriculture.[428] Atmospheric effects are described in several major reports and are active subjects of current research.[673,674,675,244,428,606,837,174] Therefore, the broad topic is but briefly introduced here, as a springboard to exploring the rich literature available. Our discussion is divided into four areas: local pollutants, regional pollutants, global pollutants, and global climate change.

Local Pollutants

Local pollutants, which exhibit effects at the local scale of adjacent ecosystems and surrounding neighborhoods, are surprisingly diverse. Solid objects, such as engines, tires, and body parts of vehicles, plus human-discarded objects are scattered along roadsides, where they tend to persist and accumulate. In contrast, many airborne materials that cause local effects persist only for seconds to hours.[428] Airborne *particulate matter* (or simply *particles*) includes the soil components, sand, silt, clay, and organic matter from roadsides and road surfaces as well as particles from the degradation of paved road surfaces.[902,277,651,74,875] (Figure 10.9). Also, particles from fuel combustion and brake linings of vehicles are distributed over local areas.

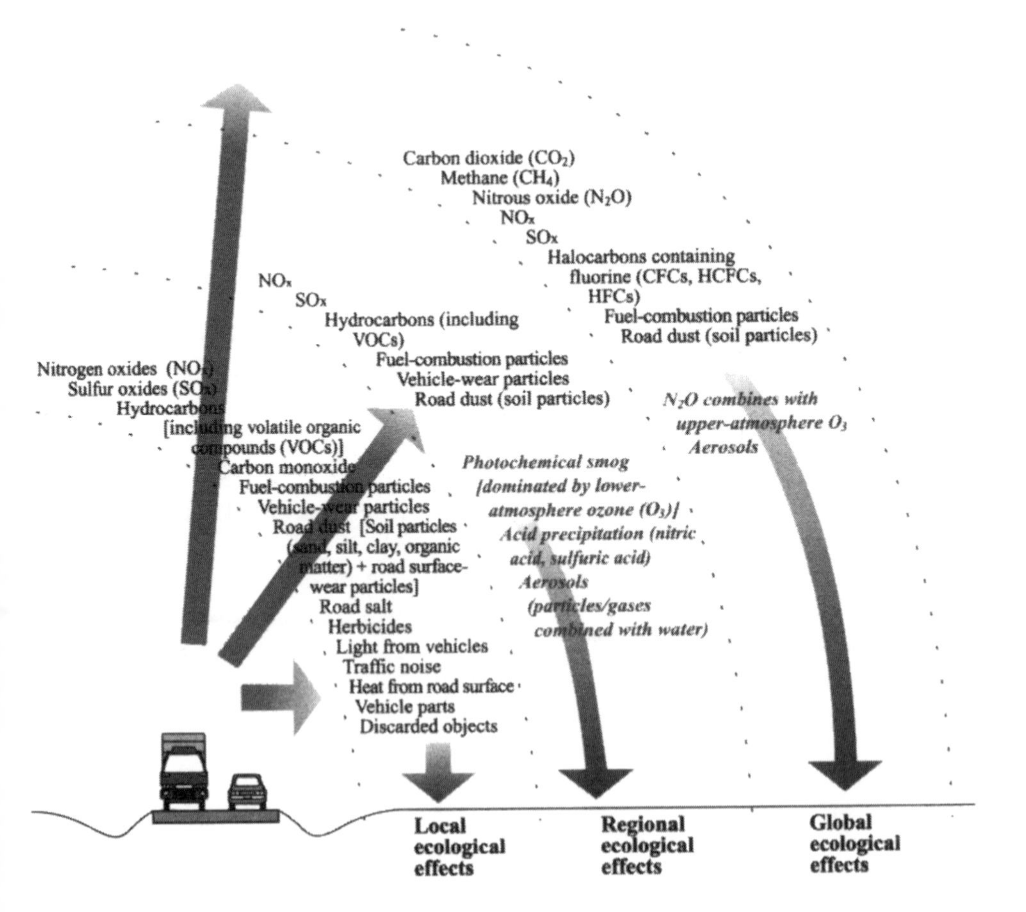

Figure 10.9. Road- and vehicle-related pollutants at local, regional, and global scales. The last pollutants (indicated by italics) in the regional and global lists are formed in the atmosphere.

Particulate matter is graded in terms of diameter as PM-10, PM-2.5, and PM-1 (less than 10, 2.5, and 1 micron particles, respectively). Finer particles are considered especially hazardous to human health.[675] Particles may carry toxic or carcinogenic compounds[277] and are most dangerous to human health in high concentrations near sources such as roads. Consequently, although research seems to be scarce, particulate matter is probably also a hazard to vertebrate wildlife near roads.

Several gases result from fuel combustion in vehicles. These include carbon monoxide (CO), hydrocarbons (organic compounds composed of only C and H), and oxides of nitrogen and sulfur (NOx and SOx).[674] Carbon monoxide probably has the most localized effect. It can asphyxiate humans and animals at high concentrations, but in circulating air CO dissipates in seconds. The other gases—NOx, SOx, and hydrocarbons, which contain VOCs (volatile organic compounds) and are emitted from fuel evaporation as well as combustion—also damage human health and, hence, probably affect vertebrate animals near roads. Catalytic converters on vehicle exhaust systems have reduced the amounts of sulfur dioxide and hydrocarbons in exhaust.

NOx emissions from vehicles on busy highways may also produce effects on adjoining ecosystems and plant communities (Figure 10.9). These emissions appear to produce a nutrient enrichment (eutrophication) effect. For instance, soil nitrogen levels were elevated to a distance of 200 m (656 ft) from busy motorways in England.[23] In areas of acid heathland, this nitrogen (nitrate) enrichment was sufficient to alter the plant community by favoring less common species at the expense of the predominant native heathland plants. High nitrogen levels and plant species favored by abundant nitrogen are present in the edges of some woods in agricultural areas.[240,302,1010]

Many other pollutants associated with roads and vehicles produce local ecological effects. Herbicides used for roadside maintenance, as well as roadsalt used to eliminate ice, are carried by wind into adjacent and nearby areas. Airborne heavy metals are mainly distributed locally, where they accumulate in soil, plants, and animals near roads. Additional local effects result from traffic noise, heat from road surfaces, and the visual stimulus of moving vehicles and headlights, as discussed earlier in this chapter. In short, airborne particles, aerosols, gases, and energy represent a unidirectional linkage between road systems and ecological effects on the local surrounding land.[573]

Regional Pollutants

Vehicles produce a range of particles and gases that are carried away from the immediate site of generation and persist for hours to years.[428] The most persistent materials find their way to the *stratosphere* (upper atmosphere), while others accumulate in the *troposphere* (lower atmosphere) and produce regional

Figure 10.10. Regional air pollution, road lights, and streams of headlights illustrating widespread but ecologically poorly understood factors. Interstate Highways 20 and 820, Fort Worth, Texas. Photo courtesy of U.S. FHWA.

effects (Figure 10.10). Dust from roads and roadsides adds more particles and chemicals to the vehicle-caused mix (Figure 10.9). Furthermore, while being carried through the air, the mixture undergoes chemical and physical transformations that produce additional toxic chemicals. The types and amounts of these diverse human-caused additions to the atmosphere at the regional scale are reasonably well understood, though portions of the picture remain shrouded.[674, 428, 174]

In contrast, the ecological effects of these regional pollutants are not well understood, though the effects are known to be widespread and important.[47] Some harm humans, while others damage plants and natural ecosystems.

For instance, Lake Tahoe in California (Chapter 9) has received an enormous amount of nitrogen from atmospheric deposition over the past century. Much of this has originated from vehicular traffic. Of equal importance to the reduced lake transparency, however, is road dust. Roadside erosion adds soil to the sand and salt mixture used for de-icing roads in the Lake Tahoe basin during winter. As vehicles grind this material into ever-finer particles, wind and passing vehicles suspend the particles, which contain phosphorus, iron, and a variety of pollutants, into the atmosphere. Contributed via both wet and dry deposition to the lake surface, these fine particles add nutrients to Lake Tahoe's water column (see Figure 9.3). The particles also absorb and scatter light, both accelerating eutrophication and decreasing transparency. Measurements indicate that most of these particles are less than 5 micrometers in diameter. Since they settle extremely slowly, these fine particles remain suspended in the lake's water column for many years, contributing significantly to the progressive loss of transparency in this uniquely clear subalpine lake.[363]

Of the total particulate matter in the atmosphere, a relatively small amount comes from vehicles. Particles from vehicles mainly originate from fuel combustion and wearing of brake linings and include nitrate and sulfate particles[674, 675, 542] (Figure 10.9). These emitted particles tend to be very small—less than 1 micrometer in diameter. Such PM-1 particles are frequently considered the most dangerous to human health and presumably also to wildlife in a region.

Three gaseous pollutants cause the greatest concern: nitrogen oxides, sulfur oxides, and hydrocarbons. About one-third of all human-caused NOx and hydrocarbons come from fossil fuel combustion in vehicles. Vehicles also emit SOx, mostly from the combustion of diesel fuel in trucks, though the proportion of the total produced by society is lower than for the other two gases. Exhaust from diesel vehicles is particularly rich in particulates and aerosols that accumulate in the atmosphere.

These atmospheric nitrates and sulfates in the presence of moisture produce nitric acid and sulfuric acid, major components of acid precipitation and deposition at the regional scale[826, 38, 47, 551] (Figure 10.9). *Acid precipitation* has widespread damaging effects on many freshwater ecosystems, mountain ecosystems, and forest trees. Damage to forests, and even lakes, in the Los Angeles region is a widely cited example of acid rain effects.

Nitrates also tend to cause nitrogen enrichment or eutrophication of aquatic and terrestrial ecosystems.[978, 453, 5, 602, 363] Such enrichment commonly alters natural communities by favoring certain nutrient-limited plants, which then outcompete the predominant species. A significant increase in NOx seems to alter or disrupt the overall nitrogen cycle[979] and may accelerate the invasion of aquatic weeds, such as the Eurasian watermilfoil *(Myriophyllum spicatum)*.[996]

Nitrogen oxides have a third major regional effect. They combine with hydrocarbons (including VOCs) in the presence of sunlight to form *photochemical smog*. The main constituent of smog is ozone, which in the lower atmosphere can damage human health.[673, 675] Ozone also damages various agricultural crops and forestry trees[408, 857, 1045] and doubtless degrades certain plant communities and natural ecosystems. Heavy metals may accumulate at the regional scale and also be transported between regions.

In short, road systems produce a range of atmospheric pollutants at the regional scale. This pollutant mixture, in turn, causes highly diverse and important effects on natural ecosystems across the region.

Global Pollutants

The global effects of road systems largely result from gases that often persist for decades to centuries in the stratosphere.[674] Global climate change associated with certain of these gases is addressed in the following section. Here we introduce the pollutants that accumulate in the upper atmosphere and produce ecological effects on the land (Figure 10.9).[428,606]

Sulfate and nitrate aerosols in the upper atmosphere scatter light, blocking the penetration of some solar radiation, and hence tend to lessen somewhat the temperature rise associated with global climate change. Other aerosols in the upper atmosphere originate from nonvehicular fossil fuel burning, plant biomass burning, and mineral dust from agriculture, roads, and other sources. Carbon dioxide, methane (CH_4), and nitrous oxide (N_2O) are the major greenhouse gases. In addition to originating from biological sources, nitrous oxide may also be a by-product of catalytic converters in vehicle exhaust systems.

Nitrous oxide, which has several effects at the regional scale, is lost in the stratosphere by reacting with ozone. Ozone in the stratosphere is the essential filter against incoming ultraviolet (UV) radiation, which tends to cause mutations and skin cancer and can be lethal. Reaction with N_2O is the natural mechanism of loss of (or a sink for) ozone. Thus increasing N_2O concentration has led to some depletion of the stratospheric ozone layer. Consequently, high UV radiation levels at the earth's surface, particularly in high southern latitudes, have been increasingly reported in recent years to be associated with the *ozone hole* (a stratosphere area with reduced O_3 levels).

A series of *halocarbons* (fluorine-containing chemicals)—with the mouthful names *chlorofluorocarbons, hydrochlorofluorocarbons,* and *hydrofluorocarbons* (or simply CFCs, HCFCs, and HFCs)—also damages the critical ozone layer (Figure 10.9). These chemicals are released from air conditioning systems, including those that have been used extensively in vehicles over the years.

Thus the upper atmosphere around the globe also accumulates human-caused chemicals, some of which are directly related to vehicles. They mainly alter global conditions, such as temperature, light penetration, and the ozone shield, which in turn affects animals, plants, and entire ecosystems. Greenhouse gases also accumulate in the upper atmosphere, as described in the following section.

Global Climate Change Effects

The globe's climate is undergoing major change.[428,606,174] During the twentieth century, the earth's average surface temperature increased by over 0.6 degree C (1 degree F), with 17 of the 18 warmest years occurring between 1980 and 2001. Global sea levels rose 15 to 20 cm (6 to 8 in), due to the added water from melting ice and snow and because warming causes water in the ocean to expand. During the century, annual precipitation increased by about 10% in the middle and higher latitudes. Curiously, recent evidence exists of some cooling occurring in the Antarctic, where many glaciers are not retreating.

A principal cause of this change is the sharp increase in greenhouse gases being produced by humans. Carbon dioxide (CO_2) is the primary greenhouse gas inducing climate change, with some additional effects contributed by methane, nitrous oxide, and other industrial gases (Figure 10.9). Since the late 1800s, the concentration of CO_2 has increased about 30%, to the highest level in 420 000 years.[595] The current rate of CO_2 increase is the highest in 20 000 years. About a quarter of this increase is due to land-use change, primarily tropical deforestation, and about three-quarters has resulted from burning of fossil fuels.[1006] The *Kyoto accord* (an international agreement to reduce greenhouse gas emissions) is an important effort to control emissions but has yet to be adopted by all of the major consumers of fossil fuel.

Motor vehicles are a major source of this CO_2, accounting for about 25% of the total CO_2 produced in the USA, and a somewhat smaller proportion worldwide. Motor vehicles are also responsible directly or indirectly for other greenhouse gases, including ozone, N_2O, and methane. Roads and vehicles are becoming so pervasive on the earth that they are affecting global climate. For example, North America alone has a quarter billion vehicles using nearly 8 million km (5 million mi) of public roads. Many of these vehicles still have relatively low fuel-burning efficiency. Typical two-cycle engines, for instance, waste into the atmosphere or water about 20% of their fuel and oil unburned. *Fuel cell technology* is still years away from any significant use, and hybrid higher-efficiency engines are appearing in only relatively small numbers on the roads (Chapter 3).

Almost all global-climate models predict a continuing temperature rise through the twenty-first century.[606, 428] The most likely types of temperature change through this upcoming period are: (1) higher minimum temperatures, (2) fewer frost days, (3) higher maximum temperature, (4) fewer cold spells (or waves), (5) more hot summer days, and (6) an increase in heat index.[235, 428] Based on the climate-change models, the likely changes in precipitation are: (1) more heavy, multiday precipitation events, (2) more heavy, one-day precipitation events, (3) more wet spells, and (4) more common El Niño–like conditions. Alteration of rainfall patterns will subject some areas to frequent droughts and yet others to increased flooding. Such major widespread alterations in temperature and precipitation are well described by the term *global climate change* and are likely to increase water shortages in some regions of the globe.

These temperature, precipitation, and other climatic shifts will produce widespread ecological changes. Likely effects include: (1) thawing of *permafrost* ("permanently" frozen Arctic soil) with increased methane release, (2) reduced ice on rivers and lakes, (3) rising sea level, (4) lengthening of growing seasons in mid- to high latitudes, (5) poleward and altitudinal shifts of plant and animal species ranges, and (6) earlier flowering of trees, emergence of insects, and egg-laying in birds.[606] These patterns have already been recorded in many aquatic and terrestrial environments. Natural systems with limited adaptive capacity are also at great risk.[606] These systems include glaciers, coral reefs and atolls, mangroves, boreal and tropical forest, polar and alpine ecosystems, prairie wetlands, and remnant native grasslands. Some species will increase in abundance or range. However, today's vulnerable species will be at greater risk of extinction. This points to a loss of biodiversity. The geographical extent of ecological damage or loss, and the number of systems affected, will increase with both the magnitude and the rate of climate change.

Plant species composition and dominance at a site will change, but ecosystems as discrete units are not expected to migrate geographically.[606] Wildlife will change in distribution, population sizes, population density, and behavior—in direct response to climate change and indirectly through changes in vegetation.[957] Freshwater fish will have more poleward distributions, with a net habitat loss for cool-water fish and gain for warm-water fish. This may alter salmonid populations as warm-water predators, such as smallmouth bass *(Micropterus dolomieu),* invade higher-latitude waters.[957] In addition, several major effects will occur in coastal zones associated with sea-level rise and in marine ecosystems.

These manifold ecological effects are well understood conceptually and have been predicted with considerable confidence at the global scale. However, climate changes at the scale of individual regions or landscapes are much less clear, and thus predictions of ecological changes at the regional scale are less certain.[428,606,174] In addition, the ecological effects just described are broad patterns based on considerable existing research. Much less understood are the more specific patterns, such as which species will go extinct, what a specific river system will look like, and what the geographical ranges of, for instance, bass, trout, and grizzlies *(Micropterus, Salmo, Ursus)* will be.

Other major changes on earth, such as human land use and habitat loss and fragmentation, will interact with climate change to produce additional ecological effects.[606] Indeed, numerous climate-change effects on human systems can be expected to spill over and alter natural ecological systems. Should water shortages and flooding occur on a repeating major scale, extensive food-producing areas of the world could be dramatically affected.

To the extent that greenhouse gas concentrations continue to increase, these effects will grow. Unless major changes are made in the production and use of energy and other materials, the rate of release of CO_2 and other greenhouse gases is expected to increase. By 2100, the human-caused greenhouse gas level may be two to three times the concentration present in the "preindustrial" mid-1800s.[428]

Conclusion

In four important ways, road systems play a central role in the landscape ecological impacts described above. First, road surfaces and roadsides are a major source of fugitive dust and associated chemical pollutants (Figure 10.4). Lake Tahoe, for example, near the crest of the Sierra Nevada between California and Nevada, has accumulated so much nitrogen nutrients from vehicle exhaust over the past 40 years that it has changed from a formerly nitrogen-deficient lake to one where algal growth is now limited by available phosphorus.

Second, the burning of fossil fuels by vehicles moving along roads is a major contributor to atmospheric pollutants and greenhouse gases. The huge fleets of diesel-burning trucks on the world's roads contribute both greenhouse gases and particulate sulfur and carbon, or soot.

Third, the concrete used in road construction contributes to greenhouse gas emissions[86] (Figure 14.2). The manufacturing of cement contributes about 3% of the global estimates of CO_2 release,[594] although not all cement is used in roads and parking lots. Roads remain a major consumer of this product as well as of petroleum-based asphalt.

Fourth, the proliferation of roads throughout the world and especially in tropical forests often leads to broad-scale forest clearing.[260, 116, 197, 198, 524] This deforestation, which results in a loss of carbon stored in trees, is a major contributor to greenhouse gas emissions. Between 1850 and 1990, changes in land use contributed about half as much carbon to the atmosphere as that released from fossil fuel combustion.[429] Most of this land-use contribution was from slash-and-burn agricultural practices and timber removal, which result in extensive tropical deforestation. Road building in the tropics invariably leads to extensive slash-and-burn agriculture, as the new roadways allow penetration of the forest by subsistence farmers and multinational lumber companies.

Finally, road systems directly and indirectly create pollution, disturbances, and climate change. More roads and more vehicles will exacerbate these adverse effects. However, a shift to cleaner, more efficient vehicles would help mitigate some of these impacts. Fewer vehicles and a reduction in vehicle distance traveled would also help. Finally, a reduced road density, as well as reduced road access to extensive forest areas, would also lessen the atmospheric impacts of roads and vehicles.

PART IV

Road Systems and Further Perspectives

CHAPTER 11

Road Systems Linked with the Land

> Grid settlement divides landscapes, ideally level ones, into equal and tradeable segments. . . . It is by design, the death of nature.
>
> —Geoff Park, *Nga Uruora: The Groves of Life,* 1995

> Traffic is audible from virtually every location in England.
>
> —Jackie E. Underhill and Penelope G. Angold, *Environmental Review,* 2000

Traveling by car is like watching two movies at once. The roadway show before us moves from serenity to danger, from curves and bumps to onrushing overlapping signs. Meanwhile, the surrounding landscape show skips from boredom to beauty, from endless development to grand vistas with towering thunderheads. At times, the movies may merge as the road becomes an attractive part of the scene, an aesthetic or even metaphorical element of the landscape. At other times, the road appears as an intrusion in a natural or pastoral scene. Occasionally, we ponder the relationship between road and landscape. Did the road lead to the surrounding development or vice versa? The juxtaposition of the road with its surroundings shapes our experience of the trip. This notion of the road in its landscape highlights the importance of understanding the specific interactions between a road network and its surrounding land.

Much of this book has emphasized ecological effects of roads at the scale of individual road locations and road segments, the focus of most engineering concerns and most ecological studies. On the other hand, some wide-ranging effects of highways and traffic, such as air pollution, are considered at much broader scales, even globally. Intermediate in scale between local and regional-to-global scales lies the landscape.[302,935,335] Here the interactions of roads with ecological and watershed conditions operate at the scale of a few to many hundreds of square kilometers. This scale of inquiry is the domain of landscape ecology, watershed science, land-use planning, highway planning, and this chapter.

Individual roads affect the ecology of their immediate surroundings, so of course, the additive effect of the total road length in a landscape or region may be considerable. More to the point here is that roads are connected in a network and that the form of the network varies and matters[898,568,129] (Figure 11.1). Furthermore, traffic volume varies across the array of segments or links in a road system.

Understanding road networks in broader landscape and land-use contexts requires consideration of the units in a landscape. Natural conditions, human land use, and interactions between the two impose a pattern on the landscape. This pattern leads directly to *landscape ecology,* which interprets landscape occupants, structures, processes, and changes.[299,316,317,302,935,441] Ecologically, the

Figure 11.1. A planned-road grid to catalyze and accommodate residential development in the desert. Note the dry streambed and tributary, a primary-secondary-tertiary road hierarchy, and roads typically two house-lots apart in the residential developments. Outskirts of Albuquerque, New Mexico. Photo by R. T. T. Forman.

landscape occupants are individuals, populations, and communities of plants and animals, in addition to people. *Landscape structures* include patches of vegetation as well as stream and road networks. *Landscape processes* include the movement of organisms, seeds, and dust in dispersed trajectories, as well as the flow of surface water with associated material and species in stream networks. *Landscape change* describes the dynamic modifications or alterations in occupants, pattern, and process over time. In effect, land-use activities, including roads and practices in areas between roads, create patterns on the land by altering landscape structures, processes, and occupants.

In adopting this landscape perspective for understanding road ecology, this chapter addresses two broad reciprocal questions: (1) How do properties of landscapes, such as topography and land use, affect road network patterns? And (2) how do road network patterns affect the ecological properties, watershed processes, and land uses of the broad landscape? In part, answering these questions involves scaling our knowledge of local, site-scale aspects of road ecology up to the landscape.

Our premise is that *land-use patterns,* the types and arrangement of human uses of the land, strongly influence the pattern of roads in a landscape. Furthermore, the interactions between roads and the ecosystems and watersheds in which they reside fundamentally shape the flows and movements across the land, in effect determining how the landscape works.[302,400,305] Ecological and watershed effects result from both the intended and the unintended functions of road systems.

The ecological effects of road systems are defined in part by the land-use context of intervening areas. In large parklands, for example, protection of native large-mammal species is a common priority, and vehicle-wildlife collisions are a public safety issue.[377,80] Therefore, road placement and management relative to wildlife travel corridors are important for successful park management. In open-range grazing lands, collisions with livestock are a safety issue, and native wildlife usually holds a lower priority than in park settings. Where road densities are high, as in suburban areas, storm runoff from hard surfaces can exacerbate flooding in downstream areas, which may degrade riparian and aquatic habitats.[283,424,825] Along mountain forestry roads, on the other hand, roadcuts often intercept groundwater and convert it to surface flow along ditches, which may deliver water to streams with the potential for greater downstream flooding.[1017] In both the suburban and the forestry example, the land uses in the surroundings may also accelerate water runoff, thus having a synergistic effect with roads by increasing peak runoff and potential downstream flooding.

In this chapter, we address the properties of road networks as a basis for examining cumulative ecological effects of road systems across landscapes and watersheds. The chapter begins with a simple theory of the properties of network systems, including roads. Next, we introduce the spatial attributes of road networks, including road density. Time is then added to understand changing networks. We then explore the ecological effects of road networks and consider the effects of surrounding lands and land uses on the road system. Chapters 12 and 13 then follow with examples of road networks and their ecological effects in different major landscape types.

Network Theory for Road Ecology

Network theory has an important root in transportation.[898,568,317] It has also been incorporated into ecology, largely in a nonspatial manner such as relationships in a food web.[727,140,653] Road ecology can bring together and gain from these two approaches and their analytic tools.

The classic "traveling salesman" or "transportation problem" is a linear optimization approach that attempts to minimize the cost or time in sending goods through a network of possible transportation routes.[898,568] Interest in the approach surged during World War II to meet a great need for rapid and efficient transport of goods. Today, ecology has begun to take notice of the sophisticated mathematics of linear optimization embedded in computer programs for a variety of applications to model flow of goods and services along networks.[558]

Network theory has been a major contribution to the ideas of ecosystem science that arose from the systems analysis approach.[692,659] Applying theoretical concepts from electrical circuitry to the network of food webs (and trophic links) takes advantage of the large body of mathematics underlying network design.[416,727] This theory describes five key features of ecosystems as network: homogenization, amplification, quantitative indirectness, synergy, and utilization efficiency.[713] These concepts incorporate energy flows, inputs, tight linkages, indirect effects, connections, and cost-benefit efficiency. The concepts may be useful in analyzing and understanding road systems relative to ecosystems.

Network theory also underlies hierarchical approaches to ecosystems, whereby interactions can be viewed at various levels within a food (trophic) chain or biological system.[690,727,939] Depending on how much of the system is taken into account, the level of understanding can vary greatly. In addition, graph theory has been used in ecology as an analytical tool to detect spatial patterns on the landscape and their degree of connectivity.[140] A more dynamic landscape ecological approach presented in the next section focuses on spatially linking a network and its flows to characteristics of its surroundings—in other words, the road system to its landscape.

Branching hierarchical networks, such as circulatory systems, roads, and streams, share common properties. The networks cover a tiny percentage of the total area or volume yet provide extensive coverage of the object occupied. Individual corridors or linkages in the network are *flow paths* that facilitate the collection and distribution of materials across the object as a whole. High densities of flow paths foster interactions between flow paths and parts of the object. These properties of networks help accomplish desired objectives and functions. At the same time, such pervasive properties may have unintended consequences with resulting costs.

Just as the circulatory system connects every cell in the human body to the vital support organs, roads provide access to and from individual units of land use throughout a landscape (Figure 11.1). The sizes and arrangement of land-use units strongly influence the structure and density of road networks and, consequently, the interactions between road networks and the landscapes in which land-use units are embedded. For example, individual residential lots in suburbs, timber harvest units in forestry areas, and farm fields in agricultural land help define the *grain* (average size or diameter of patches) of their respective landscapes.[302] Similarly, road spacing in parkland is typically determined by the dimensions of the terrain that provides access to views (or viewsheds) and recreation areas.

A sampling of landscape areas on the scale of 1000 km^2 (386 mi^2) or greater often contains several distinct types of road system (Chapter 2). The distribution of population centers typically determines the spacing and geometry of the main connector-road net[898,568] (Figure 11.2). Intervening areas are commonly filled with local land uses, and the mesh size of attendant road networks is set by the land-use patterns.[559] In effect, a fine-mesh network connects local land uses to the coarse-mesh road network which connects population centers.

To meet transportation objectives, roads generally need be no closer together than the width (or depth) of two land-use units, such as residential lots, forest harvest areas, or farm fields (Figure 11.1). For example, the arrangement of streets in suburban neighborhoods provides street frontage and short driveway access for each residential house lot. Lot size in turn is largely determined by cost, convention, and zoning, plus topography, which results in a grain size and road density typical of residential suburbia. In most areas of the midwestern USA, roads have been added within the original nineteenth-century square-mile road grid to permit ready access to individual agricultural fields. For example, a quarter-mile road-mesh size is effective for serving sixteen 40-acre (16-ha) fields per square mile (2.59 km^2). Ecological effects are sensitive to the average mesh size of a network but are probably most sensitive to variability in mesh size. A high variance implies the presence of some large roadless areas, which are especially important for sustaining many key ecological conditions.[450]

Form and function are inextricably linked.[317] Network form in roads and other systems is most suited for particular functions, principally the transfer of

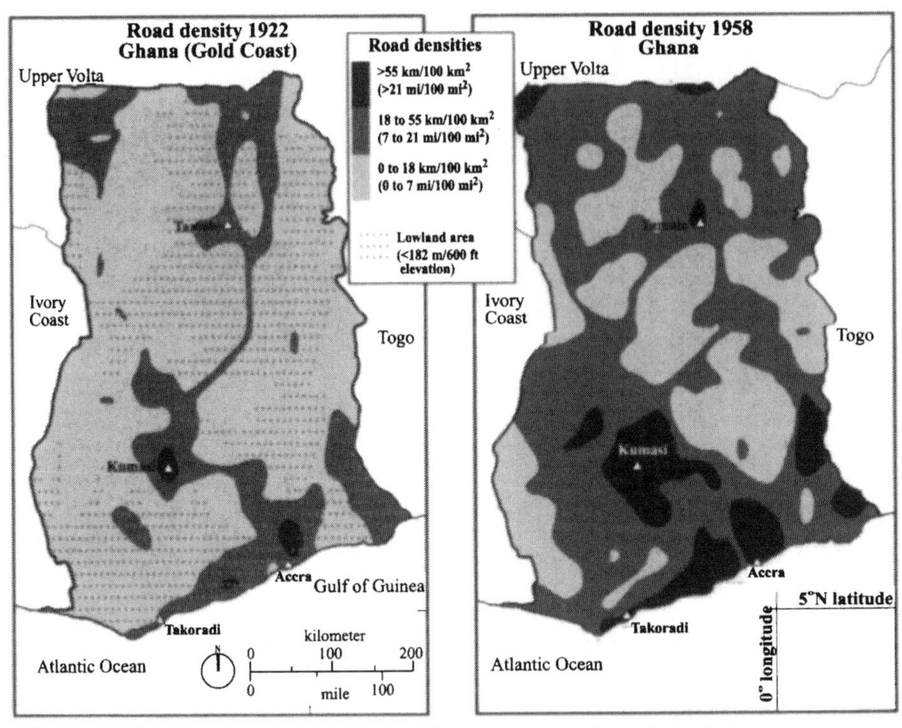

Figure 11.2. Changing road density in Ghana, as connection of population centers is followed by interconnecting access roads to dispersed resources and people. Early road penetration mainly followed the uplands and coast; later roads spread widely into lowland tropical rainforest. Adapted from Taaffe et al. (1963).

materials through the network and between flow paths and adjacent areas. In engineered and constructed networks, such as road systems, both intended and unintended functions exist. Unintended functions are illustrated by roadside plantings of non-native species that establish populations, which invade adjacent lands with undesired consequences (Chapter 4). Traffic may be the unintended vector for species dispersal,[984,563] and roadside areas may become unintended establishment sites of non-native plants. In both cases, the highly dispersed pattern of road networks exacerbates the impact.

Time is also a key to interpreting the structure and function of networks, especially road networks. Networks spread and contract in highly varied patterns depending on physical and social forces.[898,568] Changing land use, for instance, alters interactions between the road network and the landscape in which it is embedded. That improves or degrades the functional effectiveness of the network. A system that does not work well needs fixing at a cost to society. In some cases, the function of a road network may change, but the structural form remains as a legacy of former land uses.

In summary, four interlocking or feedback relationships combine to produce a simple road-network theory for interpreting the effects of road networks in landscapes:

1. The grain and arrangement of land uses surrounding roads is a major determinant of road-network structure.
2. Network structure and traffic flow strongly affect the intended societal functions of delivery of goods and services to and from accessible locations.
3. A road network has unintended effects on surrounding ecosystems, determined by road density, network structure, traffic flow, and the arrangement of ecosystems in the landscape.
4. The pattern of the surrounding land, in turn, produces intended and unintended effects on roads and traffic.

In essence, land pattern determines road-network structure, which controls flows both along and across the network, which together determine the cumulative ecological effects on the land.

Society mainly designs road networks around the first two interlocking relationships. In contrast, this chapter focuses on the latter two factors, the unintended functions of road networks.

Spatial Attributes of Road Networks

Road networks serve to connect population centers and to access and support such dispersed land-use activities as farming and forestry. The specific purposes served by roads strongly determine network form. For example, *rectilinear* (straight lines and right angles predominating) road networks serve residential uses efficiently, whereas *dendritic* (hierarchically branching) networks support mountain forestry operations. The branching patterns of road networks in mountainous terrain are shaped in part by the roads' conformance with valley floors and ridges as well as by limits on the *grade* (slope or steepness) of roads climbing the hillslopes. Flat lands and cities often contain simple road grids, whereas suburban landscapes and areas with modest topographic relief often exhibit irregular rectilinear road nets (effectively, *wavy nets*), such as in New England (USA) and eastern Canada.[302]

Simple characteristics of road networks strongly affect the intended purposes and unintended effects of road systems. Therefore, we first consider specific attributes of a network and then the integrated patterns of road density and network form, using a dual perspective that focuses on both the network and the intervening patches of land bounded by road segments.

Specific Attributes

The individual strips or corridors that combine to form a network have been much studied in landscape ecology.[815,302,73] In contrast, although numerous attributes and metrics of networks have been described in various disciplines, few have been studied ecologically.[129] The following array of specific network attributes highlights the potential importance of this subject[302]: connectivity, circuitry, directionality, corridor density, linkages per node, linkage angles, rectilinearity, converging and diverging corridors, variability in width of corridors, mesh size, enclosure size and variability, number of hierarchical levels, arrangement of intersection nodes and attached nodes, and variability in corridor context. Some attributes apply to rectilinear networks, such as circuitry and mesh size, and some to dendritic networks, including convergence and divergence. Other attributes, including corridor density and linkage angle, apply to both basic network types. The ecology of road networks is in its infancy, so only a few structural attributes of known ecological importance will be explored here.

Road density in a landscape, expressed as length of road per unit of landscape area, gives a useful first approximation of possible road effects in the various land-use types considered here.[302,817] Road networks vary greatly in density, ranging from tight grids of 40 km of road per square kilometer of area (64 mi/mi^2), typical of urban centers, to road networks of less than 0.1 km/km^2 in remote areas (Chapter 13). Suburban landscapes often have road densities of perhaps 10 km/km^2 (16 mi/mi^2), but with great variability present from place to place. The flat Central Plains agricultural landscapes of North America tend to have road densities of around 2 km/km^2 (3 mi/mi^2). Road density strongly affects spatial patterns as well as flows both along and across the network.

In contrast to road density, some processes respond to the proportion of total *road surface area* in a landscape.[825,151] The greater the road surface area, the more stormwater runs off from hardened surfaces and the more streams receiving that runoff may be degraded.

The network structures of roads take many geometric forms, ranging from simple rectangular grids to branching networks and highly irregular patterns. Flat lands permit use of simple grid road-network patterns, but steep terrain leads to branching networks. Even in gridded road structures, traffic patterns typically have a hierarchical branching pattern of traffic flows. High traffic volume on collector roads and lower volume in relatively remote areas tend to form a road-capacity hierarchy and branching network.

In steep topography, both slope steepness and river drainage patterns guide and constrain road patterns. Road systems in mountain terrain closely parallel rivers where valley floor width and slope permit. In many landscapes, roads follow ridges to avoid wet areas and more recently to respect the high ecological values of riparian areas. Where hillslope steepness exceeds road grade, the road may climb across slope, often intersecting streams nearly at right angles. These interactions among topography, roads, and streams strongly influence the effects of roads on ecosystems as well as the vulnerability of roads and travelers to landslides and snow avalanches. In relatively flat landscapes, the influence of topography on road patterns is less pronounced though still present—for example, where wetlands constrain road location and floods block travel.

Road Density and Alternative Network Forms

With an array of spatial attributes available, measuring a road network to effectively evaluate its ecological role is a challenge. Road density is the simplest spatial measure, providing an average or overview for the landscape. However, the specific form or arrangement of connected roads may be more informative.[760,918,310,319] Thus some ecological patterns related first to road density and then to *network form* are introduced here.

Road density appears to affect many species of large wildlife. For example, wolves *(Canis lupus)* in Minnesota, Wisconsin, and Michigan (USA) and mountain lions *(Felis concolor)* in Utah (USA) appear to thrive only where the road density is less than 0.6 km/km^2 (1.0 mi/mi^2).[960,618,320] Recent research further indicates that wolf packs in the western part of the Great Lakes region of the USA are most likely to occur in areas with road densities below 0.23 km/km^2 (0.4 mi/mi^2).[643] Nearly all wolves here occurred where road densities were less than 0.45 km/km^2 (0.7 mi/mi^2), and no wolf-pack territory was bisected by a major highway. The road-density effect on animals may be primarily due to road-kill, disturbance avoidance, or human access to remote areas, or to a combination, depending on the species and landscape[114,320] (Chapter 5).

Many other ecological patterns can be related to road density[318,320] (Figure 11.3). For instance, peak flows in mountain streams may increase with a higher road density.[462,106,461,725] Consider plotting a range of ecological factors—such as large predator populations, fire size, fire frequency, peak flows in streams, and disturbance due to human access—along an axis from low to high road density. Extremely different curves would be expected, suggesting that a unique combination of ecological conditions is associated with each level of road density.

Mesh size, the area or diameter of patches enclosed within a network, is inversely related to road density[302] (Figures 11.1 and 11.3). As road density increases, mesh size shrinks. As the landscape becomes more fragmented, the fragments are smaller.[396,951,827,953] Although roads are sources of effects, such as road-kills and human-caused fires, most road effects are effectively a constraining influence on the natural functioning of enclosed patches. Either road

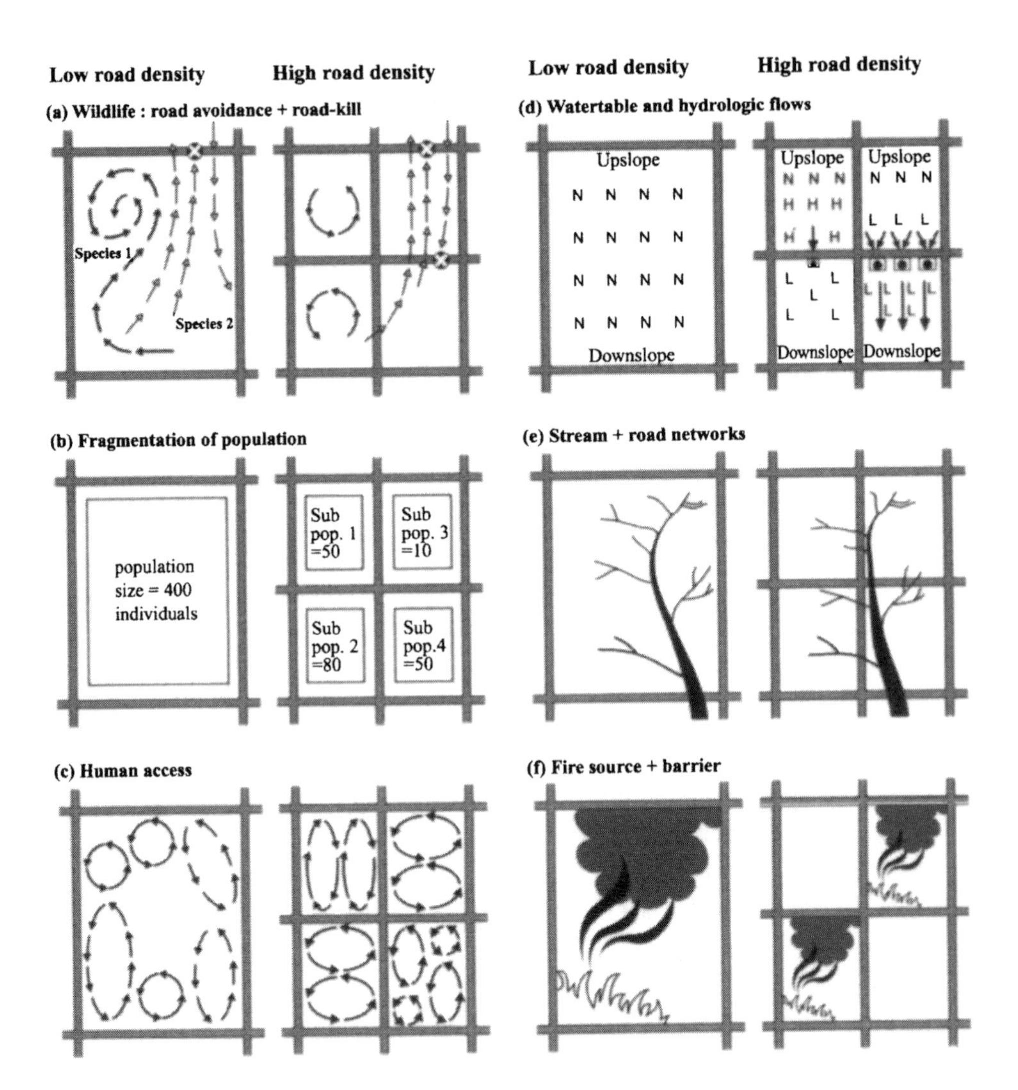

Figure 11.3. Diverse ecological effects of road density. (a) Species 1 tends to avoid the vicinity of a road; species 2 crosses roads, resulting in road-kills. (b) Species avoid the vicinity of a road; small populations fluctuate widely in size. (c) Central area on left is remote from human access and impacts. (d) N = normal, H = higher, and L = lowered watertable. Right side compares water flow through one small culvert versus flow through three large culverts. (e) Connected headwater streams on left; right (roads tending to parallel large streams and cross small streams nearly at right angles), 8 stream crossings and 1 road alongside a stream significantly affect aquatic ecosystems of the stream network. (f) Fewer and larger fires on left; more (human-caused fires) and smaller (access for fire control) fires on right. Adapted from Forman and Hersperger (1996).

density or mesh size can be used as a measure. However, the latter focuses on the enclosed ecosystems or habitat fragments or land uses, which become progressively smaller and eventually constrained (and often degraded) with the addition of roads.[450]

In a somewhat analogous case, the mesh size of fields surrounded by hedgerows in agricultural landscapes of western France appeared to be a useful overall assay of ecological conditions.[543,302] In this example, several types of species responded differently to mesh size. Predatory beetles and four herbaceous plants essentially disappeared when field size exceeded 2 ha (5 acres), owls disappeared at approximately 7 ha (17 acres), and certain shrub species disappeared at about a 100-ha (247-acre) mesh size. In addition, energy, soil, and other conditions were hypothesized to respond in distinctive ways to mesh size. Considering these responses in combination indicates that the agricultural landscape at, for instance, a 1-ha (2.5-acre) mesh size greatly differs ecologically from that at 10 ha (25 acres), which in turn differs from a landscape with 100-ha fields. Therefore, each mesh size has different ecological conditions present, just as suggested in the preceding case for road density.

The concept of *effective mesh size* for road systems provides further insight by adding two types of variables, land-use pattern within enclosures and width of the zone affected by a road.[245,450] Land-use pattern highlights the amount and arrangement of habitat or natural vegetation within each area enclosed by roads. Width of a road-effect zone highlights the fact that roads in the network have different traffic volumes. Effective mesh sizes calculated for numerous local areas in the region of Stuttgart, Germany, vary by about 10 to 30 times, and virtually all decreased progressively from 1930 to 1998.[245]

Road networks take an infinite variety of forms. Typically, natural landscapes are overwhelmingly heterogeneous and irregular, so the construction of a regular road grid would produce a dramatic contrast with nature. Indeed, as measured by fractal dimension, the pattern of landscapes with a heavy human imprint differs markedly from that in natural landscapes.[508]

Where the mesh size is large relative to the ecological flows in the landscape, it may matter little whether the network is regular or irregular. However, most landscapes are crisscrossed with relatively fine-scale road networks. Instead of perfect grids, these networks almost all contain variability. Although ecologists would generally agree that a regular grid is undesirable, little is known about what spatial arrangement would be ecologically optimum.

Consider three simple measures of variability in network density or mesh size: (1) high variance, (2) largest patch present, and (3) average size of the few largest patches. To compare these, the expected patterns for peak stream flow[462,106] and for habitat area available for wildlife that avoid the vicinity of roads[618] could be plotted against each of the three network-variability measures. The resulting curves suggest that the third option (average size of the few largest patches) may be best in providing satisfactory conditions for both of the ecological characteristics, and thus may be a useful simple measure of network variability. Of course, many alternative measures of network structure exist, as suggested in the preceding section.

Changing Road Networks

The structure of road systems is normally determined by the desire for efficiency—for instance, to minimize the construction "footprint" and travel distances—consistent with economics, aesthetics, topography, vehicle capabilities, and other factors. Thus road travel and land-use objectives interacting with topography set the grain and pattern of road networks. But often the road network is in part a legacy, in the sense that it may retain some grain and pattern related to former, rather than current, land uses. Although landscapes themselves undergo marked changes, both gradual and rapid, the focus here is on changes in road systems within landscapes.[1074,302,935,704,720,804,199]

Network Development

Form follows intended function for access to land-use units spread across the landscape (Figure 11.2). *Network development* of roads over time often reflects the sequence of diverse forces that change land use. Alternatively, network development may be a product of comprehensive planning to accomplish a specific land-use change. The initial intended functions, such as access to farm fields and pastures, are often supplanted over time as land use changes, as illustrated by suburbanization of former agricultural areas. So road networks are like stories that can be read for clues to the legacies of former land uses.

With the establishment of Concord, Massachusetts, in 1635, for example, roads were constructed to support small-town and farm uses in a previously forested landscape. In the mid-1800s, when Henry David Thoreau roamed and wrote about that land, it had been transformed to some 80% farmland and only 15% forest (Figure 11.4).[1025,323] For the century and a half thereafter, agriculture declined, forest slowly spread, and suburban town mushroomed. In short, from the seventeenth through the twentieth centuries, the landscape changed dramatically.

Meanwhile, however, the road network had a common stable backbone (Figure 11.4). It was in place by the 1660s and grew slowly in road length for three centuries. Even as suburban development began, most houses were built along or near the backbone. Only in recent decades has road density shot upward as residential development has spread over some of the enclosed patches of farmland and forest. Thus, through history, land-use changes have

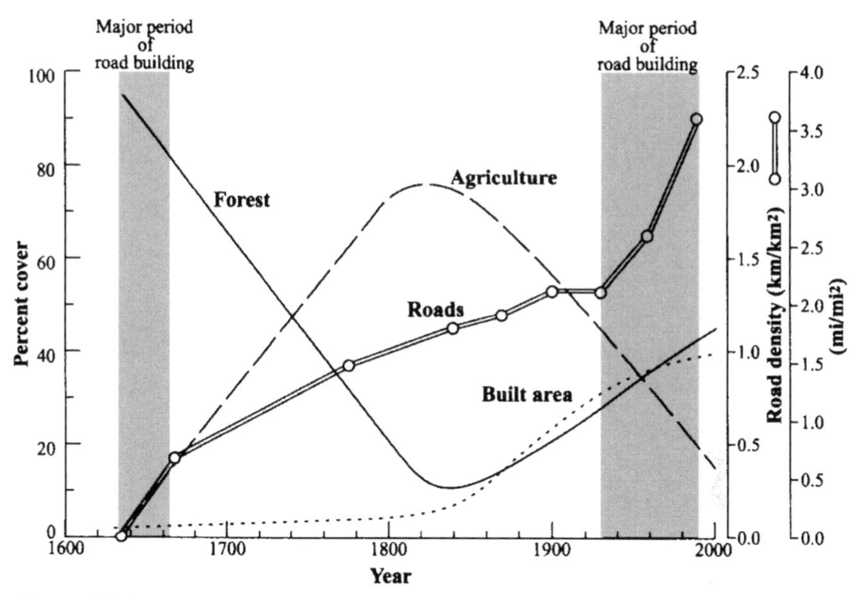

Figure 11.4. Growth of a road network as land uses and the broad landscape undergo transformations over time. A 65-km² (25-mi²) town, Concord, Massachusetts, 40 km (25 mi) west of Boston. Trains from Concord to Boston began in 1844, the year that Henry David Thoreau began a two-year stay in a cabin by Walden Pond in Concord and 400 m (0.25 mi) from the railroad. Based on Whitney and Davis (1986), Foster (2000), and Concord town records.

been dramatic, while road-network development remained on a rather steady, gradual, and different trajectory.

In contrast, recent forestry operations in the U.S. Pacific Northwest resulted in road network expansion from the 1950s through the 1980s. The native forest covering the landscape was changed by progressive clearcutting of dispersed patches of about 15 ha (37 acres) each.[396,332] In the Willamette National Forest in Oregon, approximately 1% of the land was cut each year. Roads totaling about 2 to 4 km (1.2 to 2.4 mi) long were annually and incrementally constructed over a 40-year period. Since 1990, a minor contraction occurred, as forestry uses diminished and some road segments were removed or substantially modified for watershed restoration.

Adding and Removing Roads

Network variability is also important in adding and removing roads. For example, if you were a planner with the objective of adding 10% of the road length to each of the cases in Figure 11.5, where would you place the new

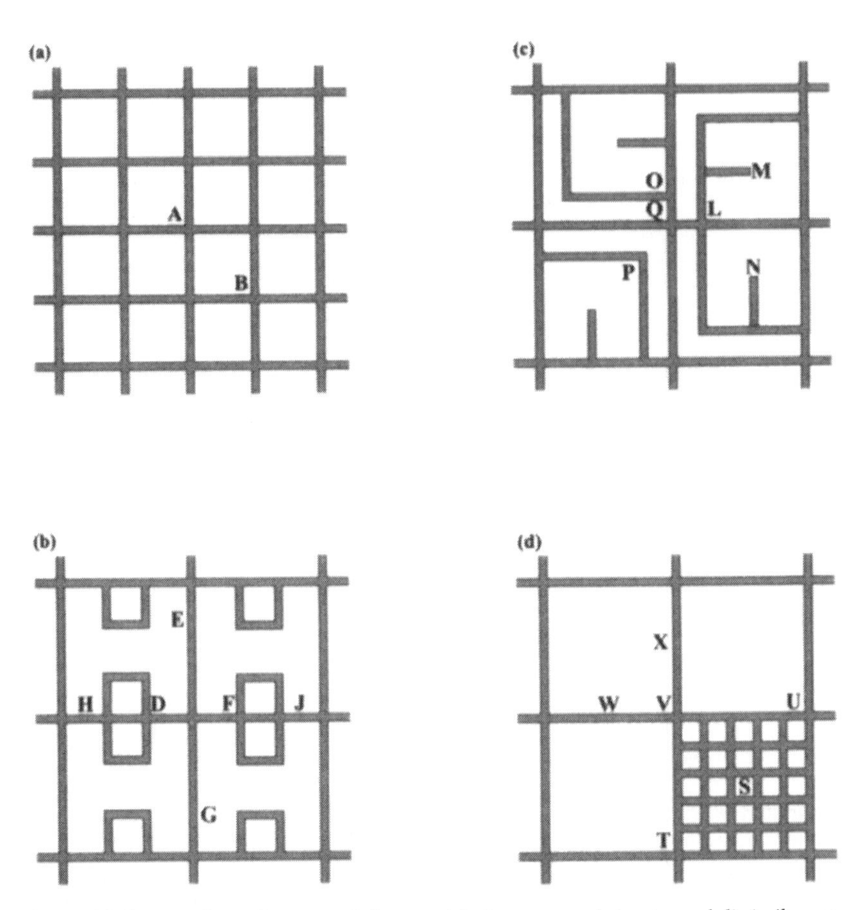

Figure 11.5. Four dissimilar network forms with the same road density and dissimilar ecological conditions. White = vegetated area. See chapter text for explanation.

roads to best provide for a range of ecological conditions? (Assume that each network is surrounded by networks identical to it and that roads can be no closer together than those in network d.) For protection of an aquifer or large–home-range species, new roads might best be built in the areas near T and U but not near W and X. For a population *(metapopulation)* separated into four large patches and occasional road-crossing movements among them, new roads might best go in the far corners of network b (which has four fairly large patches, one road apart). Such an analysis is highly relevant in suburbanizing areas with active development and in forestry landscapes for logging.

Conversely, if your task were to recommend removal of 10% of the existing road length from each road network (Figure 11.5), what roads would you remove to best enhance ecological conditions?[302] To provide the ecological benefits of large patches, the roads removed might be: (1) the cross defined by D, E, F, and G; (2) roads around L and M; and (3) roads by W and X. To enhance connectivity across the landscape, reduce traffic noise effects, restore hydrologic flows in groundwater and streams, or achieve a combination, optimal road-removal solutions may be different. This type of analysis is important in some rural landscapes with declining farm populations, as well as in forestry areas where trees have been cut and ecological restoration is a principal goal or where roadless areas are to be reestablished.

Thus the effects of both adding and removing roads strongly depend on the spatial arrangement of the road network, in many cases nearly independent of road density. In both cases, two major principles guide solutions: (1) maintain or establish large roadless areas, and (2) minimize road connections to streams and other water bodies. When adding roads (Figure 11.5), if we relax the constraint about minimum distance between roads, a third principle can be implemented: (3) add lanes to existing busy roads rather than adding new roads (see Figure 14.5 in Chapter 14).

Time and Road Systems

The flow of vehicles through the network in time and space is also a key to most ecological road effects. Traffic volume tends to vary cyclically on daily, weekly, and seasonal rhythms. Traveling to and from work or school illustrates diurnal traffic changes. Commuter flows on workdays alternating with weekend relaxation or shopping frenzy illustrate the weekly cycle. And winter weather and vacation travel highlight recurring seasonal changes in traffic. The effects of pollutants and of road crossing by wildlife are highly sensitive to these cyclic traffic flows. Some species, however, can adjust behavior or adapt over time to cyclic disturbances.

Noncyclic, irregular traffic flows also produce ecological effects. Episodic pulses of traffic associated with infrequent events occur, such as an occasional hurricane or a major rock concert. Presumably, these events mainly produce temporary, but sometimes intense, ecological effects. However, the local extinction of a rare species would typically be a long-term effect, since recolonization by the species from afar would normally be slow. A more familiar noncyclic pattern is the long-term increase or decrease in traffic on a road. For example, increasing traffic may change road-kill rates for mammals (Chapters 5 and 6) or accentuate traffic-disturbance (e.g., noise) effects on surrounding bird communities.[771,322]

The scheduling of traffic volume relative to ecological and other phenomena of interest is critical to assessing road effects. For example, the timing of traffic flow relative to animal movement affects the magnitude of interactions.[522,982] The timing of truck traffic on unpaved roads relative to the time of water runoff strongly influences the amount of fine sediment delivered to streams.[766]

In some areas, roads are closed during the migration of rare or important species (Chapter 5). Massive migrations of amphibians from overwintering woods to pond or wetland breeding sites may occur on a warm spring night. In some cases, citizens and officials have temporarily closed roads to prevent major "squishings" on roads being crossed.[522,445,982] Indeed, this process has sometimes led to the installation of amphibian tunnels for migration, as in Amherst, Massachusetts.

Finally, the link between network structure and function is a principal challenge for assessing road effects at the landscape scale. Overall, three major road-system properties determine ecological responses: (1) road density, (2) road surface area, and (3) traffic volume. With future research, network form (Figure 11.5) may be documented as a fourth major property. It appears that each ecological pattern and process is primarily related to one of the three properties. The structure of the affected ecosystems is also critical. For instance, some road effects propagate laterally into the adjacent vegetation patchwork, while other processes affect the routing of materials through stream networks interacting with road networks. Thus measures of road networks should be carefully tuned to the objective of study or environmental problem to be solved.

Road System Effects on the Surrounding Land

We begin this section by considering the distance outward that road effects extend and what that would mean on a map of the land. Then we introduce the "road-effect zone" concept as a promising assessment and planning tool that brings the careful, detailed engineer's perspective together with the broad landscape ecologist's perspective. Next, road density is discussed as a simple general measure of ecological effects in a landscape. This is followed by an examination of the intriguing question of habitat fragmentation by roads. Finally, we explore the effects of the rectilinear road network superimposed on the dendritic stream network.

The preceding chapters have provided detailed insight into the separate patterns of interactions of roads with plants, animals, water, sediment, and other ecosystem characteristics, including the many variables affecting distance effects. Now we synthesize some of the diverse results to detect overall patterns of how far road effects extend outward (Figure 11.6).[241,559,754,302] Many of the significant effects limited to short distances from a road are due to particulate and aerosol materials deposited from local air movements. Many sorts of road effects involving species and the transfer of energy and materials extend medium distances. Most effects that extend outward a long distance from a road include human-access disturbances, exotic species spread, effects in streams, the blocking of wildlife movement routes, and troposphere-stratosphere air movements (Chapter 10) (Figure 11.6). Although controversy exists about the effect of roads on caribou *(Rangifer tarandus)* (Chapter 13), the greatest avoidance zone or effect-distance reported is 5 km (3 mi) for caribou.[676,677,678]

The lateral extent of road effects (Figure 11.6) is determined by the processes involved and the transport medium, such as wind, groundwater, or animal locomotion. Some of these processes are strongly affected by topography, and others weakly. Wind transport of sand from road surfaces is more limited than transport of finer silt, clay, and salt particles. Change in vegetation structure from roadside to adjoining vegetation, such as dense forest, also affects the extent of road-effect penetration into neighboring areas. A forest canopy opening over roads may have a more spatially limited microclimatic effect on birds of the forest interior than the effect of traffic noise on these same species. Road effects typically penetrate farther into grassland ecosystems than into forests. Road effects due to suspended sediment, dissolved roadsalt, and other materials transported in water flows often extend for great distances downslope but only very limited distances upslope.

Road networks can produce cumulative effects on animal populations, hydrologic systems, stream networks, and other components of landscapes in which roads are embedded. Effects of road networks on the surrounding land depend on whether the interactions are above or below the ground surface, diffuse or concentrated, and follow or ignore gravitational flow paths. Such effects may be manifest far off site and substantially lagged in time, as in the case of the slow transport of road-related pollutants in groundwater systems.

To assess the effects of road networks on broader landscapes, we must take the one-dimensional road-segment perspective (Figure 11.6) to a two-dimensional road-network scale.[760] Mapping road network effects on a hypothetical landscape highlights several important points (Figure 11.7). In this example, the diverse factors (Figure 11.6) are divided into short-distance, medium-distance, and long-distance effects. Zone A adjacent to roads is subject to all effects, zone B receives medium- and long-distance effects, and zone C receives only long-distance effects. Zone R is not significantly affected by the road network. In this example, about 50% of the total landscape is in zone A, the heaviest impact area. The remote area R covers about 25%, and zones B and C 15% and 10% of the landscapes, respectively.

Also, since zones A, B, and C are equal width (Figure 11.7), adding to the straight-line length of a road increases the area of all three zones equally. In contrast, increasing road length by progressively adding intersecting roads, as in a network, increases the area of zone A more than the area of zone C. Ultimately, in creating a dense network, as in some forestry land, R may disappear, followed in turn by the disappearance of C and then B.

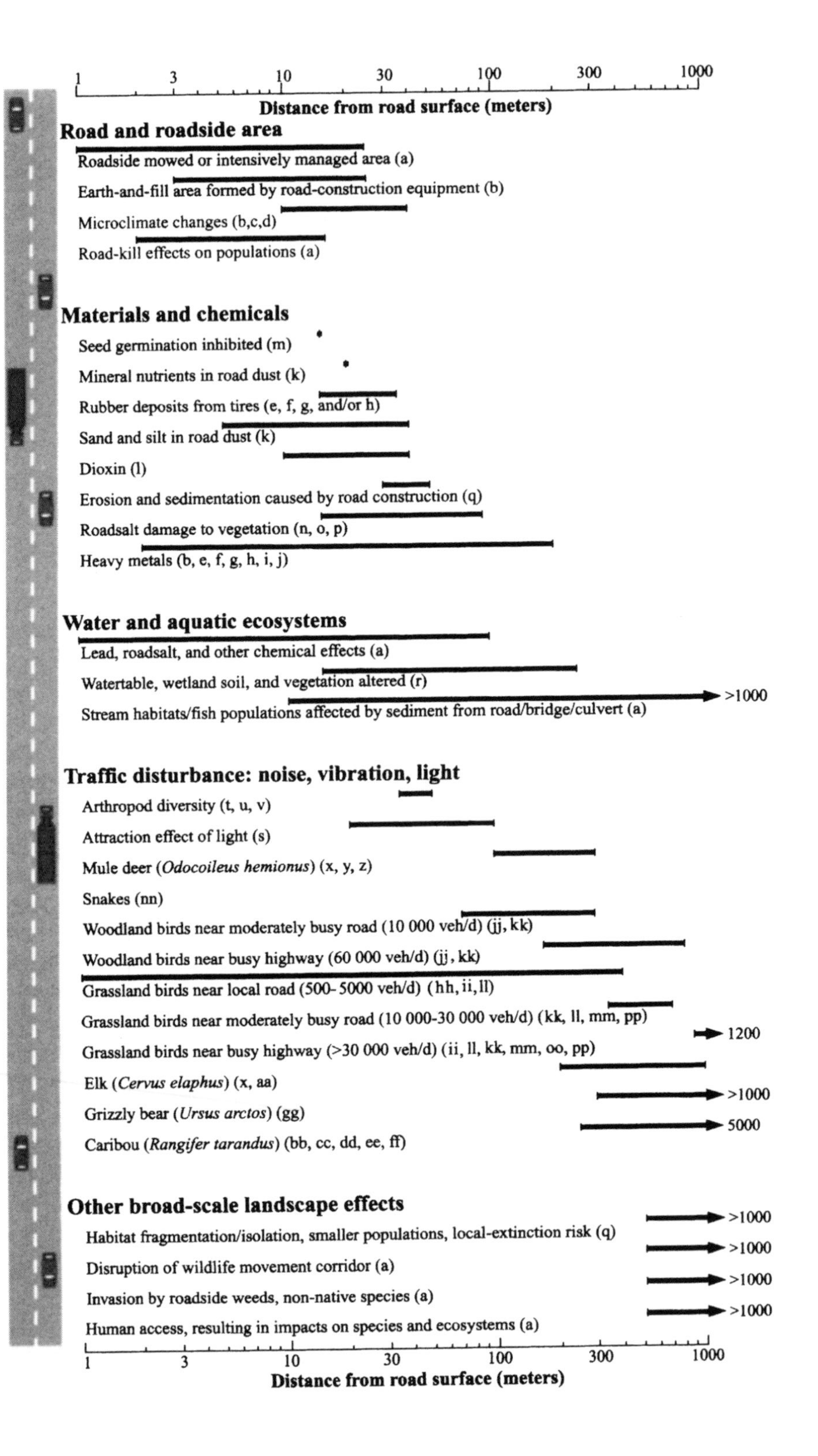
1
3
10
30
100
300
1000
Distance from road surface (meters)
Road and roadside area
Roadside mowed or intensively managed area (a)
Earth-and-fill area formed by road-construction equipment (b)
Microclimate changes (b,c,d)
Road-kill effects on populations (a)
Materials and chemicals
Seed germination inhibited (m)
Mineral nutrients in road dust (k)
Rubber deposits from tires (e, f, g, and/or h)
Sand and silt in road dust (k)
Dioxin (l)
Erosion and sedimentation caused by road construction (q)
Roadsalt damage to vegetation (n, o, p)
Heavy metals (b, e, f, g, h, i, j)
Water and aquatic ecosystems
Lead, roadsalt, and other chemical effects (a)
Watertable, wetland soil, and vegetation altered (r)
Stream habitats/fish populations affected by sediment from road/bridge/culvert (a)
>1000
Traffic disturbance: noise, vibration, light
Arthropod diversity (t, u, v)
Attraction effect of light (s)
Mule deer (Odocoileus hemionus) (x, y, z)
Snakes (nn)
Woodland birds near moderately busy road (10 000 veh/d) (jj, kk)
Woodland birds near busy highway (60 000 veh/d) (jj, kk)
Grassland birds near local road (500-5000 veh/d) (hh, ii, ll)
Grassland birds near moderately busy road (10 000-30 000 veh/d) (kk, ll, mm, pp)
1200
Grassland birds near busy highway (>30 000 veh/d) (ii, ll, kk, mm, oo, pp)
Elk (Cervus elaphus) (x, aa)
>1000
Grizzly bear (Ursus arctos) (gg)
5000
Caribou (Rangifer tarandus) (bb, cc, dd, ee, ff)
Other broad-scale landscape effects
>1000
Habitat fragmentation/isolation, smaller populations, local-extinction risk (q)
>1000
Disruption of wildlife movement corridor (a)
>1000
Invasion by roadside weeds, non-native species (a)
>1000
Human access, resulting in impacts on species and ecosystems (a)
1
3
10
30
100
300
1000
Distance from road surface (meters)

The spatial arrangement of the zone types also has potential ecological importance (Figure 11.7). Zone A is connected in a network encompassing the road network. Zones B and C are only in the form of thin rectangular rings or donuts. In fact, there are three of each equidistant from one another. In contrast, the unaffected zone R appears only in isolated rectangular patches. One large remote patch is well separated from the two tiny ones (by twice the width of A + B + C), which has important implications for some remote-patch species (or metapopulations).

Several aspects of the network-scale ecological effects of roads emerge from this simple analysis. First, road effects can extend over a large part of a landscape where the scale of distance effects approaches half the average distance between road segments (a measure of network grain).[760] Second, road effects can saturate (penetrate throughout) a landscape even with moderate road densities where long-distance effects are a factor. Third, road-effect patterns can create isolated habitat patches free of road effects (Figure 11.7). The size and arrangement of these habitat features may limit their function. If these patches are too widely dispersed and interpatch dispersal is too difficult, for example, populations of organisms relegated to these patches may eventually decline or disappear, despite having otherwise sufficient suitable habitat in the landscape as a whole.

Finally, roads tend to avoid certain habitats, such as wetlands and steep slopes, which therefore may be strongly represented in the remote R zone.[881,817] This means that overall road effects would be less in such habitats. However, it also means that impacts would be concentrated on certain other habitat types, which would tend to decrease the diversity of natural communities in the landscape.

Figure 11.6. Effect-distances from roads for diverse ecological factors. A horizontal bar indicates the approximate range of average and maximum distances from a road that significant ecological effects have been recorded (outliers, if present, are typically excluded). Letters in parentheses indicate sources: (a) authors' estimate; (b) Ellenberg et al. 1981; (c) ⋆Mader 1981; (d) ⋆Pauritsch et al. 1985; (e) ⋆Keller and Preis 1967; (f) ⋆Fidora 1972; (g) ⋆Hoffmann et al. 1989; (h) ⋆Reinirkens 1991; (i) Santelmann and Gorham 1988; (j) Ministry of Transport, Public Works and Water Management 1994c; (k) Tamm and Troedsson 1955; (l) ⋆Unger 1991; (m) ⋆Fluckiger et al. 1978; (n) Hofstra and Hall 1971; (o) ⋆Evers 1976; (p) ⋆Wentzel 1974; (q) Forman and Deblinger 2000; (r) ⋆Adam 1992; (s) ⋆Meier 1992; (t) ⋆Port and Hooton 1982; (u) ⋆Maurer 1974; (v) ⋆Przybylski 1979; (x) Rost and Bailey 1979; (y) Dorrance et al. 1975; (z) Singer and Beattie 1986; (aa) Edge and Marcum 1985; (bb) James and Stuart-Smith 2000; (cc) Dyer et al. 2001; (dd) Nellemann et al. 2001; (ee) Nellemann and Cameron 1996; (ff) Nellemann and Cameron 1998; (gg) Gibeau 2000; (hh) Clark and Karr 1979; (ii) Reijnen et al. 1996; (jj) R. Reijnen et al. 1995; (kk) M. Reijnen et al. 1995; (ll) Forman et al. 2002; (mm) Green et al. 2000; (nn) Rudolph et al. 1999; (oo) van der Zande et al. 1980; (pp) Raty 1979. ⋆ = cited by Reck and Kaule 1992.

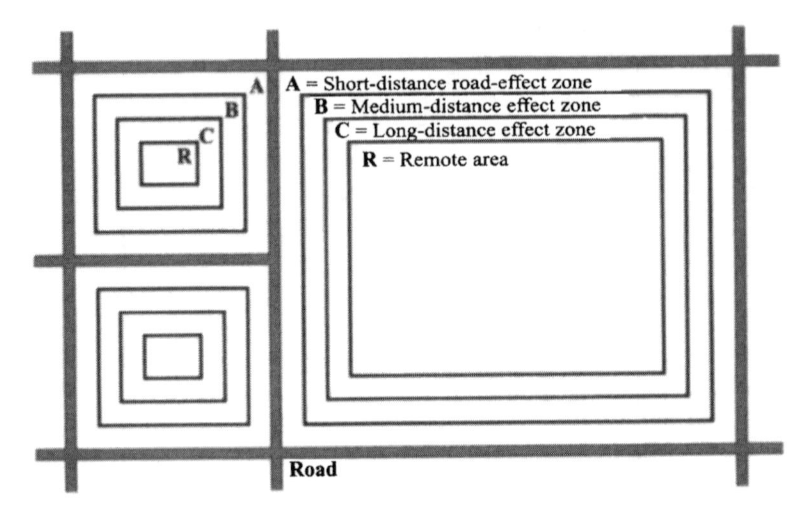

Figure 11.7. Zones of ecological road effects within a network.

Road-Effect Zone and Terrestrial Systems

Highway engineers carefully design roadbeds, surfaces, ditches, roadsides, and immediately adjoining areas according to rigorous design criteria and extensive experience. Meanwhile, landscape ecologists and colleagues in related fields document the major patterns and processes of wildlife movement, plant communities, rare species, surface and groundwater flows, and fish populations across the broad landscape. Conservation planners, transportation planners and, indeed, society need to integrate these critical fine-scale and broad-scale perspectives. The following model may be a promising common ground for creating beneficial solutions in managing real processes, road systems, and landscapes.

Consider driving along a road as it slices through a heterogeneous landscape. Clearly, the pair of land uses or local ecosystems on opposite sides of the road changes, reminiscent of a movie. Also, in passing intersections and perhaps communities, traffic volume changes. This sequence of adjacent lands and traffic volumes certainly creates a variable width zone of road effects to the right and the left. But let us look more closely at the mechanisms.

In addition to changes in adjacent land type and traffic volume, many different types of road effects extend outward both to the right and to the left for different distances (Figure 11.8). Consider three processes or mechanisms that help determine the distance that road effects extend[320,310,305]:

1. Wind—for example, blowing from left to right—causes asymmetric effects (Figure 11.8). Road dust, nitrogen from tailpipes, and traffic noise are carried farther downwind than upwind, thus producing greater effects.

2. Water—for example, flowing from left to right—carries sediment and dissolved chemicals, rises to flood levels, and disappears during droughts (Figure 11.8). The effects are greater and extend farther downslope than upslope.
3. The behavioral attraction of animals and people to more suitable habitat

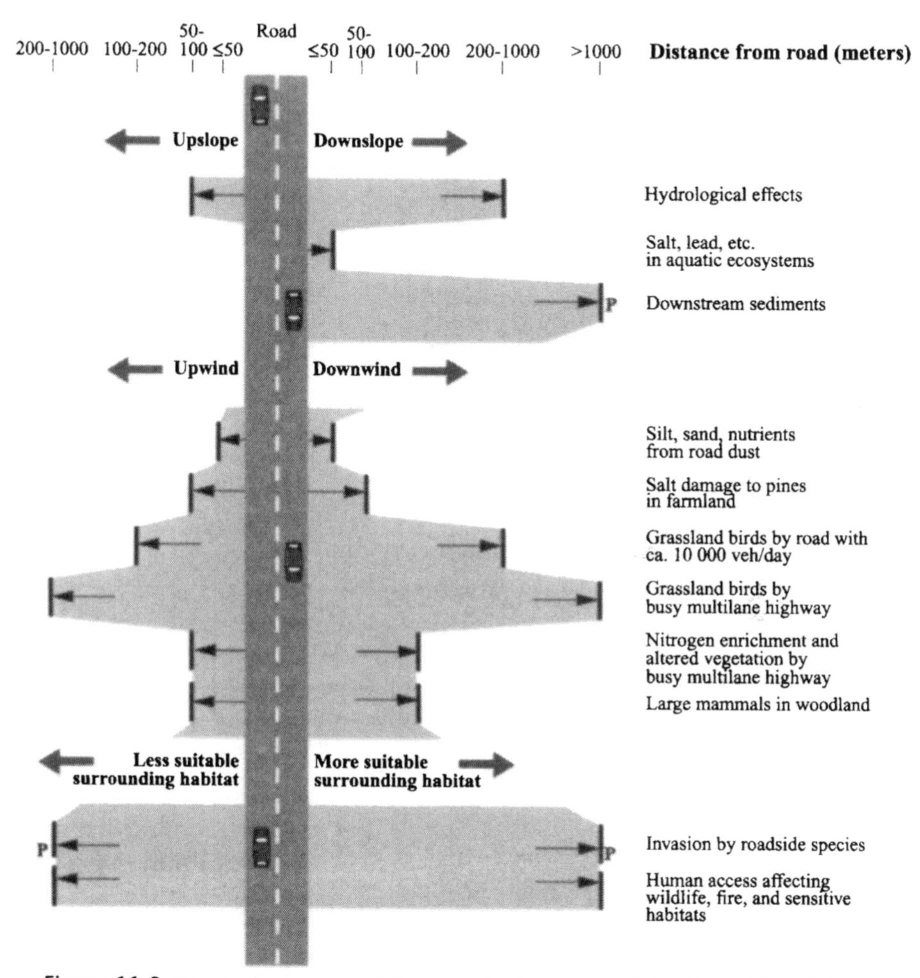

Figure 11.8. Road-effect zone and three mechanisms determining its width and form. Illustrative effects listed on right based on one or more studies. Three mechanisms—gravity (upslope/downslope), wind (upwind/downwind), and behavior or habitat suitability (less/more)—in addition to walls or hills near the road, produce greater effect-distances on one side of the road than on the other. With a scarcity of data, in general distances for the examples are approximately halved on the left side. Shaded area = road-effect zone. Each effect typically extends outward along a stretch of road or road segment; P = an effect extending from a point on the road. Adapted from Forman and Alexander (1998).

or places, or avoidance of less suitable habitat, leads to different effects on opposite sides of the road. Deer browse more, non-native species invade more, and humans picnic or hunt more in one direction than the other.

In essence, a road system crossing a land mosaic alters material, energy, and species patterns under the influence of wind, water, and behavioral processes. The result is to produce a zone of influence along a road, a *road-effect zone,* over which significant ecological effects extend (Figure 11.8). The zone has asymmetric convoluted margins. Effects that extend far or relatively far from the road surface, such as B and C in Figure 11.7, normally define the margin of the road-effect zone.

An example of the road-effect zone model was mapped for a 25-km (16-mi) stretch of a multilane divided highway with about 50 000 veh/d extending westward from Boston (USA).[315,306] Nine ecological factors that extend more than 100 m (328 ft) from the road—wetlands, streams, roadsalt, exotic plants, moose *(Alces)*, deer *(Odocoileus)*, amphibians, forest birds, and grassland birds—were measured or estimated. A few road effects extended more than 1 km (0.6 mi). Most factors were detected at two to five specific points or locations, whereas traffic noise effects apparently were evident along much of the road length. Mapping these road effects indicated that the road-effect zone in this landscape averages approximately 600 m (0.4 mi) in width (300 m each side), is asymmetric, and has convoluted boundaries with a few long "fingers" extending farther outward (Figure 11.9).

The fingers generally represent locations where the highway crosses a different type of corridor, such as a river or railroad corridor. The highway passes through a slightly undulating forested and residential landscape. Occasional roadside barriers and hills inhibited the propagation of traffic noise, whereas an elevated roadbed adjacent to fields doubtless enhanced it. Ecological effects were measured or estimated only for natural and agricultural areas. Ignoring effects on residential areas therefore effectively narrowed the average width of the road-effect zone.

The margin of a road-effect zone does not imply that there are no ecological effects beyond, such as in the remote zone R of Figure 11.7. It simply means that, for example, based on measurements along lines perpendicular to a road, effects beyond the margin are not (statistically) significant based on present evidence.

Extending the road-effect zone model to a two-dimensional road system in a landscape suggests a *network-effect zone*, as in Figure 11.7. In the form of a lattice or net, its area combines the individual road-effect zones present, together with any area due to their interactions. For instance, the road-effect zones surrounding a small enclosed patch might interact to affect the central area of the patch. Thus wind might carry vehicular pollutants across the patch,

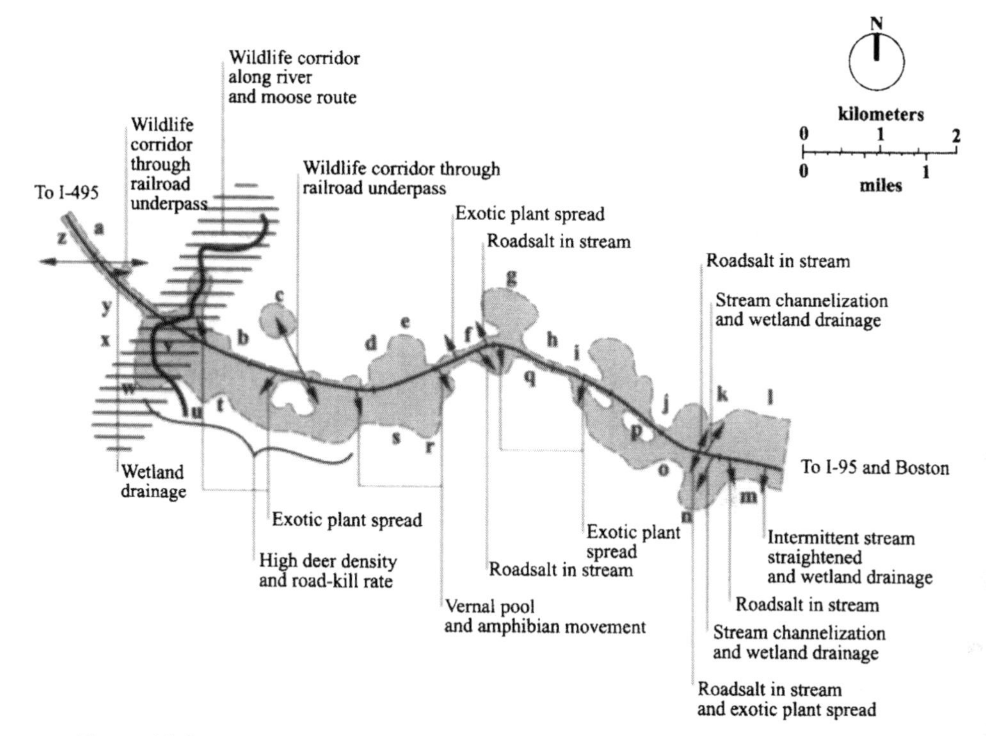

Figure 11.9. Road-effect zone for a 10-km (6.2-mi) segment of a multilane highway. Arrows indicate distances of estimated or measured ecological effects. Dashed lines indicate areas of natural vegetation estimated to be affected by traffic disturbance or noise. Letters refer to factors such as built area, road embankment, and open grassland that affect the width of the road-effect zone (see Forman 2000). Route 2 with 50 000 veh/d in the Lincoln/Concord area west of Boston. Adapted from Forman and Deblinger (1998, 2000) and Forman (2000).

or non-native roadside animals might move across it. Wind, water, and behavioral movements tend to be highly nonrandom and directional, which offers promise for modeling and understanding network-effect zones.

The importance of a network-effect zone is suggested by a study of grassland birds in open patches of a "typical" outer-suburban landscape west of Boston (Chapter 10).[315,322] In the 240-km^2 (93-mi^2) landscape area, the network of roads with different traffic volumes was mapped along with all open patches apparently suitable for grassland birds. Road-effect zones from a few hundred meters to over a kilometer wide, depending on traffic volume, were found where the presence and regular breeding of the birds were inhibited (see Figure 10.7). Half of the open patches and half of their total area fell within this network-effect zone. This suggests that road effects—in this case, probably mainly due to traffic noise—extend over a surprisingly large portion of the suburban landscape.

Equally interesting was an estimate of change over time. If the effect distances were decreased by 10%, perhaps mimicking an earlier time period with less traffic, the network-effect zone dropped only 2% to 3%.[322] However, if the effect distances were increased by 10%, analogously mimicking a future with more traffic, the zone increased by 7%. This difference in zone size highlights the curvilinear increase in area of the network-effect zone that occurs as either road density or traffic volume increases linearly.[245] The increase in network-effect zone area from 2% to 3% to 7% also suggests that the zone has reached its rapidly expanding phase in this typical outer-suburban landscape. Enclosed relatively natural patches (zone R in Figure 11.7) without significant road-traffic effects on bird communities are probably rapidly disappearing. Since traffic has increased proportionally much more than road length has in the past two or three decades, the rapid loss of remote unaffected natural habitat may be largely related to increased traffic. Indeed, this effect of growing traffic may be among the most pronounced ecological impacts of road systems.

These patterns point to the idea of a *saturation effect,* which may appear when roads reach a density at which additional roads have little added effect. A saturation effect is suggested by a curvilinear or threshold response, as in the case of the width of the avoidance or road-effect zone for grassland birds as traffic increases[322] (Chapter 10). The curvilinear response was illustrated in the outer-suburban landscape, where a sixfold increase in traffic corresponded with a threefold increase in road-effect–zone area. In an extreme, once a wildlife population has disappeared from an area due to roads with traffic, adding more roads cannot increase the effect on that species.

In addition to these limited research results available, we can gain insights from spatial models (Figure 11.7). Zones A, B, and C together compose the road-effect zone for the schematic road system in a homogeneous landscape. That is the minimum area required for environmentally sensitive transportation planning, design, construction, maintenance, mitigation, and management. Clearly, to accomplish this objective would require the combined expertise of at least ecologists, engineers, and planners. Solutions may differ markedly in zones A, B, and C. This approach also suggests that zone R is largely peripheral to transportation planning, though extremely important to landscape and regional planning.

Road Network Effects on Stream Systems

Road networks can affect stream systems in many ways, including by serving as sources and conduits of surface water runoff to the natural stream system. In watershed science and management, these phenomena are referred to as off-site *cumulative effects,* the spatially and temporally dispersed changes in watershed processes that result in effects far removed from sites of direct effects (and which must be considered to fulfill NEPA regulations in the USA—see Chapter 3). These processes have been observed in urban, suburban, and mountain forestry landscapes where floods in downstream areas increase by several mechanisms.[1017] Hardened road surfaces prevent infiltration of rain and snowmelt, thus speeding delivery of water to streams and potentially increasing flood levels. Roadcuts on steep hillslopes may intercept subsurface water flow and divert it to surface water flowing via ditches to the natural stream network. This process effectively extends the stream network, increasing *drainage density* (total stream length per unit area of land).[1017,106] Also, network length and the degree of connectivity of stream and road networks, rather than road surface area, primarily determine downstream flows in steep terrain. The increased runoff can result in channel incision or deepening by streams, thus altering aquatic and riparian ecological conditions, including fish populations, in the lower parts of stream systems.[234]

Simple measures of effects of road networks on streams begin with an assessment of stream drainage density relative to road density.[1044,1017,106,463] In addition, the degree of connectivity between the two network types is ecologically important. The two major types of connectivity between the two network types are: (1) road-stream crossings on bridges and culverts, and (2) stretches where roads and streams run alongside each other.

A third type of connectivity between the road and stream networks occurs where water from roadside ditches connects the two. In a mountain forestry case, for instance, approximately 57% of the road network functioned as an extension of the stream network, thereby increasing drainage density by 21% to 50%.[1017] Most of this road-stream connectivity occurred by the connection of roadside drainage ditches to streams. However, some occurred where culverts carried water from ditches onto soil below the road, which resulted in gully formation and unintentional connection of ditches to native streams.

Road systems also provide *human access,* which may result in direct and extensive alteration of stream and riparian ecosystems.[405] Diverse, often subtle factors, such as legal and illegal fishing, hunting of species common to riparian zones, and removal of wood from streams, result in environmental degradation due to human access. The potential for such alterations of stream and streamside ecosystems to occur may be proportional to the frequency of crossings of road networks and stream networks. Or such impacts may relate to the proximity of a single road to streams, such as recreational impacts catalyzed by a spur road into a remote area.[495,547,405]

Groundwater systems may also be altered by road networks, which serve as sources of pollutants or structures that limit water infiltration (Chapter 7). Chemicals intentionally or unintentionally applied to roads, such as roadsalt and spills of hazardous materials, can degrade groundwater systems. Where road networks have a substantial surface area with extensive water runoff into storm drainage systems instead of into local groundwater systems, groundwater recharge may be reduced.

Effects of Surrounding Land on Road Systems

Many conditions of the adjacent land affect road networks and travelers on the roads. This section sequentially considers the effects of adjacent lands and land uses on the road structure itself, on traffic flow, on the roadside habitat and associated species, and on how adjacent land use and objectives affect animal movement interacting with road networks.

Adjacent lands and land uses affect the road structure where movement of water, sediment, and landslides damages roads. Rivers flood roads, but roads may aggravate the effects of floods in several ways (Chapter 7). Poor design may cause roads to block drainage, thus promoting inundation of roadways. Blocked drainage may also concentrate flow and scour on the downstream side of roads, resulting in erosion of the road fill and road surface. Rapid runoff from hardened surfaces (e.g., rooftops) in highly developed land promotes flooding of roads in downstream areas. Bare ground in agricultural landscapes may similarly promote flooding with sediment-charged water, potentially inundating and damaging roads. Also, gravel extraction from riverbeds, commonly for road construction and maintenance, can lead to river down-cutting and the undermining of bridge piers and approaches.

Land-use conditions and activities in areas adjacent to roads can affect traffic flow. Dust and smoke impede travel and even lead to highway fatalities in multi-vehicle pileups, where smoke from field or forest burning, for example, obscures visibility. Park visitors stopping to view elk *(Cervus)*, bison *(Bison)*, and bear *(Ursus)* create traffic hazards in settings such as Yellowstone and Banff National Parks[377] (see Figure 13.3 in Chapter 13). Landslides from recent forestry operations on steep slopes have knocked vehicles off highways into rivers, causing fatalities. Thus traveler safety is determined by local land-use effects in quite diverse ways.

The roadsides between roadway and adjacent land uses are a distinctive habitat within many landscapes. Interactions between terrestrial animal populations and road networks are strongly determined by adjacent land uses. But animals are also affected by roadside habitats, which may be quite distinctive, with important consequences for both ecological concerns and traveler safety (Chapters 4 and 5). Where the adjacent lands are wild and protection of native species is a prime objective, traveler safety may involve the risk of collision with large animals.[377,443]

Adjacent land-use and management objectives affect animal movement in ways that influence road-network structure and use (Chapter 5). These land uses as well as natural landscape features can intentionally or inadvertently guide animal movement toward some highway segments and away from others. This process affects the distribution of hazards posed by animals crossing roads. It also affects the ability of highway managers to deal with the hazards by altering road structure, animal behavior, or human behavior.

Lastly, the density of residential and commercial development strongly determines the magnitude of traffic on adjacent roadways. Traffic volumes help determine the classification of road types. This in turn affects planning and construction changes in existing roads, as well as the probability and location of future roads.

Conclusion

Evaluation of road systems linked with adjacent lands or land uses suggests the presence of an overarching system with feedback mechanisms. Rather than road systems simply affecting the land or the land only affecting road systems, both apply for many ecological factors. In other words, road systems and the surrounding land represent a tightly linked feedback system. Thus roads may contribute to flooding by blocking drainage or extending the stream network system, while flooding in turn can scour and undermine roads. Without designing road networks to mesh harmoniously with natural landscape processes, significant human costs are required to maintain a functioning road network. Similarly, a highway with traffic commonly alters adjacent habitat and concentrates large-mammal crossing to certain locations along the road. The concentration of wildlife crossings, in turn, often leads to more animal-vehicle collisions and more associated human activity at these locations, with consequent disruptions of traffic flow. In some landscapes, increased road density leads to more human access and, consequently, more fires. Fires in turn can temporarily block traffic on the roads. Furthermore, the higher road density provides more access for fire control, resulting in smaller fires that have less effect on roads and traffic. Road systems and the land are indeed tightly intertwined.

Finally, to address the question of an ecologically optimum road network, consider two extreme cases with the same overall road density. The first is an evenly distributed road network with moderate impacts over the entire landscape. The second concentrates almost all of the impacts in a small portion of the landscape with a large roadless area in the rest of the landscape. The information presented in this chapter points to the second case as being less ecologically damaging. For example, clean groundwater and many large–home range animals are most effectively maintained in large vegetated areas. Added insight comes from evidence that as traffic increases, the road-effect zone widens but at a progressively slower rate. This suggests that if more traffic is to be added to a landscape, it will have the smallest impact if added to the roads that currently have the highest traffic volume—that is, by widening those roads (see Figure 14.5 in Chapter 14).[241,255]

Almost all road networks in real landscapes represent gradations between the extreme cases considered. Two principles or guidelines for designing or enhancing these gradations emerge from this brief analysis. Road networks with the smallest ecological impact are those that not only (1) maintain large roadless areas, but also (2) concentrate traffic onto a small number of large roads.

Furthermore, if small areas with concentrated roads are present, they are analogous to *point sources* of pollutants, the finite locations that can be targeted for mitigation solutions. Similarly, if most traffic goes on a few roads rather than being distributed over many, mitigation can be targeted to the busy roads, with large benefit to society.

CHAPTER 12

The Four Landscapes with Major Road Systems

> Two roads diverged in a wood, and I—
> I took the one less traveled by,
> And that has made all the difference.
>
> —Robert Frost, "The Road Not Taken," 1916

> In those brevities just before dawn and a little after dusk—times neither day nor night—the old roads return to the sky some of its color . . . a mysterious cast of blue, and it's that time when the pull of the blue highway is strongest, when the open road is beckoning, a strangeness, a place where a man can lose himself.
>
> —William Least Heat-Moon, *Blue Highways: A Journey into America,* 1982

Imagine standing on a hillside overlooking a familiar suburban landscape. A sea of residential house lots engulfs some scattered parks and commercial areas. Seemingly everywhere lie highways and streets, driveways and parking places. They must cover at least 10% of the land surface, maybe twice that. The roads are all connected, much like the stream system crossing the land beneath them. Suddenly, the thought arises that the road system is connected to the stream system at every crossing, and that indeed the road network is really an extension of the stream network, like a huge set of extra fingers reaching everywhere throughout the scene. A heavy rain would send water gushing down the streets and cascading through storm-drains into streams. The water would wash every street clean of chemicals and debris, pouring them into the narrow stream channels. What would happen to the rich aquatic habitats and the fish in streams in the face of this enormous surge of water and pollutants? Indeed, what about the habitats over the landscape that are deprived of the fast-disappearing water? The big picture before you reflects, in essence, the cumulative effect of a road network on a whole landscape.

In this chapter, we build on concepts from Chapter 11 to explore road ecology in major widespread landscape types. Focusing on the USA and Canada sets some limits to the cases represented. For example, in North America, agriculture is normally practiced in landscapes with slopes of less than a few percent, whereas mountain agriculture is common in many nations from the Alps and Andes to the Himalayas. In developed countries, roads overwhelmingly access land uses, but in some parts of the world, foot travel provides effective access to sites well removed from road systems. In some nations, rail and transit are major components of human travel, so housing and road systems are more organized around access to public transportation. And in some cases, villages and a sense of place are deeply ingrained in the culture, and correspondingly, sprawl and its road-system effects are scarce.

Four broad landscape types, based on the predominant land use present, are explored in this chapter: (1) built land, (2) forestry land, (3) agricultural land, and (4) grazing-arid land. (Natural land is explored in Chapter 13, and both chapters provide partial syntheses of diverse concepts in preceding chapters.) Roads and vehicles range from dominant features of the landscape in many built areas to minor components in some grazing-arid land (or desert-grassland). For each landscape type, we begin with characteristics linking the terrain, land use, and road system and then consider ecological effects. The focus is on the road system, the network of roads and vehicles, interacting with the landscape.

Built Land and Road Systems

In a sense, cities are great for nature. By squeezing together into large aggregations, people effectively concentrate rather than disperse their impacts on the natural resources on which they depend. Resources distributed across the land thus are channeled along transportation corridors to cities. Cities provide for a specialization of human activities. They offer a distinctive urban and cosmopolitan culture as an alternative to traditional local and regional culture. Keeping a city vibrant may slow the dispersal of people sprawling across the land. Yet, irrespective of the quality of life in cities, people will live in, and move to, suburbs and outer areas of sprawl. A focus on urban regions, especially the outer non-city portions that compose most of their area, is a key to the future of natural resources and society.

Three types of *built land* (or built-up landscape) are considered[188,576,68,70,715]:

(1) city, (2) suburban landscape or suburbia, and (3) sprawl area. In brief, a *city* is a large or important municipality, and *suburbs* are just outside the city, with suburban towns normally dominated by residential areas. Together, suburbs form a *suburban landscape,* or simply *suburbia,* which contains small pieces of *open space* (nonbuilt area). *Inner suburbs* are close to the city, have fairly dense populations and roads, and have relatively little total open space. *Outer suburbs,* with less dense populations and roads and more open space, differ quantitatively in these attributes. *Sprawl areas,* characterized by dispersed, low-density residential housing, contain both large and small pieces of non-built area, such as agriculture, forest, or desert. An *urban region,* or *metropolitan area,* includes city, suburbia and, if present, sprawl area. Rural towns may appear similar to outer suburbs.

Landscapes change by (1) shifting land uses within a landscape, (2) becoming more dense, and (3) spreading over another one. A commercial area may replace a residential area, and vice versa. Residential areas often become denser in population over time. Spread is directional: city expands over inner suburbs, inner suburbs over outer suburbs, outer suburbs over adjacent nonbuilt or sprawl area, and sprawl over more distant nonbuilt area.[188,976,68,70,217] Other types of spread may also occur, such as along transportation corridors (e.g., growth along a rail line, or strip development) and growth of satellite cities plus infilling.

Terrain, Land Use, and Road System

Today's cities generally originated where a water body and good agricultural soil met.[188,507] Some or all of that soil is under city now. Most downtown areas (or central business districts) exhibit a relatively regular grid of streets with a road density of about 40 km/km^2 (64 mi/mi^2).[576,2] The non-downtown city area tends to have a mixture of regular and irregular (wavy net) grid areas (Chapter 11) serving commercial and industrial districts, residential areas differing in population density, and mixed-use areas. *Through streets* generally connect the dense local networks present in these districts and areas. The non-downtown area in North America often has a surprisingly large hard-surface area devoted to car parking.

Most open spaces in cities are *woodlawn parks,* the proverbial grass-tree-bench ecosystem that provides space for reflection or intensive recreation.[612,1028,356,186,732,427,870,890,113,458] Small patches of relatively natural vegetation tend to survive in areas that escaped development or are overlooked by society, such as a wet spot or a site surrounded by highways or railroads. Major green corridors generally parallel rivers, rail lines, and highways (Figure 12.1), whereas thin green strips are mainly limited to planted shrub lines and street trees.[870,853,60]

Figure 12.1. Major highway corridor and interchange in an urban residential and commercial area. The below-grade (sunken) highways reduce traffic disturbance and noise effects on the surroundings (but note the vertical reflecting walls and elevated traffic ramps). A walkway overpass and 10 road bridges in the distance provide connectivity across the highway (although road alignments suggest that the upper and lower neighborhood was split). Homogenizing earth material in construction, smoothing of its surface, and intensive mowing of the grassy areas have created a large area in the city almost devoid of ecological value (1985 photo), representing a challenge to society to discover alternative approaches. Interstate Highway 630, Little Rock, Arkansas. Photo courtesy of U.S. FHWA.

City traffic flows never stop. The 10-times-a-week, big traffic jams reflect the arrival and departure of commuters. The vehicle saturation of parking lots and the generation of vehicle-caused pollutants also track this combination of never-ending and commuter-pulsing traffic. A stormwater-drainage piping system is designed to take as much of the natural precipitation provided to the city as fast as possible out of the city and, in the process, clean the streets of pollutants. In general, former streams are in pipes, former wetlands have been filled, and major water bodies receive the stormwater pulses and pollutants.

The suburban landscape composed of suburban towns is blanketed by inputs from both the adjacent city and the surrounding rural area. Heat and chemicals from the city plus dust and species from the country cover suburbs.

Yet suburbs are also sources of effects, as commuters head for the city and weekend recreationists head for the country. Moreover, the city tends to push outward, converting inner suburb to city, while the outer suburb expands by rolling over rural land.

In suburbia, land uses form distinct patches: the town centers; commercial shopping areas; industrial and office parks; high-, medium- and low-density residential areas; and often a matrix of mixed-use areas.[732,217,68,70] The overall road network has a wavy net form with a strong hierarchy, from local roads to multilane major arterial highways. Road density typically averages about 10 to 30 km/km^2 (16 to 48 mi/mi^2). Fine networks providing access to and from homes in housing developments and residential neighborhoods are especially distinctive in suburbs. Cul-de-sac spur roads may be common in some residential neighborhoods. Varied collector and arterial roads with traffic volumes from a few hundred to a few tens of thousands of vehicles per day connect major land uses covering suburbia.[2,3] Cutting through the suburban landscape are multilane highways, sometimes going around the city but often connecting the country to the city. The suburban road network, in general, is variable and complex.

Nature in suburbs usually includes a few large, relatively natural green areas.[87,732,217,305,68,70,322,458] At the other end of the scale are the shreds of nature surviving by the thousands in suburban house lots.[104] Between house lots and large natural-vegetation patches is a complexity of somewhat degraded nature, including cemeteries, golf courses, farm fields, school yards, town parks, corporate parks, river corridors, stream corridors (in stretches where the water is not underfoot in pipes), bike paths, power lines, rail lines, and lines of street trees.

The daily pulses of commuters between city and suburb (and country) and between places within suburbia are pronounced. Since the majority of North America's cars are housed in suburbs, traffic flows on the complex suburban wavy net are complex indeed. Trucks carrying goods to the city are especially concentrated on certain highways at certain times.

Sprawl is more difficult to describe. The varied concepts generally agree that sprawl has areas of low-density houses, groups of houses tend to be separated, and strip development is usually present.[788,383,68,70] Most houses are relatively new, representing recent or ongoing growth. Although the process of sprawl occurs in parts of some existing suburban towns, we focus on the characteristic zone of sprawl outside of the suburban landscape.

Some people observe that sprawl is simply a product of market forces without the guiding hand of government. Thus, despite inevitable environmental problems, sprawl provides a range of benefits to society.[715] These include linking homeowners with nature, providing for self-reliance and private space, and offering an alternative to life in city, suburb, or isolated rural community. In fact, sprawl may simply be today's term or manifestation for a process that in earlier eras gave rise to our present cities and suburban towns.

Alternatively, some people liken sprawl to a cancer or a process out of control. Modern North American cities sprawl at a rate "unhealthy" in proportion to what they support. Between 1970 and 1990, for example, Chicago's population increased by 4% while its surface area expanded by 50%[217] (see Figure 2.4 in Chapter 2). The average ratio in the USA for growth in area versus population is 2.3. In other words, the developed land area expands by 23% with a population increase of 10%.[17] Some cities, such as Baltimore and St. Louis, have experienced five to seven times faster growth in area than in population.

As metropolitan regions sprawl outward, road infrastructure is needed to connect communities to services and institutions as well as to one another. Yet this observation raises a long-standing (some say chicken-and-egg) controversy in the transportation community.[674,208] Do roads catalyze development, or does development lead to roads? No doubt both cases exist. Some roads are built to encourage economic development,[197,198,524] while from a state transportation perspective, most are built or expanded to serve existing development.

Irrespective, the road network in sprawl areas is mainly the preexisting network serving rural land uses and people. This tends to be supplemented by long driveways and local roads to groups of new houses. Traffic volumes overall are low but cause consternation because the rural road network was not built for the added vehicles from many new homes. Surprisingly, no landscape ecology analysis of sprawl areas seems to have been done yet.

Ecological Effects

Nature in cities is tersely described as somewhere between extremely important and absolutely absent. For some urban residents, the geranium on a windowsill, the tree by a park bench, or the cluster of energetic sparrows is a delight, an inspiration, or a tangible link to a place in the country or where one grew up. In contrast, urban ecologists recognize with scientific interest the diversity of native plants, the migrating birds that stop by a cemetery, and the high densities of squirrels, raccoons, and proverbial insects in cities.[356,186,732,890,88,113] Horticulturalists, landscape designers, schoolteachers, and meteorologists see other pieces of nature in cities.

City roads and vehicles produce many varied effects on human health.[675] Ecologically, streets radiate heat, parking lots provide microsites for plant growth, vehicles emit pollutants, and traffic keeps particles in the air, all of which affect urban organisms. Indeed, streets funnel rainwater directly into stormwater drains, and raccoons *(Procyon)* love stormwater drains. But overall, the road system has a relatively minor ecological effect in the city.

However, like suburbs themselves, the ecological effects of roads and vehicles in the suburban landscape are complex. Suburban growth and supporting surface-transportation systems reduce vegetative cover and produce large areas of relatively impervious or hard surface, including buildings, roads, driveways, and parking lots. Hard surfaces effectively alter watershed hydrology, producing rapid, high-volume, water-runoff patterns[424,1070] (see Figure 7.2 in Chapter 7). Hard surfaces typical of suburban development reduce the infiltration of rainwater into soils.[283]

In terrain with natural vegetation, most precipitation water commonly infiltrates into the ground. Studies in several suburban landscapes suggest that when *hard surface* (hardened or impermeable) covers 5% to 10% of the area, about 10% of precipitation water becomes surface runoff that flows into receiving waters, such as streams and lakes, and the rest infiltrates into the ground[283,151] (Figure 12.2). When hard surface reaches 30% to 40% of the area, about 30% of precipitation water becomes surface runoff. When hard surface

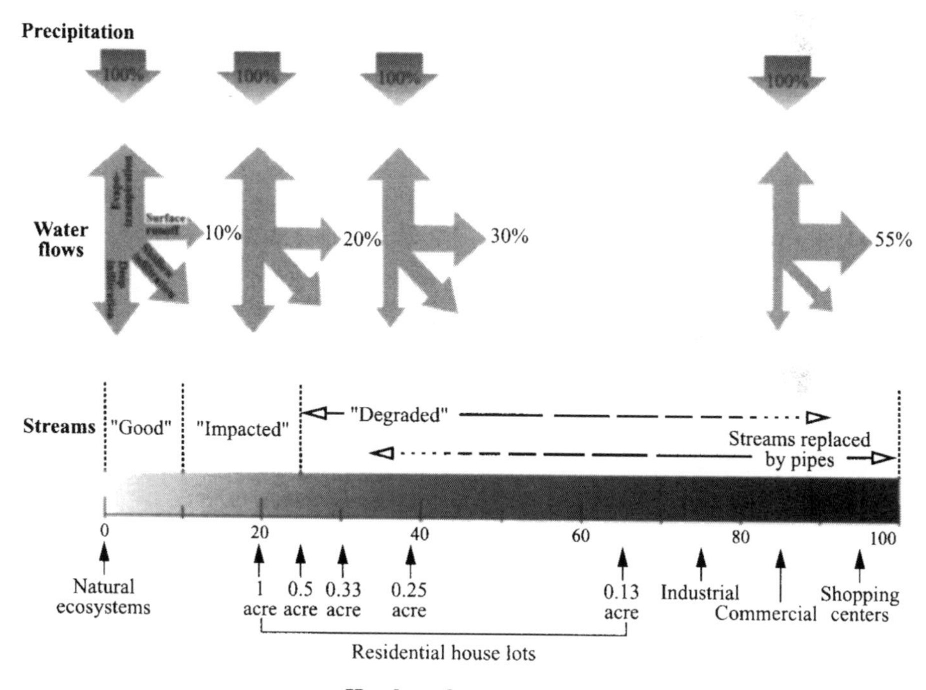

Figure 12.2. Generalized model of water flows and stream quality related to the amount of hard surface in a watershed area. Hard or impervious surface is largely composed of roads, parking areas, and buildings. For surface runoff and shallow infiltration into the soil, the amount of water flow and the pollutants carried in it have major effects on the overall quality of streams in an area. General water flow diagrams in the upper portion are for 5%–10%, 15%–25%, 30%–40%, and 80%–90% hard surface, respectively. Adapted from USDA Soil Conservation Service (1975), Schueler (1992, 1995), Environmental Protection Agency (1993), Arnold and Gibbons (1996), and Center for Watershed Protection (1998).

Table 12.1. Generalized relationship between hard-surface (impervious-surface) cover and stream quality. Changes expected in most streams once hard-surface cover of the contributing watershed land area exceeds about 10%. Adapted from Schueler (1995).

Water—Hydrology
- Increased flood peaks
- More frequent bank-full flows
- Lower stream flow during dry-weather periods

Stream Habitat Structure
- Widening and/or deepening of the stream channel
- Increased streambank and channel erosion
- Increase in artificial channels relative to natural ones
- Fewer logs and branches in streams
- Loss of pool-and-riffle structure
- Reduction in stream habitat diversity
- Decline in streambed quality due to sediment deposition

Water Quality
- Increase in stream temperature
- Higher stormwater pollutant loads and concentrations
- Increased risk of coastal shellfish-bed closure
- High probability that fecal-coliform levels exceed water-contact recreation standards

Aquatic Invertebrates and Fish
- Decline in aquatic insect and freshwater mussel diversity
- Decrease in fish habitat quality
- Decline in total number of fish species, but increase in non-native species
- Loss of sensitive coldwater fish (trout and salmon)
- Reduced spawning of migratory (anadromous) fish
- Higher potential for fish barriers where stream crossings are more frequent

Floodplain
- Fragmentation of riparian forest corridor
- Decline in wetland plant and animal diversity

reaches 80% to 90% (approximating the boundary between city and suburb), more than half (about 55%) of precipitation water washes away as surface runoff. New suburban development typically has 30% to 60% of the area as hard surface, composed largely of roads and buildings. As a result, groundwater supplies may not be fully recharged, streams tend to degrade, and flooding often increases (Table 12.1).[825]

As the water runoff moves over hard surfaces, it collects contaminants, which leads to increasing pollution in water bodies such as streams and lakes.[424,907] In addition, hard surfaces absorb considerable solar energy. This heats the air and the water, which in turn negatively affects aquatic communities. An increasing body of scientific research, conducted in many geographic areas, supports the theory that the percent of hard-surface land cover is a reliable overall indicator of the condition or degradation of streams in an area[825] (Figure 12.2 and Table 12.1).

Road density (road length per unit area) typically correlates closely with the amount of hard-surface cover, which is typically dominated by buildings, roads, and parking areas.[151] In one example from the Seattle (USA) region, a road density of 5 km/km^2 (8 mi/mi^2) corresponded to a hard-surface cover of about 20% in the watershed. A *hard-surface model* (or *impervious-cover model*), developed after reviewing hundreds of studies from across the USA, suggests that stream networks are "impacted" at levels of hard surface as low as 10% coverage[151] (Figure 12.2). Although many stream attributes are affected (Table 12.1), reduced populations of sensitive native fish are typically an effective overall measure.[470,159] Once hard-surface coverage exceeds 25%, streams in the area tend to be *degraded,* a condition characterized by unstable channel morphology, polluted water, and highly altered or impoverished fish communities.[151,1070] Empirical relationships between the percentage of hard-surface cover and the diversity of fish species and of aquatic invertebrates have been developed, and are useful for planning and management.

Thus roads play a major role in aquatic pollution in suburban settings. Roads reduce water infiltration into groundwater, cause heat buildup, and accelerate flow of water and chemical pollutants through a storm-drain system. Water flow is accentuated by the presence of the curbs and gutters associated with traditional street architecture. Curbs provide an effective trap for accumulating leaves, trash, snow, roadsalt, sand, particles from pavement deterioration, deposits of airborne sediment, and vehicular particles and drippings. If not removed by a road-sweeper, this material collects along the curbside and in cool climates may be mixed with snowmelt. Typically, this array of material and pollutants is carried downward into a storm drain and then to a water body during the first flush of rainfall, as water courses off the road.

The flush of contaminated stormwater from streets into receiving streams and lakes is made worse in suburban settings by two contributions associated with residential homes. First, due to widespread nutrient fertilization of lawns, the levels of phosphorus are higher in water from residential driveways (average 1.16 mg/liter) and streets (0.63 mg/l) than from either urban highways or commercial parking lots (0.42 mg/l).[825] Second, because of conspicuous piles of dog feces along suburban roadsides, the levels of fecal coliform bacteria (some pathogenic) are considerably higher on residential streets (average 37 000 counts/100 ml) than on commercial streets and parking lots (2500 to 12 000 counts/100 ml). Together, these nutrient and pollution sources associated with suburban road networks contribute to freshwater eutrophication with nuisance algae blooms.

Suburban—and, to a lesser extent, sprawl area—road systems also exert many significant effects on plants and wildlife.[732,88,68,70,315] The mostly narrow roadsides are strewn with non-native plants.[890,113,393] Although some of these non-native species probably invade relatively natural areas (Chapter 4), the abundance of house lots and town centers forms a much larger source of exotic species. Heavy traffic volumes mean that roadsides and adjoining space are heavily coated with chemical pollutants, including nutrients, toxic substances, and roadsalt in cold climates (Chapter 8). Traffic movement keeps particles with attached pollutants mobilized in the air. The abundance of culverts and bridges (Chapter 9) in suburbs means that streams are channelized and disrupted at frequent intervals along their length. Frequent roadbeds themselves alter hydrological flows and wetlands present (Chapter 7).

Wildlife is also strongly affected (Chapter 5). Road-kills are especially conspicuous in suburbs. Although the bulk of the animals killed are common species, such as domestic cats, squirrels, and raccoons, the abundance of traffic means that some uncommon species are killed in relatively high numbers. Local roads with little traffic are not barriers to most wildlife movement, but such roads in suburbs are mainly limited to residential areas often overrun with cats and dogs. In the somewhat natural areas of suburbia, roads tend to be through routes with relatively large traffic volumes. Such roads, especially multilane highways, can be formidable barriers to animals (Chapter 5).

High traffic volumes mean that traffic noise effects extend outward long distances[770,771,322] (Chapter 10). Because the road density is high, few areas in suburbia escape the effects of traffic noise. An example for an entire urban region is instructive. The total area estimated to be directly affected by the road system in the Quebec City region was mapped using road-effect zones for roads with different traffic volumes (Chapter 10) (R. Balej and M. Lee-Gosselin, personal communication). The overall image resembles a scan of a heart, with major wide arteries radiating outward, surrounded by a system of progressively smaller and "squigglier" arteries. Wide road-effect bands run along both sides of a major river. On a limestone plain to the south, small rectangular-like enclosures are separated by medium-width bands. On rolling granitic forestland to the north, large amoeboid enclosures are separated by narrow road-effect zones. Traffic-disturbance-sensitive species would have difficulty crossing the region in any direction, and most would find only certain enclosures large enough to inhabit.

But the ecological effects of road systems need to be viewed in the context of nature in suburbs. The relatively high human-population density produces an array of ecological effects, which means that only a few large natural-vegetation patches remain in relatively natural condition.[305] These important large areas are surrounded by smaller corridors and patches, which are green but represent varied stages of degraded nature. Various rare state-listed species are present or disperse into suburbia, relatively few thrive there, and probably very few can be sustained there for the long term. Few aquatic ecosystems in suburbia are even close to natural. Superimpose the suburban road system on this pattern of nature, and human impacts are accentuated. Fewer and smaller areas remain natural, chemical pollutants are much more dispersed, water flows are acutely disrupted, aquatic ecosystems are highly degraded, and significant barriers are seemingly everywhere. Yet an increasing percentage of the human population lives in suburbs. People in suburban towns see, learn about, and may depend on the nature around them. Caring for that nature, and understanding their dependence on it, translates into local, state, and national policies affecting nature everywhere.

Low-Impact Development and Clustered Development

A combination of *preemptive planning* and *retrofitted design* can reduce the deleterious effects of suburban roads on aquatic ecosystems. Both approaches address current suburban stormwater management that focuses on a carefully engineered conveyance system and is often costly to construct and maintain. Normally, rainwater and snowmelt from roads are carried as fast as possible out of a built area in a piping system to minimize the flooding of basements and septic systems and to maintain road stability. The water and pollutants transported then cause environmental problems beyond the piping system. Two alternative but related approaches, low-impact development and clustered development, are sentinels for how to reduce the effects of roads on stormwater quantity and quality.

Low-Impact Development

Low-impact development disperses effects widely to avoid creating places with severe environmental alteration or degradation. How we design road networks is of cardinal importance in affecting the watersheds of which they are a part. The title of a report, "Start at the Source: Residential Site Planning and Design Guidance Manual for Stormwater Quality Protection," highlights an important dimension of modern suburban stormwater-management thinking.[775] Comparative economic analyses have shown that the higher up in a watershed that stormwater controls are placed, the less expensive the overall costs will be to reduce the effects of hard surfaces, including roads, on water quality. Centralized *outlet management* is the most expensive option, and smaller *enroute management* is intermediate. Large detention ponds illustrate outlet management, and stormwater wetlands illustrate enroute management. Underground rain barrels, "rain gardens" designed to hold water from heavy rainstorms, and terracing are examples of *headwater management.* In short, *starting at the source* with dispersed headwater or micromanagement solutions appears to be best both ecologically and economically.

Residential streets are at the center in the design of suburban communities in North America (Figure 12.1), often comprising 10% to 20% of the suburban land surface area as well as over half of the total impervious land coverage. Thus it has been said, "This makes street design the single greatest factor in a residential development's impact on environmental quality."[775] In the past, most residential streets served multiple uses, such as playgrounds and walking and meeting places. Now, for safety reasons, roads are restricted almost solely to vehicular traffic plus a limited amount of bicycling. The paved area associated with roads has commonly increased by up to 50% in the past 50 years, due to new residential streets, their progressive widening, and the addition of paved sidewalks.[775]

Traditional urban planning in the USA is based on a set of municipal standards for street design corresponding to vehicle usage in a roadway classification system (local, collector, arterial).[2] In addition, the width and arrangement of neighborhood streets are commonly linked to their use by increasingly large fire trucks. A modification of this approach matches street widths to anticipated traffic volumes, allowing "headwater streets" with little usage to be much narrower than downstream streets.[2] So, just as stream widths increase the lower one gets in a watershed in order to accommodate increased water volumes, with this approach most streets across a neighborhood would be of minimal width. Unlike traditional practices, in which residential streets tend to be uniformly wide, here wider streets would be present only at specific locations to support greater traffic volumes. Under such a strategy, less hard surface would typically be required, and therefore the deleterious effects of stormwater runoff on aquatic ecosystems would be reduced.

Although headwater streets in residential areas receive relatively little traffic flow, they commonly represent 50% to 65% of the length, and surface area, of the road network in a residential area.[825] A variety of additional methods can be employed in low-impact development to reduce the amount of hard-surface coverage.[740] For example, in an area with 50 homes, by narrowing headwater streets from the conventional 36-foot (11-m) width to a 26-foot (8-m) width, over 12 000 ft^2 (1115 m^2) of paved-road coverage can be eliminated.[775] Assuming construction costs of about US$10 per square foot, this impervious surface reduction would save $125 000, or $2500 per house lot.

Many other low-impact development approaches can be planned or retrofitted to suburban environments to mitigate the effects of roads on aquatic ecosystems.[1070] All are based on reducing the amount of hard surface and on finding ways for the surrounding landscape to retain stormwater runoff, allowing it to infiltrate to the groundwater. For instance, in many locations, sidewalks need to be present only on one side of a street and can sometimes be constructed of porous pavement.[283,825] Curbs can be removed or modified to allow surface flow into *bioretention swales* (small "pocket wetlands") planted with attractive species that will also enhance the diversity of wildlife (Chapter 8). Most parking lots and home driveways can be made hydrologically sensitive by using porous construction materials, minimizing areal coverage, and planting diverse vegetation.

In addition to shrinking the size of neighborhood streets, the form of their layout on the landscape can have a substantial influence on total imperviousness and site hydrology.[740] For example, shifting from the typical right-angled gridiron layout to one characterized by "loops and lollipops" can result in a total reduction in hard surface of roughly 25%. Even attention to small details can help, as demonstrated, for example, by the turnarounds on dead-end streets in residential subdivisions. By decreasing the turning radius from 40 to 30 feet (12 to 9 m) and/or including a drainage island in the middle, impervious coverage can be reduced by almost half.[825] Another alternative to traditional design is to replace the circular turning radius with a T-shaped (or hammerhead) option, which requires only about one-third the paved surface area for turnarounds.

Furthermore, how roads are placed on the suburban landscape will have just as strong an influence on hydrologic processes and aquatic ecology as either the amount of roads or their layout design.[775] The typical approach, in which conventional curb-and-gutter streets with storm sewers are placed in low areas, concentrates hydrologic surges from storms. This pattern promotes pulse flooding and erosive scouring in the downstream aquatic ecosystems (Chapters 7 and 9). Alternatively, a suburban road-development project working with, rather than against, site hydrology permits natural water pathways in the land to mainly determine the location of vehicular pathways. For example, residential roads placed along ridgelines preserve the natural drainage patterns. This approach helps protect the lower areas where water converges, which are then addressed with the best management practices (Chapter 8) that retain natural hydrologic conditions.

To *disconnect hard surfaces* is yet another stormwater-management guideline for reducing impacts.[740] Instead of creating hard surface or pipes from home driveway to water body, the water route is broken up as often as possible. Each break then uses techniques to get water back into the soil, where it may infiltrate to groundwater or be pumped vertically in evapo-transpiration by natural vegetation. Disconnecting hard surfaces reduces water flow velocity and flood level, as well as the amount of pollutants reaching stream and lake ecosystems.

Finally, an important planning message emerges. Every road has a distinctive water flow and also a unique location in the network of flows in a watershed.

Clustered Development

One of the most efficient ways to reduce impervious surface cover is *clustered development,* in which conventional subdivision lot dimensions are relaxed to allow for a more compact building footprint, which also saves undeveloped land in reserve.[825,1070] The more the buildings are clustered, the less the road length that is needed (Table 12.2).Various economic advantages should be evident. Depending on the original lot size and road layout, hard surfaces can be reduced by up to 50%. With clustered development, the same length of road can serve a far greater number of people than with dispersed developments. This building strategy also allows for the protection of large open spaces to be available for stormwater detention or water quality improvement. Moreover, the open spaces resulting from clustered development will provide other

Table 12.2. Generalized benefits of clustering rather than dispersing residential development. Adapted from Schueler (1995).

Water, Erosion, and Runoff

- Reduces on-site hard-surface cover by about 10% to 50%, depending on original property size and road network
- Reduces soil erosion potential, because about 25% to 60% of site is not cleared of vegetation
- Reduces stormwater runoff and pollutant loads
- Reduces the size and cost of stormwater quantity and quality controls
- Concentrates runoff where it can be most effectively treated
- Provides a wider range of feasible sites to locate "stormwater best management practices"

Economics

- Reduces capital cost of development by an estimated 10% to 33%
- Reduces the cost of future public services needed by the development
- May increase future residential property values
- Provides partial or total compensation for land devoted to natural resource protection

Open Space

- Reserves about 25% to 50% of site as green space, unlike dispersed development
- Reserves about 15% of site in open space dedicated to passive recreation
- Creates larger habitat patches for wildlife and biodiversity

Broader Community Goals

- Increases sense of community and enhances pedestrian movement
- Reduces potential pressure to encroach on nearby natural resource areas
- May enhance farmland preservation, affordable housing, architectural diversity, and so forth

non-stormwater benefits, including wildlife, recreation, and aesthetics,[27] in cases where the spaces remain open space rather than sites for future development.

Strip Development

As distinctive fingers or protrusions extending along major roads from towns,[715] *strip development* is probably effective at blocking the movement of wildlife across the landscape, especially in open landscapes with relatively little vegetation cover. In addition, the strip is commonly colonized by aggressive non-native, generalist, or domestic species, and traffic noise and disturbance are prominent. Both factors inhibit some wildlife. Furthermore, in some areas the shrinking distance between the tips of fingers from adjacent towns (combined with sprawl) increasingly threatens to subdivide the land, causing a new broad-scale fragmentation. Strip development thus is especially a threat to certain large-wildlife species, which presumably require foraging, dispersal, and migration movements for survival.[400]

Forestry Land and Road Systems

Forestry is practiced in moist areas from the tropics to cool regions. Roads provide access to the forest and thus support a variety of management activities and forest uses. The U.S. Forest Service manages nearly 10% of the length of the public road system of the United States (Chapter 2), and in some states forestry roads on private lands are widespread. Roads are essential for many forestry operations, including the delivery of logging equipment to individual cutting units and the transport of logs to mills. Road networks in forestry lands are also used for recreation, fire detection and suppression, hunting, fishing, and a host of other activities.

Steep terrain characterizes forestry lands in many parts of western North America; gentler terrain, in central, southern, and eastern parts of the continent. Terrain sets the stage for many road network characteristics, traffic volumes, and associated ecological issues. Thus forestry lands in both mountainous and somewhat flat terrain are considered here, with a substantial emphasis on the effects of topography.

Terrain, Land Use, and Road System

Topography, distribution of forest resources, and concerns for environmental protection strongly influence the geographic distribution of forestry roads to serve forestland uses, especially in mountain landscapes.[378] In landscapes of gentle slope, roads may be placed across the landscape in simple grid patterns, but steeper slopes impose a series of constraints on locations and forms of road networks. A simple comparison of three mountain landscapes shows the diversity of these interactions (Figure 12.3).[331] In steep areas of the Oregon Cascade Mountains, the primary forestry-access roads are on valley floors, and secondary roads climb hillslopes to reach ridges, providing access for logging operations. Many areas on the western slopes of the Sierra Nevada Mountains in California have broad ridges where the primary roads are located, interspersed with deep narrow canyons (Figure 12.3). Secondary roads drop downslope from the ridges onto the upper slopes of the canyons. In parts of Southeast Alaska, such as Prince of Wales Island, high-value forests are located on steep lower hillslopes and broad, marshy valley floors long ago carved by glaciers. Here primary forestry roads follow the edges of the valley floors, and secondary roads may extend short distances up lower hillslopes. These contrasting patterns of roads across landscapes can have quite different effects on ecosystems, even for similar road densities. For example, the Cascades case may result in greater proximity of forestry roads to stream ecosystems and hence greater ecological impact than in the Sierra Nevadas.

Much of the steep U.S. Pacific Northwest and Northern Rockies land shares similar historic patterns of forestry-road development. In the late 1800s and early 1900s, logs were mainly transported by river and railroad. By the mid-1900s, road access was developed predominantly up the main valley floors

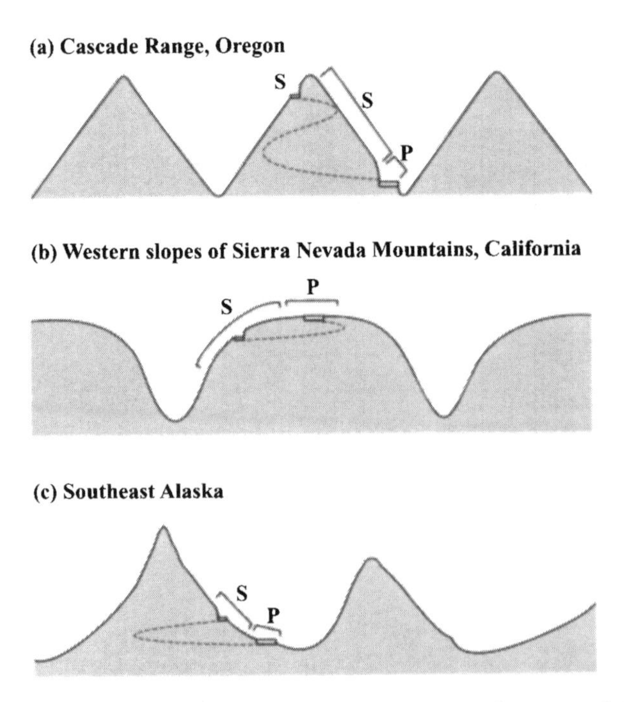

Figure 12.3. Typical locations of primary (P) and secondary (S) forestry roads in different mountain landscapes.

and then up the hillslopes to the ridges. Where possible, ridge roads are used to access *landings* (flat areas of 0.4 ha [1 acre] or so), where logging equipment is positioned so that a tower-and-cable system can be used to move logs upslope to the open landings. In gentler terrain, off-road tractors and rubber-tired equipment may be used to transport logs to landings. Logs are collected at the landings and placed on trucks for transport to mills (Figure 12.4). To some extent, the road network mirrors the trail systems used by earlier native people, who followed both river and ridgeline networks, to capitalize on diverse resources associated with these two most-separated parts of the landscape.

The location of roads in this landscape, therefore, depends on several key factors: the desired location of landings; the size of *cutting units,* which is constrained to a certain extent by the capabilities of the logging systems; effects of topography on logging systems; and environmental considerations, such as protection of streams and riparian zones (Figure 12.4).[332,373,175,667,376] Logging

Figure 12.4. Log-hauling truck and logging system on adjacent slope as important determinants of forest-cutting-unit size, road density, and network form. Douglas-fir *(Pseudotsuga menziesii)* and tamarack *(Larix)* being harvested in 1976. Note the cutbank, where, during moist periods, groundwater seeps out to become surface water exposed to solar heating and with erosion potential. Washington, near Potlach, Idaho. Photo courtesy of USDA Soil Conservation Service.

equipment developed in the second half of the twentieth century had the capability of harvesting trees from individual units of about 10 to 30 ha (25 to 75 acres) each. Therefore, these dimensions of cutting units set the grain of the road network necessary to support forestry operations. Other uses of forest roads, including access for recreation and fire detection and suppression, are generally fully served by the road network developed for forestry operations.

Constraints on road grade (steepness) and turning radius, especially considering the hauling of long, heavy loads of logging equipment and logs, also significantly influence road network design. Forestry roads can follow relatively flat valley floors, but as roads climb steep slopes, they traverse diagonally across the slopes until they reach ridges. This pattern creates several different arrangements of road segments relative to the nearest stream. Roads parallel rivers on the valley floor but cross streams in traversing steep slopes. Roads along ridges are above the highest tips of streams and receive little runoff from higher slopes, so the amount of water intercepted by these roads is normally minimal. These relationships strongly influence the interaction between forestry roads and streams, which is mediated by the movement of water, sediment, and landslides[1018] (Chapter 7).

Development of the cumulative road imprint on federal forestland over time changes in response to shifts in policy, management objectives, and technology.[175,376,405] During the mid-1900s, when protection standards for stream and riparian ecosystems were lower, the main forestry-access road segments were generally constructed in valley floors and in mid-hillslope locations, where environmental impacts are most acute. A common history in many parts of the western USA shows that road systems constructed on steep hillslopes contributed to the occurrence of landslides and other erosion problems triggered by storms.[843] Recognition of these relationships led to the strengthening of standards for road location, design of road-drainage systems, and road maintenance, in an effort to reduce environmental impacts in the future.[843] By the end of the 1970s, additions to the road system were focused on ridges to minimize environmental impact. However, this shift occurred rather late in development of the networks, and most high-impact portions of road networks remain.

These factors set the broad geographic and temporal patterns of road networks that support mountain-forestry operations.[1017,760,918] Constraints of steep, complex topography commonly contribute to road networks becoming a hierarchical branching pattern. The average size of cutting units helps establish an overall grain size for the landscape, which results in road densities typically of about 1.2 to 2.5 km/km^2 (2 to 4 mi/mi^2) in both federal and private forest lands sampled in the U.S. Pacific Northwest.[1017]

Locally, road density is much higher where early logging systems with limited reach fostered construction of closely spaced roads. On gentle terrain, road densities may be lower where heavy off-road equipment is involved in harvest operations.[175,319,817] Over time, various efforts have attempted to balance the environmental effects of off-road logging equipment, which can cause forest soil compaction, against the environmental effects of engineered forestry roads. However, even if modern equipment makes it possible to conduct forestry operations with a less-extensive road system, commonly the road system from a former era would remain and cause some ecological effects.

Traffic patterns are believed to generally correlate with the type of segment in the forestry road network, with highest traffic volumes and most diverse uses along the main segments of the road network. Traffic patterns may be quite irregular in time, with use most intensive during periods of log hauling and rather light during other periods. A rhythm of recreational traffic may be superimposed on certain forestry road segments.

At the beginning of the twenty-first century, after decades of expanding forestry roads in North America, many forestry lands, both public and private, are essentially fully "roaded." Forestry roads are expensive to maintain, and timber harvest levels are down in many areas. A change is under way in some forestry lands.[405] The U.S. Forest Service is beginning a process of *road system reduction* in response to a declining budget for maintenance and a drastic decrease in harvest levels, which reduces the need to transport logs. An additional strong motivation for reducing the extent of the road network is the renewed emphasis on protecting biodiversity and water resources. Cutting levels are down on private forestland in many eastern states of the USA, and forest cutting levels have also decreased in some areas of Canada. Thus reducing road networks is of widespread interest and has both economic and ecological importance.

Road network contraction is a slow, costly process. It has begun most vigorously in areas where benefits are most obvious and endangered species need protection, including areas of Idaho and the Pacific Coast, where road-related landslides have a high potential to damage salmon habitat.[376] Reduction in the extent of forestry roads has the added ecological benefit of reducing human access into more remote areas or large blocks of natural habitat. Road reduction may significantly decrease direct human effects—for instance, of recreational activities—on sensitive habitats, rare species, and aquatic ecosystems.[495,547,405] Reducing potential indirect impacts, such as unregulated hunting and fishing, provides benefits for maintaining wildlife and fish populations.[114,1050] And, of course, people-management costs are reduced by closing and removing forestry roads.

A great variety of methods, ranging from temporary road closure to entire road removal, is used to reduce the level of road impact on terrestrial and aquatic ecosystems.[376,405] A simple road treatment to reduce human access is to place locked gates at entrances to roads. This road-closure technique is normally temporary. Various types of blockages and roadbed removal near the road entrance tend to be more effective and long term. Another approach to limit travel uses *waterbars,* diagonal trenches crossing a road that divert road-surface or roadside-ditch water onto hillslope areas at numerous points, thereby not concentrating water at specific sites that might be prone to gully or landslide erosion. Waterbars can be constructed either to permit or to block vehicle access, depending on how deeply they are cut into the road surface.

A more intensive and expensive road-closure approach involves removal of roadbed fill along the general road right-of-way or specifically at stream crossings.[582] Finally, the entire removal of roads, as done in certain forestry areas of Alberta and in Redwood National Park, California, involves reconstructing the pre-road hillslope topography. In flattish areas, former earthen road surfaces may be *decompacted* (e.g., by bulldozing, plowing, or tilling) to produce a porous substrate for rapid vegetation growth. These approaches, which effectively modify road fill and reconstruct hillslope or surface topography, normally require the movement of earth by heavy equipment, creating temporary environmental degradation. The effects of this process can be expected to last until vegetation with or without other erosion-control measures stabilizes the newly established earthen surface.[582]

The simplest part of a road network to close is the small, short dead-end road, or *spur road.* Short spur roads commonly carry low traffic volume and may be located in parts of the landscape, such as near ridges, with relatively low impact on watershed resources. Longer spur roads often provide access to a more remote watershed and may traverse a valley floor, with associated effects on riparian and stream ecosystems. Typically, an overriding benefit of closing spur roads is the reduction of human access into a remote area. This is a highly effective and cost-efficient way to protect biodiversity, aquatic ecosystems, and fish and wildlife populations. Of course, measures of success are more appropriately focused directly on reducing ecological effects than on reduction in road length.

Forestry road systems in areas of low relief, such as in the southeastern USA, can be laid out on a simpler grid. Although landslides are not an issue, concentrations of drainage water may aggravate gully erosion (Chapter 7). Vegetation types and rare animal habitats are heterogeneously arranged across the forested landscape.[300,378] Irregularities in the grid are often imposed by wetlands and flooding associated with floodplains. In addition, land-use history, a mosaic of land ownerships, and multiple access points often produce highly irregular forestry road network forms.

Ecological Effects

Roadsides along forestry roads generally have limited impact on adjacent native vegetation except where dust is a problem (Chapter 10) and where non-native species are introduced (Chapter 4). The establishment of non-native plants, insect pests, and pathogens in forestry landscapes can be facilitated by roads.[703,376] Roadsides penetrating forest lands provide favorable habitat for weedy species, especially on bare soil and in sunny areas. Vehicles and passengers may serve as dispersal vectors for the spread of seeds. Despite the small total surface area, road networks create the potential for extensive impact on the landscape due to their wide distribution. Management actions or natural processes in adjacent areas may alter roadside vegetation, which facilitates the spread of non-native species from roadsides into neighboring land. In some areas, non-native pest species may reduce forest productivity as well as wildlife habitat value.

Other pests and pathogens can be dispersed across forestry lands via vehicles. Spores of root decay fungi, such as *Phytophthora lateralis,* which attacks and kills Port Orford cedar *(Chamaecyparis lawsoniana),* can be dispersed in soil attached to vehicles such as log trucks and road construction machinery.[1073,376] Gypsy moths *(Porthetria dispar)*, which periodically defoliate large areas of forest, are dispersed by vehicles both locally and regionally.

Forestry roads can also harbor uncommon or rare species. In Britain, these very lightly traveled 5 to 15 m (16 to 49 ft) wide strips or "rides" in woodland contain plants and butterflies that would thrive in meadows.[720] However, with intensive agricultural cultivation surrounding woodlands, the roads used for selective logging are among the few habitats remaining that support these species.

Terrestrial animals, including game and nongame species, abound in forestry landscapes. Many vertebrates are subject to disturbance by road networks and traffic, and the human access they provide.[1050] However, road-kill is generally not a big issue, because traffic volumes are low and speeds are restricted by narrow, winding roadways. Road-kill rates for mammals are probably highest on two-lane highways through forests where visibility is limited for both animal and driver. Amphibians and reptiles are frequently road-killed near wetlands and other water bodies along these highways. Forestry roads with very low traffic tend to be well-used travel corridors for many wildlife species, including large predators.[302]

Forestry roads also provide important human access for animal viewing as well as both legal hunting and poaching. In recent decades, road access has sometimes been regulated seasonally to protect animal populations from harassment by humans (Chapter 6). In the case of elk, for example, road closures may be instituted to limit interruption during the breeding season or to enhance the hunting experience by eliminating vehicle traffic and associated disturbance in selected hunting zones. Thus, in a general sense, managing road networks can be a form of managing the relationship of humans with large mammals.

Poorly designed and maintained forestry roads represent one of the most important sources of nonpoint pollution in the landscape. In forestry landscapes of the western USA, road damage to aquatic resources, especially where salmon *(Oncorhynchus)* run, has been a greater management issue than the damage to terrestrial wildlife.[376] Several key issues affect the protection of fish and fish habitat as well as the recovery of migratory movement patterns by salmon[666,376]: (1) increase and/or decrease of peak storm-flow levels; (2) concentration of suspended sediment in stream water; (3) extent of fine sediment that covers, or chokes, the gravel important for spawning; (4) alteration of the beneficial effects of streamside vegetation on stream ecosystems; (5) impacts of road-related landslides; and (6) blockage by poorly designed, constructed, or maintained culverts and bridges. Forestry roads have been implicated as culprits in all of these issues, though other factors are typically also involved.

Alteration of stream-flow regimes is a critical and complex issue in river management, so attention has been focused on the full distribution of flows rather than only on high or low flows.[733] Roads have mainly been a concern in terms of potentially increasing peak flows (Chapter 7). Excessive peak flows, for example, can persistently disrupt aquatic habitat and dislodge eggs or fish from the streambed. On the other hand, reduced peak flows can limit the periodic removal of fine sediment from spawning gravel, an important process for maintaining fish spawning habitat.

The concentration of *suspended sediment* in streams affects the quality of both water and streambed habitat. Muddy or turbid water benefits some fish species at certain times when, for instance, it provides cover and protection from predation. On the other hand, excessive fine sediment may have direct physiological impacts on fish or may disrupt their migratory patterns. It also plugs spaces between gravel particles, impeding the flow of oxygenated water to fish eggs incubating in spawning beds. Traffic produces fine sediment by pulverizing roadbed material and by mixing the sediment with rainwater. Both processes increase sediment runoff to streams.[766]

Streamside vegetation has many beneficial effects on stream ecosystems, including providing shade.[331,373,667] It also provides leaves and twigs, which serve as a food source for aquatic invertebrates, and large wood, which creates structure in the aquatic habitat[470] (Chapter 9). By removing vegetation in riparian zones, the presence of roads can directly alter these important vegetation functions. Indirectly, roads provide human access that results in alteration of streamside vegetation. For example, roadside trees may be removed from riparian zones where the trees pose a safety hazard to travelers or campers, or trees may be illegally removed for firewood.

The negative impacts of road-related landslides on stream channels, riparian zones, and associated aquatic communities are widely recognized.[376] On the other hand, landslides are also sources of large wood and boulders, which help form complex aquatic habitat to the benefit of fish.[619]

The effect of forestry roads on stream systems is disproportionate to the total area of roads in watersheds. In an Oregon Cascade Range watershed, for instance, roads occupy only about 3.1% of a drainage basin studied, but they have been observed to increase drainage density (channel length per unit area) by 20% to 50%,[1017] potentially leading to increased peak flows and floods (Chapter 7).[462,106] Despite their limited area, roads also play a major role in patterns of soil and landslide movement through watersheds,[891,1018] with road location strongly influencing road effects. Erosion events associated with forest roads involve either the removal from or deposition on a road of soil and sediment. During a major flood the frequency of such events (>10 m^3 of material) was higher for roads on valley floors than for roads on adjacent steep hillslopes.[1018] The lowest frequency of erosion events occurred near ridges where roads were sources of sediment to downstream areas. Hillslope roads trapped a large volume of sediment delivered from upslope, but twice that amount moved downslope from the roads. Road segments on the valley floor were sites of net accumulation of sediment from both upslope areas and tributary streams.

In summary, forestry roads in steep mountain slope areas affect aquatic ecosystems mainly by altering flows of water and sediment and by replacing vegetation in riparian zones. Depending on climate, soil, topography, and land-use practices, road effects may be major or minor. Nevertheless, roads are foreign or exotic structures in the landscape and have no counterpart in the natural functioning of watersheds (as in the case of attempting to replace suppressed wildfire by prescribed fire). Therefore, the environmental effects of roads are likely to be viewed as negative, rather than as replacing natural processes inhibited by other human activities.

Agricultural Land and Road Systems

Irrespective of the region, farmlands in North America tend to exhibit a common set of attributes. They are mostly on flattish, sloping, or hilly land and have sufficient rainfall during the growing season for crop production, though in some areas irrigation is used to increase production. Although dairy farms are included here, most farmland has a rectilinear form of *tilled* (cultivated) fields. These fields form a matrix around farmsteads, woods, streams, villages, and towns which are usually frequent in this landscape. Fields also surround farm roads, local roads, and two-lane highways. Uncommon features, such as cities, rivers, multilane highways, and sand/gravel extraction areas, are typically scattered but important in agricultural landscapes.

Terrain, Land Use, and Road System

Road networks in farmland take varied forms reflecting topography, era of construction, and type of agriculture at the time of initial road development. Roads in rural, hilly New England landscapes once served agriculture in small fields and pastures long before mechanization permitted farming of large fields.[323] Forest and house lots have now overtaken much of this landscape, although parts of the eighteenth- and nineteenth-century road system persist to serve residential and forestry uses (see Figure 11.4 in Chapter 11). Evidence of the agricultural roads remains, including adjoining parallel stone fences or walls.

In contrast, the thoroughly regular road network of the Great Plains, so evident on flights across the Midwest, marks the section lines forming a 1-mi (1.6-km) grid.[812,827] These roads provide access to the agricultural fields, which range up to a full square mile (2.6 km^2) in extent and necessitate highly mechanized management practices (Figure 12.5).

Figure 12.5. Farmland, highway, and wetland in competition for space. The former rectangular grid of roads enclosing fields (note the scattered tree patches in the distance, mostly associated with farmsteads) was irregular in this natural wetland area. Farmers drained some of the wetland for cultivation (note the dark moist lines in the mottled field on right), and remaining wetland was protected as a state wildlife-management area. Multilane highway was then built through fields and wetland, and new ponds, ditches, and dikes were created, in part for waterfowl. A diversity of habitats with irregular boundaries promotes high plant and animal diversity. In the absence of culverts along the highway, diking and bridges have fixed the stream location and eliminated stream migration (note the former stream channel on the right of the small road). Natural wildlife movement between wetlands on right and left is blocked or subject to road-kills, especially of amphibians and reptiles. Interstate Highway 90, South-central Minnesota. Photo courtesy of U.S. FHWA.

Farms and fields are the two basic repeated units across the agricultural landscape, with farmers (and tractors) the organizing forces molding the soil types present. Farms, the larger unit, may be bisected or crisscrossed by roads, though the original coarse network of roads mainly connected farmhouses. Fields, the smaller unit, act as the key structuring mechanism for the road system, since in North America normally roads reach every field but do not cross fields. Traditionally, farm size in a local area tended to vary within a relatively narrow range, reflecting the area needed to support a family. Local areas vary greatly in topography, soil, and precipitation, so farm size in different parts of a landscape or region varies accordingly. Today, farmland often has a mixture of agribusiness enterprises and family farms, so two or more typical farm sizes tend to be present in the landscape. Farm management, appearance, and ecology normally vary more than farm size.[671,670] Field size also varies more than farm size, so variability in mesh size and network structure is considerable, even for the road system in farmland flat as a pancake.

Contour plowing and associated vegetation strips on slopes are related to erodible soils and rainfall patterns.[921,726] In parts of Northern Europe, hedgerows in such areas tend to have ditches in them to drain off excessive water flows.[311,129]

The road network tends to be regular, varying from the Jeffersonian government-imposed grid in many agricultural parts of the USA to a wavy network in topographically diverse areas of eastern Canada. Road density varies from area to area but may average about 2 km/km^2 (3.2 mi/mi^2). Where farm size is relatively constant, variability in field size may be low. In this case, access to farms and fields would tend to make a regular road network. Furthermore, farmers have a typical home range (e.g., including equipment repair, shopping, social interactions, and neighbor concerns), which could also lead to a regular road network. In the agricultural landscape, the home range of town residents may be mostly around town, though both farmers and town residents travel to other towns and cities periodically.

Traffic is relatively low on most roads in the agricultural landscape, except on highways connecting towns and on through streets in the vicinity of towns.[995] The proportion of truck traffic is relatively high, reflecting the transport needs of a farming economy.

Ecological Effects

In some rolling farmland, a striking gridlike hedgerow network partially encloses many fields.[734,311,128,129,624] Two basic hedgerow types are present: double strips sandwiching roads and single strips without roads. The roadsides usually contain ditches, often with wet soil and associated species, and may contain intensively mowed strips between the ditches and adjoining vegetation.

Fencing and gaps are common in the hedgerow strips. In a hedgerow network in France, species richness was highest along farmers' lanes, where there was a double line of trees.[128]

Often contrasting with and connected to the rectilinear hedgerow network is a dendritic stream network.[587,812,827] The smallest fingers of the dendritic network are often *swales* (linear open vegetated depressions) or simply traces left in fields. The road grid crosses the stream network with culverts and bridges in many places (Figure 12.5).

Also conspicuous in farmland are patches of natural vegetation, usually composed of large and small woods plus scattered trees and shrubs.[321,311,392,695,812,827] Most large woods are typically in isolated locations in the agricultural landscape, whereas small woods and scattered trees may be near or far from roads. Thus the threat to large woods from non-native species in roadsides seems minimal in this landscape.

Since farmers try to eliminate wet spots from fields, and because roadside ditches are kept maintained for drainage, these ditches may often be the only areas of wet soil present. Ditches provide two parallel linear strips of habitat for wetland or even streamside plants. Roadside ditches, which are often sunny and contain marsh species, also may connect with wetlands, ponds, or streams (Figure 12.5).

Roadside maintenance along local roads tends to be less intensive than along highways, so the farmland landscape may have a distinctive network of native wildflowers and other fallow plants (Chapter 4). However, herbicides may be sprayed on roadsides, sometimes at the request of farmers. Little seems to be documented about the spread of seeds, pollinators, and wildlife along roads in farmland.

In intensive farmland in The Netherlands, highway roadsides periodically cut across sandy soils.[1,63] Plant communities on these sandy roadside stretches are sometimes protected as the only places in the landscape where the plant species thrive. Road intersections do not seem to be promising habitats for protection, due to relatively intensive human usage and the unlikelihood of long-term care.

The most important habitats for woodland animals are the large woods and the stream corridors in farmland (Figure 12.5).[812,827] The relatively connected hedgerows, stream corridors, roadside strips, and rows of small-patch stepping stones are probably effective for movement of some of these species, though not the most sensitive ones, across the landscape.[311,628,624,815,129]

Hedgerows intersecting roads channel wildlife movement to roadsides, where road-kills may be concentrated.[418] However, except for lightly used roads, natural strips along roads are probably less used by animals of conservation interest than are equivalent hedgerows at a distance. Worldwide evidence suggests relatively little movement of animals along roads or roadsides or parallel to roads with traffic (Chapter 4).[302] Thus the roadside natural strip network, though prominent in the landscape, may not function much as part of the movement route of key animals across the landscape. An important exception is where the roadside natural strip is almost the only relatively natural corridor near roads in the agricultural landscape.[815,72,73]

Most roads in farmland are local, with low traffic volumes of 400 veh/d or fewer,[3] so traffic noise effects overall are limited. Yet the open nature of farmland means that noise effects from highways extend a long distance, from hundreds of meters to over a kilometer (Chapter 10). Hence, highways in farmland may form significant avoidance zones and barriers to animal movement. Visual disturbance by vehicles may be important for a narrow zone around many farmland roads.

Wetland animals, such as amphibians, which function at a finer scale, doubtless readily move along roadside ditches[812] (Figure 12.5). Ditches may provide connectivity between small populations in nearby wetlands and streams. Indeed, where the road network is superimposed on a stream network, the ditches may provide connections between tributaries, forming loops or alternative routes for movement.[302]

Since roadside ditches (and, similarly, roadside natural strips) are separated from each other only by the road itself, there is doubtless much crossing of the road by amphibians, reptiles, mammals, and birds. This suggests that road-kill rates, at least per vehicle passing, could be especially high along farmland roads. This may be ecologically important where road ditches are almost the only wet areas.

Animals of agricultural fields may predominate in roadsides. If road-kills are abundant, scavengers feeding on road-kills are also abundant and move along roads. For animals moving across an agricultural landscape,[494,302] the farmland road may not be very effective as a barrier.

During road construction, adjacent wetlands may have been partially or entirely drained (Figure 12.5) (Chapter 9). Where streams and roads cross, stream channelization—upstream, downstream, or both—together with various levees and pilings, are often prominent. The road network that repeatedly crosses a stream tends to eliminate stream migration, pool-and-riffle sequences, and other key aquatic habitats. These processes thus eliminate much scarce natural habitat remaining in a farming landscape.

Extreme water runoff events can damage roads in farmland (Chapter 7). The road network with vehicles also is a known source of heavy metals, hydrocarbons, roadsalt, herbicides, and so forth entering the farmland landscape. Much would directly enter the ditches, with groundwater and streams as typical next stops. However, overall, this landscape has a low percent of hard surface and low traffic volumes.

Roadside ditches and roadbanks by bridges and culverts are common sources of eroded sediments entering streams (Chapter 9). In farmland, however, fields must be much greater sources of sediment. Nevertheless, stream restoration remains one of the major challenges of the agricultural landscape.[812,827] The road network repeatedly crosses a stream network, and roadside ditches funnel water and sediment a considerable distance to a stream. Thus a landscape ecological rather than road segment or stream segment approach is critical for successful stream restoration.

Bridges and infrequent culverts tend to prevent stream migration across a floodplain (Chapter 9). They also constrain flows during flood periods, cause unnatural scouring, provide habitat for certain fish, and much more. Constrained floodwaters may wash out bridges and culverts.

Road-stream crossing areas in farmland offer long-term societal opportunities. Stretches of stream, especially upstream of crossings, may be protected in relatively natural condition, and mitigation can help control downstream flooding. Fish reserves can be maintained and can even capitalize on the shade from bridges.[572,597] Wildlife habitat can be enhanced, and attractive local recreation spots can be established. Indeed, the process of restoring the absorptive capacity and biodiversity value of headwater channels and streams can begin at road-stream crossings.

Grazing Land, Arid Land, and Road Systems

The extensive arid lands of the western USA provide the central focus here.[4,607,812,566] Most of these areas have been grazed by livestock.[921,726] Overall livestock density has varied over time, and in a scarce rainfall area the effects of overgrazing are widely visible. Moister grazing land extends into the eastern portion of the Great Plains and is present in Florida and elsewhere in the southeastern USA.[812,827] Landscape ecology analysis may show that road ecology patterns are similar in other dry regions of the world.[994,921,726,571]

Terrain, Land Use, and Road System

What comes to mind when imagining roads in arid grazing lands? Most people visualize isolated stretches of mostly two-lane, undivided highway over dry, hostile, and often unforgiving country.[995] Newer roads may be wider, often divided four-lane interstate highways. Sparse traffic, except in and near larger cities, and light local traffic around smaller towns and municipalities are usually related to ranching and other agricultural activities. As the population of the American West increases, however, there is a concomitant increase in traffic of both recent residents and nonlocal, seasonal visitors.

Primary roads are most often found in flat valley bottoms surrounded by foothills and often linear mountain ridges. Occasionally, the land resembles

Figure 12.6. Road and roadsides through desert grassland with some cattle grazing. Note the black road surface, which is solar heated. Roadsides irrigated by road runoff, plus soil disturbance and larger shrubs removed by maintenance activity, are subject to invasion by non-native plants. North of Monument Valley, southeastern Utah. Photo by R.T.T. Forman.

bajada terrain, with wide, open gentle slopes that stretch to the horizon and with mountain-sized rocky outcrops scattered like huge lumps across the sparsely vegetated surface (Figure 12.6). Western arid lands tend to be dominated by arid bunch grasses as well as by arid and saline-tolerant shrubs—including creosotebush *(Larrea),* sagebrush *(Artemisia),* tarbush *(Flourensia),* and saltbush *(Atriplex)*—that are widely but sparsely distributed.[607] Western soils are typically highly alkaline, with basins having the most saline properties. Waterways are fed by mountain watersheds and may produce flash floods after snowmelt in the spring or after large precipitation events. Water rights for irrigation often result in much-reduced stream flow, and smaller streams are commonly ephemeral.

The topography of the landscape certainly affects the nature of the road system. Two patterns are prominent. First, small towns and municipalities tend to be spaced widely in arid grazing country, with primary roads connecting these settlements. The emerging, growing economies of many small western towns, driven by increasing demand for recreation and second homes, have sharply increased seasonal traffic and fostered localized development in many parts of the Arid West. For example, such towns as Jackson (Wyoming), Boise

(Idaho), Park City (Utah), West Yellowstone (Montana), and Flagstaff (Arizona) experienced dramatic growth during the 1990s. New service roads, improvement of access, and increased housing development have had a major impact on the road network and ecological conditions around these municipalities.[788] Strip development extending outward along highways is often conspicuous and doubtless strongly affects wildlife movement patterns.

The second pattern reflects the increasing growth of suburbs around major cities, which is often guided by topographic features.[995] For example, development in the corridor from Provo through Salt Lake City to Ogden, Utah, left little open space separating the cities and towns. Known as the Wasatch Front, this corridor is home to about 62% of the population of Utah, which drives demand for increased road and suburban development. Similarly, the developing corridor from Denver north to Fort Collins, Colorado, has grown quickly in the recent past.

Elsewhere in the Arid West, even smaller towns have become permanent destinations for a U.S. population on the move.[995] However, the regionwide road network that provides connectivity between populated areas has remained much the same, save for widening or improvement of existing highways. Few new roads appear to have been built at the regional level, although much highway improvement occurs around populated areas.

Road networks across the Arid West have similar characteristics, although certain differences reflect the diversity of each state. Total road length varies from 43 087 km (26 757 mi) in Wyoming to twice that in Arizona, compared with 268 658 km (166 836 mi) in California (see Figure 2.8 in Chapter 2). Rural roads comprise 91.4% of the total in Wyoming, 68.3% in Arizona, and 49.8%, in California. Overall road density in the Arid West is less than 0.3 km/km^2 (0.5 mi/mi^2).

Two federal-aid highway systems are present—the National Highway System and the Surface Transportation Program—in addition to state-, county-, and city-maintained roads (Chapter 2). The departments of transportation in the western states as a rule provide maintenance for a majority of the federal-aid highways, although non-federal-aid roads may comprise most of the road length in a state. Thus the State of Nevada provides maintenance for about 13% of all roads in the state. As in all western states, federal-aid highways carry most of the traffic.

During the 1990s, state and municipal governments became increasingly concerned with the loss of open space in urban areas. Suburban or rural communities that were distinctive two decades ago often are hardly recognizable anymore. Similar to the other western states, the population of Utah is expected to grow from 2.1 million in 2000 to 3.7 million by 2030. Population increases will both exacerbate the open space problem and fuel the demand for wider, higher-speed roads. Between 1982 and 1996, the Salt Lake City area experienced the second-greatest increase in highway congestion in the USA (Utah Quality Growth Commission, 2001 website). In response, Utah embarked on the largest road-building project since the state was settled 150 years ago and, in the summer of 2001, completed the rebuilding of Interstate Highway 15 through the Salt Lake Valley. Utah is not unique.

The State of Nevada has had an ambitious road-building program to accommodate the state's rapidly increasing population. Road and interchange building has especially occurred around Las Vegas, Reno, and Sparks. In Las Vegas, an interchange aptly named "Spaghetti Bowl," which previously could accommodate 330 000 veh/d, now can handle half a million. Typically, interchanges in the region cover as much as 32 ha (80 acres) of land.

In these two examples of major road construction, transportation is serving the public by "meeting demand" (Chapter 3). With this approach, road system expansion is likely to abate only when population pressures abate, which is not on the horizon here. Yet in some other regions, road construction has been constrained as meeting demand is gradually replaced by a palette of societal and transportation approaches to using and developing the existing road system more wisely (Chapter 3).

"Boom-and-bust" towns in the brief history of the American West make one ponder the long-term impact of the primary western road network. More than a millennium and a half after the Roman Empire fell, the rectilinear grid of former Roman roads is still detectable from Turkey to Britain. In North America, the Oregon Trail was used by wagon trains only from 1836 to 1869, yet the ruts left by the wagon wheels are still present for all to see today. For example, at Guernsey, Wyoming, the Oregon Trail crossed soft sandstone, where the wagon wheels reputedly gradually carved a spectacular trough 5 ft (1.5 m) deep. The lessons learned from the imprint of Roman roads and Oregon Trail wagon ruts suggest that the transportation services provided by roads must be balanced with the ecosystem services provided when roads "tread lightly" through the land, especially in arid country.

Ecological Effects

Roads and road networks can exert positive ecological effects in dry grazing land. Most roads deliver water from rainfall and snowmelt to the roadsides (rights-of-way) (Figure 12.6). This often results in more-productive roadside vegetation. Furthermore, in many parts of the American West where intensive grazing or cropland agriculture is the predominant land use, roadsides maintain some of the only natural or seminatural vegetation present.

Large and small mammals are often attracted to roadsides because of the greenness of the vegetation and for cover.[431] Electric power poles, which trace so many miles of the rural road network, provide excellent perches for birds.[88] Raptors, especially, hunt the roadsides and adjoining land for small mammals.

Town and city dwellers in the West use recreational areas at an increasing rate.[495,566] For the most part, use is driven by road access and the presence of water. In effect, these areas are being targeted in much greater proportion than their availability, resulting in a greater localized riparian degradation, which is particularly prominent and important in a dry region. In addition, off-road vehicles may damage fragile ecosystems characteristic of the region.[1012,1013] As riparian areas are degraded, local impacts and declines of birds, mammals, water quality, and stream integrity become evident. As these impacts are repeated in the scattered riparian areas across the landscape, concomitant effects are seen with regionwide declines in the abundance of wildlife species.

Water, a key resource in the arid-grazing land, dictates where settlements develop. As a result, roads proliferate around water and riparian areas, with corresponding and usually negative effects on stream channel integrity, water quality, and instream flow.[921,726,566] Riparian trees are often reduced in number, resulting in higher water temperature and less habitable stream length for many fish populations. Fish then often seek stretches near remaining riparian vegetation and bridges for shade.[812] Higher water temperatures also reduce bottom-dwelling invertebrates that are the base of the food chain for fish.

A large portion of rural roads in the Arid West are unpaved. As described in Chapter 10, fugitive dust from vehicular travel has important human health and ecological effects. This problem is especially pronounced downwind of unpaved roads and worsens as traffic increases.

Conclusion

The four landscape types considered here contain the bulk of the human population, the road network, and the traffic flows on it. Road density varies by at least two orders of magnitude over the range of land-use types included. In some forest land and arid-grazing land, animals can travel a few kilometers before encountering a road, whereas in suburban landscapes, that distance is little more than a few tenths of a kilometer.

Traffic volume determines a large part of the barrier effect of a road on wildlife. Therefore, with increasing traffic, animal populations are increasingly being isolated in smaller areas within the polygons of a road network. This is especially important in the suburban landscape.

Stream systems and hydrologic flows are especially sensitive to the road network, which tends to cross or bump into the flowing water multiple times. At each interaction, the aquatic ecosystem is usually degraded, so with frequent interactions a cumulative degradation occurs.

Several ecological effects of road systems occur at much greater levels than the proportion of road surface in the landscape would suggest. Prime examples are road effects on riparian zones, on stream drainage systems and potential flood levels, and on wildlife habitats in dry regions.

CHAPTER 13

Roads and Vehicles in Natural Landscapes

We have agreed not to drive our automobiles into cathedrals, concert halls, art museums, legislative assemblies, private bedrooms and other sanctums of our culture; we should treat our national parks with the same deference, for they, too, are holy places.

—Edward Abbey, *Desert Solitaire,* 1971

Rural roads promote economic development, but they also facilitate deforestation.

—K. M. Chomitz and D. A. Gray, *World Bank Economic Review,* 1996

The four major landscape types that make up most of the USA and Canada were explored in the preceding chapter: built land, forestry land, farmland, and grazing-arid land. These landscapes overwhelmingly contain most of the roads, vehicles, and people. However, other landscape types in other parts of the world, as well as in North America, are also extremely important to transportation, society, and the environment.

We therefore highlight four additional landscape types in this chapter: (1) remote land, (2) parkland, (3) the Arctic, and (4) the tropics. These four landscapes are distinctive for having few roads, few vehicles and few people. Nature and natural processes predominate. Unlike in agricultural and built areas, here natural ecosystems surround nearly every vehicle and every kilometer of road. Consequently, the ecological effects of roads and vehicles have a special importance. Moreover, with a limited-size road system, each road in the network is important for travel and for its ecological effects. Changes and policies for the road system offer big problems and big opportunities.

Natural landscapes are a heterogeneous lot, varying widely depending on region and terrain. Overlap exists in the patterns and principles for the four landscape types, yet considering the types separately and sequentially is quite informative. Furthermore, exploring natural landscapes and their road systems provides a partial synthesis of the diverse important concepts in Chapters 1 through 12. The landscape ecology lens helps bring these large areas "alive."

Remote Land, Roads, and Vehicles

Remote land refers to extensive areas—for example, the size of whole landscapes—with extremely low densities of roads, buildings, and people. Remote land contributes three major values to society: natural systems, scenic beauty, and recreation.[495,547,405] Typically it has few known economic resources and has little long-term protection. It can change rapidly and radically. Remote land ranges from desert to tundra, from boreal forest to tropical forest. Depending on geography, it may be coastal or interior, bisected by rivers or rugged mountain ranges, and have a continental or maritime climate. Terms like "wilderness" and "wildlife refuge" describe some remote areas. Here we consider first the terrain and road system and then the ecological effects.

Terrain and Road System

The major dispersed human activities occurring in remote lands will affect how road networks are developed and how they interact with the ecosystems present. For this discussion, road networks in flat lowlands or plains will be contrasted with those in mountainous areas. The impacts of road systems in hilly or rolling landscapes will likely fall somewhere in between.

In remote lowlands or plains, roads tend to be simple from an engineering point of view. They seem uneventful to the motorist, and some travelers would consider them downright boring. Flatland roads are distinguished by their long, straight stretches. The occasional curve is noticeable, and the gradual descent to cross a river bottomland is almost memorable.

Roads in mountainous areas are the polar opposite. They are complex and play out their full potential on horizontal and vertical planes. Remote mountain roads are often narrow, winding, gut-wrenching routes that may have either paved or rough earthen surfaces. The roads swing around big trees, stubborn outcrops, and other natural impediments that early trailblazers, and subsequent road builders, smartly opted to avoid. Like riding a goat, mountain roads pitch back and forth. Once the rider is convinced that a road has "settled down," the route abruptly becomes nearly vertical, climbing steep slopes in endless switchbacks. Finally, the road attains a mountain pass or summit, only to dive down the squiggles on the other side.

Some remote natural areas are an anomaly and tend to be inhospitable badlands or "malpais." In essence, these are the unproductive and unwanted leftovers of a massive land grab. Remote natural areas are extreme in that they are hot and dry, or cold and ice-covered, or have shallow soils or deep permafrost. Nearly always, remote areas have been labeled by outsiders as infertile lands, where little can be profitably grown and where few people want to live. Residents, however, whether native or immigrant, have often developed a distinctive culture.

Terrain or topographical features of the land often dictate where roads can go and therefore strongly affect road-network structure and pattern.[532] Road access or construction can be prevented because of steep slopes, deep canyons, lakes, rivers, and wetlands. Road systems in remote lands therefore may take several forms. Most roads in mountainous areas today are believed to overlay earlier wagon tracks and, prior to that, trail systems used by native people and most probably game animals.[532] For that reason, many roads follow rivers and streams as well as ridgeline networks.

Human settlements and associated roads in remote landscapes tend to be near resources, including rivers, especially the junction of major tributaries. Rivers provide a route for travel as well as an ending point for some roads.[532] Ports at the end of roads often become transformed into important human settlements, with associated bottomlands providing fertile agriculture area. Therefore, remote-land road networks are frequently situated adjacent to, and follow the course of, rivers and streams. Small roads in the network are often old, originating in early travel routes of native people, explorers, and pioneers.

Recently constructed roads that traverse remote lands tend to trace a more straight-line trajectory—for instance, connecting city A to city B—and therefore follow topography less rigorously. Such roads may only cross, rather than follow, a meandering river system.

Where large lakes or cliffs prevent road construction, the routes around them may act as funnels with considerable traffic. This often leads to development along such road stretches. For example, roads built along water bodies for viewing and recreation can concentrate tourists and their vehicles. Parking areas at popular trailheads, and the roads that lead to them, can acquire a substantial amount of traffic and parked vehicles (Figure 13.1). Roads and motorized vehicles often must avoid, or are restricted in, designated wilderness areas, sanctuaries, and reservations belonging to native people.

Compared to the more developed landscapes discussed in Chapter 12, remote lands have relatively small roads, low road densities, low traffic volumes, and low vehicle speeds.[598] Depending on the amount of vehicle traffic and weather conditions, remote-land roads may be paved or unpaved. In cold

Figure 13.1. Visitor center at an isolated location on a spur road in parkland. Sawgrass *(Cladium jamaicense)* marsh with scattered tree hammocks elongated in the direction of surface water flow is relatively impassable, so human damage to habitats is constricted and concentrated. During water-runoff season, the road acted as a dam and was regularly inundated, so it was raised 30 cm (1 ft) and 150 culverts were installed to provide for a continuous sheet of water flow (as well as species movement). Note the buses to transport tourists and reduce vehicle use, the boardwalk into a hammock for nature viewing, and the invisibility of (underground) utility lines. Shark Valley Road, Everglades National Park, Florida. Photo courtesy of U.S. FHWA.

climates, roads are often seasonal, with winter closures due to snow. In warm regions, roads may be closed during heavy-rain seasons, when floods and road washouts occur. Human settlements in remote natural areas tend to be sparse and dispersed over the land. Typical road networks exhibit low circuitry, contain few loops, and have numerous spur roads.[898,302]

Modern engineering methods and equipment enable roads to be placed almost anywhere and to conform to any given landscape. In general, however, road alignments today are built with a least-cost or least-resistance principle in mind—that is, minimize cut-and-fill and follow relatively level valley bottoms. In lowlands and plains, roads generally circumnavigate large wetlands and lakes to avoid hydrologic problems. Some road alignments may exhibit a coalescence of least cost and least ecological impact. With today's flexibility in engineering technology, nearly all roads could probably be built with rather little ecological impact.

Ecological Effects

Vehicles traveling in remote road networks are at least partially responsible for the gradual introduction of non-native plant species into areas historically free of such species (Chapter 4). However, the relative scarcity of disturbed areas in remote landscapes tends to limit the subsequent spread of exotic species. Also, the near-absence of grazing land—commonly an area of non-native plant colonization and spread—limits this problem. Overall, the number of both non-native and invasive species is probably low in remote land.

The effects of vehicles in remote areas are somewhat limited because of the low road density and sparse traffic. Wildlife may be habituated to the possible disturbance caused by passing vehicles. The roads are permeable for crossing by most animals and thus have little effect on movement patterns except in locations where human development and road density are relatively great. Wildlife collisions with vehicles are usually not a problem, in part due to low vehicle speeds. Noteworthy exceptions are where animals such as reptiles are attracted to the road surface, particularly at night.

Remote landscapes harbor some of the last remaining intact ecosystems on the earth. Wildlife communities generally thrive in remote areas. Especially characteristic are wary and road-sensitive large predator species, such as grizzly bears, wolverines, lynx, cougars, wolves, jaguars, and tigers. However, increasing road density to the threshold level for a species can result in its disappearance from an area (Chapter 5).

Changes related to active or former resource-extraction operations such as mining, quarrying, and logging in remote areas may have a large influence on road-network development. Resource-extraction roads often are primary access roads. When closed, they provide ready access to legal and illegal hunters and to off-road vehicles. The indirect impact of former resource-extraction roads on other natural resources, such as fish and wildlife, can be significant unless measures are taken to remove the roads. Roads in remote areas also often provide access for off-road vehicle use that may be poorly controlled and produce many effects on soil, vegetation, and wildlife.[1013,547,405] Overall, however, the impacts roads and vehicles have in remote areas appear to be considerably less than in other landscapes.

Parkland, Roads, and Vehicles

Parkland and remote land have many patterns and principles in common relative to roads and vehicles. Therefore, this section generally highlights the distinctive characteristics of parklands and their road systems. *Parkland* refers to large, landscape-size areas that are protected for the public by government, mainly for nature, scenic beauty, and passive or *nature-based recreation* (such as hiking and birdwatching). Small parks as patches within a landscape are not included.[302]

Terrain, Land Use, and Road System

Parkland is designed to showcase natural treasures by providing access to visitors. At the same time, this function usually increases exposure of park interiors to road-related impacts.[823] Preserving the natural environment and maintaining its ecological integrity are paramount in many parklands. Some national parks, including Yosemite (California), Great Smoky Mountains (Tennessee/North Carolina), and Yellowstone (Wyoming/Montana), are faced with a major dilemma: how to adequately handle the volume of traffic entering the park yet maintain ecological-integrity goals and a quality experience for the visitor.[495,547,405] As time passes, road-network patterns often change little, while the number of vehicles traveling on the roads generally increases. This usually leads almost inexorably to road-widening and lane-expansion projects, rather than reconfiguration of the road system to protect the essential ecological integrity of the park.

Practically all visitors arrive by car or bus, so roads become obvious mitigation devices. In some parks with high visitation rates and traffic, such as Banff National Park (Alberta), efforts are under way to minimize road impacts on the park environment by using temporary road closings. Thus roads are closed during calving, post-denning, or winter periods when wildlife may be more sensitive to road disturbance. Because these measures are relatively new, little information is yet available regarding their effectiveness.

The distribution and relative importance of attraction points for visitors in parkland will influence road-network pattern and the extent of its impact on the land. Viewing areas or vista points tend to be strategically located and dispersed in the landscape. Some roads are built on high ridges and near cliffs to provide spectacular views. However, these roads often create undesirable engineering and visual impacts. Other highlighted points of attraction, including natural features and visitor centers, are also normally dispersed, rather than clustered in parkland. The network structure of park roads can also be affected by cultural values, such as archaeological sites and sacred sites of native people.

Some visitor centers receive the bulk of the tourists and therefore are intentionally placed at park entrances to help protect the essential interior of the park. Other visitor facilities at more remote locations attract relatively few visitors and cause few ecological impacts. Visitor impacts themselves lead to road-network changes. For example, facilities originally constructed within a giant sequoia grove *(Sequoia gigantea)* in Sequoia–Kings Canyon National Park (California) were moved down the road to an area of lesser

Figure 13.2. Narrow, winding park-like road, which has low traffic volume, vehicle speed, traffic noise, pollutant level, and road-kill rate. Coastal redwoods *(Sequoia sempervirens)* in California. Photo courtesy of USDA Forest Service and Jerry F. Franklin.

ecological importance, because of high tourist visitation and harmful impacts on the stunning grove.

Most parklands have narrow, winding, two-lane roads that provide opportunities for visitors to sightsee while traveling in the park (Figure 13.2). The abundance of curves in park roads helps reduce speeding and collisions with wildlife. Interest in sightseeing is so great in some parks that scenic loop drives are designed to provide a broad coverage of popular sections of the parks, as in Yosemite, Yellowstone, and Saguaro (Arizona) National Parks.

Few large parks have major highways running through them. One that does is Banff National Park in Alberta. The Trans-Canada Highway bisects the park and has peak tourist-season traffic volumes of approximately 35 000 veh/d (see Figure 2.5 in Chapter 2) (Parks Canada Highway Service Centre, unpublished data). As a comparison, the average daily traffic volume on the Tioga Pass road in Yosemite National Park during the peak summer months is 4000 veh/d (Caltrans, unpublished data). The Trans-Canada Highway through Banff was nicknamed the "meat-maker" because of the extensive carnage of road-killed wildlife, largely resulting from high traffic volume and vehicle speeds. Other parklands bisected by highways include Parc de la Gaspesie (Québec), Algonquin Provincial Park (Ontario), Olympic National Park (Washington State), and Great Smoky Mountains National Park.

To protect the especially valuable natural resources in the interior of a park, roads, access to attractive features, and visitor facilities can be concentrated in the extensive edge area of the park.[301,302] However, in an extreme case in Saguaro National Park, park boundary roads heavily used by early morning and evening commuters to the nearby city of Tucson had exceptionally high road-kill rates because busy travel times coincided with peak periods of animal activity.[492] Road-kill surveys in Saguaro National Park indicate that paved roads have higher mortalities than unpaved roads by a 13-to-1 ratio.[492] Also, fewer animals are killed on internal loop roads that are closed at night than on open through-traffic roads. Road-effect zones, barrier effects, and habitat fragmentation effects may be ecologically more important than road-kills (Chapter 5) but are yet little studied in parkland.

Maintenance cost or motorist safety may cause the seasonal closure of parkland roads. An example is the Tioga Pass road in Yosemite and Going-to-the-Sun Road in Glacier National Park (Montana). In winter, these roads traverse high mountain passes covered with snowdrifts 3 to 6 m (10 to 20 ft) deep. Permanent road closure, such as in Canyonlands National Park (Utah), may be a difficult process because people are accustomed to using the road for access, yet it could lead to road removal and impressive restoration of biological diversity and natural process in the large area previously affected.

Ecological Effects

Plant species composition along roadsides in parks is affected over time by traffic. Vehicles introduce non-native plant species (Chapter 4).[937,563] Also, areas disturbed by construction of park facilities or road widening doubtless encourage the spread and dispersal of exotic species.

The low road density and sparse traffic characteristic of most parks allow wildlife to become habituated to traffic on park roads. Roads with low traffic volumes and low vehicle speeds are for the most part permeable for animal crossing.[115,565] Heavy traffic even at low speed, however, can block animal movement.[352] In the expansive Lamar Valley of Yellowstone National Park, there are so many binocular-clad and spotting scope–equipped tourists following the recently reintroduced wolves that the resulting line of cars sometimes forms a virtual barrier to wolf movement. The park has now created "no-stopping zones" for cars so the wolves can get across the road.

Extensive park landscapes contain some of the last remaining intact ecosystems on the continent. Large predators and prey species generally thrive in large parks (where, in many countries, hunting is prohibited), such as Yellowstone, Great Smoky Mountains, and the Banff area, and include such wary and road-sensitive species as grizzly bears, cougars, and wolves[617,350] (Figure 13.3). Over time, increasing vehicle traffic on park roads may alter behavior of individual

Figure 13.3. Grizzly bear *(Ursus arctos)* on a 52 m (171 ft) wide wildlife overpass over a multilane highway in a park. A large–home range species thriving in remote land or large blocks of parkland with few people. Note the savanna-like cover and the 2 m (6.5 ft) high soil berm along the edge of the overpass. Probably an adult female, this bear shows light-colored "grizzled" hair. The first grizzly bear crossing of an overpass occurred in the second summer after construction, and by the third year, female grizzlies crossed regularly.[350, 352] Trans-Canada Highway (see Figure 2.5 in Chapter 2) with peak tourist season traffic of 35 000 vehicles per day, Banff National Park, Alberta. Photo by A. P. Clevenger.

animals and decrease habitat quality.[350] Highway mitigation measures, such as wildlife fencing and passages, have been installed along roads with high traffic volumes in such parks as Everglades (Florida), Glacier, and Banff[845,324,167,170] (Figure 13.3). Overall, these structures have proven effective (Chapter 6). As park popularity and visitation rates increase in the future, such measures may become more common in parkland.

Animals such as black bear, coyotes, and even deer may become dependent on human food sources and handouts from passing vehicles. In 1997 alone, there were more than 80 bear jams (traffic snarls caused by motorists stopping to look at bears) reported along roads in Banff National Park.[351] This exposure to people—and often their food—results in bears becoming food conditioned, habituated to humans, or subject to road-kills, or being removed from the park.[351]

The impacts of roads in park landscapes are considerably less than in most other landscapes, due to the strict conservation measures and relatively sparse road system. Relatively intact wildlife communities, including large predator and prey species, are found in large park landscapes and are little affected by low-volume roads and the vehicles that travel on them. Park roads tend to be near major points of attraction, including large water bodies and rivers and along ridge tops. Diversity of non-native plant species is lower than in most other landscapes. Smaller parks tend to have higher road densities and be divided into more fragments than larger parks.[823]

Today, few new roads are being built in parks. But the looming problem ahead is how to handle the motorized masses of people entering parkland and still provide a quality experience for the visitor and meet essential ecological quality goals. The road network itself is a key to accomplishing this. Improvements in the network, such as mitigation measures (Chapter 6), public transit, and road closures, should provide the answer to that dilemma.

The Arctic and Its Roads and Vehicles

In a landscape where the remains of camps left by the first European polar explorers appear nearly unchanged after more than a century, it is no surprise to learn that many human alterations are conspicuous for a long time.[991,296] This special region easily captures the imagination of visitors—Barry Lopez (*Arctic Dreams,* 1988) referred to the Arctic as "the ground floor of creation." Yet it is one of the most environmentally sensitive areas on earth, as will become clear in considering road systems and nature. Perhaps nowhere else does Thoreau's statement (*A Week on the Concord and Merrimack Rivers,* 1849) that "roads do some violence to Nature" resonate more aptly and loudly.

In the arctic portion of Alaska, the presence of roads is largely a result of exploration for petroleum[97,96] (Figure 13.4) and is therefore an issue of renewed interest today. Early roads left scars that persisted for decades. Vegetation recovered slowly, largely because of permafrost and hydrologic conditions.[991,296,966,297]

Alaska maintains 2400 km (1500 mi) of paved roads and 3200 km (2000 mi) of gravel roads (Alaska Department of Transportation, unpublished data). The state has 1 km of road for every 109 km^2 of land area (1 mi/175 mi^2), far below the U.S. road-density average of 0.6 km/km^2 (1 mi/mi^2) (see Figure 2.8 in Chapter 2). Despite low population and low road density, Alaska has nearly 37 km (23 mi) of road per 1000 inhabitants, which is 50% higher than the national average (see Figure 2.4 in Chapter 2). Nearly 30% of Alaska's population is not connected by road or ferry to the road network across North America. Indeed, less than 20% of Alaska's roads are paved, compared with 91% of the roads in the other 49 states (see Figure 2.6 in Chapter 2).

Figure 13.4. Arctic road in summer with vehicle dust, dust-deposit zones, and flooded (thermokarst) areas adjoining road. Natural surface-water areas and "ice-wedge" polygons characteristic of the flat tundra are visible.

In this section, we begin with road construction techniques, including unusual ones. Then earth, water, vegetation, and wildlife follow in somewhat overlapping sequence.

Road Construction in the Arctic

Continuous areas of permafrost greatly complicate planning, construction, and use of paved roads in arctic environments.[531,96,1039] Seasonal freezing and thawing of the ground surface result in the breaking of road pavement surfaces. Hence, gravel roads are widely used in the Arctic because they resist permafrost damage and are easily maintained. However, gravel roads and the vehicles that travel on them can significantly alter drainage and snowdrift patterns, change vegetation composition and soil properties, and lead to more off-road-vehicle trails.

Several road-building techniques have been used in arctic environments. The old and now-prohibited practice of bulldozing and laying down a gravel road base resulted in melted permafrost plus severe, long-lasting damage to the surrounding environment.[992,991,966,297] Considerably less environmentally damaging are the two main types of winter roads used today: ice roads and snow roads.[6,95,992] In addition, recent engineering solutions have been developed for gravel road construction over permafrost, using such materials as geotextiles, thermosyphons, and insulating hard-fiber board (Alaska Department of Transportation, unpublished data). However, for budgetary reasons, these materials are employed in less than 1% of the road length in Alaska.

The use of temporary winter roads and construction pads (at oil-well sites) that are built of ice and snow has become increasingly popular. This is because surface travel and use of heavy equipment on the unprotected tundra have become severely restricted due to environmental impacts and regulations.[992,993,818] In some areas of Alaska, *ice roads* are used to traverse rivers and normally soft ground. When built on existing areas of frozen water, these roads require little more than plowing to keep the snow clear. However, in other areas, water is pumped and spread on the surface to freeze and build structural strength. Ice roads are considered an acceptable alternative to the damage caused by permanent roads in sensitive landscapes. Such roads are also commonly used in the arctic areas of Canada and Russia.

Snow roads are basically composed of compacted snow. They are used extensively in areas where seasonal access to remote areas would otherwise be difficult or impossible for wheeled vehicles. Snow roads and construction pads have two objectives: (1) to protect the underlying vegetation and upper layers of the ground, and (2) to provide a hard, smooth surface for travel and operating equipment.[6,992] Forestry operations in Scandinavia and Canada, petroleum operations in Alaska and Canada, and almost all activities in Antarctica make extensive use of this snow technology. Many techniques of preparing snow roads and runways have been used and studied, but at least two problems remain unsolved: (1) how to extend the service life of the road as the warm season approaches, and (2) how to bridge damaged sections.[220] Other problems associated with snow roads include parked vehicles that sink, damage to heavily trafficked areas, damage caused by fluid spills, and use being limited to vehicles with low tire pressures.

Research has addressed some of these problems, with promising preliminary results.[220] *Geocells,* plastic containers normally designed for use with sand or gravel, were filled with pack snow and used to construct a test section of road. The resulting surface was reported to be hard, stable, and resistant to damage by vehicles. *Paving blocks* were also prepared by converting snow directly to ice using high-compaction pressures in a hydraulic press. The resulting blocks placed in a snow base were reported to be strong and resistant to the infiltration of fluids of all kinds tested. Geocell and paving block techniques greatly reduced most of the problems encountered in using snow roads.

Snow pads—built of snow hauled to the site, compacted, and watered—also withstood heavy traffic and use by heavy construction equipment in several Alaskan test sites.[460] Some common problems were lack of snow, unseasonably warm spring weather, and inexperience on the part of contractors and construction personnel.

One study of the effects of construction and wheeled-vehicle use on ice-capped snow roads and ice roads at a test site near Norman Wells, Northwest Territories, Canada, found that the underlying peat was compressed as a result of vehicle use.[6] The test roads were used only by vehicles during winter, and natural revegetation was monitored during summer. The proportion of the test roads covered by live plants was about 12% during the first summer after construction and increased to about 35% the following summer. The authors concluded that such ice-capped snow roads and ice roads, if properly constructed and maintained, were capable of withstanding the traffic and loads to be expected during possible pipeline construction.

Another study in a sensitive wildlife area of northern British Columbia created a road up a dry creek bed using snow and ice as a temporary roadbase.[818] The road was created with layers of geotextile matting, artificial snow, and a cap of shale. In spring, all road materials were removed, separated, and stockpiled outside the area. A summer inspection reported a relatively low impact on the vegetation, and good vegetation recovery on areas previously covered by matting and roadbase material. Ice road technology is considered most appropriate in areas of wet meadow or willow *(Salix),* but not in wetland (muskeg) areas of low forest because of long-term impacts to the vegetation. These limited studies await research evaluation.

Snow roads and ice roads are both common means of winter travel, and both affect the vegetation due to the plowing and consequent decapitating of tussocks of arctic plants.[992,297] Nevertheless, the ecological damage is less than that for summer off-road-vehicle usage or gravel pads. Much of the northern coniferous forest area consists of wetland and rivers, and for that reason little overland travel by vehicles during the summer is possible.

In contrast, Alaska becomes a relatively open and widely traveled region in winter. Ice road and river travel is essential to communication between villages and to subsistence activities. However, global warming could reduce the extent of these activities by melting extensive areas of permafrost and shortening the available winter months during which rapid travel can occur.[1039,297] For oil- and gas-related activities, without the transport of heavy equipment over oil-lease areas, geophysical seismic surveys would have to be performed by helicopter or other means. Though possibly more expensive, that could be accomplished with less environmental impact to the arctic tundra.

One improvement in arctic road design that has become the norm in many areas is to elevate summer roads atop a 1.5- to 2-m (5- to 7-ft) gravel base. This approach provides adequate insulation to minimize "thermokarst" development (described on the next page) and aids the wind in natural snow removal from road surfaces.[993] Initially the natural vegetation groundcover provides insulation for the permafrost beneath. Gravel fill that also has insulating properties is then placed right over undisturbed vegetation, instead of the older practice of bulldozing the vegetation away and then adding gravel on top of the permafrost. This new procedure, though certainly an improvement over the old way of building arctic roads, still has an environmental effect due to the need for large gravel-extraction sites. Such gravel or *borrow* pits are known to be "hostile" locations for plant growth.[992]

Without adequate insulation against heat from the sun in warmer months, permafrost simply melts, leaving a depression covered with water. The incorporation of adequate drainage culverts under elevated roads is essential to prevent the accumulation of flood water on the upslope side of the road as well as to prevent road washout (Figure 13.4). To accommodate and prevent flooding during the roughly two-week spring thaw, extra-large culverts are needed.[780] Furthermore, the retrofitting of culverts can be an important strategy to facilitate the draining of old impoundments of water. In all situations, the culverts should be regularly checked to prevent ice blockage.

In summary, gravel roads, snow roads, and ice roads are all widely used in the Arctic, and all have environmental impacts on the fragile tundra. Because snow and ice roads only last for months, the winter road system is much more extensive and serves different areas than the limited summer road system. Consequently, ecological impacts on wildlife often occur in different areas in winter and summer.

Earth, Dust, Ice, and Water

Drainage patterns on the flat tundra can be extremely complex as a result of the presence of many dispersed, unconnected flowing-water systems.[993] A major effect of arctic roads is the blockage of surface drainage during spring snowmelt, thereby allowing water to accumulate beside a road in the relatively flat terrain (Figure 13.5). This standing water, in turn, tends to accelerate the summer thaw or melting of permafrost in ground, thereby threatening the structural integrity of the road.[992,96,1039] Such meltwater tends to move alongside roads and, with the increased flow from culverts, can produce thermal and hydraulic erosion by undercutting the road structure. (In thermal erosion, water from melted snow absorbs solar radiation, heats up, and accelerates melting of permafrost beneath it.[1039,966]) In short, interrupted drainage due to roads results in flooded areas or impoundments and sometimes road washouts.[993]

One major impact of roads, unique to the Arctic, relates to alterations in the thermal regime of the tundra. During summer, a portion of the permafrost may melt, a process called *thermokarst,* which can be greatly accelerated by human activity[531,993,297] (Figure 13.4). Often, the developing thermokarst process beside arctic roads leads to the sinking or collapse of the ground surface due to deep melting of permafrost ice beneath. This sunken

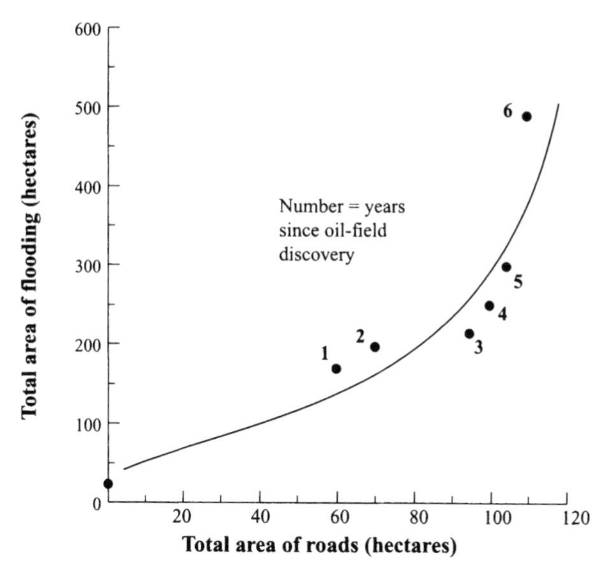

Figure 13.5. Area of arctic thermokarst flooding relative to time since oil-field discovery and total road area. Ten hectares = 24.7 acres. Prudhoe Bay, Alaska. Adapted from Walker et al. (1987b).

ground surface may extend outward from the road for meters, sometimes up to 100 m (328 ft), on each side of the road.[992,993] The resulting depressions fill with water. The low solar reflection *(albedo)* of the water surface results in energy absorption by the water, with some of the absorbed energy then conducted horizontally as heat to the adjacent strip of terrain. The heat absorbed there then increases permafrost melting, which tends to accelerate development of the thermokarst process.[1039] Once the thermokarst process begins, it is difficult to stop. The total amount of thermokarst area is likely to correlate with the density of summer roads in the landscape (Figure 13.5).

In addition, during dry periods, dust originates from gravel roads crossing the Arctic[990] (Figure 13.4) (Chapter 10). Dust fallout in one study was found to extend outward in significant amounts to 20 m (66 ft) on either side of a 16 m (52 ft) wide gravel road.[993] Dust is a summer road phenomenon that is highly dependent on traffic volume.

A covering of road dust on natural soil and vegetation reflects more of the incoming solar heat, which in turn is carried by wind into adjoining natural ecosystems. This additional heat tends to melt the fragile "ice-wedge" polygons that are such a distinctive surface feature of the tundra landscape (Figure 13.4). In some situations, spring snowmelt can occur as much as two weeks earlier, and extend up to 100 m (328 ft) outward, near heavily traveled gravel roads.[992]

Vegetation and Wildlife

The preceding section has pinpointed vegetation in addition to earth, dust, ice, and water. Here we focus more directly on vegetation, and then on wildlife, relative to arctic roads with vehicles.

Vegetation and Arctic Roads

Road development in the Arctic can potentially affect large areas of tundra, particularly where dense road networks exist.[459,547] Soil characteristics and vegetation have been studied in vehicle tracks (some as ruts created by vehicles rolling over the tundra, and others where vegetation was removed for vehicular travel) compared with adjacent undisturbed tundra in northern Alaska, and environmental characteristics have been found to differ markedly.[153] Vehicle tracks have higher temperatures, deeper thaw of permafrost, and higher concentrations of soil phosphate compared to undisturbed tundra. The vegetation is quite different too. Tracks have fewer species of plants, more evergreen shrubs, and greater dominance by grasslike species. Thus, in addition to summer gravel roads, even vehicle tracks across the tundra produce significant ecological effects.

Both acidic and nonacidic tundra, and the different dominant species associated with each, are common in arctic environments.[988] After 15 years of chronic road and dust disturbance, the effects on acidic and nonacidic areas were found to differ.[990,32] Overall, the effects on species richness and composition of vegetation were more severe in acidic tundra. However, the invasion of non-native species is uncommon along arctic roads.[992,296]

Heavy dusting near roads altered the depth of the thaw layer in the ground, compared with areas far from roads. Also near roads, total plant biomass (particularly lichens and *Sphagnum* peat mosses) was lower, and plant communities were species-impoverished, possibly due to increased soil acidity caused by road dust. These findings underscore the importance of developing sensitivity maps based on substrate (soil chemistry) and vegetation composition when planning arctic roads and mitigating their negative ecosystem effects.[32]

Many attempts to restore disturbed arctic sites, including gravel roads, have included plantings of native and non-native species, often with limited success.[459,992,990,547,991,296,966] One study allowing gravel roads to revegetate naturally resulted in relatively little vegetation cover and a low diversity of native plants.[90] These findings were attributed to several factors, including limited seed dispersal, lack of water, and slow growth due to low nutrient levels. Other studies have also found extremely slow natural recovery.[296,297]

One interesting finding in the gravel road–revegetation study was the prevalence of nitrogen-contributing *(N-fixing)* legumes colonizing gravel roads.[90] This suggests that a lack of nitrogen on arctic roads may inhibit the growth of some or many plant species. Legumes can increase soil nitrogen that is then available to other plants on disturbed soils. Therefore, sowing or planting legumes to accelerate plant succession on gravel roads may be a useful restorative approach. The effects of experimentally adding fertilizer are as yet little studied.[992,296] Most restoration research has been carried out in *low-arctic* (low-latitude) environments. In *high-arctic* (high-latitude) areas, patterns and processes of vegetation recovery may be slower and different.[295] Overall, though, arctic roads have major effects on the vegetation, and recovery or restoration efforts to date have met with limited success.

Wildlife and Arctic Roads

Relatively little research has been done on the effects of roads on arctic wildlife, with one notable exception. Several studies have examined the effects of roads and other infrastructure on caribou *(Rangifer tarandus)*. An early study in Scandinavia concluded that winter caribou used roads preferentially because the snow compaction on roads facilitated travel.[491] This resulted in high numbers of road-killed caribou during winter. On the other hand, caribou also were disturbed by traffic (including snowmobile traffic). Thus roads with relatively high traffic volumes were avoided by caribou as travel routes. Furthermore, these roads served as filters or barriers, obstructing normal movement across the land.

The only wild caribou currently remaining in Scandinavia occur in southern Norway.[678] Over the past 30 years, an increase in the number of roads and railways and in traffic volume in this area has subdivided the original caribou population into its present 26 separate herds. Areas within an impressive 5 km (3 mi) of roads and power lines are found to be essentially avoided by the caribou.[678] Due to continued development, the long-term survival of these populations is currently in doubt (see Figure 11.6 in Chapter 11).

Research on the Arctic Coastal Plain of Alaska suggested that the presence of a well-traveled road along a pipeline reduced the probability of caribou crossing the pipeline by over 50%.[858,194] Caribou have also shown altered behaviors, such as reduced feeding and resting time, in response to traffic.[662]

Caribou density is inversely related to road density.[677] At 0.3 km/km^2 (0.5 mi/mi^2), the population declined by 63%. At 0.9 km/km^2 (1.4 mi/mi^2), it was 86% lower. Areas with over 0.9 km/km^2 of road length have no cow-calf pairs, and maternal females show a displacement of 4 km (2.5 mi) away from roads.[676,677] Also, insufficient spacing of roads in the Prudhoe Bay, Alaska, area has apparently depressed overall calving activity.[136] Although road density is important, construction of the first road into an area has the largest relative impact on caribou distribution.[677] This important point is probably widely applicable to numerous species in diverse landscapes.

In Northern Alberta, caribou are found to avoid areas within 250 m (820 ft) of roads[451,233] (see Figure 11.6 in Chapter 11). The distance effect is greatest during times of higher traffic in late winter.[233] In addition, predation on caribou by wolves throughout the year is significantly higher near roads than far from roads.[451]

In contrast, another set of studies reported minor or no effects of infrastructure on caribou distribution and suggested that caribou do not avoid roads.[76,187] Similarly, in Denali National Park, Alaska, no evidence was found for caribou or grizzly bear avoidance of a road, though moose apparently did avoid the road.[1061] Taking all of the research results together, it appears that although caribou are highly sensitive to traffic, they seem to be relatively insensitive to the road itself. Therefore, roads that are used only infrequently may not be a large threat to caribou, but roads with regular traffic can have major effects on caribou density and distribution. Caribou were observed to commonly hesitate for about 10 minutes before crossing a road and pipeline.[194] In contrast, a passing vehicle usually caused an animal to retreat. Therefore, as traffic levels increase, the opportunities for caribou to cross roads decrease.

Numerous other wildlife and fish species are doubtless affected by arctic roads but are largely unstudied[931] (W. J. Wailand, personal communication). Grizzly bears are often attracted to human activities because of the presence of food and may end up being shot. Numerous bird species, some in huge number, breed across the Arctic where roads may be built.[75] A few species, such as the Pacific loon (*Gavia arctica*), may be favored by the water impoundments from thermokarst development adjoining roads.[484] Others, including the black brandt *(Branta bernicla nigricans)* and lesser snow goose *(Chen caerulescens),* may be sensitive to traffic disturbance, either visual or noise.[75,835]

Finally, various freshwater fish species move through and breed in the network of surface waters in the Arctic.[75] Culverts that are too small or poorly located may leave fish stranded in shallow pools that freeze solid. Also, the effectiveness of fish spawning grounds can be readily destroyed by high sediment loads from roadbed erosion.

The Tropics and Its Roads and Vehicles

Studies of the ecology of road systems in the wet tropics illustrate several additional important patterns. Most recent literature comes from the Amazon and from South Asia, which therefore are the focus here.

The Amazon

The establishment and paving of roads into a remote region often open up the area to development and major changes in land use[524,526] (Figure 13.6). Before

Figure 13.6. New road into a remote area, stimulating forest resource extraction, population growth, transformation of culture and land use, traffic, and additional roads. This road was part of a government plan to resettle urban immigrants onto small forest plots, which in turn led to cattle ranches and loss of animal diversity (Dale et al. 1994a, 1994b). Rondonia, Brazil. Photo courtesy of Elizabeth L. Taylor.

a road system is in place, native people of the region are relatively separate from one another and from the modern world. The livelihood of the people is often considered to be a careful balancing act of using the natural resources in a sustainable manner. For instance, the previously widespread slash-and-burn agriculture practiced by some native tribes involves small agricultural plots cut into the forest plus frequent movement of the plots.

Usually, roads are built into undeveloped areas for the extraction of resources such as timber or oil. Concurrently, the new roads change almost all native practices. Diseases, such as influenza, malaria, and yellow fever, are often introduced and spread by people from the outside who build or travel the new roads. Traditional, sustainable slash-and-burn agriculture is often transformed into a type of shifting land tenure whereby families burn several acres of land instead of small plots, grow crop monocultures and, after a few years, abandon the land when it is largely depleted of nutrients.[327] An increasing population pressure means that less land is available when soil nutrients on a farm plot run low after a few years. The end result is broad-scale deforestation with little sustainability for the farmers.[197,198,261] As people literally move down the road, more and more land is cleared and largely depleted.

With the introduction of roads and access to developed civilizations, native people also seek basic improvement in their lifestyles. Many want health, education, good jobs, and a bright future for their children. Roads are seen as a means of obtaining these opportunities. In Peru and Guatemala, for example, when oil companies exploring for oil decided not to build roads into the remote locations where the wells were located, native people demanded the roads. Indeed, in northern Guatemala, the local people held some oil workers hostage until an agreement was reached whereby the road adjacent to the pipeline would be paved.

Roads have provided access, and thus land-use change, for as long as people have moved with carts from place to place over land. The introduction of roads into natural landscapes usually increases the rate of turnover of adjoining forestland into agriculture and built land.[559,197,198] This pattern of loss of natural landscapes with road penetration has occurred around the world—first in Asia, Northern Africa, and Europe; then in North America and Australia; and recently in Latin America and Central Africa. In the Amazon, road construction is rapidly resulting in large areas of selective logging,[680] deforestation[680,146,525,527,526] and forest fragmentation.[848]

Concurrently, key wildlife populations decrease and in some cases disappear[197,198,1042] (Susan Laurance, personal communication). Road density has an important effect on wildlife, the largest relative effect results from construction of the first road into an area.[677]

This pervasive effect of roads on ecological systems occurred recently on a major scale in Rondonia, Brazil (Figure 13.6), where a herringbone pattern of roads was constructed that could be observed from space.[588] Once road access was available, a massive migration of people entered the area via automobiles, motorcycles, trucks, buses, bicycles, and walking. The roads led to extensive deforestation, planting of small farm plots, later plot abandonment, and often the spread of low-quality grassland with livestock.[260,196,197] The projected effects of these roads on continued deforestation and on sustainability of the ecological system are expected to persist long into the future.[527,526] Furthermore, associated environmental alterations, such as increasing habitat for mosquitoes that transmit malaria, can affect human health.

The widespread establishment and paving of such roads lead to soil and vegetation degradation. Hence, there is a need for careful planning of roadways into newly settled areas so that the ecological system can be maintained and the people can be supported by the land.[146] Often, incompatible uses for the land occur because there is little control of actions once the roads are in place. Environmentally aware citizens worldwide increasingly put pressure on international banks and other nongovernmental organizations to spend more of their resources on environmental concerns. Such pressure applied to agencies or companies introducing roads into new areas would pay rich environmental dividends. Society would benefit greatly by funding environmental protection or land management at the same time that a road project is being funded.

The importance of roads in the Amazon is illustrated by a recent study projecting future conditions for the Amazon Basin.[527] One of the key differences in the "optimistic" and "non-optimistic" scenarios presented was the role of roads. In the non-optimistic projection, the degraded zones near highways, roads, and infrastructure projects were more widespread. When projected 20 years into the future, both alternatives showed heavy forest degradation along the southern and eastern areas of the Amazon Basin. Roads into the interior of the basin are a major source of impact (Figure 13.6). The non-optimistic scenario suggested that few pristine areas will remain outside of the western portion of the basin. When the model was run without the planned highways, waterways, and other projects, the projected deforestation rate for the optimistic scenario dropped to 269 000 ha (664 000 acres) per year and for the non-optimistic scenario to 506 000 ha (1 250 000 acres) per year. Without roads and other access projects, these projected rates of deforestation are 14% to 27% of the current deforestation rate of 1.89 million ha (4.67 million acres) per year (based on data from 1995 to 1999), an enormous difference and expected ecological impact.[527]

The conclusion of this analysis of the future impacts of roads in the Amazon Basin highlights the importance of planning roads that minimize deforestation and subsequent changes in ecosystem services (loss of water quality, habitat, stored carbon, and so forth). The scenario of lost forests and lost ecosystem services due to road and infrastructure development is being repeated elsewhere in Latin America and Africa.[439,679,586]

South Asia

In South Asia, most national road networks are between 7700 km (4800 mi) (in Nepal) and 225 000 km (140 000 mi) (in Pakistan) total length.[750] The smallest country, Bhutan, has less than 1000 km (620 mi) of roads, whereas the largest country, India, has 3 million km (1.9 million mi) of roads, of which half are paved. In 1999, India announced plans to construct by 2010 a 13 000-km (8100-mi) national expressway system extending from north to south and east to west across the nation. The new network is expected to conflict with many of the important wildlife sanctuaries and national parks protected in India. Other South Asian nations are also undertaking major road construction programs.

Road-killed wildlife is of particular concern in this region, which has many large, famous animals. For example, a 1997–98 study in six important protected areas of India and Nepal reported that vehicles had road-killed 16 rare and endangered large predators (lion, leopard, and tiger; *Panthera leo, P. pardus, P. tigris*).[750] One of the reserves had a "forest road," three had state highways, and two had national highways. Approximately 88% of the large predators were killed on state highways and 12% on a national highway. Prey species of these predators were road-killed about equally on the three types of road. A few species, such as mongoose, civet, and langur (Paradoxurinae, Viverridae, *Semnopithecus*), were especially subject to road-kill on the forest road.

A closer look at an 80-km (50-mi) stretch of highway along one protected area, the Pench Tiger Reserve in India, for 1996 to 1999 provides a fuller picture of the types of species road-killed.[26 (cited in 750)] The 52 road-killed animals reported included 2 tigers, 37 langurs, 3 rhesus macaques (*Macaca mulatto*), 4 large birds (3 species), 4 large snakes (rat snakes and python), and two other mammals. A study in the Indira Gandhi Wildlife Sanctuary in Tamil Nadu, India, recorded 371 road-killed amphibians and reptiles of 20 species (in 171 "km-days" of sampling).[471 (cited in 750)] The authors concluded that most road-kills occurred at night. Also, road-kills tended to be higher where large canopy-trees and woody vegetation adjoined the road, in contrast to adjoining areas that were barren or deforested.

A highway bisecting a national park in Nepal illustrates one of the attempted methods of mitigation for road-kill.[750] The Royal Bardia National Park reportedly contains the largest and finest combination of woodland and grassland, and the highest biodiversity levels, in the Indian subcontinent. A ban on night driving (when apparently much truck travel occurs) was implemented from 1992 to 1995, after which it was removed. In the following three years, 1995 to 1998, annual road-kill mortality was six times higher than when the night-driving ban was in place.

Roads as barriers to wildlife movement is also considered to be ecologically important in South Asia.[750] In Bhutan, a newly approved road is expected to create a 15 m (49 ft) wide slice through a continuous evergreen-forest canopy on slopes of the Himalayas.[457 (cited in 750)] An arboreal golden langur *(Semnopithecus geei)* is an endemic species found in three groups in this forest. In order to cross the road, the animals would have to descend to the ground, where they are subject to competitors and predators. In effect, such a slice out of the canopy would isolate the golden langurs into two separate, smaller subpopulations. The authors recommend retaining large canopy-trees that maintain contact over the road every few kilometers. Such clusters of canopy-trees would serve as corridors that maintain the connectivity of the landscape for golden langurs and numerous other arboreal species. The frequency and locations of these canopy corridors can be tailored to the home-range sizes of key species and to the topography. In this hilly or mountainous terrain, narrow roads apparently tend to have the same amount of nearby tree cover on the upslope and downslope sides. After road widening, nearby cover on the downslope side is expected to decrease sharply.[750]

Roads as linear barriers illustrate other effects on wildlife. Thus major roads are of great concern for blocking the traditional movement and vital genetic exchange of Asian elephants *(Elephas maximus)* among three reserves aligned in a row in Uttaranchal, India.[750] Along linear coastal areas, roads are considered to be barriers to wildlife movement as well as threats to sea turtles due to human access. Roads along mangrove zones that parallel a coastline also tend to disrupt water flows and therefore alter salinity patterns, which strongly determine the survival and zonation of coastal vegetation.

Two other road ecology issues are highlighted in the South Asia research. First, where areas of limestone caves and associated *karst* topography (limestone surface with numerous depressions) are limited or scarce, they tend to be regionally important for biodiversity.[975 (cited in 750)] However, for the same reason of scarcity, they are targeted and exploited for cement production for roads and other uses. This results in a major ecological and economic conflict. Second, on Lantau Island in Hong Kong, a protected pitcher-plant species *(Nepenthes mirabilis)* was discovered during clearing of vegetation for a new road.[750] The species was then transplanted to another site for protection.

Finally, in the planning of new roads in South Asia, it appears that an impressive diversity of types of mitigation is proposed for minimizing ecological impacts. Some of the mitigation approaches are to be implemented in road construction and some incorporated in existing roads.

Conclusion

Roads in most natural landscapes are channels for development, harbingers of the future. Natural landscapes are particularly subject to new road construction. However, modern engineering technology now has the capability to construct roads that the environmental community would declare to be ecologically sensitive or enlightened.

Because roads are a major way that development is introduced into an area, environmental impacts of a budding infrastructure need early evaluation in a project. The Land Use Committee of the Ecological Society of America developed several guidelines or rules of thumb for incorporating ecological principles into land-use decision making.[195,199] These guidelines emphasize that land managers should take the following steps:

1. Examine impacts of local decisions in a regional context.
2. Plan for long-term change and unexpected events.
3. Preserve rare landscape elements and associated species.
4. Avoid land uses that deplete natural resources.
5. Retain large contiguous or connected areas that contain critical habitats.
6. Minimize the introduction and spread of non-native species.
7. Avoid or compensate for the effects of development on ecological processes.

8. Implement land-use and management practices that are compatible with the natural potential of the area.

Except in parklands, roads in natural land tend to receive less maintenance than do those in more populated areas. Therefore, such natural processes as heavy rains, landslides, and snow accumulation tend to damage and temporarily close roads.

Temporary roads and road closures are common, but permanent road closures permit the restoration of biological diversity and natural processes in large areas previously impacted by road effects and human access. Such mitigation measures as underpasses and overpasses for animal movement are particularly appropriate in parkland, where wildlife protection is a major objective and land on both sides of a highway is in government ownership.

Except for human-access impacts, the effects of roads in remote areas overall are relatively low. Yet, even with no more roads, vehicles, and people, impacts are likely to increase because of technology. People go farther and more often off road, using off-road vehicles, GIS images at the fingertip, geographic positioning systems, cellular phones, high-tech hiking and camping gear, and other computer-run technology.

Population and economic pressures combined with technological advances are likely to lead to the proliferation of arctic and tropical roads with associated ecological impacts. In the Arctic permafrost, thermokarst, flooding, and global warming pose one category of difficult, major issues that warrant further research. In the tropics deforestation, global carbon cycling, soil erosion and degradation, and human settlement and land-use patterns are key issues. In all natural landscapes vegetation, fish and wildlife habitat, and road avoidance and migration patterns represent another major set of problems and research needs.

Although road density has many ecological effects, the first road into a natural landscape area has the largest relative effect. It represents a threshold after which effects rapidly cascade and multiply. Avoiding the threshold is the prime objective to maintain nature, natural processes, and natural ecosystems for society long term. If the threshold is crossed, three new ecological goals take over: (1) prevent or minimize the construction of branch and spur roads, (2) establish mitigation techniques along all road segments, and (3) attempt to close and remove road segments, especially spur roads.

CHAPTER 14

Further Perspectives

We can't extend the model of . . . the United States to the rest of the world. A world with three billion or more private cars is unthinkable. . . . So we must think of something else.

—Jose Antonio Lutzenberger, quoted in the *New York Times,* 30 April 1991

Ecological design is the careful meshing of human purposes with the larger patterns and flows of the natural world. . . . When human artifacts and systems are well designed, they are in harmony with the larger patterns in which they are embedded.

—David W. Orr, *Earth in Mind,* 1995

The four broad areas of road ecology—(1) roads, vehicles, and ecology; (2) vegetation and wildlife; (3) water, chemicals, and atmosphere; and (4) road systems and the land—have been explored here in 13 focused chapters. Each pulls together much of the key literature; develops and articulates principles, models, and concepts; and illustrates these with diverse examples, applications, and solutions. An extremely rich body of emerging principles can be put right to work for transportation and society. Yet, also, we the authors are humbled by the frontiers awaiting research, as well as the extreme importance of road ecology. This effort is more a beginning than an end.

As fourteen authors with diverse expertise and perspectives, we have chosen not to close this volume with a summary of the preceding chapters, or even with tidy conclusions about the way the world should be. Rather, we offer an array of perspectives—both questions and thoughts—to stimulate the reader to go beyond the previous pages. Recall that we are not endorsing or recommending any specific application, mitigation, or solution in this book, but rather are using them to illustrate concepts and to suggest possible opportunities for action. Road ecology is in the early stage of emerging rapidly. This is the moment when one can have the greatest impact.

The chapter begins by providing a sustainability context to road ecology. Next, to illustrate some possible policy and planning solutions, we briefly highlight actions and aspirations of the major agencies responsible for road systems in the USA. Finally, an array of ideas is introduced for the reader to evaluate, augment, and/or implement.

Road Ecology and Sustainability

Road ecology is an integral component of sustainable development and sustainable transportation. Roads and vehicles provide many benefits—allowing economies to function efficiently and providing people and businesses with access to a variety of critical services, goods, and activities—but they also threaten the earth's biological and physical systems. They facilitate greenhouse gas buildup, acid rain, deforestation, erosion of water quality, and loss of habitats and species. The challenge is to retain the benefits while eliminating or mitigating the threats.

A recent task force appointed by the National Research Council's Transportation Research Board (of the National Academy of Sciences) reviewed the influence of U.S. transportation on critical natural resources and environmental systems and identified three components that threaten sustainable development[674]: (1) the U.S. transportation sector contributes about 5% of all the carbon dioxide generated worldwide and accounts for 30% of all carbon dioxide emissions in the USA; (2) other transportation emissions have cumulative and long-lasting adverse effects on the biodiversity of the nation's ecosystems; and (3) the extensive system of roads and other transportation infrastructure occupies large amounts of land, fragments habitats, and disrupts natural processes in ways that reduce biological diversity and diminish ecosystem functions.

Sustainability has a time horizon that extends over human generations and focuses on conservation of natural resources while providing for basic social and economic conditions.[1056,739] Road ecology is one element of a sustainability strategy because it is closely related to the attributes above: greenhouse gas emissions from vehicles, habitat fragmentation by new and existing infrastructure, and the impacts of roads and vehicles on biodiversity. Carbon dioxide accumulation in the atmosphere (Chapter 10) and tree planting and maturation in roadsides both illustrate the time scale of sustainability (Figure 14.1).

Figure 14.1. Tree-planting program by volunteer groups and a state department of transportation for the extensive area of roadsides (rights-of-way). Spruce (*Picea* sp.) being planted in an area where native spruce is absent or rare, though this species may not invade surrounding land. Route 29, Howard County, Maryland. Photo courtesy of U.S. FHWA.

Road ecology is a key to enhancing the process of transportation planning and project development for a sustainable future.

Yet road ecology exists in a social and economic context. To effectively implement its principles requires an appreciation of the value and uses of land in addition to the environmental dimensions. Two perspectives seem especially useful: (1) sustainable development and (2) respecting the land that nurtures us.

The President's Council on Sustainable Development published a statement in 1996 that lists as its first principle[739]: "We believe, to achieve our vision of sustainable development, some things must grow—jobs, productivity, wages, capital and savings, profits, information, knowledge, and education—and others—pollution, waste, and poverty—must not." This statement highlights growth of the human enterprise as the central goal and adds that the current level of environmental degradation should not worsen. Policy and planning for sustainable development—and its inherent social, economic, and equity dimensions—are thus especially appropriate for governments at national and international scales.[674]

The second key perspective for sustainability is, effectively, respecting the land that nurtures us.[541,1041,834] This perspective highlights the fact that land and nature have their own important attributes that require attention, and that society fundamentally depends on the land and its water. This dependence, as home and as source of diverse natural resources, underlies essentially all human activities. Caring for the land thus becomes critical. In effect, the first perspective views nature as a background framework for humans on center stage, whereas the second views nature as the major enduring force on earth, within which humans must fit.

Where does this leave the transportation planner seeking a perspective on sustainability that can be applied at various levels of government?[1072] Ultimately, sustainability-related transportation planning at state and metropolitan levels seems essential. Scientific results, and scientists, emphasize that several environmental trends, including global climate change, are serious phenomena[674,675,159,428,606,174] (Figure 14.2) requiring a response from the transportation profession. Most components of transportation give off CO_2 and other greenhouse gases to the atmosphere, whereas roadside shrub and tree plantings absorb and sequester CO_2. Although more scientific research is needed (Chapter 10), the absence of research does not justify inaction. Effective integration of ecological thinking into transportation planning and action can proceed. Pilot projects, use of new models, technology advances, public transit innovations, public education, and so forth could rapidly become widespread. In so doing, they would cumulatively produce large benefits.

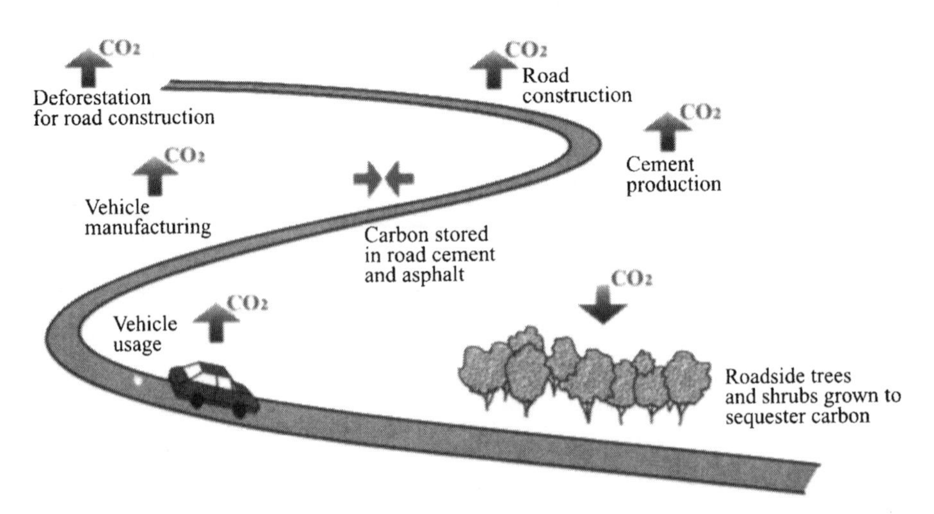

Figure 14.2. Carbon, the road system, and roadside trees.

Consider, for example, the potential impact of global warming on the infrastructure of the New York City region. Estimates were made of how deep and how often important transportation facilities would be under water as a result of sea level rise and increased storm frequency.[1071] The magnitude of the results brought home the reality of global warming to participants and readers of the case study. Such a study could be done for roads in many regions. Another study in Delaware concluded that global warming was a major problem for the state and that, if a consensus on the seriousness of the problem could be developed, the measures to address the problem would be feasible.[150]

Toward Policy and Planning Solutions

North America boasts a massive road network and a huge fleet of vehicles. The application of road ecology principles to mitigate and restore degraded nature and natural processes will never return the land to a pretransportation state. However, a clear, feasible objective stands before us. Implementation of research results and principles should produce significant visible improvements in natural systems along the entire road network, as well as reestablishment and protection of near-natural conditions in large areas of nature especially important to society and the future.[675]

Three factors have hindered the incorporation of road ecology concepts into transportation planning, construction, and management. First, *natural resource management*—occasionally labeled "ecosystem management" in the USA—is based on a fluid group of principles from which practitioners or agencies draw to serve their objectives. Second, no implementing laws or statutes are available that serve to structure public policy. Third, oversight of natural resources is highly fragmented at both state and federal levels.

An illustration of fragmentation at the national level is the large number of federal laws under the National Environmental Policy Act "umbrella" that bear upon transportation project development. Proposed federal regulations published on May 25, 2000, in the U.S. Congress' *Federal Register* listed 54 statutes that constitute the NEPA umbrella (Chapter 3). The statutes are administered by at least 10 federal departments and agencies. Most of the statutes predate the advent of broad ecosystem or landscape ecology thinking. Laws often work at cross-purposes when values and interests conflict, as in the case of a project involving both historic resources and wetlands. One recurring conflict, for example, occurs when an unsafe and congested older road is proposed for traffic-flow improvement. In many cases, widening the road may affect historic resources, while moving to a new location may damage wetlands and other natural resources. The basic statutes do not consider such conflicts, and no effective mechanism exists for priority setting or resolution of such difficult trade-offs.

To better understand the opportunities and constraints for applying road ecology concepts, we now turn to four major "players" that determine the nature of road systems on the land in the USA: (1) the U.S. Federal Highway Administration (FHWA), (2) the USDA Forest Service, (3) agencies for other federal lands, and (4) the states. Policy or management approaches, particularly focused on the present and future rather than the past, are briefly highlighted for each. A gap and a lag always separate articulated policy from practice on the ground. Nonetheless, the trajectory, the size of the gap, and the expected rate of closure are useful keys to gauge the future.

The FHWA and the Environment

The enactment of NEPA in 1969 confronted highway builders with a major challenge (Chapter 3). With hundreds of interstate highway projects under way and no "grandfathering" provisions provided by the U.S. Congress, many projects were challenged in court. More often than not, the courts determined that the projects would have to begin the project development process anew, in full compliance with NEPA. After a number of difficult years, the FHWA and the state departments of transportation developed a successful NEPA compliance process.

Until the mid-1990s, it had been FHWA policy to compensate for environmental damage and to attempt to return a negatively affected environmental resource to its prior condition. In 1994, the FHWA articulated a major environmental policy statement[266] that included the following: "To be an environmental leader FHWA must go beyond compliance and strive for environmental excellence. The widest possible range of both traditional and innovative measures to protect and enhance the environment must be pursued."

The section headings of the policy are an indication of its scope and emphasis: Complete Integration of Environmental Concerns; Active Protection and Enhancement of Our Environment; Vigorous Research; Technology Transfer and Training; and Effective Development and Promotion of Environmental Expertise. Such a policy framework, which for the first time used the term "enhance," creates an opportunity for implementing ecological goals in highway planning and development. In the past, it might have been argued that to create wildlife crossings where none previously existed was not an eligible expenditure of highway funds. Now such an expenditure is much easier to justify, since such crossings can clearly enhance the environment. Of course, such expenditures are still not guaranteed; they must compete for funding with other transportation priorities.

Congressional funding of transportation planning, projects, and construction has also evolved so that more funding became available to directly address environmental concerns. Such federal funding noticeably increased in the

1991 Intermodal Surface Transportation Equity Act (ISTEA), and then more so in the 1998 Transportation Equity Act for the 21st Century.[275] Despite this progress, widespread remediation opportunities and mitigation needs demand attention for an extensive road system largely built over a 200-year period. Moreover, major new threats continue to emerge, especially as an expanding population and greater vehicle use increasingly impact local natural systems and extend into ecologically sensitive areas.

USDA Forest Service Land

The Forest Service manages a huge road network in the USA (Chapter 2).[175] These roads have increasingly been drawn into the center of debates about the management of forests, both within the agency and more broadly. With budgets limited, logging reduced, and greater importance given to protecting watersheds and biodiversity, the Forest Service has been reassessing how it manages this extensive road network.[573] The agency recently completed a national assessment of the science of road effects on ecosystems[376] as part of a review of its policies on roads.

This assessment is related to and was inspired in part by the parallel development of *watershed analysis* as a formal process on federal,[763] state,[1003] and private[672] forestry lands. This watershed analysis approach provides a broad integrated perspective on the biological, geophysical and, in some cases, social characteristics of a watershed for the purpose of informing management. The approach also provides a framework for planning modifications of road systems and for watershed restoration. The Forest Service has used the watershed analysis approach to carry out an extensive evaluation of its road system with the intent of[945] (1) identifying needed and unneeded roads; (2) identifying road-associated environmental and public safety risks; (3) identifying site-specific priorities and opportunities for improving or removing roads; (4) identifying areas of special sensitivity, unique resource values, or both; and (5) developing other specific information that may be needed to support project-level decisions. Although these analyses are carried out mainly on federal forestry lands, they provide an informative example of comprehensive road system assessment.

Road engineering and road removal practices are especially important because they have a large effect on watershed conditions (Chapters 7 and 9), such as those needed for salmon recovery.[376,405] Improved practices include fitting roads to the landscape better and altering water-drainage structures. Thus, when major storms occur, roads not only suffer less damage but also can withstand hydrologic impacts with minimal alteration of natural processes.

A long-term goal is to "stormproof" watersheds instead of following the standard "defensive" approach of constructing roads with the intent of withstanding disturbances. *Stormproofing* involves reducing landslide potential by removing road fills and landings after cutting. More culverts are installed to disperse water collected in ditches, rather than delivering it directly to stream channels. Stream crossings are constructed so that flood waters or landslides can more easily cross over or under a road, thus reducing damage to both road and stream system. Although these mainly hydrologic based approaches cannot solve many of the water quality and fish population issues, they can provide numerous ecological benefits in a watershed. The approaches also provide economic benefits where low traffic volume and modest commercial-resource value do not justify the financial resources needed to support a large, highly engineered road network.

With a massive 650 000-km (400 000-mi) infrastructure composed of somewhat separate road networks in nearly 200 national forests scattered across America, visible change will inevitably come slowly.[175] Nonetheless, the recent history of directed policy reviews, road-network assessments, enhanced policy and planning, and initiation of more environmentally focused practices could result in a bright future for the ecology of forestry land.

Other Federal Lands

Roads on most U.S. public lands are used in a distinctive way, typically providing access only to that property rather than serving as thoroughfares. Hence, access is provided to specific locations within a relatively protected landscape. Examples include access for wildlife viewing and management on Fish and Wildlife Service refuges, residences on Indian reservations, military activities on Department of Defense reservations, and scenic spots in National Park Service parks. As the largest custodian of U.S. public lands, the Bureau of Land Management has both human-access roads and many highways for through traffic on its land. Traffic and human impacts may also be concentrated around a single road bisecting a public land, such as in the Great Smoky Mountains National Park, Glacier National Park, Banff National Park, and Parc de la Gaspesie (Québec).

By typically being large and relatively protected, such public lands tend to abound in natural resources. Also, in most cases, limited access is provided for recreational purposes, such as hiking, fishing, hunting, and skiing.[405] Road systems on federal lands exhibit several other common characteristics. A relative abundance of spur roads provides human access. Most roads are two lanes, and many are earthen surfaced. Four-wheel-drive vehicles are common, traffic volumes tend to be low, and traffic peaks are often associated with weekend recreation and tourist seasons.

These roads permeate and affect public lands that are generally subject to specific laws and regulations, such as the Endangered Species Act in the USA.

NEPA applies specifically to activities on federal lands or to other actions that use federal funds. As a result, public lands tend to have more environmental protection than elsewhere, which in turn means that they harbor numerous unique natural resources.

However, management and funding for environmental protection vary widely on federal lands. Road systems on military reservations illustrate several useful patterns. Property owned by the U.S. Department of Defense contains more federally listed endangered species per unit area than lands owned by the Forest Service, National Park Service, Fish and Wildlife Service, or Bureau of Land Management.[544] Similarly, former nuclear weapons facilities of the Department of Energy may have considerable natural resource value. Some military reservations for military training and testing are quite proactive in the management of rare species.[544] Thus areas with rare species adjacent to roads may be designated as sensitive areas, with signs posted to limit access.

On the other hand, military reservations tend to have high road densities for communication and training, and use of off-road vehicles is normally extensive. Use of explosives and fires tends to be common or widespread. A history of pollutants and toxic substances associated with military transportation has plagued many sites. Moreover, soil erosion can be quite severe. Heavy-tracked vehicles, such as tanks, churn up soil and extensively damage vegetation during maneuvers, though such intense training normally occurs in designated areas.[636]

Overall, military lands provide large areas that contain both protected sites and areas of intense mechanical disturbance. Species that survive in the face of regular disturbance tend to flourish on these lands and may be uncommon in surrounding areas. For instance, the endangered red-cockaded woodpecker *(Dendrocopos borealis)* in the southeastern United States apparently requires mature longleaf pine *(Pinus palustris)* for breeding. This pine species thrives on military reservations where fire is fairly frequent and reduces competition from hardwoods or other pine species.[358] Furthermore, controlled-burning programs have helped maintain several of the best remaining longleaf pine forests.

Resource managers on some military reservations alter the timing and intensity of roadside mowing and management to enhance conditions on ecological systems. For example, at Fort McCoy in central Wisconsin, the beautiful, endangered Karner blue butterfly *(Lycaeides melissa samuelsis)* is obligated to spend its life as a larva on wild blue lupine *(Lupinus perennis),* but lupine often grows along the roadside, where it is susceptible to mowing (Figure 14.3). The Department of Defense, working with many partners, is committed to protecting the plant because of its essential role in the life cycle of the rare butterfly. A procedure has been implemented whereby roadside populations of wild blue lupine are marked each year, and mower operators are instructed to lift their blades as they come to a marked population of lupine.[1037] As a result

Figure 14.3. Road in a military reservation where roadside vegetation is managed to protect a population of endangered butterflies. Mowing is restricted along both sides of the road to protect wild blue lupine *(Lupinus perennis)* plants used by the Karner blue butterfly *(Lycaeides melissa samuelsis)*. Fort McCoy, Wisconsin. Photo by V. H. Dale.

of selective mowing, light disturbance, and other management actions at Fort McCoy, the installation currently supports significant populations of the Karner blue butterfly.[950,855]

Overall, environmental management on military reservations is difficult, but some managers demonstrate ecological leadership in developing environmental management plans and also join forces with other partners for environmental protection. The costs of environmental challenges on military installations are significant and can even jeopardize mission activities. Thus natural resource management is implemented, based on 10 guiding principles spelled out in a 1994 memorandum from the Office of the Deputy Secretary of Defense for Environmental Security[367]: (1) maintain and approve sustainability and native biodiversity of ecosystems; (2) administer in accordance with ecological units and time frames; (3) support sustainable human activities; (4) develop a vision of ecosystem health; (5) develop priorities and reconcile conflicts; (6) develop coordinated approaches to working toward ecosystem health; (7) rely on the best science available; (8) use benchmarks to monitor and evaluate; (9) use adaptive management; and (10) implement through installation plans and programs. The widespread implementation of this ecological and adaptive management approach[544] should further strengthen military reservations as ecological gems, commonly in a matrix of urban and agricultural land.

Managers of all public lands commonly face many challenges in limiting human overuse or impacts on natural resources. Yet, despite this hurdle, the large chunks of landscape under public ownership are crown jewels to guard for a sustainability future for society (Chapter 13). Roads represent a key handle for management of public lands. The limitation and removal of access roads represent a particularly effective and cost-efficient way to limit the proliferation of human impacts on the natural environment in large federal lands.

The States

Most roads in Canada and the USA are owned and managed by local, state, and provisional governments (Chapter 2) (Figure 14.4). No strong, top-down, statutory approach to the environment exists, so state and local environmental protection is quite variable. The best examples often occur where environ-

Figure 14.4. Highway bridge where state engineers addressed concerns for visual quality and avalanche danger. Motorist views from the highway, plus views of the bridge from a campground and the distant road, affected the design (see Chapter 4). Wildlife, hikers, water, soil, and avalanches cross under the bridge. Interstate 90 near Snoqualmie Pass summit, Washington. Photo courtesy of U.S. FHWA.

mentally committed governors personally require good stewardship on the part of their cabinet secretaries and insist on good coordination. Additionally, some transportation secretaries have shown great leadership by insisting that all key people in their departments of transportation (DOTs) be properly trained in environmental stewardship and put mechanisms in place to monitor performance. Also, within DOTs there are outstanding environmental units that accomplish high-quality work.

Those governmental bodies more committed to environmental protection employ some form of strategic management with goal setting, benchmarking, and performance measurement. For instance, an excerpt from promotional materials recently distributed by the New York State DOT (2002 website) to state agencies entitled "Steps to Stewardship" indicates a commitment to environmental stewardship:

1. *Lead—show it is real.* Put environment in your mission statement. Publish an environmental policy. Require everyone to do a project.
2. *Institutionalize—make it happen.* Hold workshops and meetings. Develop action plans. Issue official instructions. Conduct training and workshops. Revise procedures and guidance. Develop promotional publications. Try to leave project areas better than you found them. Maintain lists of projects and enhancements. Adopt context sensitive design. Develop best practices. Track progress of your plan.
3. *Partner—reach out to agencies and groups.* Meet at different levels. Build environmental enhancements. Pilot new technologies. Streamline permit procedures.
4. *Promote—bring others along.* Publish accomplishments. Present awards to outstanding teams and projects. Establish a website. Keep spreading the word!

As many as ten state DOTs have shown special leadership in the environmental area. Other states, while not as far along, have made vast strides in moving their transportation programs from an environmental compliance mode to proactive protection of environmentally sensitive areas. At the other end of the spectrum are some states still in the compliance mode, doing only what is required to get permits and approvals. The individual steps and projects are important, but the big picture is more important. Maintaining viable wetland functions in a large area, salmon populations in a river system, or biodiversity in an outer suburban landscape requires more than a sequence of detailed steps.

Given the absence of strict rules and guidelines, state and local governments have considerable discretion. Consider for a moment that you have been appointed to a new position in a state DOT. You are committed to introducing road ecology concepts. How might you proceed? The following brief list of ideas highlights the opportunities:

1. Determine the degree to which environmental approaches currently use ecology principles.
2. Consult and partner with state and federal natural resource agencies and appropriate nongovernmental groups.
3. Reorganize and consolidate existing organizational units that break up environmental problems into separate resource areas.
4. Charge planning units with strengthening environmental considerations in the planning process.
5. Encourage state and federal natural resource agencies to delineate and establish pilot ecosystem or landscape areas and to assist in the development of cross-cutting protection or restoration strategies for each area.
6. Introduce ecology principles and set ecology goals in ongoing performance-based strategic management programs.
7. Reward and highlight projects and personnel that implement road ecology principles.

An Array of Perspectives for Future Policy, Planning, and Projects

As highlighted at the beginning of this chapter, instead of simply summarizing material or listing conclusions, the authors chose to enrich the opportunities on the table in hopes that the reader will explore further and create a better future. Readers are invited to pick and choose from the following selections. Each reader will have additional ideas of what should be on the table. We have grouped the authors' brief nuggets into eight overlapping categories: (1) roads and vehicles, (2) fuel and ecology, (3) travel and traffic, (4) nature, wildlife, and water, (5) communities and neighborhoods, (6) economics, (7) planning, and (8) policy. The nuggets build on and are inspired by, but rarely repeat, the preceding chapters. All are designed to stimulate the maturation and application of road ecology in the years ahead.

Roads and Vehicles

The role of ecology has too often been kept on the sidelines in designing roads and vehicles, but how might or should this change? Consider road surfaces. Solid wastes from society are recycled into road-surface pavements, and surfaces continually degrade, giving off particles and chemicals. Apparently no one knows whether this is ecologically significant and therefore whether zoning road segments[851] for different types of material is desirable. Evaluation near sensitive ecosystems and drinking water supplies would help gauge whether the effect may be large or can be forgotten.

In response to growing traffic, is it ecologically better to add lanes on an existing highway or to build a new road?[241] Seemingly, adding lanes is better, as suggested by Figure 14.5. Yet a fuller, more-informed answer depends on how the options fit into a road network and how that network, in turn, ecologically interacts with patterns and processes in the landscape.

The concept of "life cycle" planning for vehicles accounts for the disposition of all components from design to disposal so that virtually all parts are recycled. This approach, if widely embraced as an analytical and policy framework, would reduce natural resource consumption and should provide a clearer picture of pollutant flows through ecosystems over time.

Is it possible to envision a future with few vehicles and few roads? What might it look like, and how would it work? What ecological benefits would it provide? What would be the benefits and shortcomings for society? Students preparing for the real world and faced with these questions sometimes outline intriguing solutions. Their designs for the future are often based on addressing the problem of global climate change, providing flexible, alternative means of transportation linked to housing and jobs, increasing "smart growth" initiatives, pursuing enlightened planning with political muscle, and viewing the emergence of road ecology issues as appealing and compelling.

Fuel and Ecology

The link between fuel use and air quality is well documented. But what about the broader ecological gain of improved fuel efficiency and new fuels? What are the effects on soil erosion, habitat disturbance, and water pollution resulting from less fuel use and a switch to non-petroleum fuels?[866,865] Oil extraction in ecologically sensitive areas and transport over long distances, with occasional spills, can cause huge environmental damage. Extracting heavy oils, tar sands, oil shale, and coal to produce fuels may well cause even greater ecological damage than from petroleum extraction. The widespread use of biomass would also cause major environmental problems, but of a very different nature. Would a switch to battery electric, hybrid gasoline-electric, biomass fuels, or hydrogen lead to significant ecological improvements? What are the spatial and temporal dimensions of such changes? These questions have never been answered in a rigorous or comprehensive fashion.

Furthermore, consider carbon and greenhouse gases. Carbon sequestered in growth and biomass of trees along roadsides could help mitigate the global increase in atmospheric concentrations of CO_2 (Figures 14.1 and 14.2). In the same way that native flowers are planted in spots to beautify roadways, trees and shrubs can be planted or allowed to develop naturally in many locations to enhance carbon storage. Not only would carefully designed shrub-and-sapling vegetation capture more CO_2, in many locations it would also be

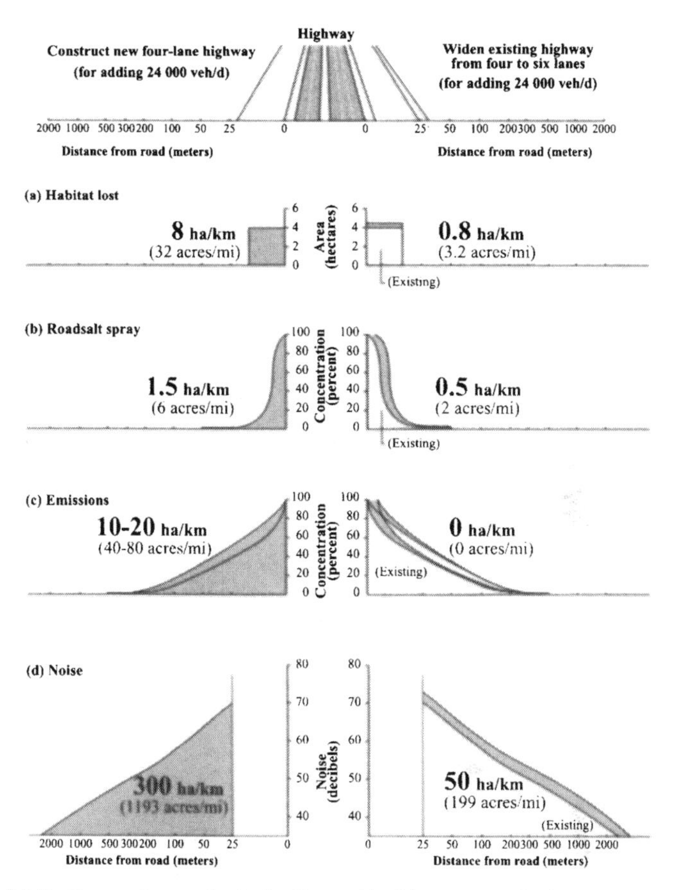

Figure 14.5. Comparing ecological effects of building a new highway versus widening an existing highway. The shaded portions with associated area (in hectares per kilometer length of highway) indicate the effect of a change for each of the four ecological factors illustrated. In (c) the upper curve represents average maximum, and the lower curve average minimum emissions levels. Based on studies of German highways. Adapted from Ellenberg et al. (1981).

useful for safety by slowing errant vehicles and blocking headlight glare across median strips. The millions of miles of roadsides (rights-of-way) (Chapters 2 and 4) include extensive largely unused areas for plantings in woodland or savanna form to store carbon.

Travel and Traffic

Despite the long-term trend and perception of continued growth in travel and traffic in the USA (see Chapter 3 and Figure 1.6 in Chapter 1), evidence periodically surfaces suggesting a reduction in travel.[510] It is unclear how widely applicable such reports are or whether they represent normal variation within a trend or the onset of a new trend. Irrespective, if the evidence is real, it suggests that changing the perceived juggernaut of increasing VMT or VKT may be easier than we thought.

"Remove it and they will disappear" summarizes the findings of a series of 60 studies worldwide of roads that were temporarily or permanently closed; on average, the total amount of traffic decreased by 20% rather than use alternative vehicular routes.[510] The highest reductions in vehicle trips occurred where the public had flexibility with alternative means of travel, from trains to walkways.

For ecological protection, less vehicle traffic is better than more, and traffic concentrated on a few major thoroughfares is better than that dispersed onto roads in more ecologically sensitive areas.[241] Clearly, infrastructure, pricing, and financing decisions are continually being made that bear upon how much travel will occur and where. Indeed, no one fully exercises his or her travel desires. One of the California authors of this book asserts that, left unconstrained by cost or time, he would be in Paris nightly to dine. But what is the optimal amount of transportation infrastructure and services for society, and how should it be provided and priced? In the USA, where land is relatively abundant and the value of individual freedom is salient, society has been willing to accommodate more and bigger vehicles and more travel.

As the local, national, and global implications of these decisions, including road ecology costs, become better understood, society exercises more restraint. Decisions are made to enhance transit services, reduce vehicle emissions and energy use, design more aesthetic and ecologically sensitive roads, and limit vehicle use. Singapore, Denmark, and many other countries have imposed 100% sales taxes on vehicles for many years. Many cities ban vehicle use in certain downtown areas. Most countries impose fuel taxes of 300% or more on gasoline. In the aftermath of the September 11, 2001 attacks in the USA, only vehicles with two or more occupants were allowed to enter southern Manhattan, New York. In wartime, fuel is rationed. In Mexico City and Santiago (Chile), only some cars can be driven on especially polluted days. Many governments design road networks to direct traffic away from areas with ecological and cultural heritage values.[131]

Choices and investments are made every day. Long-distance rail and bus systems are subsidized, light-rail systems are built in suburbs, and bikeways and walkways are built (and removed). New technologies and new facilities and services will continue to emerge. The desire for more personalized travel will continue unabated. The challenge is to inform and guide travel choices, vehicle and road designs, and government policy and investment so that they balance the benefits and costs of mobility. Ecology, including its investment, cost, and benefit dimensions, is a growing part of that equation.

Nature, Wildlife, and Water

Nature protection in many areas of Europe includes maintaining ecosystems or nature that is "traditional," a product of intensive human management and resource extraction—in a sense, human-created nature or natural systems.[719,339] To enhance nature in The Netherlands, road corridors are perforated with numerous types of tunnels and overpasses that facilitate wildlife movement across the landscape[63] (Chapter 6). In contrast, the Australian approach to protecting wildlife and biological diversity in intensive agricultural landscapes is to maintain strips of somewhat natural vegetation along roads.[815] Both approaches are officially dictated, funded, and managed by government. Neither approach is common in North America, where protection of a perceived nearly pristine nature with natural processes is a prime policy goal.

For alleviating problems of wildlife-vehicle accidents, a combination of recent detection technology, modifying motorist driving behavior, modifying animal behavior, and improving public awareness of the road-kill problem offers promise.[785,904] Crossing-structure projects and analyses tend to focus on a single species, yet multispecies movements typical of natural communities may be more important.[760,41,845,168] How to engineer low-cost wildlife crossing structures using animal behavior principles, a coating of soil, and tough, drought-resistant plants—perhaps using technology of existing lightweight, pedestrian-friendly designs—remains a little-explored opportunity.

A litany of ecological difficulties is familiar when considering road systems and water and aquatic ecosystems (Chapters 7–9). Are there any simple answers or obvious places to start in solving such a complex problem? Water quantity and water quality are the two problems, with the first also affecting the second. Redesigning and enlarging culverts and bridges where roads directly affect streams might produce the biggest improvement in water quantity and quality issues. But across the land, roadbeds block sheets of surface water from storms and in some places block groundwater flows. Maybe simply increasing the number of water pipes and culverts along roadways—perhaps by 5, 10, or 20 times—would produce the greatest gain for hydrology, erosion, sedimentation, streams, and wetlands (see Figure 13.1 in Chapter 13). Small animals would benefit as well. The ongoing schedule of "upgrading," roadbed maintenance, and resurfacing projects appears to be a cost-effective mechanism to accomplish needed water and aquatic ecosystem improvements.

Places, Communities, and Neighborhoods

The sense of place implies a deep connection between people and a location with distinctive characteristics and meaning. The characteristics vary widely from heritage,[642,598,704,670] to beauty,[237,670,1041,876] to ecology,[1041,876,834,704,598,199] to story,[122,4,808,834,642] to experience.[4,704,808,642] Roads and vehicles typically are also characteristics of a place and its meaning. Yet change in the broader road system may threaten the deep connections with a place.

Some highways effectively create new communities for people by opening up new resource locations or providing new transportation centers. In contrast, highways may also cause communities to wither, as in the case of a bypass route around a town or the development of a new shopping or strip-development area. Both creation and destruction have huge ecological implications.

Moreover, neighborhoods and communities can be sliced in two by a highway.[675] This division typically leads to two different, smaller neighborhoods or to the withering of one. Over time, public services, including transit, walkways, park areas, and security, tend to differ in the two neighborhoods. Environmental and social justice impacts[675] highlight not just the physical presence of a bisecting or adjacent highway but the consequent environmental effects. Thus air pollutants, traffic noise, increased flooding, and so forth cause major effects on the human community.

Traffic calming reduces and slows traffic within a community[2] and might especially reduce travel if applied at a regional scale. Local aquatic ecosystems and nature reserves, for example, should benefit from an associated decrease in road-kills, pollutants, and traffic noise. Since slower traffic may also, for instance, produce more engine idling, acceleration time, and consequent pollutant output, a broad, rigorous ecological evaluation of traffic calming should be highly informative.

Economics

Economists and ecologists treat each other warily and often skeptically. Despite their common "eco" origin, the disciplines are rooted in a very different set of values and beliefs. Yet, the less they talk past each other, the more likely progress will be made in resolving road ecology issues. In general, cost estimates of problems and solutions exist for only a few pieces of the puzzle—such as the cost of wildlife crossing structures (Chapter 6), different types of winter de-icing agents (Chapter 8), and insurance costs of wildlife-vehicle accidents (Chapter 6). What about the costs of alternative roadside management practices for native wildflower plantings, natural wildlife habitat, and carbon sequestration? Similarly, it is not "rocket ecology" (or economics) to jointly develop cost estimates for alternative approaches to restore streams and fish populations degraded by roads or for alternative ecologically sound approaches to highway planning and construction.

In the USA, the TEA-21 Act provides tens of billions of dollars per year for highways, public transit, rail, and safety investments.[275] A Highway Trust Fund with revenues from gasoline and other taxes on automobiles and trucks is the source of funds for most of the programs in the act, including a large portion of the expenditures for transit. In 1995, only 2% of all personal travel was by transit (91% was by private vehicles and the remainder by plane, train, and so forth), yet about 22% of all federal highway and transit expenditures were for transit.[281,278,280] Still, the bulk of the funding is for highways, and it remains an economic and cultural puzzle how to increase alternatives to the personal vehicle in the USA, especially in the face of so many impressive time-tested public transportation systems in Europe.

The TEA-21 Act, when coupled with NEPA, requires mitigation for damage to the environment caused by new projects. Remediation of past damage is eligible for funding but must be justified and compete with other priorities for funds. Transportation enhancement funding has helped spur activity in bikeways, historic preservation, enhanced landscaping, and native wildflower use, and recently projects to reduce vehicle-caused wildlife mortality have become eligible for funding.[275,279] The trends for road ecology investments are slowly moving in the right direction, but they also lead to further questions. What is the economic and societal rationale for the use, or non-use, of transportation funds for ecology, how should those funds be spent, and what is an appropriate funding level?

Planning

Environmental planning is often directed at road segments. Broadening it spatially to road networks at the regional or state level should provide new, more sustainable cost-effective solutions. For example, in Florida, a statewide environmental priority–setting project for roads is currently based on nine factors (riparian zone, rare species, wildlife corridor, and so forth), of which road-kill (especially of black bears, *Ursus americanus*) is particularly important[851] (Figure 14.6). This regional road-network approach highlights priority road segments for ecological mitigation by the state department of transportation. Alternative assays might focus on stream habitat structure, on the degree of disruption of animal-movement patterns,[12] on population sizes (linked to extinction probabilities) of key species, or directly on biodiversity. Addressing landscape ecological patterns and processes relative to road networks gets beyond the site-by-site and species-by-species approach—a big gain for society.

Sprawl over two or three decades in North America represents one of the great challenges to society, especially because of its road ecology dimensions[68,70,715] (Chapter 12). Aghast, ecologists have largely moved their research farther away

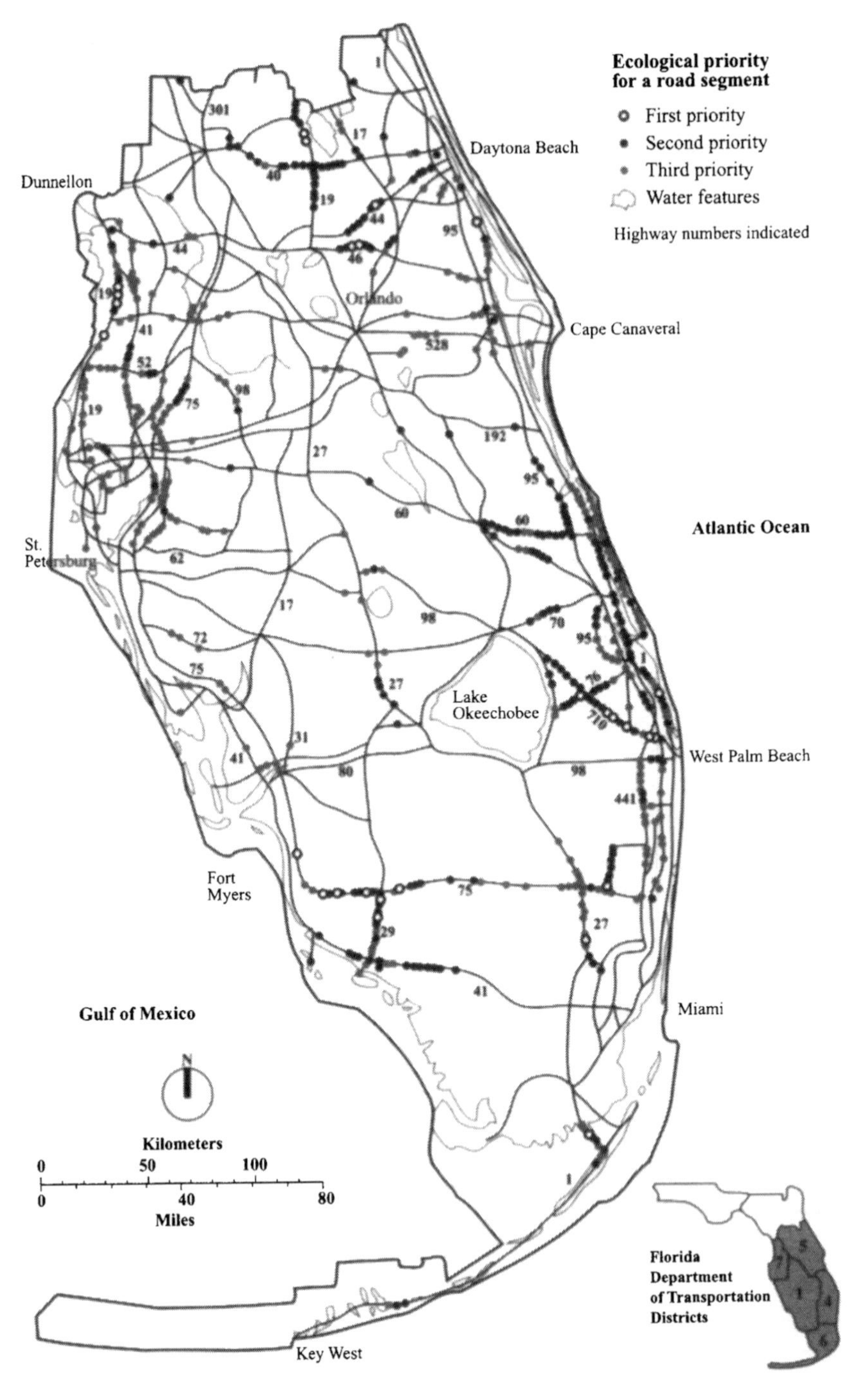

Figure 14.6. Priority stretches of road for nature protection and mitigation based on a regional analysis of several key ecological attributes in the vicinity of a road. Many wildlife underpasses have been built on Highways 75, 29, 46, and elsewhere in the flat state (Chapter 6 and Figure 6.7). See text for explanation. Florida. Adapted from D. Smith (1999).

rather than grasp the mantle of sprawl ecology. Yet ecologists have the tools to study and model, for example, alternative road networks, changing land-use patterns, generic spatial patterns of nature, ecological flows across the land, and combinations thereof. Working hand in hand with the transportation and planning community, ecologists could produce great ecological and societal benefits.

Finally, to emphasize that many approaches and solutions can be simple, consider the following road ecology approach (which was approved in law and funded by Parliament in The Netherlands).[652,951,318] Develop a map of the primary ecological network of the land (basically large vegetation areas plus connecting major water and wildlife corridors).[302,697,417,305] Then overlay the road network onto the ecological network (Figure 14.7). This simple process is used

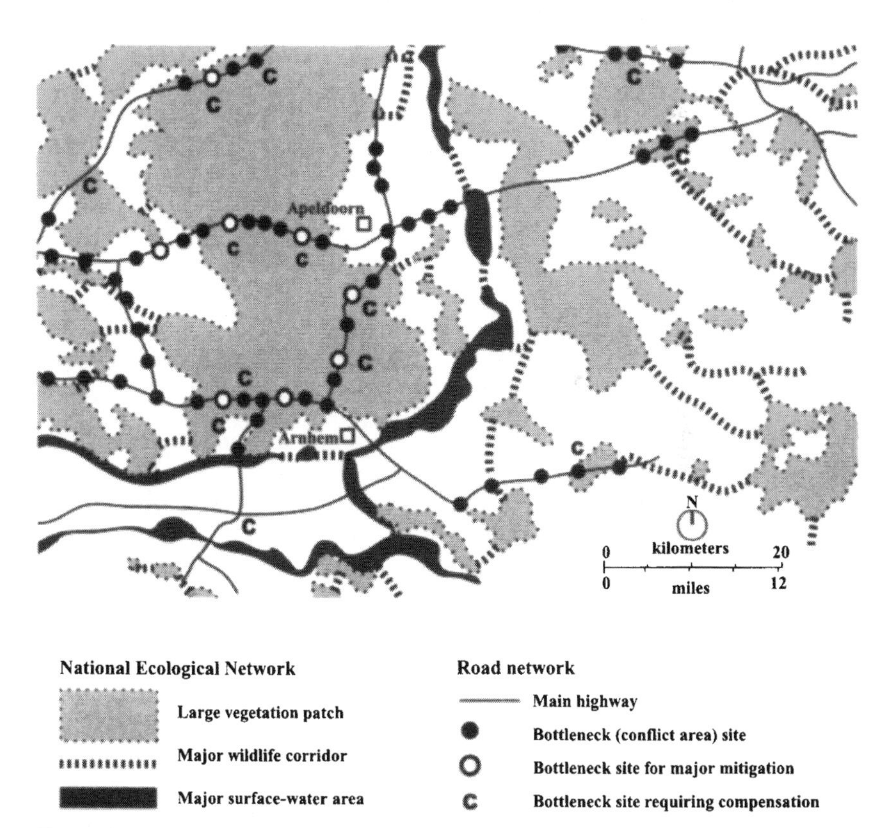

Figure 14.7. Main road network superimposed on a national ecological network to identify conflict areas for ecological mitigation or compensation. Most major mitigation sites include a wildlife overpass. Compensation, where mitigation on site is apparently not possible, is accomplished off site (Chapter 6). Vegetation patches are technically soil types that support vegetation, plus a few planned revegetation spots; major wildlife corridors are known or probable. East-central portion of The Netherlands. Based on Morel and Specken (1992) and van Bohemen et al. (1994) and adapted from Forman and Hersperger (1996).

nationally in The Netherlands to identify ecology-transportation conflict areas (or bottlenecks) for mitigation and compensation.[63,191,192,193,697]

Policy

Six broad policy initiatives are suggested by road ecology[308] (Figure 14.8):

1. Perforate road corridors with underpasses and overpasses for frequent wildlife and water crossings to reduce the road-barrier effect and habitat fragmentation.
2. Use soil berms and vegetation and depress roads to reduce traffic disturbance and noise effects on wildlife (as well as people).
3. Decrease and divert traffic, including trucks, and channel it onto primary roads to reduce the dispersion of both noise and barrier effects.
4. Improve engineering designs of road surface, tires, motors, and vehicles (aerodynamics) to reduce the ecological effects of noise.
5. Use cleaner fuel and "life-cycle" vehicular materials (by designing essentially all vehicle parts to be recycled) to reduce greenhouse gases as well as pollutants of soil, water, and air.
6. Close or remove remote roads to reduce human access and disturbance.

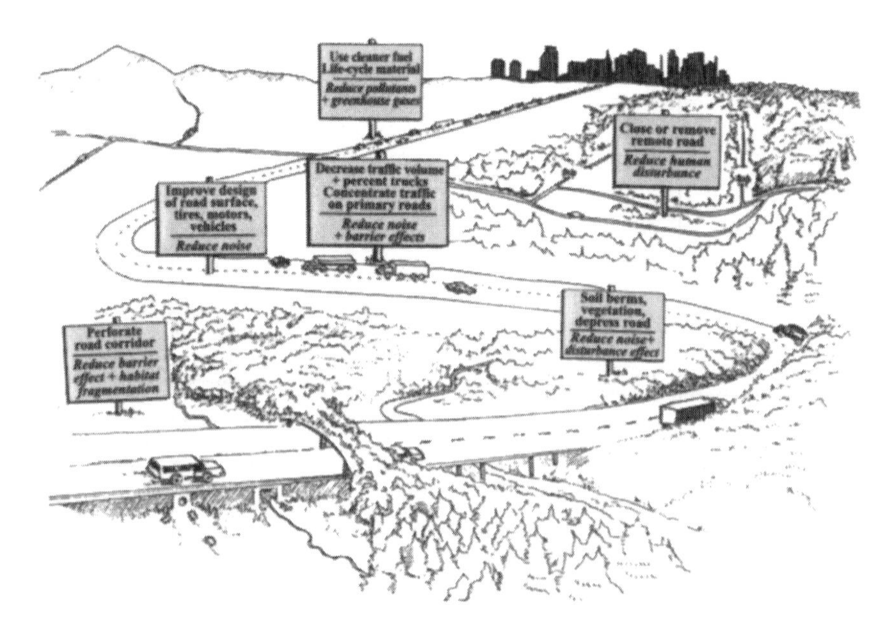

Figure 14.8. Six broad policy changes that would reduce major ecological impacts of road systems. The changes address effects that extend the farthest outward (usually > 100 m [328 ft]) from roads. Adapted from Forman (2002); original drawing courtesy of Lisa Giersbach.

If more of the highway in Figure 14.8 were visible, the driver might also see signs related to (7) roadside design and management, (8) disrupted hydrologic flows, (9) streams and other aquatic ecosystems, and even (10) the policy process itself, to understand the needed interdisciplinarity, the key temporal and spatial scales, how to deal with negative side effects, and so forth.

These simplified broad initiatives may be considered as goals and guidelines. Some are directly applicable, but more often the solutions involve complex trade-offs. Just as there is no magic answer for reducing pollutants in stormwater runoff (Chapter 8), most of the specific issues will be effectively ameliorated with a combination of actions producing success.

In practice, most policy initiatives will have four components in parallel, like a fork with four tines: (1) pilot projects, (2) rigorous research, (3) project evaluations, and (4) public education. For success in accomplishing initiatives, effective leadership of course is needed. In addition, schoolteachers, environmental groups, community groups, and local governments, for instance, should know the pilot projects and receive the periodic monitoring reports. Policy initiatives begin as seedlings, which are nurtured and then rapidly spread throughout a region. What can happen? Within a few years, what was politically impossible becomes compelling, and political leaders are eager to step forward and take credit.

Whither Road Ecology?

This book is wide ranging in its subject matter, and intentionally so. In some ways, it represents a distillation of knowledge relevant to the ecological study and management of roads and vehicles. Our goal has been to provide the scientific basis for road ecology.

At this early stage of coalescence, the essential core and the priority areas remain open, up for grabs by the upcoming generation of road ecologists. Where should researchers and practitioners focus their efforts to be most effective? Road ecology is highly relevant in many applications and situations, from new roads being built on the urban fringe to forestry roads, rural roads crossing fragile lands, and mountain areas coming under new intense development pressure, such as in the Rocky Mountains and Sierra Nevadas. Road ecology is critical in all of these.

But some focus is needed. Where are the problems most threatening and demand the most attention? Where can answers and solutions emerge in a time-effective manner? Are the roads and light development characteristic of sprawl better or worse ecologically than intensive agriculture or forestry? Should ecological inquiry mainly focus on roads in large parks and remote fragile areas where few people live?

Are these even important questions? Ecologists have been little involved in such questions. Answers are also scarce because little is known of what the public thinks and values about the ecological impacts of roads. Sophisticated survey research and valuation studies would provide policy makers some sense of the relative priorities. Such research could also highlight the need for and focus of public education relative to road ecology.

In a larger sense, how might road ecology evolve as a body of knowledge and field of study? From a spatial and professional perspective, road ecology lies at the intersection of traditional metropolitan-focused land-use and transport planning, and traditionally rural- and remote-area-focused ecologists. So far, the latter have been mainly interested in large, relatively natural areas without roads or with lightly traveled roads and with a minimal human population. In contrast, land-use and transport planners focus on the built environment, often teeming with people.

On a professional level, the two groups are quite distinct. Land-use and transport planners employ a large variety of transport-network and travel-forecasting models. These professionals are part of a deeply entrenched and institutionalized process responsible for allocating transportation funds. Natural-land rural ecologists, in contrast, are typically rooted in the biological sciences, and their activities are not nearly as codified or institutionalized. They deal with wildlife, plants, aquatic ecosystems, landscape ecology, and the like. Is road ecology simply an overlap of the two groups?

As an "-ology," or area of study, road ecology draws heavily on science but reaches into the fields of engineering, design, and social science. Indeed, big-picture solutions require expertise from several fields beyond transportation and ecological science. Road ecology could remain simply the overlap area of these diverse disciplines, or it could develop its own body of theory. At present, spatial models, networks, and populations appear to be the most promising areas to explore in developing a potential body of theory[304] (Chapters 5 and 11). However, concepts, principles, models, and theories unraveled through the pages of this book are grist for thoughtful people and future theory.

What are the next steps for road ecology professionals and scientists? To what extent should or might road ecology remain a "scientific community" that serves other professional communities? Studying road systems in parks, forestry areas, and remote lands is ecologically important, but it does not engage with the mainstream transportation community, which deals with the central issues of a nation's road infrastructure and traffic. One obvious opportunity for collaboration and common interest is metropolitan-fringe and sprawl areas. Here population growth overall is most rapid, and most new roads are likely to be built in the foreseeable future. Here ecologically important habitats, species, and processes are present but are undergoing rapid change and degradation. And here road ecology research and application could have a noticeable impact quickly.

Certainly, expansion of the scientific and research base is important. How might further research on road ecology be inspired, and how might resources be generated to create necessary tools and methods for planners and engineers? Possible steps include holding meetings and symposia, solidifying funding mechanisms, broadening the focus of existing quality journals and launching a scholarly journal, developing effective linkages with state and federal transportation agencies, and creating standing committees with the American Association of State Highway and Transportation Officials, the National Research Council's Transportation Research Board of the National Academy of Sciences, and elsewhere.

Perhaps more important, road ecology itself, with its inevitable attraction of ecologists and related scientists, will greatly benefit from closer alignment with transportation and land-use planners. Ecologists have more of a scientific bent, are politically less entangled, and have scientific credibility. But the transportation community manages vast sums of money and is the locus of road decision making and management. The transportation community also has an important research core, an obvious entrée for scientific linkages with ecological scientists.

Both groups have much to gain in collaboration and synergy. Transportation will be noticeably enhanced. The land will be more species—and people—friendly. Society will be the winner. In essence, the real achievement will be safe and efficient transport elegantly embedded in nature's biological diversity and natural process.

Looking ahead, we should all be able to say, "Remember what it was like back at the turn of the century? Since then, the ecological transformation of our land and our transportation system is so conspicuous, like an enormous green flag waving in the wind, that even space travelers are impressed."

Bibliography

1. Aanen, P., W. Alberts, G. J. Bekker, H. D. van Bohemen, P. J. M. Melman, J. van der Sluijs, G. Veenbaas, H. J. Verkaar, and C. F. van de Watering. 1991. *Nature Engineering and Civil Engineering Works.* Wageningen, Netherlands: Pudoc.
2. AASHTO. 2001a. *A Policy on Geometric Design of Highways and Streets: 2001a.* Washington, D.C.: American Association of State Highway and Transportation Officials.
3. ———. 2001b. *Guidelines for Geometric Design of Very Low-Volume Local Roads (ADT ≤ 400): 2001.* Washington, D.C.: American Association of State Highway and Transportation Officials.
4. Abbey, E. 1971. *Desert Solitaire: A Season in the Wilderness.* New York: Ballantine Books.
5. Aber, J. D., A. Magill, S. G. McNulty, R. D. Boone, K. J. Nadelhoffer, M. Downs, and R. Hallett. 1995. Forest biogeochemistry and primary production altered by nitrogen saturation. *Water, Air and Soil Pollution* 85:1665–70.
6. Adams, K. M., and H. Hernandez. 1977. Snow and ice roads: Ability to support traffic and effects on vegetation. *Arctic* 30:13–27.
7. Adams, L. W. 1984. Small mammal use of an interstate highway median strip. *Journal of Applied Ecology* 21:175–78.
8. Adams, L. W., and A. D. Geis. 1983. Effects of roads on small mammals. *Journal of Applied Ecology* 20:403–15.
10. Adams, T. C. 1964. Salt migration to the northwest body of Great Salt Lake, Utah. *Science* 143:1028–30.
11. Ahrens, C. D. 1991. *Meteorology Today: An Introduction to Weather, Climate, and the Environment.* St. Paul, Minn.: West Publishing.
12. Alexander, S., and N. Waters. 2000. The effects of highway transportation corridors on wildlife: A case study of Banff National Park. *Transportation Research Part C: Emerging Technologies* 8:307–20.
13. Allen, R. E., and D. R. McCullough. 1976. Deer-car accidents in southern Michigan. *Journal of Wildlife Management* 40:317–25.
14. Althouse, J. 2001. Dust, washboards, and deep stabilization. *Erosion Control Magazine,* March/April, 44–49.
15. AMBS Consulting. 1997. *Fauna Usage of Three Underpasses beneath the F3 Freeway*

between Sydney and Newcastle. Final Report to New South Wales Roads and Traffic Authority. Sydney, Australia.

16. American Public Transit Association. 2001. *Public Transportation Fact Book.* Washington, D.C.: American Public Transit Association.
17. American Society of Landscape Architects. 1996. *Integrating Stormwater into the Urban Fabric.* Portland, Ore.: Portland State University.
18. Amor, R. L., and P. L. Stevens. 1976. Spread of weeds from a roadside into schlerophyll forests at Dartmouth, Australia. *Weed Research* 16:111–18.
19. Anderson, P. G. 1998. Sediment generation from forestry operations and associated effects on aquatic ecosystems. In *Forest-Fish Conference: Land Management Practices Affecting Aquatic Ecosystems,* edited by M. K. Brewin and D. M. Monita, 491–508. Ottawa: Canadian Forest Service.
20. Anderson, R. 1999. Disturbance as a factor in the distribution of sugar maple and the invasion of Norway maple into a modified woodland. *Rhodora* 101:264–73.
21. Anderson, R., and D. Asquin. 1986. *Culvert Sizing for Stream Crossings: A Handbook of Three Common Methods.* Edmonton, Alberta: Forestry, Land, and Wildlife.
22. Andrews, A. 1990. Fragmentation of habitat by roads and utility corridors: A review. *Australian Zoologist* 26:130–42.
23. Angold, P. G. 1997. The impact of road upon adjacent heathland vegetation: Effects on plant species composition. *Journal of Applied Ecology* 34:409–17.
24. Appleyard, D., K. Lynch, and J. R. Myer. 1964. *The View from the Road.* Cambridge: MIT Press.
25. Apps, C. D. 1999. Space-use, diet, demographics, and topographic associations of lynx in the southern Canadian Rocky Mountains: A study. In *Ecology and Conservation of Lynx in the United States,* edited by L. F. Ruggiero, K. B. Aubry, S. W. Buskirk, G. M. Koehler, C. J. Krebs, K. S. McKelvey, and J. R. Squires, 351–72. General Technical Report RMRS-GTR-30WWW. Fort Collins, Colo.: USDA Forest Service.
26. Areendran, G., and M. K. S. Pasha. 2000. *Gauer Ecology Project.* Dehradun, India: Wildlife Institute of India.
27. Arendt, R. G. 1996. *Conservation Design for Subdivisions: A Practical Guide to Creating Open Space Networks.* Washington, D.C.: Island Press.
28. Arnold, C. L., Jr., and C. J. Gibbons. 1996. Impervious surface coverage: The emergence of a key environmental indicator. *Journal of the American Planning Association* 62:243–58.
30. Ashley, E. P., and J. T. Robinson. 1996. Road mortality of amphibians, reptiles and other wildlife on the Long Point Causeway, Lake Erie, Ontario. *Canadian Field-Naturalist* 110:404–12.
31. Ashley, M., and C. Wills. 1987. Analysis of mitochondrial DNA polymorphisms among Channel Island deer mice. *Evolution* 41:854–63.
32. Auerbach, N. A., M. D. Walker, and D. A. Walker. 1997. Effects of roadside disturbance on substrate and vegetation properties in arctic tundra. *Ecological Applications* 7:218–35.
33. Bache, D. H., and I. A. MacAskill. 1984. *Vegetation in Civil and Landscape Engineering.* London: Granada.

34. Backman, L., and L. Folkeson. 1995. The influence of de-icing salt on vegetation, groundwater and soil along highways E20 and 48 in Skaraborg County during 1994. *VTI meddelande Nr* 775A:45.
35. Bagley, S. 1998. *The Road-Ripper's Guide to Wildland Road Removal.* Missoula, Mont.: Wildlands Center for Preventing Roads.
36. Bagnold, R. A. 1954. *The Physics of Blown Sand and Desert Dunes.* London: Methuen.
37. Baker, H. G. 1986. Patterns of plant invasion in North America. In *Ecology of Biological Invasions of North America and Hawaii,* edited by H. A. Mooney and J. A. Drake, 43–57. New York: Springer-Verlag.
38. Baker, L. A., A. T. Herlihy, P. R. Kaufmann, and J. M. Eilers. 1991. Acid lakes and streams in the United States: The role of acidic deposition. *Science* 252:1151–54.
39. Baker, R. H. 1998. Are man-made barriers influencing mammalian speciations? *Journal of Mammalogy* 79:370–71.
40. Bakowski, C., and M. Kozakiewicz. 1988. The effect of forest road on bank vole and yellow-necked mouse populations. *Acta Theriologica* 33:345–53.
41. Ballon, P. 1985. Premières observations sur l'efficacité des passages à gibier sur l'autoroute A36. In *Routes et faune sauvage,* 311–16. Bagneaux, France: Service d'Etudes Techniques de Routes et Autoroutes.
42. Bangs, E. E., T. N. Bailey, and M. F. Portner. 1989. Survival rates of adult female moose on the Kenai Peninsula, Alaska. *Journal of Wildlife Management* 53:557–63.
43. Banks, J. E. 1997. Do imperfect trade-offs affect the extinction debt phenomenon? *Ecology* 78:1597–1601.
44. Bannister, D. 2001. Transport planning. In *Handbook of Transport Systems and Traffic Control,* edited by K. J. Button and D. A. Hensher, 9–20. New York: Pergamon.
45. Barbosa, A. E., and T. Hvitved-Jacobsen. 1999. Highway runoff and potential for removal of heavy metals in an infiltration pond in Portugal. *Science and the Total Environment* 235:151–59.
46. Barbour, E. H. 1895. Bird fatality along Nebraska railroads. *Auk* 12:187.
47. Barker, J. R., and D. T. Tingey, eds. 1991. *Air Pollution Effects on Biodiversity.* New York: Van Nostrand Reinhold.
48. Barnett, J. L., R. A. How, and W. F. Humphreys. 1978. The use of habitat components by small mammals in eastern Australia. *Australian Journal of Ecology* 3:277–85.
50. Barret, M. E., R. D. Zuber, E. R. Collins, J. F. Malina, R. J. Charbeneau, and G. H. Ward. 1993. *A Review and Evaluation of Literature Pertaining to the Quantity and Control of Pollution from Highway Runoff and Construction.* Austin: Center for Research in Water Resources, Bureau of Engineering Research, University of Texas.
51. Bartoldus, C. C., and E. W. Garbish. 1994. *Evaluation for Planned Wetlands (EPW): A Procedure for Assessing Wetland Functions and a Guide to Functional Design.* St. Michaels, Md.: Environmental Concern Inc.
52. Bashore, T. L., W. M. Tzilkowski, and E. D. Bellis. 1985. Analysis of deer-vehicle collision sites in Pennsylvania. *Journal of Wildlife Management* 49:769–74.
53. Batanouny, K. H. 1979. Vegetation along the Jeddah-Mecca road: Pattern and process as affected by human impact. *Journal of Arid Environments* 2:21–30.

54. Bates, G. H. 1937. The vegetation of wayside and hedgerow. *Journal of Ecology* 25:469–81.
55. Baumgartner, F. M. 1934. Bird mortality on the highways. *Auk* 51:537–38.
56. Baur, A., and B. Baur. 1990. Are roads barriers to dispersal in the land snail *Arianta arbustorum? Canadian Journal of Zoology* 68:613–17.
57. Bauske, B., and D. Goetz. 1993. Effects of de-icing salts on heavy metal mobility. *Acta Hydrochimica et Hydrobiologica* 21:38–42.
58. Bayley, P. B. 1995. Understanding large river-floodplain ecosystems. *BioScience* 45:153–58.
59. Bazzaz, F. A. 1996. *Plants in Changing Environments: Linking Physiological, Population, and Community Ecology.* New York: Cambridge University Press.
60. Beatley, T. 2000. *Green Urbanism: Learning from European Cities.* Washington, D.C.: Island Press.
61. Beier, P., and R. H. Barrett. 1991. *The Cougar in the Santa Ana Mountain Range, California. Final Report.* Berkeley, Calif.: Orange County Co-operative Mountain Lion Study. Department of Forestry and Resource Management.
62. Beier, P., and S. Loe. 1992. A checklist for evaluating impacts to wildlife movement corridors. *Wildlife Society Bulletin* 20:434–40.
63. Bekker, H., B. van den Hengel, H. van Bohemen, and H. van der Sluijs. 1995. *Natuur over Wegen* (Nature across motorways). Delft, Netherlands: Ministry of Transport, Public Works and Water Management.
64. Bekker, H. G. J., and K. Canters. 1997. The continuing story of badgers and their tunnels. In *Habitat Fragmentation & Infrastructure,* edited by K. Canters, 344–53. Delft, Netherlands: Ministry of Transport, Public Works and Water Management.
65. Bellis, E. D., and H. B. Graves. 1971. Deer mortality on a Pennsylvania interstate highway. *Journal of Wildlife Management* 35:232–37.
66. Bender, D. J., T. A. Contreras, and L. Fahrig. 1998. Habitat loss and population decline: A meta-analysis of the patch size effect. *Ecology* 79:517–33.
67. Benfenati, E., S. Valzacchi, G. Maniani, L. Airoldi, and R. Farnelli. 1992. PCDD, PCDF, PCB, PAH, cadmium and lead in roadside soil: Relationship between road distance and concentration. *Chemosphere* 24:1077–83.
68. Benfield, F. K., M. D. Raimi, and D. D. T. Chen. 1999. *Once There Were Greenfields.* New York: Natural Resources Defense Council.
70. Benfield, F. K., J. Terris, and N. Vorsangeret. 2001. *Solving Sprawl: Models of Smart Growth in Communities across America.* New York: Natural Resources Defense Council.
71. Bennett, A. F. 1988. Roadside vegetation: A habitat for mammals at Naringal, southwestern Victoria. *Victorian Naturalist* 105:106–13.
72. ———. 1991. Roads, roadsides and wildlife conservation: A review. In *Nature Conservation 2: The Role of Corridors,* edited by D. A. Saunders and R. J. Hobbs, 99–117. Chipping Norton, Australia: Surrey Beatty.
73. ———. 1999. *Linkages in the Landscape: The Role of Corridors and Connectivity in Wildlife Conservation.* Gland, Switzerland and Cambridge, U.K.: IUCN—The World Conservation Union.
74. Bennett, S. L. 2000. Dust: Invisible, dangerous, avoidable. *Erosion Control Magazine,* March, 42–50.

75. Berger, T. R. 1977. *Northern Frontier, Northern Homeland: The Report of the Mackenzie Valley Pipeline Inquiry*, Vols. 1 and 2. Ottawa: Supply and Services Canada.
76. Bergerud, A. T., R. D. Jakimchuk, and D. R. Carruthers. 1984. The buffalo of the North: Caribou *(Rangifer tarandus)* and human developments. *Arctic* 37:7–22.
77. Bergin, T. M., L. B. Best, K. E. Freemark, and K. J. Koehler. 2000. Effects of landscape structure on nest predation in roadsides of a midwestern agroecosystem: A multiscale analysis. *Landscape Ecology* 15:131–43.
78. Beringer, J. J., S. G. Seibert, and M. R. Pelton. 1990. Incidence of road crossing by black bears on Pisgah National Forest, North Carolina. *International Conference on Bear Research and Management* 8:85–92.
79. Bernardino, F. S., Jr., and G. H. Dalrymple. 1992. Seasonal activity and road mortality of the snakes of the Pa-hay-okee wetlands of the Everglades National Park, USA. *Biological Conservation* 72:71–75.
80. Bertwistle, J. 1999. The effects of reduced speed zones on reducing bighorn sheep and elk collisions with vehicles on the Yellowhead highway in Jasper National Park. In *Proceedings of the Third International Conference on Wildlife Ecology and Transportation,* edited by G. L. Evink, P. Garrett, and D. Zeigler, 89–97. FL-ER-73-99. Tallahassee: Florida Department of Transportation.
81. Beschta, R. L. 1978. Long-term patterns of sediment production following road construction and logging in the Oregon Coast Range. *Water Resources Research* 14:1011–16.
82. Bevers, M., and C. H. Flather. 1999. Numerically exploring habitat fragmentation effects on populations using cell-based coupled map lattices. *Theoretical Population Biology* 55:61–76.
83. Bider, J. R. 1968. Animal activity in uncontrolled terrestrial communities as determined by a sand transect technique. *Ecological Monographs* 38:274–91.
84. Bilby, R. E. 1985. Contributions of road surface sediment to a western Washington stream. *Forest Science* 31:827–38.
85. Bilby, R. E., K. Sullivan, and S. H. Duncan 1979. The generation and fate of road-related sediment in forested watersheds in southwestern Washington. *Forest Science* 35:453–68.
86. Bilodeau, A., and M. Malhotra. 2000. High-volume fly ash system: Concrete solution for sustainable development. *ACI Mater Journal* 97:41–48.
87. Binford, M. W., and M. J. Buchenau. 1993. Riparian greenways and water resources. In *Ecology of Greenways: Design and Function of Linear Conservation Areas,* edited by D. S. Smith and P. C. Hellmund, 69–104. Minneapolis: University of Minnesota Press.
88. Bird, D., D. Varland, and J. Negro. 1996. *Raptors in Human Landscapes: Adaptations to Built and Cultivated Environments.* San Diego, Calif.: Academic Press.
90. Bishop, S.C., and F. S. Chapin. 1989. Patterns of natural vegetation on abandoned gravel pads in arctic Alaska. *Journal of Applied Ecology* 26:1073–81.
91. Bisson, P. A., R. E. Bilby, M. D. Bryant, C. A. Dolloff, R. A. House, M. L. Murphy, K. V. Koski, and J. R. Sedell. 1987. Large woody debris in forested streams in the Pacific Northwest: Past, present, and future. In *Streamside Management and Fishery Interactions,* edited by E. O. Salo and T. W. Cundy, 143–90. Seattle: Institute of Forest Resources, University of Washington.

92. Bissonette, J. A., and M. Hammer. 2000. Effectiveness of earthen return ramps in reducing big game highway mortality in Utah. *UTCFWRU Report Series 2000* no. 1: 1–29.
93. Bissonette, J. A., and I. Storch, eds. 2002. *Landscape Theory and Resource Management: Linking Theory to Management*. Washington, D.C.: Island Press.
94. Blais, J. M., R. L. France, L. E. Kimpe, and R. J. Cornett. 1998. Climatic changes in northwestern Ontario have had a greater effect on erosion and sediment accumulation than logging and fire: Evidence from ^{210}Pb chronology in lake sediments. *Biogeochemistry* 43:235–52.
95. Bliss, L. C. 1983. Modern human impacts in the Arctic. In *Man's Impact on Vegetation,* edited by W. Holzner, M. Werger, and I. Ikusima, ch. 13. London: Dr. W. Junk Publishers.
96. ———. 1995. Industrial development, Arctic. *Encyclopedia of Environmental Biology* 2:277–88.
97. Bliss, L. C., and E. B. Peterson. 1973. *The Ecological Impact of Northern Petroleum Development*. Fifth International Conference on Arctic Oil and Gas. Le Havre, France: Fondation Française d'Etudes Nordiques.
98. Blocher, A. 1926. More fatalities. *Oölogist* 43:66–67.
99. ———. 1927. A summer's record of the automobile. *Oölogist* 44:96.
100. ———. 1936. Fatalities. *Oölogist* 53:19–22.
101. Blomqvist, G., and E.-L. Johansson. 1999. Airborne spreading and deposition of de-icing salt—a case study. *Science and the Total Environment* 235:161–68.
102. Bonnett, X., N. Guy, and R. Shine. 1999. The dangers of leaving home: Dispersal and mortality in snakes. *Biological Conservation* 89:39–50.
103. Bormann, F. H., and G. E. Likens. 1979. *Pattern and Process in a Forested Ecosystem*. New York: Springer-Verlag.
104. Bormann, F. H., D. Balmori, and G. T. Geballe. 1993. *Redesigning the American Lawn*. New Haven, Conn.: Yale University Press.
105. Bowles, A. E. 1997. Responses of wildlife to noise. In *Wildlife and Recreationists: Coexistence through Management and Research,* edited by R. L. Knight and K. J. Gutzwiller, 109–56. Washington, D.C.: Island Press.
106. Bowling, L. C., and D. P. Lettenmaier. 1997. *Evaluation of the Effects of Forest Roads on Streamflow in Hard and Ware Creeks, Washington*. Water Resources Series Technical Report 155, Department of Civil Engineering. Seattle: University of Washington.
107. Brandenburg, D. M. 1996. Effects of roads on behavior and survival of black bears in coastal North Carolina. Master's thesis, University of Tennessee, Knoxville.
108. Brandes, D. 1983. Flora und Vegetation deer bahnhofe Mitteleuropas. *Phytocoenologia* 11:31–115.
110. Brandle, J. R., D. L. Hintz, and J. W. Sturrock, eds. 1988. *Windbreak Technology*. Amsterdam: Elsevier. Reprinted from *Agriculture, Ecosystems and Environment* 22/23 (1988).
111. Braudel, F. 1981. *Civilization and Capitalism 15th–18th Century*. Vol. 1, *The Structures of Everyday Life: The Limits of the Possible*. New York: Harper and Row.
112. Brett, M. T., and D. C. Mueller-Navarra. 1997. The role of highly unsaturated fatty acids in aquatic food-web processes. *Freshwater Biology* 38:483–99.

113. Breuste, J., H. Feldmann, and O. Uhlmann, eds. 1998. *Urban Ecology.* New York: Springer-Verlag.
114. Brocke, R. H., J. P. O'Pezio, and K. A. Gustafson. 1990. A forest management scheme mitigating impact of road networks on sensitive wildlife species. In *Is Forest Fragmentation a Management Issue in the Northeast?* 13-17. General Technical Report NE-140. Radnor, Pa.: USDA Forest Service.
115. Brody, A. J., and M. R. Pelton. 1989. Effects of roads on black bear movements in western North Carolina. *Wildlife Society Bulletin* 17:5–10.
116. Brothers, T. S. 1997. Deforestation in the Dominican Republic: A village-level view. *Environmental Conservation* 24:213–23.
117. Bryens, O. M. 1931. A recovery and a return from the highway. *Bird-Banding* 2:85–86.
118. Bubeck, R. C., W. H. Diment, B. L. Deck, A. L. Baldwin, and S. D. Lipton. 1971. Runoff of deicing salt: Effect on Irondequoit Bay, Rochester, New York. *Science* 172:1128–32.
119. Buchanan, B. W. 1993. Effects of enhanced lighting on the behaviour of nocturnal frogs. *Animal Behaviour* 43:893–99.
120. Buchanan, J. B. 1987. Seasonality in the occurrence of long-tailed weasel road-kills. *Murrelet* 68:67–68.
121. Buckler, D., and G. Granato. 1999. *Assessing Biological Effects from Highway-Runoff Constituents.* Northborough, Mass.: U.S. Department of Interior, Geological Survey.
122. Buell, L. 2001. *Writing for an Endangered World: Literature, Culture, and Environment in the U.S. and Beyond.* Cambridge: Belknap Press of Harvard University Press.
123. Bugg, R. L., C. S. Brown, and J. H. Anderson. 1997. Restoring native perennial grasses to rural roadsides in the Sacramento Valley of California: Establishment and evaluation. *Restoration Ecology* 5:214–28.
124. Buhlmann, K. A. 1995. Habitat use, terrestrial movements, and conservation of the turtle, *Deirochelys reticularia,* in Virginia. *Journal of Herpetology* 29:173–81.
125. Bunyan, R. 1990. *Monitoring Program of Wildlife Mitigation Measures: Trans-Canada Highway Twinning—Phase II.* Final Report to Parks Canada, Banff National Park.
126. Bureau of Transportation Statistics. 1999. *National Transportation Statistics 1999.* Publication BTS99-04. Washington, D.C.: U.S. Department of Transportation.
127. ———. 2001. *National Transportation Statistics 2000.* Publication BTS01-01. Washington, D.C.: U.S. Department of Transportation.
128. Burel, F., and J. Baudry. 1990. Hedgerow network patterns and processes in France. In *Changing Landscapes: An Ecological Perspective,* edited by I. S. Zonneveld and R. T. T. Forman, 99–120. New York: Springer-Verlag.
129. ———. 1999. *Ecologie du paysage: Concepts, méthodes et applications.* Paris: Editions TEC & DOC.
130. Burnham, K. P., and D. R. Anderson. 1998. *Model Selection and Interference: A Practical Information Theoretic Approach.* New York: Springer.
131. Button, K., P. H. Nijkamp, and H. Priemus, eds. 1998. *Transport Networks in Europe: Concepts, Analysis and Policies.* Cheltenham, U.K.: Edward Elgar.
132. Byrd, D. S., J. T. Gilmore, and R. H. Lea. 1983. Effect of decreased use of lead in gasoline on the soil of a highway. *Environmental Science and Technology* 17:121–23.
133. Cale, P., and R. J. Hobbs. 1991. Condition of roadside vegetation in relation to

nutrient status. In *Nature Conservation 2: The Role of Corridors,* edited by D. A. Saunders and R. J. Hobbs, 353–62. Chipping Norton, Australia: Surrey Beatty.

134. Calvo, R., and N. Silvy. 1996. Key deer mortality, U.S. Highway 1 in the Florida Keys. In *Trends in Addressing Transportation Related Wildlife Mortality,* edited by G. L. Evink, P. Garrett, D. Zeigler, and J. Berry, 287–296. Report FL-ER-58-96. Tallahassee: Florida Department of Transportation.
135. Camby, A., and C. Maizeret. 1985. Perméabilité des routes et autoroutes vis-à-vis des mammifères carnivores: exemple des études menées dans les Landes de Gascogne par radio-poursuite. In *Routes et faune sauvage,* 183–96. Bagneaux, France: Service d'Etudes Techniques des Routes et Autoroutes.
136. Cameron, R., D. J. Reed, J. R. Dau, and W. T. Smith. 1992. Redistribution of calving caribou in response to oil field development on the arctic slope of Alaska. *Arctic* 45:338–42.
137. Camp, M., and L. B. Best. 1994. Nest density and nesting success of birds in roadsides adjacent to rowcrop fields. *American Midland Naturalist* 131:347–58.
138. Campbell, C. S., and M. H. Ogden. 1999. *Constructed Wetlands in the Sustainable Landscape.* New York: John Wiley.
139. Canters, K., ed. 1997. *Habitat Fragmentation & Infrastructure.* Delft, Netherlands: Ministry of Transport, Public Works and Water Management.
140. Cantwell, M. D., and R. T. T. Forman. 1994. Landscape graphs: Ecological modeling with graph theory to detect configurations common to diverse landscapes. *Landscape Ecology* 8:239–55.
141. Capel, S. W. 1988. Design of windbreaks for wildlife in the Great Plains of North America. *Agriculture, Ecosystems and Environment* 22/23:337–47. (Reprinted in J. R. Brandle, D. L. Hintz, and J. W. Sturrock, eds., *Windbreak Technology,* Amsterdam: Elsevier, 1988.)
142. Carbaugh, B. J., J. P. Vaughan, E. D. Bellis, and H. B. Graves. 1975. Distribution and activity of white-tailed deer along an interstate highway. *Journal of Wildlife Management* 39:570–81.
143. Carr, L. W., and L. Fahrig. 2001. Impact of road traffic on two amphibian species of different vagility. *Conservation Biology* 15:1071–78.
144. Carr, L. W., L. Fahrig, and S. E. Pope. 2002. Impacts of landscape transformation by roads. In *Applying Landscape Ecology in Biological Conservation,* edited by K. J. Gutzwiller, 225–43. New York: Springer.
145. Carsignol, J. 1993. *Passages pour la grande faune.* Bagneaux, France: Service d'Etudes Techniques des Routes et Autoroutes, Centre de la Sécurité et des Techniques Routières.
146. Carvalho, G., A. C. Barros, P. Moutinho, and D. C. Nepstad. 2001. Sensitive development could protect the Amazon instead of destroying it. *Nature* 409:131.
147. Case, R. M. 1978. Interstate highway road-killed animals: A data source for biologists. *Wildlife Society Bulletin* 6:8–13.
148. Caughley, G., and A. R. E. Sinclair. 1994. *Wildlife Ecology and Management.* Cambridge, Mass.: Blackwell Science.
149. Center for Energy and Environmental Policy. 2000. *Delaware Climate Change Action Plan.* Newark: University of Delaware.
150. Center for Transportation and the Environment. 2002. *International Conference*

on Ecology and Transportation 2001. Proceedings:*A Time for Action*. North Carolina State University, Raleigh.

151. Center for Watershed Protection. 1998. *Rapid Watershed Planning Handbook: A Comprehensive Guide for Managing Urbanizing Watersheds.* Ellicot City, Md.: Center for Watershed Protection.
152. Chadwick, D. H., and R. Gehman. 2000. *National Geographic Destinations:Yellowstone to Yukon*. Washington, D.C.: National Geographic Society.
153. Chapin, F. S., and G. R. Shaver. 1981. Changes in soil properties and vegetation following disturbance of Alaskan arctic tundra. *Journal of Applied Ecology* 18:605–17.
154. Chen, J., J. F. Franklin, and T. A. Spies. 1992. Vegetation responses to edge environments in old-growth Douglas-fir forests. *Ecological Applications* 2:387–96.
155. ———. 1995. Growing-season microclimatic gradients from clearcut edges into old-growth Douglas-fir forests. *Ecological Applications* 5:74–86.
156. Chepil, W. S., and N. P. Woodruff. 1963. The physics of wind erosion and its control. *Advances in Agronomy* 15:211–302.
157. Chollar, B. H. 1990. An overview of deicing research in the United States. In *The Environmental Impact of Highway Deicing,* edited by C. R. Goldman and G. J. Malyj, 1–6. Institute of Ecology Publication No. 33. Davis: University of California.
158. Chomitz, K. M., and D. A. Gray. 1996. Roads, land use, and deforestation: A spatial model applied to Belize. *World Bank Economic Review* 10:487–512.
159. Chu, E. W., and J. R. Karr. 2001. Environmental impact, concept and measurement of. *Encyclopedia of Biodiversity* 2:557–77.
160. Clark, C. F., P. G. Smith, G. Neilson, et al. 2000. Chemical characterisation and legal classification of sludges from road sweepings. *Journal of Chart Institute Water E* 14:99–102.
161. Clark, D. R. 1979. Lead concentrations: Bats vs. terrestrial small mammals collected near a major highway. *Environmental Science and Technology* 13:338–40.
162. Clark, W. D., and J. R. Karr. 1979. Effect of highways on red-winged blackbird and horned lark populations. *Wilson Bulletin* 91:143–45.
163. Clarke, G. P., P. C. L. White, and S. Harris. 1998. Effects of roads on badger *Meles meles* populations in south-west England. *Biological Conservation* 86:117–24.
164. Clarke, H. 1930. Birds killed by automobiles. *Bird-Banding* 32:271.
165. Clevenger, A. P. 1998. Permeability of the Trans-Canada Highway to wildlife in Banff National Park: Importance of crossing structures and factors influencing their effectiveness. In *Proceedings of the International Conference on Wildlife Ecology and Transportation,* edited by G. L. Evink, P. Garrett, D. Zeigler, and J. Berry, 109–119. FL-ER-69-98. Tallahassee: Florida Department of Transportation.
166. Clevenger, A. P. 2001. *Highway Effects of Wildlife.* Progress Report 6 prepared for Parks Canada. Banff, Alberta.
167. Clevenger, A. P., and N. Waltho. 1999. Dry drainage culvert use and design considerations for small- and medium-sized mammal movement across a major transportation corridor. In *Proceedings of the Third International Conference on Wildlife Ecology and Transportation,* edited by G. L. Evink, P. Garrett, and D. Zeigler, 263–77. FL-ER-73-99. Tallahassee: Florida Department of Transportation.

168. ———. 2000. Factors influencing the effectiveness of wildlife underpasses in Banff National Park, Alberta, Canada. *Conservation Biology* 14:47–56.
169. Clevenger, A. P., B. Chruszcz, and K. Gunson. 2001a. Drainage culverts as habitat linkages and factors affecting passage by mammals. *Journal of Applied Ecology* 38:1340–49.
170. ———. 2001b. Highway mitigation fencing reduces wildlife-vehicle collisions. *Wildlife Society Bulletin* 29:646–53.
171. Clevenger, A. P., F. J. Purroy, and M. A. Campos. 1997. Habitat assessment of a relict brown bear *Ursus arctos* population in northern Spain. *Biological Conservation* 80:17–22.
172. Clevenger, A. P., J. Wierzchowski, B. Chruszcz, and K. Gunson. 2002. GIS-generated expert based models for identifying wildlife habitat linkages and mitigation passage planning. *Conservation Biology* 16:503–14.
173. Clifford, H.T. 1959. Seed dispersal by motor vehicles. *Journal of Ecology* 47:311–15.
174. *Climate Change Impacts on the United States: The Potential Consequences of Climate Variability and Change.* 2001. Report of National Assessment Synthesis Team, U.S. Global Change Research Program. New York: Cambridge University Press.
175. Coghlan, G., and R. Sowa. 1998. *National Forest Road System and Use.* Washington, D.C.: USDA Forest Service.
176. Coker, J. E. 2000. Optical water quality of Lake Tahoe. Master's thesis, University of California, Davis.
177. Cole, C. A., R. P. Brooks, and D. H. Wardrop. 1998. Building a better wetland—a response to Linda Zug. *Wetland Journal* 10(2): 8–11.
178. Cole, M. A., A. J. Rayner, and J. M. Bates. 1997. The environmental Kuznets curve: An empirical analysis. *Environment and Development Economics* 2:401–16.
179. Coleman, J. S., and J. D. Fraser. 1989. Habitat use and home ranges of Black and Turkey Vultures. *Journal of Wildlife Management* 53:782–92.
180. Comins, H. N., W. D. Hamilton, and R. M. May. 1980. Evolutionary stable dispersal strategies. *Journal of Theoretical Biology* 82:205–30.
181. Conover, M. R. 1997. Monetary and intangible valuation of deer in the United States. *Wildlife Society Bulletin* 25:298–305.
182. Conover, M. R., W. C. Pitt, K. K. Kessler, T. J. DuBow, and W. A. Sanborn. 1995. Review of human injuries, illnesses, and economic losses caused by wildlife in the United States. *Wildlife Society Bulletin* 23:407–14.
183. Cordone, A. J., and D. W. Kelly. 1961. The influence of sediment on aquatic life in streams. *California Department of Fish and Game Journal* 47:1047–80.
184. Costa, J. E., and V. R. Baker. 1981. *Surficial Geology: Building with the Earth.* New York: John Wiley.
185. Cowlishaw, G. 1999. Predicting the pattern of decline of African primate diversity: An extinction debt from historical deforestation. Conservation Biology 13:1183–93.
186. Craul, P. J. 1992. *Urban Soil in Landscape Design.* New York: John Wiley.
187. Cronin, M. A., S. C. Amstrup, G. M. Durner, L. E. Noel, T. L. McDonald, and W. B. Ballard. 1998. Caribou distribution during the post-calving period in relation to infrastructure in the Prudhoe Bay oil field, Alaska. *Arctic* 51:85–93.
188. Cronon, W. 1991. *Nature's Metropolis: Chicago and the Great West.* New York: Norton.

189. Cruden, D. M., and D. J. Varnes. 1996. Landscape types and processes. In *Landslides: Investigation and Mitigation,* edited by A. N. Turner and R. L. Schuster, 36–75. National Research Council Special Report 247. Washington, D.C.: Transportation Research Board.
190. Cummins, K. W. 1992. Invertebrates. In *The River Handbook. I. Hydrological and Ecological Principles,* edited by P. Calow and G. E. Pitts, 234–50. Philadelphia: W. B. Saunders.
191. Cuperus, R., M. G. G. J. Bakermans, H. A. Udo de Haes, and K. J. Canters. 2001. Ecological compensation in Dutch highway planning. *Environmental Management* 27:75–89.
192. Cuperus, R., K. J. Canters, and A. A. G. Piepers. 1996. Ecological compensation of the impacts of a road. Preliminary method for the A50 road link (Eindhoven-Oss, Netherlands). *Ecological Engineering* 7:327–49.
193. Cuperus, R., K. J. Canters, H. A. Udo de Haes, and D. S. Friedman. 1999. Guidelines for ecological compensation associated with highways. *Biological Conservation* 90:41–51.
194. Curatolo, J. A., and S. M. Murphy. 1986. The effects of pipelines, roads, and traffic on the movements of caribou, *Rangifer tarandus. Canadian Field-Naturalist* 100:218–24.
195. Dale, V. H., S. Brown, R. A. Haeuber, N. T. Hobbs, N. Huntly, R. J. Naiman, W. E. Riebsame, M. G. Turner, and T. J. Valone. 2000. Ecological principles and guidelines for managing the use of land. *Ecological Applications* 10:639–70.
196. Dale, V. H., R. V. O'Neill, M. A. Pedlowski, and F. Southworth. 1993. Causes and effects of land-use change in central Rondonia, Brazil. *Photogrammetric Engineering and Remote Sensing* 59:997–1005.
197. Dale, V. H., R. V. O'Neill, F. Southworth, and M. Pedlowski. 1994a. Modeling effects of land management in the Brazilian Amazonian settlement of Rondonia. *Conservation Biology* 8:196–206.
198. Dale, V. H., S. M. Pearson, H. L. Offerman, and R. V. O'Neill. 1994b. Relating patterns of land-use change to faunal biodiversity in the Central Amazon. *Conservation Biology* 8:1027–36.
199. Dale, V. H., and R. A. Haeuber, eds. 2001. *Applying Ecological Principles to Land Management.* New York: Springer-Verlag.
200. Daly, H. E. 1991. *Steady-State Economics.* Washington, D.C.: Island Press.
201. D'Antonio, C. 1993. Mechanisms controlling invasion of coastal plant communities by the alien succulent, *Carpobrotus edulis. Ecology* 74:83–95.
202. Dargay, J., and D. Gately. 1997. Vehicle ownership to 2015: Implications for energy use and emissions. *Energy Policy* 25:14–15.
203. Davis, M. B. 1975. Erosion rates and land use history in southern Michigan. *Environmental Conservation* 3:139–48.
204. Davis, S. 2000. *Transportation Energy Data Book, Edition 20.* Report ORNL-6959. Oak Ridge, Tenn.: Center for Transportation Analysis, U.S. Department of Energy, Oak Ridge National Laboratory.
205. Davis, S., L. Truett, and P. Hu. 1999. *Fuel Used for Off-Road Recreation.* Report ORNL/TM-1999/100. Oak Ridge, Tenn.: Oak Ridge National Laboratory.
206. Davis, W. H. 1934. The automobile as a destroyer of wild life. *Science* 79:504–05.
207. Dawson, B. L. 1991. South African road reserves: valuable conservation areas? In

Nature Conservation 2: The Role of Corridors, edited by D. A. Saunders and R. J. Hobbs, 119–29. Chipping Norton, Australia: Surrey Beatty.

208. Deakin, E. A. 1990. Jobs, housing, and transportation: Theory and evidence on interactions between land use and transportation. In *Transportation, Urban Form, and the Environment,* 25–39. Washington, D.C.: National Research Council.
210. Dearing, J. A. 1991. Lake sediment records of erosional processes. *Hydrobiologia* 214:99–106.
211. Decker, D. J., K. M. Loconti Lee, and N. A. Connelly. 1990. *Incidence and Cost of Deer-related Vehicular Accidents in Tompkins County, New York.* Human Dimensions Research Group 89-7. New York: Cornell University.
212. DeFerrari, C. M., and R. J. Naiman. 1994. A multi-scale assessment of the occurrence of exotic plants on the Olympic Peninsula, Washington. *Journal of Vegetation Science* 5:247–58.
213. Delumyea, R., and R. Petal. 1977. *Atmospheric Input of Phosphorus to Southern Lake Huron April–October 1975.* Washington, D.C.: Office of Research and Development, U.S. Environmental Protection Agency.
214. DeMaynadier, P. G., and M. L. Hunter Jr. 1995. The relationship between forest management and amphibian ecology: A review of the North American literature. *Environmental Review* 3:230–61.
215. DeMers, M. N. 1993. Roadside ditches as corridors for range expansion of the western harvester ant (*Pogonomyrmex occidentalis* Cresson). *Landscape Ecology* 8:93–102.
216. Dhindsa, M. S., J. S. Sandhu, P. S. Sandhu, and H. S. Toor. 1988. Roadside birds in Punjab (India): Relation to mortality from vehicles. *Environmental Conservation* 15:303–10.
217. Diamond, H. L., and P. F. Noonan. 1996. *Land Use in America.* Washington, D.C.: Island Press.
218. Diamond, J. 1999. *Guns, Germs, and Steel: The Fates of Human Societies.* New York: Norton.
219. Dickerson, L. M. 1939. The problem of wildlife destruction by automobile traffic. *Journal of Wildlife Management* 2:104–16.
220. Diemand, D., R. Alger, and V. Klokov. 1996. Snow road enhancement. *Transportation Research Record* 1534:1–4.
221. Dionne, M., F. T. Short, and D. M. Burdick. 1998. Fish return quickly to restored salt marshes. *Coastlines* 10:35–38.
222. Doncaster, C. P. 1999. Can badgers affect the use of tunnels by hedgehogs? A review of the literature. *Lutra* 42:59–64.
223. Dorrance, M. F., P. J. Savage, and D. E. Huff. 1975. Effects of snowmobiles on white-tailed deer. Journal of Wildlife Management 39:563–69.
224. Douglas, I. 1969. Man, vegetation and the sediment yield of rivers. *Nature* 215:925–28.
225. Downes, S. J., A. Handasyde, and M. A. Elgar. 1997. The use of corridors by mammals in fragmented Australian eucalypt forests. *Conservation Biology* 11:718–26.
226. Dramstad, W., J. D. Olson, and R. T. T. Forman. 1996. *Landscape Ecology Principles in Landscape Architecture and Land-Use Planning.* Washington, D.C.: Harvard Uni-

versity Graduate School of Design, American Society of Landscape Architects, and Island Press.

227. Drapper, D., R. Tomlinson, and P. Williams. 2000. Pollutant concentrations in road runoff: Southeast Queensland case study. *Journal of Environmental Engineering* 126:313–20.

228. Dreyer, W. A. 1935. The question of wildlife destruction by the automobile. *Science* 82:439–40.

230. Driscoll, E., P. E. Shelley, and E. W. Strecker. 1990. *Pollutant Loadings and Impacts from Highway Stormwater Runoff.* Vols. 1–4. FHWA/RD-88-006-9. Oakland, Calif.: Federal Highway Administration and Woodward-Clyde Consultants.

231. Dunne, T., and L. B. Leopold. 1978. *Water in Environmental Planning.* San Francisco: W. H. Freeman.

232. Dunning, J. B., B. J. Danielson, and H. R. Pulliam. 1992. Ecological processes that affect populations in complex landscapes. *Oikos* 65:169–75.

233. Dyer, S. J., J. P. O'Neill, S. M. Wasel, and S. Boutin. 2001. Avoidance of industrial development by woodland caribou. *Journal of Wildlife Management* 65:531–42.

234. Eaglin, G. S., and W. A. Hubert. 1993. Effects of logging and roads on substrate and trout in streams of the Medicine Bow National Forest, Wyoming. *North American Journal of Fisheries Management.* 13:844–46.

235. Easterling, W. E., L. O. Mearns, C. J. Hays, and D. Marx. 2001. Comparison of agriculture impacts of climate change calculated from high and low resolution climate change scenarios. Part II. Accounting for adaptation and carbon dioxide direct effects. *Climate Change* 51:173–97.

236. Easterlings, K. 1999. *Organizational Space: Landscapes, Highways, and Houses in America.* Cambridge: MIT Press.

237. Eaton, M. M. 1997. The beauty that requires health. In *Placing Nature: Culture and Landscape Ecology,* edited by J. I. Nassauer, 85–106. Washington, D.C.: Island Press.

238. Edge, W. D., and C. L. Marcum. 1985. Movements of elk in relation to logging disturbances. *Journal of Wildlife Management* 49:926–30.

239. Ehmann, H., and H. G. Cogger. 1985. Australia's endangered herpetofauna: A review of criteria and policies. In *The Biology of Australasian Frogs and Reptiles,* edited by G. Grigg, R. Shine, and H. Ehmann, 435–447. Sydney, Australia: Surrey Beatty and Royal Zoological Society of New South Wales.

240. Ellenberg, H. 1986. *Vegetation Ecology of Central Europe.* 4th ed. Cambridge: Cambridge University Press.

241. Ellenberg, H., K. Muller, and T. Stottele. 1981. Strassen-Okologie: Auswirkungen von Autobahnen und Strasse auf Okosysteme deutscher Landschaften. In *Okologie und Strasse,* 19–122. Bonn, Germany: Broschurenreihe der Deutschen Strassenliga, Ausgabe 3.

242. Environmental Protection Agency. 1993. *Guidance Specifying Management Measures for Sources of Nonpoint Source Pollution in Coastal Waters.* Office of Water, Report 840-B-92-002. Washington, D.C.: U.S. Environmental Protection Agency.

243. ———. 1995. *Controlling Nonpoint Source Runoff Pollution from Roads, Highways,*

and Bridges. Report EPA-841-F-95-008a. Washington, D.C.: U.S. Environmental Protection Agency.

244. ———. 2001. *National Air Quality and Emissions Trend Report 1999.* EPA454/R-01-004. Research Triangle Park, N.C.: U.S. Environmental Protection Agency.
245. Esswein, H., J. Jaeger, H.-G. Schwarz-von Raumer, and M. Muller. 2002. Landschaftszerschneidung in Baden-Wurttemberg: Zerschneidungsanalyse zur aktuellen Situation und zur Entwicklung der letzten 70 Jahre mit der effektiven Maschenweite. Nr. 214. Stuttgart, Germany: Akademie fur Technikfolgenabschatzung in Baden-Wurttemberg.
246. Eversham, B. C., and M. G. Telfer. 1994. Conservation value of roadside verges for stenotopic heathland *Carabidae:* Corridors or refugia? *Biodiversity and Conservation* 3:538–45.
247. Evink, G. L. 1996. Florida Department of Transportation initiatives related to wildlife mortality. In *Trends in Addressing Transportation Related Wildlife Mortality,* edited by G. L. Evink, P. Garrett, D. Zeigler, and J. Berry, 278–86. Publication FL-ER-58-96. Tallahassee: Florida Department of Transportation.
248. Evink, G. L., P. Garrett, and D. Zeigler, eds. 1999. *Proceedings of the Third International Conference on Wildlife Ecology and Transportation.* Publication FL-ER-73-99. Tallahassee: Florida Department of Transportation.
250. Evink, G. L., P. Garrett, D. Zeigler, and J. Berry, eds. 1996. *Trends in Addressing Transportation Related Wildlife Mortality.* Publication FL-ER-58-96. Tallahassee: Florida Department of Transportation.
251. ———.1998. *Proceedings of the International Conference on Wildlife Ecology and Transportation.* Publication FL-ER-69-98. Tallahassee: Florida Department of Transportation.
252. Ewel, J. J. 1986. Invasibility: Lessons from South Florida. In *Ecology of Biological Invasions of North America and Hawaii,* edited by H. A. Mooney and J. A. Drake, 214–30. New York: Springer-Verlag.
253. Fahrig, L. 1991. Simulation models for developing general landscape-level hypotheses of single species dynamics. In *Quantitative Methods in Landscape Ecology,* edited by M. G. Turner and R. H. Gardner, 417–42. New York: Springer-Verlag.
254. ———. 1997. Relative effects of habitat loss and fragmentation on species extinction. *Journal of Wildlife Management* 61:603–10.
255. Fahrig, L., J. H. Pedlar, S. E. Pope, P. D. Taylor, and J. F. Wegner. 1995. Effect of road traffic on amphibian density. *Biological Conservation* 74:177–82.
256. Falk, N. W., H. B. Graves, and E. D. Bellis. 1978. Highway right-of-way fences as deer deterrents. *Journal of Wildlife Management* 42:646–50.
257. Farina, A. 2000. *Principles and Methods in Landscape Ecology.* Boston and Dordrecht: Kluwer.
258. Farmer, A. M. 1993. The effects of dust on vegetation: A review. *Environmental Pollution* 79:63–75.
259. Fay, R. R. 1988. *Hearing in Vertebrates: A Psychophysics Databook.* Winnetka, Ill.: Hill-Fay Associates.
260. Fearnside, P., and D. Ferreira. 1984. Roads in Rondonia—highway construction

and the farce of unprotected reserves in Brazil's Amazonian forest. *Environmental Conservation* 11:358–60.

261. Fearnside, P. M. 2001. Land-tenure issues as factors in environmental destruction in Brazilian Amazonia: The case of southern Para. *World Development* 29:1361–72.
262. Federal Highway Administration. 1979. *America's Highways 1776–1976: A History of the Federal Aid Program*. Washington, D.C.: U.S. Department of Transportation.
263. ———. 1990. *Fish Passage through Culverts*. Publication FHWA-FL-90-006. Washington, D.C.: U.S. Department of Transportation.
264. ———. 1995a. *Highway Traffic Noise Analysis and Abatement Policy and Guidance*. Washington, D.C.: FHWA Office of Environment and Planning.
265. ———. 1995b. *Our Nation's Highways: Selected Facts and Figures*. Publication FHWA-PL-95-028. Washington, D.C.: U.S. Department of Transportation.
266. ———. 1995c. *Environmental Policy Statement 1994: A Framework to Strengthen the Linkage Between Environmental and Highway Policy*. Publication FHWA-PD-95-006. Washington, D.C.: U.S. Department of Transportation.
267. ———. 1995d. *Best Management Practices for Erosion and Sediment Control*. Report FHWA-FLP-94-005. Washington, D.C.: U.S. Department of Transportation.
268. ———. 1996a. *Evaluation and Management of Highway Runoff Water Quality*. Publication FHWA-PD-96-032. Washington, D.C.: U.S. Department of Transportation.
270. ———. 1996b. *Highway Snowstorm Countermeasure Manual: Snowbreak Forest Book* (translated from the Japanese original). Publication FHWA-PL-97-010. Washington, D.C.: U.S. Department of Transportation.
271. ———. 1996c. *Highway Statistics 1995*. Publication FHWA-PL-96-017. Washington, D.C.: U.S. Department of Transportation.
272. ———. 1996d. *Highway Traffic Noise Barrier Construction Trends*. Washington, D.C.: Office of Environment and Planning Noise Team, U.S. Department of Transportation.
273. ———. 1997a. *Highway Statistics Summary to 1995*. Publication FHWA-PL-97-009. Washington, D.C.: U.S. Department of Transportation.
274. ———. 1997b. *Pavement Recycling Guidelines for State and Local Governments: Participant's Reference Book*. Publication FHWA-SA-98-042. Washington, D.C.: U.S. Department of Transportation.
275. ———. 1998. *TEA-21 Transportation Equity Act for the 21st Century: Key Information*. Publication FHWA-PL-98-043. Washington, D.C.: U.S. Department of Transportation.
276. ———. 2000a. *Our Nation's Highways*. Publication FHWA-PL-00-014. Washington, D.C.: U.S. Department of Transportation.
277. ———. 2000b. *Recycled Materials in European Highway Environments: Uses, Technologies, and Policies*. Publication FHWA-PL-00-025. Washington, D.C.: U.S. Department of Transportation.
278. ———. 2000c. *1999 Status of the Nation's Highways, Bridges, and Transit: Conditions and Performance*. Publication FHWA-PL-00-017. Washington, D.C.: U.S. Department of Transportation.

279. ———. 2000d. *Critter Crossings: Linking Habitats and Reducing Roadkill.* Publication FHWA-EP-00-004. Washington, D.C.: U.S. Department of Transportation.
280. ———. 2002. *Our Nation's Highways.* Publication FHWA-PL-00-014. Washington, D.C.: U.S. Department of Transportation.
281. Federal Highway Administration and Federal Transit Administration. 1997. *Transportation Conformity: A Basic Guide for State and Local Officials.* Washington, D.C.: U.S. Department of Transportation.
282. Feldhammer, G. A., J. E. Gates, D. M. Harman, A. J. Loranger, and K. R. Dixon. 1986. Effects of interstate highway fencing on white-tailed deer activity. *Journal of Wildlife Management* 50:497–503.
283. Ferguson, B. K. 1994. *Stormwater Infiltration.* New York: Lewis Publishing.
284. Ferreras, P., J. J. Aldama, J. F. Beltran, and M. Delibes. 1992. Rates and causes of mortality in a fragmented population of Iberian lynx *Felis pardini* Temminck, 1824. *Biological Conservation* 61:197–202.
285. Ferris, C. R. 1979. Effects of Interstate 95 on breeding birds in northern Maine. *Journal of Wildlife Management* 43:421–27.
286. Fiedler, P., and P. M. Kareiva, eds. 1998. *Conservation Biology for the Coming Decade.* New York: Chapman and Hall.
287. Findlay, C. S., and J. Bourdages. 2000. Response time of wetland biodiversity to road construction on adjacent lands. *Conservation Biology* 14:86–94.
288. Findlay, C. S., and J. Houlahan. 1997. Anthropogenic correlates of species richness in southeastern Ontario wetlands. *Conservation Biology* 11:1000–1009.
290. Finnis, R. G. 1960. Road casualties among birds. *Bird Study* 7:21–32.
291. Fisher, G. L., A. G. H. Locke, and B. C. Northey. 1985. *Stream Crossing Guidelines: Operational Guidelines for Industry.* Edmonton: Alberta Department of Energy and Natural Resources.
292. Flint, W. P. 1926. The automobile and wild life. *Science* 63:426–27.
293. Florida Department of Environmental Protection. 1996. *Surface Water Quality Standards: Florida Administrative Code, Chapter 62-302.* Tallahassee: Florida Department of Environmental Protection.
294. Foppen, R., and R. Reijnen. 1994. The effect of car traffic on breeding bird populations in woodland. II. Breeding dispersal of male willow warblers *(Phylloscopus trochilus)* in relation to the proximity of a highway. *Journal of Applied Ecology* 31:95–101.
295. Forbes, B. C. 1992. Tundra disturbance studies, I: Long-term effects of vehicles on species richness and biomass. *Environmental Conservation* 19:48–58.
296. Forbes, B. C., and R. L. Jefferies. 1999. Revegetation of disturbed arctic sites: Constraints and applications. *Biological Conservation* 88:15–24.
297. Forbes, B. C., J. J. Ebersole, and B. Strandberg. 2001. Anthropogenic disturbance and patch dynamics in circumpolar ecosystems. *Conservation Biology* 15:954–69.
298. Forcella, F., and S. J. Harvey. 1983a. Eurasion weed infestation in western Montana in relation to vegetation and disturbance. *Madrono* 30:102–09.
299. Forman, R. T. T. 1979a. The New Jersey Pine Barrens, an ecological mosaic. In *Pine Barrens: Ecosystem and Landscape,* edited by R. T. T. Forman, 569–85. New York: Academic Press.

300. ———, ed. 1979b. *Pine Barrens: Ecosystem and Landscape.* New York: Academic Press. (Reprint, New Brunswick, N.J.: Rutgers University Press, 1998.)
301. ———. 1989. Landscape ecology plans for managing forests. In *Proceedings of the Society of American Foresters 1988 National Convention,* 131–36. Bethesda, Md.: Society of American Foresters. (Reprinted 1990 in *Is Forest Fragmentation a Management Issue in the Northeast?*, compiled by R. M. DeGraaf and W. H. Healy, 27–32. General Technical Report NE-140. Radnor, Penn.: U.S. Forest Service.)
302. ———. 1995. *Land Mosaics: The Ecology of Landscapes and Regions.* New York and Cambridge, England: Cambridge University Press.
303. ———. 1998. Road ecology: A solution for the giant embracing us. *Landscape Ecology* 13(4): iii–v.
304. ———. 1999a. Spatial models as an emerging foundation of road system ecology and a handle for transportation planning and policy. In *Proceedings of the Third International Conference on Wildlife Ecology and Transportation,* edited by G. L. Evink, P. Garrett, and D. Zeigler, 118–23. Publication FL-ER-73-99. Tallahassee: Florida Department of Transportation.
305. ———. 1999b. Horizontal processes, roads, suburbs, societal objectives, and landscape ecology. In *Landscape Ecological Analysis: Issues and Applications,* edited by J. M. Klopatek and R. H. Gardner, 35–53. New York: Springer-Verlag.
306. ———. 2000. Estimate of the area affected ecologically by the road system in the United States. *Conservation Biology* 14:31–35.
307. ———. 2002a. Roadsides and vegetation. In *Proceedings of the Fourth International Conference on Ecology and Transportation: A Time for Action,* 7–8. Raleigh, N.C.: Center for Transportation and the Environment, North Carolina State University.
308. ———. 2002b. The missing catalyst: Design and planning with ecology roots. In *Ecology and Design: Frameworks for Learning,* edited by B. R. Johnson and K. Hill, 85–109. Washington, D.C.: Island Press.
310. Forman, R. T. T., and L. E. Alexander. 1998. Roads and their major ecological effects. *Annual Review of Ecology and Systematics* 29:207–31.
311. Forman, R. T. T., and J. Baudry. 1984. Hedgerows and hedgerow networks in landscape ecology. *Environmental Management* 8:495–510.
312. Forman, R. T. T., and S. K. Collinge. 1995. The "spatial solution" to conserving biodiversity in landscapes and regions. In *Conservation of Faunal Diversity in Forested Landscapes,* edited by R. M. DeGraaf and R. I. Miller, 537–68. New York and London: Chapman Hall.
313. ———. 1997. Nature conserved in changing landscapes with and without spatial planning. *Landscape and Urban Planning* 37:129–35.
314. Forman, R. T. T., and R. D. Deblinger. 1998. The ecological road-effect zone for transportation planning, and a Massachusetts highway example. In *Proceedings of the International Conference on Wildlife Ecology and Transportation,* edited by G. L. Evink, P. Garrett, D. Zeigler, and J. Berry, 78–96. Publication FL-ER-69-98. Tallahassee: Florida Department of Transportation.
315. ———. 2000. The ecological road-effect zone of a Massachusetts (USA) suburban highway. *Conservation Biology* 14:36–46.

316. Forman, R. T. T., and M. Godron. 1981. Patches and structural components for a landscape ecology. *BioScience* 31:733–40.
317. ———. 1986. *Landscape Ecology.* New York: John Wiley.
318. Forman, R. T. T., and A. M. Hersperger. 1996. Road ecology and road density in different landscapes, with international planning and mitigation solutions. In *Trends in Addressing Transportation Related Wildlife Mortality,* edited by G. L. Evink, P. Garrett, D. Zeigler, and J. Berry, 1–22. Publication FL-ER-58-96. Tallahassee: Florida Department of Transportation.
319. Forman, R. T. T., and A. D. Mellinger. 1999. Road networks and forest spatial patterns: Comparing cutting-sequence models for forestry and conservation. In *Nature Conservation 5: Nature Conservation in Production Environments: Managing the Matrix,* edited by J. L. Craig, N. Mitchell, and D. A. Saunders, 71–80. Chipping Norton, Australia: Surrey Beatty.
320. Forman, R. T. T., D. S. Friedman, D. Fitzhenry, J. D. Martin, A. S. Chen, and L. E. Alexander. 1997. Ecological effects of roads: Toward three summary indices and an overview for North America. In *Habitat Fragmentation & Infrastructure,* edited by K. Canters, 40–54. Delft, Netherlands: Ministry of Transport, Public Works and Water Management.
321. Forman, R. T. T., A. E. Galli, and C. F. Leck. 1976. Forest size and avian diversity in New Jersey woodlots with some land-use implications. *Oecologia* 26:1–8.
322. Forman, R. T. T., B. Reineking, and A. M. Hersperger. 2002. Road traffic and nearby grassland bird patterns in a suburbanizing landscape. *Environmental Management* 29:782–800.
323. Foster, D. R. 1999. *Thoreau's Country: Journey through a Transformed Landscape.* Cambridge: Harvard University Press.
324. Foster, M. L., and S. R. Humphrey. 1995. Use of highway underpasses by Florida panthers and other wildlife. *Wildlife Society Bulletin* 23:95–100.
325. Foufopoulos, J., and A. R. Ives. 1999. Reptile extinctions on land-bridge islands: Life-history attributes and vulnerability to extinction. *American Naturalist* 153:1–25.
326. Fowle, S. C. 1990. The painted turtle in the Mission Valley of western Montana. Master's thesis, University of Montana, Missoula.
327. Fox, J., D. M. Truong, A. T. Rambo, N. P. Tuyen, C. T. Cuc, and S. Leisz. 2000. Shifting agriculture: A new paradigm for managing tropical forests. *BioScience* 50:521–28.
328. France, R. L. 2002a. Factors influencing sediment transport from logging roads near boreal trout lakes (Ontario, Canada). In *Handbook of Water Sensitive Planning and Design,* edited by R. L. France, 635–44. Boca Raton, Fla.: Lewis Publishers.
330. ———. 2002b. *Creating Wetlands: Design Principles and Practices for Landscape Architects and Land-Use Planners.* New York: Norton.
331. Franklin, J. F., and C. T. Dyrness. 1988. *Natural Vegetation of Oregon and Washington.* Corvallis: Oregon State University Press.
332. Franklin, J. F., and R. T. T. Forman. 1987. Creating landscape patterns by forest cutting: Ecological consequences and principles. *Landscape Ecology* 1:5–18.
333. Franklin, W. 1988. Boats, trains, cars and the popular eye. *The North American Review,* March, 15–19.

334. Fraser, D., and E. R. Thomas. 1982. Moose-vehicle accidents in Ontario: Relation to highway salt. *Wildlife Society Bulletin* 10:261–65.

335. Freemark, K., D. Bert, and M.-A. Villard. 2002. Patch-, Landscape-, and Regional-Scale Effects on Biota. In *Applying Landscape Ecology in Biological Conservation,* edited by K. J. Gutzwiller, 58–83. New York: Springer.

336. Freeze, R. A., and J. A. Cherry. 1979. *Groundwater.* Englewood Cliffs, N.J.: Prentice-Hall.

337. Frenkel, R. E. 1970. *Ruderal Vegetation along Some California Roadsides.* Berkeley: University of California Press.

338. Freund, P., and G. Martin. 1993. *The Ecology of the Auto.* New York: Black Rose Books.

339. Friedman, D. S. 1997. *Nature as Infrastructure: The National Ecological Network and Wildlife-Crossing Structures in The Netherlands.* Report 138. Wageningen, Netherlands: DLO Winand Staring Centre.

340. Fuller, T. 1989. Population dynamics of wolves in north-central Minnesota. *Wildlife Monographs* 105:1–41.

341. Garland, T., Jr., and W. G. Bradley. 1984. Effects of highway on Mojave desert rodent populations. *American Midland Naturalist* 111:47–56.

342. Gaston, K. J., and T. M. Blackburn. 1995. Birds, body-size and the threat of extinction. *Philosophical Transactions of the Royal Society of London, Series B—Biological Sciences* 347:205–12.

343. Geiger, R. 1965. *The Climate near the Ground.* Cambridge: Harvard University Press.

344. Gere, W. M. 1977. South Dakota's harvesting of crops in highway rights-of-way. *Transportation Research Record* 647:25–26.

345. Gerlach, G., and K. Musolf. 2000. Fragmentation of landscape as a cause for genetic subdivision in blank roles. *Conservation Biology* 14:1066–74.

346. Gese, E. M., O. J. Rongstad, and W. R. Mytton. 1989. Changes in coyote movements due to military activity. *Journal of Wildlife Management* 53:334–39.

347. Getz, L. L., F. R. Cole, and D. L. Gates. 1978. Interstate roadsides as dispersal routes for *Microtus pennsylvanicus. Journal of Mammalogy* 59:208–13.

348. Getz, L. L., L. Verner, and M. Prather. 1977. Lead concentrations in small mammals living near highways. *Environmental Pollution* 13:151–57.

349. Gibbs, J. P. 1998. Distribution of woodland amphibians along a forest fragmentation gradient. *Landscape Ecology* 13:263–68.

350. Gibeau, M. L. 2000. A conservation biology approach to management of grizzly bears in Banff National Park, Alberta. Ph.D. dissertation, University of Calgary, Alberta.

351. Gibeau, M. L., and S. Herrero. 1998. Roads, rails and grizzly bears in the Bow River Valley, Alberta. In *Proceedings of the International Conference on Wildlife Ecology and Transportation,* edited by G. L. Evink, P. Garrett, D. Zeigler, and J. Berry, 104–08. Publication FL-ER-69-98. Tallahassee: Florida Department of Transportation.

352. Gibeau, M. L., A. P. Clevenger, S. Herrero, and J. Wierzchowski. 2002. Grizzly bear response to human development and activities in the Bow River watershed, Alberta. *Biological Conservation* 103:227–36.

353. Gifford, J. L. 2002. *Flexible Urban Transportation.* Oxford, England: Elsevier.

354. Gilbert, J., D. L. Danielopol, and J. A. Stanford. 1994. *Groundwater Ecology.* San Diego, Calif.: Academic Press.
355. Gilbert, J. J., and D. C. Froehlich. 1987. *Simulation of the Effect of U.S. Highway 90 on Pearl River Floods of April 1980 and April 1983 near Slidell, Louisiana.* U.S.G.S. Water-Resources Investigations Report 85-4286. Baton Rouge, La.: U.S. Geological Survey.
356. Gilbert, O. L. 1991. *The Ecology of Urban Habitats.* London: Chapman and Hall.
357. Gilfedder, L., and J. B. Kirkpatrick. 1993. Germinable soil seed and competitive relationships between a rare native species and exotics in a semi-natural pasture in the Midlands, Tasmania. *Biological Conservation* 64:540–51.
358. Gilliam, R. S., and W. J. Platt. 1999. Effects of long-term fire exclusion on tree species composition and stand structure in an old-growth *Pinus palustris* (longleaf pine) forest. *Plant Ecology* 140:15–26.
359. Gloyne, C. C., and A. P. Clevenger. 2001. Cougar use of wildlife crossing structures on the Trans-Canada Highway in Banff National Park, Alberta. *Wildlife Biology* 7:117–24.
360. Gobster, P. H. 1999. An ecological aesthetic for forest landscape management. *Landscape Journal* 18:54–64.
361. Goff, K. 1999. Fugitive dust. *Erosion Control Magazine,* March, 86–95.
362. Goldman, C. R. 1960. Molybdenum as a factor limiting primary productivity in Castle Lake, California. *Science* 132:1016–17.
363. ———. 2000. Baldi lecture. Four decades of change in two subalpine lakes. *Verh. Internat. Verein. Limnol.* 27:7–26.
364. Goldman, C. R., and G. J. Malyj. 1990. *The Environmental Impact of Highway Deicing.* Institute of Ecology Publication 33. Davis: University of California.
365. Goldman, C. R., F. Lubnow, and J. Elser. 1990. Environmental effect of calcium magnesium acetate on natural phytoplankton populations in ten Sierra Nevada and Klamath mountain lakes. In *The Environmental Impact of Highway Deicing,* edited by C. R. Goldman and G. J. Malyj, 9–19. Institute of Ecology Publication 33. Davis: University of California.
366. Goldsmith, C. D., and P. F. Scanlon. 1977. Lead levels in small animals and selected invertebrates associated with highways of different traffic densities. *Bulletin of Environmental Contamination and Toxicology* 17:311–16.
367. Goodman, S. W. 1996. Ecosystem management at the Department of Defense. *Ecological Applications* 6:706–7.
368. Goodwin, C. R. 1987. *Tidal-Flow, Circulation, and Flushing Changes Caused by Dredge and Fill in Tampa Bay, Florida.* U.S.G.S. Water-Supply Paper 2282. Washington, D.C.: U.S. Geological Survey.
370. Gover, A., J. M. Johnson, and L. J. Kuhns. 2001. *Roadside Vegetation Management Research Report—Fourteenth Year Report.* Report No. PA 00-4620 and 85-08. University Park: Pennsylvania Department of Transportation and Pennsylvania State University.
371. Green, R. E., G. A. Tyler, and C. G. R. Bowden. 2000. Habitat selection, ranging behaviour and diet of the stone curlew *(Burhinus oedicnemus)* in southern England. *Journal of Zoology, London* 250:161–83.
372. Greenberg, C. H., S. H. Crownover, and D. R. Gordon. 1997. Roadside soils: A

corridor for invasion of xeric scrub by nonindigenous plants. *Natural Areas Journal* 17:99–109.

373. Gregory, S.V., F. J. Swanson, W. A. McKee, and K. W. Cummins. 1991. An ecosystem perspective of riparian zones. *BioScience* 41:540–51.
374. Grover, K. E., and M. J. Thompson. 1986. Factors influencing spring feeding site selection by elk in the Elkhorn Mountains, Montana. *Journal of Wildlife Management* 50:466–70.
375. Grue, C. E., D. J. Hoffman, W. N. Beyer, and L. P. Franson. 1986. Lead concentrations and reproductive success in European starlings *Sturnus vulgaris* nesting within highway roadside verges. *Environmental Pollution* (Series A) 42:157–82.
376. Gucinski, H., M. J. Furniss, R. R. Ziemer, and M. H. Brookes. 2001. *Forest Roads: A Synthesis of Scientific Information.* General Technical Report PNW-GTR-509. Portland, Ore.: USDA Forest Service.
377. Gunther, K. A., M. J. Biel, and H. L. Robison. 1998. Factors influencing the frequency of road-killed wildlife in Yellowstone National Park. In *Proceedings of the International Conference on Wildlife Ecology and Transportation,* edited by G. L. Evink, P. Garrett, D. Zeigler, and J. Berry, 32–42. Publication FL-ER-69-98. Tallahassee: Florida Department of Transportation.
378. Gustafson, E. J., and N. Diaz. 2002. Landscape pattern, timber extraction, and biological conservation. In *Applying Landscape Ecology in Biological Conservation,* edited by K. J. Gutzwiller, 244–65. New York: Springer.
379. Gutzwiller, K. J., ed. 2002. *Applying Landscape Ecology in Biological Conservation.* New York: Springer.
380. Guyot, G., and J. Clobert. 1997. Conservation measures for a population of Hermann's tortoise, *Testudo hermanni,* in southern France bisected by a major highway. *Biological Conservation* 79:251–56.
381. Hadley, A. H. 1927. Wild life and automobiles. *Bird-Lore* 32:391.
383. Hansen, A. J., and J. J. Rotella. 2001. Nature reserves and land use: Implications of the "place" principle. In *Applying Ecological Principles to Land Management,* edited by V. H. Dale and R. A. Haeuber, 54–72. New York: Springer.
384. Hansen, K., and J. Jensen. 1972. The vegetation on roadsides in Denmark: Qualitative and quantitative composition. *Dansk Botanisk Arkiv* 28(2):1–61.
385. ———. 1974. Edaphic conditions and plant-soil relationships on roadsides in Denmark. *Dansk Botanisk Arkiv* 28, no. 3: 1–143.
386. Hanski, I., and M. Kuussaari. 1995. Butterfly metapopulation dynamics. In *Population Dynamics: New Approaches and Synthesis,* edited by N. Capuccino and P. W. Price, 149–71. San Diego, Calif.: Academic Press.
387. Hanski, I., A. Moilanen, and M. Gyllenberg. 1996a. Minimum viable metapopulation size. *American Naturalist* 147:527–41.
388. Hanski, I., A. Moilanen, T. Pakkala, and M. Kuussaari. 1996b. The quantitative incidence function model and persistence in an endangered butterfly metapopulation. *Conservation Biology* 10:578–90.
389. Hanski, I., T. Pakkala, M. Kuussaari, and G. Lei. 1995. Metapopulation persistence of an endangered butterfly in a fragmented landscape. *Oikos* 72:21–28.
391. Harmon, J. M., and J. F. Franklin. 1995. *Seed Rain and Seed Bank of Third- and*

Fifth-Order Streams on the Western Slope of the Cascade Range. Research Paper PNW-RP-480. Portland, Ore.: USDA Forest Service.

392. Harms, B., and P. Opdam. 1990. Woods as habitat patches for birds: Application in landscape planning in The Netherlands. In *Landscape Change: An Ecological Perspective,* edited by I. S. Zonneveld and R. T. T. Forman, 73–97. New York: Springer-Verlag.
393. Harper-Lore, B., ed. 1999. *Roadside Use of Native Plants.* Washington, D.C.: U.S. Federal Highway Administration and Island Press.
394. Harrington, J. A. 1991. Survey of landscape use of native vegetation on Midwest highway rights-of-way. *Transportation Research Record* 1326:19–30.
395. ———. 1994. Roadside landscapes: Prairie species take hold in Midwest rights-of-way. *Restoration and Management Notes* 12:8–15.
396. Harris, L. D. 1984. *The Fragmented Forest: Island Biogeography Theory and the Preservation of Biotic Diversity.* Chicago: University of Chicago Press.
397. ———. 1988. Edge effects and conservation of biotic diversity. *Conservation Biology* 2:330–39.
398. Harris, L. D., and K. Atkins. 1991. Faunal movement corridors in Florida. In *Landscape Linkages and Biodiversity,* edited by W. E. Hudson, 117–34. Washington, D.C.: Island Press.
399. Harris, L. D., and J. Scheck. 1991. From implications to applications: The dispersal corridor principle applied to the conservation of biological diversity. In *Nature Conservation 2: The Role of Corridors,* edited by D. A. Saunders and R. J. Hobbs, 189–220. Chipping Norton, Australia: Surrey Beatty.
400. Harris, L. D., T. S. Hoctor, and S. E. Gergel. 1996. Landscape processes and their significance to biodiversity conservation. In *Population Dynamics in Ecological Space and Time,* edited by O. Rhodes Jr., R. Chesser, and M. Smith, 319–47. Chicago: University of Chicago Press.
401. Harrison, S. 1994. Metapopulations and conservation. In *Large-Scale Ecology and Conservation Biology,* edited by P. J. Edwards, R. M. May, and N. R. Webb, 111–28. London: Blackwell Scientific Publications.
402. Harrison, S., and L. Fahrig. 1995. Landscape pattern and population conservation. In *Mosaic Landscapes and Ecological Processes,* edited by L. Hansson, L. Fahrig, and G. Merriam, 293–308. London: Chapman and Hall.
403. Haugen, A. O. 1944. Highway mortality of wildlife in southern Michigan. *Journal of Mammalogy* 25:177–84.
404. Hautala, E.-L., R. Rekila, J. Tarhanen, and J. Ruuskanen. 1995. Deposition of motor vehicle emissions and winter maintenance along roadside assessed by snow analyses. *Environmental Pollution* 87:45–49.
405. Havlick, D. G. 2002. *No Place Distant.* Washington, D.C.: Island Press.
406. Haxton, T. 2000. Road mortality of snapping turtles, *Chelydra serpentina,* in central Ontario during their nesting period. *Canadian Field-Naturalist* 114:106–10.
407. Heath, B.A., J. A. Maughan, A. A. Morrison, I. W. Eastwood, I. B. Drew, and M. Lofkin. 1999. The influence of wooded shelterbelts on the deposition of copper, lead and zinc at Shakerley Mere, Cheshire, England. *Science and the Total Environment* 235:415–17.
408. Heck, W. W., O. C. Taylor, and D. T. Tingey, eds. 1988. *Assessment of Crop Loss from Air Pollutants.* New York: Elsevier.

410. Heindl, B., and I. Ullmann. 1991. Roadside vegetation in mediterranean France. *Phytocoenologia* 20:111–43.
411. Heisler, G. M., and D. R. DeWalle. 1988. Effects of windbreak structure on wind flow. *Agriculture, Ecosystems and Environment* 22/23:41–69. (Reprinted in J. R. Brandle, D. L. Hintz, and J. W. Sturrock, eds., *Windbreak Technology,* Amsterdam: Elsevier, 1988.)
412. Hels, T., and E. Buchwald. 2001. The effect of road kills on amphibian populations. *Biological Conservation* 99:331–40.
413. Herben, T., H. Rydin, and L. Söderström. 1991. Spore establishment probability and the persistence of the fugitive invading moss, *Orthodontium lineare:* A spatial simulation model. *Oikos* 60:215–21.
414. Hewitt, D. G., A. Cain, V. Tuovila, D. B. Shindle, and M. E. Tewes. 1998. Impacts of an expanded highway on ocelots and bobcats in southern Texas and their preferences for highway crossings. In *Proceedings of the International Conference on Wildlife Ecology and Transportation,* edited by G. L. Evink, P. Garrett, D. Zeigler, and J. Berry, 126–34. Publication FL-ER-69-98. Tallahassee: Florida Department of Transportation.
415. Hicks, B. 1996. *A Historic Look at the Federal Government's Involvement in Highway Infrastructure.* Paper No. TP-12799E. Ottawa: Transport Canada.
416. Higashi, M., and T. P. Burns, eds. 1991. *Theoretical Studies of Ecosystems: The Network Approach.* Cambridge, England: Cambridge University Press.
417. Hobbs, R. J. 2002. Habitat networks and biological conservation. In *Applying Landscape Ecology in Biological Conservation,* edited by K. J. Gutzwiller, 150–170. New York: Springer.
418. Hodson, N. L., and D. W. Snow. 1965. The road deaths enquiry, 1960–61. *Bird Study* 12:90–99.
419. Hofstra, G., and R. Hall. 1971. Injury on roadside trees: Leaf injury on pine and white cedar in relation to foliar levels of sodium and chloride. *Canadian Journal of Botany* 49:613–22.
420. Hoffman, R. W., C. R. Goldman, S. Paulson, and G. R. Winters. 1981. Aquatic impacts of deicing salts in the central Sierra Nevada mountains, California. *Water Resources Bulletin* 17:280–85.
421. Hole, F. D., and J. B. Campbell. 1985. *Soil Landscape Analysis.* Totowa, N.J.: Rowman and Allanheld.
422. Holzapfel, C., and W. Schmidt. 1990. Roadside vegetation along transects in the Judean desert. *Israel Journal of Botany* 39:263–70.
423. Horne, A. J., and C. R. Goldman. 1994. *Limnology.* New York: McGraw-Hill.
424. Horner, R. R., J. J. Skupien, E. H. Livingstone, and H. E. Shaver. 1994. *Fundamentals of Urban Runoff: Technical and Institutional Issues.* Washington, D.C.: Terrene Institute.
425. Hornig, D., ed. 1994. *State of the Cape 1994: Progress Toward Preservation.* Boston: Association for the Preservation of Cape Cod.
426. Horvath, A., and C. Hendrickson. 1999. Comparison of environmental implications of asphalt and steel-reinforced concrete pavements. *Transportation Research Record* 1626:105–13.

427. Hough, M. 1984. *City Form and Natural Processes: Towards an Urban Vernacular.* New York: Van Nostrand Reinhold.
428. Houghton, J. T., Y. Ding, D. J. Griggs, M. Noguer, P. J. van der Linden, X. Dai, K. Maskell, and C. A. Johnson, eds. 2001. *Climate Change 2001: The Scientific Basis.* Contribution of Working Group I to the Third Assessment Report of the Intergovernmental Panel on Climate Change. New York: Cambridge University Press.
429. Houghton, R.A. 1999. The annual net flux of carbon to the atmosphere from changes in land use 1850–1990. *Tellus B* 51:298–313.
430. Hubbard, M. W., B. J. Danielson, and R. A. Schmitz. 2000. Factors influencing the location of deer-vehicle accidents in Iowa. *Journal of Wildlife Management* 64:707–12.
431. Huey, L. M. 1941. Mammalian invasion via the highway. *Journal of Mammalogy* 22:383–85.
432. Huijser, M. P., and P. J. M. Bergers. 2000. The effect of roads and traffic on hedgehog *(Erinaceus europaeus)* populations. *Biological Conservation* 95:111–16.
433. Hunt, A., H. J. Dickens, and R. J. Whelan. 1987. Movement of mammals through tunnels under railway lines. *Australian Zoologist* 24:89–93.
434. Hunter, M. L. Jr. 1990. *Wildlife, Forests, and Forestry.* Englewood Cliffs, N.J.: Prentice Hall.
435. Husakova, J., and M. Guzikowa. 1979. Flora and vegetation of road-side in western part of the Czech Krkonose (Giant Mountains). *Opera Corcontica* 16:87–112.
436. Hutchinson, G. E., and U. Cowgill. 1970. An account of the history and development of the Lago di Monterosi, Latium, Italy. XII. The history of the lake: a synthesis. *Transactions of the American Philosophical Society* 60(4):163–70.
437. Hvitved-Jacobsen, T., N. B. Johansen, and Y. A. Yousef. 1994. Treatment systems for urban and highway runoff in Denmark. *Science and the Total Environment* 146/147:499–506.
438. Hynes, H. B. N. 1970. *The Ecology of Running Waters.* Liverpool: University of Liverpool Press.
439. Imbernon, J. 2000. Deforestation and population pressure in the state of Rondonia, Brazil. *Bois et Forêts de Tropiques* 266:23–33.
440. Inbar, M., and R. T. Mayer. 1999. Spatio-temporal trends in armadillo diurnal activity and road-kills in central Florida. *Wildlife Society Bulletin* 27:865–72.
441. Ingegnoli, V. 2002. *Landscape Ecology: A Widening Foundation.* New York: Springer-Verlag.
442. Ingersoll, C. A., and M. V. Wilson. 1990. Buried propagules in an old-growth forest and their response to experimental disturbances. *Canadian Journal of Botany* 68:1290–92.
443. Iverson, A. L., and L. R. Iverson. 1999. Spatial and temporal trends of deer harvest and deer-vehicle accidents in Ohio. *Ohio Journal of Science* 99(4):84–94.
444. Iverson, R. M., B. S. Hinckley, and R. M. Webb. 1981. Physical effects of vehicular disturbances on arid landscapes. *Science* 212:915–17.
445. Jackson, S. D. 1996. Underpasses for amphibians. In *Trends in Addressing Transportation Related Wildlife Mortality,* edited by G. L. Evink, P. Garrett, D. Zeigler, and J. Berry, 240–44. Publication FL-ER-58-96. Tallahassee: Florida Department of Transportation.

446. ———. 1999. Overview of transportation related wildlife problems. In *Proceedings of the Third International Conference on Wildlife Ecology and Transportation,* edited by G. L. Evink, P. Garrett, and D. Zeigler, 1–4. Publication FL-ER-73-99. Tallahassee: Florida Department of Transportation.
447. Jackson, S.D., and T. Tyning. 1989. Effectiveness of drift fences and tunnels for moving spotted salamanders *Ambystoma maculatum* under roads. In *Amphibians and Roads,* edited by T. E. S. Langton, 93–100. Shefford, Bedfordshire, England: ACO Polymer Products.
448. Jacobson, R. B., and K. A. Oberg. 1997. *Geomorphic Changes on the Mississippi River Floodplain at Miller City, Illinois, as a Result of the Flood of 1993.* U.S.G.S. Circular 1120-J. Washington, D.C.: U.S. Geological Survey.
450. Jaeger, J. A. G. 2000. Landscape division, splitting index, and effective mesh size: New measures of landscape fragmentation. *Landscape Ecology* 15:115–30.
451. James, A. R. C., and A. K. Stuart-Smith. 2000. Distribution of caribou and wolves in relation to linear corridors. *Journal of Wildlife Management* 64:154–59.
452. Jassby, A. D., C. R. Goldman, J. E. Reuter, R. C. Richards, and A. C. Heyvaert. 2001. Lake Tahoe: diagnosis and rehabilitation of a large mountain lake. In *The Great Lakes of the World (GLOW): Food-web, Health, and Integrity,* edited by M. Munawar and R. E. Hecky, 431–54. Leiden, Netherlands: Backhuys Publishers.
453. Jassby, A. D., J. E. Reuter, R. P. Axler, C. R. Goldman, and S. H. Hackley. 1994. Atmospheric deposition of nitrogen and phosphorus in the annual nutrient load of Lake Tahoe (California–Nevada). *Water Resources Research* 30:2207–16.
454. Jeffries, D. J., and M. C. French. 1972. Lead concentrations in small mammals trapped on roadside verges and field sites. *Environmental Pollution* 3:147–56.
455. Jenny, H. 1980. *The Soil Resource: Origin and Behavior.* New York: Springer-Verlag.
456. Jensen, W. F., T. K. Fuller, and W. L. Robinson. 1986. Wolf, *Canis lupus,* distribution on the Ontario-Michigan border near Sault Ste. Marie. *Canadian Field-Naturalist* 100:363–66.
457. Jimba, K., H. Adhikarey, P. D. Wangdi, L. Sherpa, P. Tshering, and D. Dorji. 1998. *Dakpai—Buli Road: Environmental Impact Assessment Report.* Zhemgang, Bhutan: ISDP.
458. Johnson, B. R., and K. Hill, eds. 2002. *Ecology and Design: Frameworks for Learning.* Washington, D.C.: Island Press.
459. Johnson, L. A. 1981. *Revegetation and Selected Terrain Disturbances along the Trans-Alaska Pipeline, 1975–1978.* Report 81-12. Hanover, N.H.: Cold Regions Research and Engineering Laboratory.
460. Johnson, P. R., and C. M. Collins. 1980. *Snow Pads for Pipeline Construction in Alaska, 1976.* Report 80-17. Hanover, N.H.: Cold Regions Research and Engineering Laboratory.
461. Jones, J. A. 2000. Hydrologic processes and peak discharge response to forest removal, regrowth, and roads in 10 small experimental basins, western Cascades, Oregon. *Water Resources Research* 36:2621–42.
462. Jones, J. A., and G. E. Grant. 1996. Peak flow responses to clearcutting and roads in small and large basins, Western Cascades, Oregon. *Water Resources Research* 32:959–74.

463. Jones, J. A., F. J. Swanson, B. C. Wemple, and K. U. Snyder. 2000. Effects of roads on hydrology, geomorphology, and disturbance patches in stream networks. *Conservation Biology* 14:76–85.

464. Jonsen, I. D., and P. D. Taylor. 2000. Fine-scale movement behaviors of calopterygid damselflies are influenced by landscape structure: An experimental manipulation. *Oikos* 88:553–62.

465. Joselow, M. M., E. Tobias, R. Koehler, S. Coleman, J. Bogden, and D. Gause. 1978. Manganese pollution in the city environment and its relationship to traffic density. *American Journal of Public Health* 68:557–60.

466. Joyce, T. L., and S. P. Mahoney. 2001. Spatial and temporal distributions of moose-vehicle collisions in Newfoundland. *Wildlife Society Bulletin* 29:281–91.

467. Judd, J. H. 1976. Lake stratification caused by runoff from street deicing. *Water Research* 4:521–32.

468. Junk, W. J., P. B. Bayley, and R. E. Sparks. 1989. The flood pulse concept in river-floodplain systems. *Canadian Special Publication of Fisheries and Aquatic Sciences* 106:110–27.

469. Kameyama, A. 1977. Succession of the slope vegetation of expressways. *Journal of the Japan Institute of Landscape Architects* 41:23–33.

470. Karr, J. R., and E. W. Chu. 1999. *Restoring Life in Running Waters: Better Biological Monitoring.* Washington, D.C.: Island Press.

471. Karthikeyan, V., S. P. Vijay Kumar, and N. M. Ishwar. 1999. *Western Ghat Habitat Fragmentation Project.* Dehradun, India: Wildlife Institute of India.

472. Keeley, B., and M. Tuttle. 1999. Bats in American bridges. In *Proceedings of the Third International Conference on Wildlife Ecology and Transportation,* edited by G. L. Evink, P. Garrett, and D. Zeigler, 167–72. FL-ER-73-99. Tallahassee: Florida Department of Transportation.

473. Keen, C. L., and B. Lonnerdal. 1986. Manganese toxicity in man and experimental animals. In *Manganese in Metabolism and Enzyme Function,* edited by V. L. Schramm and F. C. Wedler, 35–49. New York: Academic Press.

474. Keller, V. 1999. *The Use of Wildlife Overpasses by Mammals: Results from Infra-Red Video Surveys in Switzerland, Germany, France and The Netherlands.* Report to Infra Eco Network Europe (IENE), Fifth IENE Meeting. Budapest, Hungary.

475. Keller, V., and H. P. Pfister. 1997. Wildlife passages as a means of mitigating effects of habitat fragmentation by roads and railway lines. In *Habitat Fragmentation & Infrastructure,* edited by K. Canters, 70–80. Delft, Netherlands: Ministry of Transport, Public Works and Water Management.

476. Keller, V., H.-G. Bauer, H.-W. Ley, and H. P. Pfister. 1996. Bedeutung von Grunbrucken uber Autobahnen fur Vogel. *Der Ornithologische Beobachter* 93:249–58.

477. Keller, V. E. 1991. The effect of disturbance from roads on the distribution of feeding sites of geese *(Anser brachyrhynchus, A. anser),* wintering in north-east Scotland. *Ardea* 79:229–32.

478. Kelley, D. W., and E. A. Nater. 2000. Historical sediment flux from three watersheds into Lake Pepin, Minnesota. *Journal of Environmental Quality* 29:561–68.

479. Kellman, M. 1974. Preliminary seed budgets for two plant communities in coastal British Columbia. *Journal of Biogeography* 1:123–33.

480. Kelsey, P. D., and R. G. Hootman. 1992. Deicing salt dispersion and effects on

vegetation along highways, case study: Deicing salt deposition on the Morton Arboretum. In *Chemical Deicers and the Environment,* edited by F. M. D'Itri, 253–81. Lewis, Mich: Chelsea.

481. Kent, R. L. 1993. Determining scenic quality along highways: A cognitive approach. *Landscape and Urban Planning* 27:29–45.
482. Kerri, K. D., J. A. Racin, and R. B. Howell. 1985. Forecasting pollutant loads from highway runoff. *Transportation Research Record* 1017:39–46.
483. Kershaw, K. A., and J. H. H. Looney. 1985. *Quantitative and Dynamic Plant Ecology.* London: Edward Arnold.
484. Kertell, K. 2000. Pacific loon. In *The Natural History of an Arctic Oil Field: Development and the Biota,* edited by J. C. Truett and S. R. Johnson, 181–96. San Diego, Calif.: Academic Press.
485. Kiester, A. R. 1995. Aesthetics of biological diversity. *Human Ecology Review* 3:151–63.
486. King, R. S., K. T. Nunnery, and C. J. Richardson. 2000. Macroinvertebrate assemblage response to highway crossings in forested wetlands: Implications for biological assessment. *Wetlands Ecology and Management* 8:243–56.
487. Kivilaan, A., and R. S. Bandurski. 1981. The one-hundred-year period for Dr. Beal's seed viability experiment. *American Journal of Botany* 68:1290–92.
488. Kiviniemi, K., and O. Eriksson. 1999. Dispersal, recruitment and site occupancy of grassland plants in fragmented habitats. *Oikos* 86:241–53.
490. Kjensmo, J. 1997. The influence of road salts on the salinity and the meromictic stability of Lake Svinsjoen, Norway. *Oecologia* 347:151–58.
491. Klein, D. R. 1971. Reaction of reindeer to obstructions and disturbance. *Science* 173:393–98.
492. Kline, N. C., and D. E. Swann. 1998. Quantifying wildlife road mortality in Saguaro National Park. In *Proceedings of the International Conference on Wildlife Ecology and Transportation,* edited by G. L. Evink, P. Garrett, D. Zeigler, and J. Berry, 23–31. Publication FL-ER-69-98. Tallahassee: Florida Department of Transportation.
493. Klinkhamer, P. G., T. J. de Jong, J. A. J. Metz, and J. Val. 1987. Life history tactics of annual organisms: The joint effects of dispersal and delayed germination. *Theoretical Population Biology* 32:127–56.
494. Knaapen, J. P., M. Scheffer, and B. Harms. 1992. Estimating habitat isolation in landscape planning. *Landscape and Urban Planning* 23:1–16.
495. Knight, R. L., and K. J. Gutzwiller, eds. 1995. *Wildlife and Recreationists: Coexistence through Management and Research.* Washington, D.C.: Island Press.
496. Knight, R. L., and J. Y. Kawashima. 1993. Responses of raven and red-tailed hawk populations to linear right-of-ways. *Journal of Wildlife Management* 57:266–70.
497. Knobloch, W. 1939. Death on the highway. *Journal of Mammalogy* 20:508.
498. Knutson, M. G., D. J. Leopold, and R. C. Smardon. 1993. Selecting islands and shoals for conservation based on biological and aesthetic criteria. *Environmental Management* 17:199–210.
499. Kobringer, N. P. 1984. *Sources and Migration of Highway Runoff Pollutants—Executive Summary.* Vol. 1. FHWA/RD-84/057. Milwaukee, Wis.: Federal Highway Administration and Rexnord EnviroEnergy Technology Center.
500. Koehler, G. M., and J. D. Brittell. 1990. Managing spruce-fir habitat for lynx and snowshoe hares. *Journal of Forestry* 88:10–14.

501. Koelman, M., V. D. Janssen, W. Laak, and H. Ietswaart. 1999. Dispersion of PAH and heavy metals along motorways in The Netherlands: An overview. *Science and the Total Environment* 235:347–49.

502. Koford, E. J. 1993. Assessment and mitigation for endangered vernal pool invertebrates. In *Conference Proceedings for the 20th Anniversary Conference on Water Management in the 90's,* 839–41. New York: Water Resource Planning and Management of Urban Water Resources, ASCE.

503. Komura, J., and M. Sakamoto. 1991. Short-term oral administration of several manganese compounds in mice: Physiological and behavioral alterations caused by different forms of manganese. *Bulletin of Environmental Contamination and Toxicology* 46:921–28.

504. Kopecky, K. 1988. Einfluss der Strassen auf die Synanthropisierung der Flora und Vegetation nach Beobachtungen in der Tschechoslowakei. *Folia Geobotanica et Phytotaxonomica* 23:145–71.

505. Kozel, R. M., and E. D. Fleharty. 1979. Movement of rodents across roads. *Southwestern Naturalist* 24:239–48.

506. Kreithen, M. L., and D. B. Quine. 1979. Infrasound detection by the homing pigeon: A behavioral audiogram. *Journal of Comparative Physiology* (Series A) 129:1–4.

507. Krieger, A., D. Cobb, and A. Turner, eds. 2001. *Mapping Boston.* Cambridge: MIT Press.

508. Krummel, J. R., R. H. Gardner, G. Sugihara, R. V. O'Neill, and P. R. Coleman. 1987. Landscape patterns in a disturbed environment. *Oikos* 48:321–24.

510. Kruse, J. 1998. Remove it and they will disappear: Why building new roads isn't always the answer. *New England Planning* (June 1998): 1, 4, and 8.

511. Kucera, T. E., and R. H. Barrett. 1993. The Trailmaster camera system for detecting wildlife. *Wildlife Society Bulletin* 23:110–13.

512. Kuemmel, D. A., R. C. Sonntag, J. Crovetti, Y. Becker, J. R. Jaeckel, and A. Satanovsky. 2000. *Noise and Texture on PCC Pavements—Results of a Multi-state Study.* Madison: Wisconsin Department of Transportation.

513. Kuennen, T. 1989. New Jersey's I-78 preserves mountain habitat. *Roads and Bridges* (February 1989): 69–73.

514. Kuitunen, M., E. Rossi, and A. Stenroos. 1998. Do highways influence density of land birds? *Environmental Management* 22:297–302.

515. Kusler, J. A., D. E. Willard, and H. C. Hull. 1998. *Wetlands and Watershed Management: Science Applications and Public Policy.* Berne, N.Y.: Association of State Wetland Management.

516. Lal, R., ed. 1994. *Soil Erosion: Research Methods.* Ankeny, Iowa: Soil and Water Conservation Society.

517. La Marche, J. L., and D. P. Lettenmaier. 2001. Effects of forest roads on flood flows in the Deschutes River, Washington. *Earth Surface Processes and Landforms* 26:115–34.

518. Land, D., and M. Lotz. 1996. Wildlife crossing designs and use by Florida panthers and other wildlife in Southwest Florida. In *Trends in Addressing Transportation Related Wildlife Mortality,* edited by G. L. Evink, D. Zeigler, P. Garrett, and

J. Berry, 323–28. Publication FL-ER-58-96. Tallahassee: Florida Department of Transportation.

519. Lande, R. 1987. Extinction thresholds in demographic models of territorial populations. *American Naturalist* 130:624–35.
520. ———. 1988. Genetics and demography in biological conservation. *Science* 241:1455–60.
521. Lane, D. 1976. The vegetation of roadsides and adjacent farmland of the Mornington Peninsula, Victoria, Australia. *Weed Research* 16:385–89.
522. Langton, T. E. S., ed. 1989. *Amphibians and Roads.* Shefford, Bedfordshire, England: ACO Polymer Products.
523. Laurance, W. F. 1991. Edge effects in tropical forest fragments: application of a model for the design of nature reserves. *Biological Conservation* 57:205–19.
524. ———. 2000. Mega-development trends in the Amazon: Implications for global change. *Environmental Monitoring and Assessment* 61:113–22.
525. Laurance, W. F., A. K. M. Albernaz, and C. Da Costa. 2001a. Is deforestation accelerating in the Brazilian Amazon? *Environmental Conservation* 28:305–11.
526. Laurance, W. F., A. K. M. Albernaz, G. Schroth, P. M. Fearnside, S. Bergen, E. M. Venticinque, and C. Da Costa. 2002. Predictors of deforestation in the Brazilian Amazon. *Journal of Biogeography.* 29:737–48.
527. Laurance, W. F., M. A. Cochrane, S. Bergen, P. M. Fearnside, P. Delamonica, C. Barber, S. D'Angelo, and T. Fernandes. 2001b. The future of the Brazilian Amazon. *Science* 291:438–39.
528. Laursen, K. 1981. Birds on roadside verges and the effect of mowing on frequency and distribution. *Biological Conservation* 20:59–68.
530. Lausi, D., and P. L. Nimis. 1985. Roadside vegetation in boreal South Yukon and adjacent Alaska. *Phytocoenologia* 13:103–38.
531. Lawson, D. E. 1986. Response of permafrost terrain to disturbance: A synthesis of observations from Northern Alaska, U.S.A. *Arctic and Alpine Research* 18:1–17.
532. Lay, M. G. 1992. *Ways of the World.* New Brunswick, N.J.: Rutgers University Press.
533. Lee, C. S. Y., and G. G. Fleming. 1996. *Measurement of Highway-Related Noise.* Report FHWA-PD-96-046. Washington, D.C.: Federal Highway Administration, U.S. Department of Transportation.
534. Lee, C. S. Y., G. G. Fleming, and J. Burstein. 1998. *FHWA Traffic Noise Model: Look-Up Tables.* Report FHWA-PD-98-047. Washington, D.C.: U.S. Department of Transportation, Federal Highway Administration.
535. Leeson, B. 1996. Highway conflicts and resolutions in Banff National Park, Alberta. In *Trends in Addressing Transportation Related Wildlife Mortality,* edited by G. L. Evink, P. Garrett, D. Zeigler and J. Berry, 91–96. Report FL-ER-58-96. Tallahassee: Florida Department of Transportation.
536. Legret, M., and C. Pagotto. 1999. Evaluation of pollutant loadings in the runoff waters from a major rural highway. *Science and the Total Environment* 235:143–50.
537. Leharne, S., D. Charlesworth, and B. Chowdhry. 1992. A survey of metal levels in street dusts in an inner London neighborhood. *Environment International* 18:263–70.
538. Lehnert, M. E., and J. A. Bissonette. 1997. Effectiveness of highway crosswalk structures at reducing deer-vehicle collisions. *Wildlife Society Bulletin* 25:809–18.

539. Lehnert, M. E., J. A. Bissonette, and J. W. Haefner. 1998. Deer (Cervidae) highway mortality: Using models to tailor mitigative efforts. *Gibier Faune Sauvage, Game Wildlife* 15 (Hors Série Tome 3): 835–41.

540. Leopold, A. 1933. *Game Management*. New York: Charles Scribner's Sons.

541. ———. 1949. *A Sand County Almanac and Sketches Here and There*. New York: Oxford University Press.

542. Lerdau, M. T., J. W. Munger, and D. J. Jacob. 2000. The NO_2 conundrum. *Science* 289:2291–93.

543. *Les Bocages: Histoire, Ecologie, Economie*. 1976. Rennes, France: Institut National de la Recherche Agronomique, Centre National de la Recherche Scientifique, et Université de Rennes.

544. Leslie, M., G. K. Meffe, J. L. Hardesty, and D. L. Adams. 1996. *Conserving Biodiversity on Military Lands: A Handbook for Natural Resources Managers*. Arlington, Va.: The Nature Conservancy.

545. Leung, P. L., and R. M. Harrison. 1999. Roadside and in-vehicle concentrations of monoaromatic hydrocarbons. *Atmosphere and Environment* 33:191–204.

546. Levin, S. A., D. Cohen, and A. Hastings. 1984. Dispersal strategies in patchy environments. *Theoretical Population Biology* 26:165–91.

547. Liddle, M. 1997. *Recreation Ecology: The Ecological Impact of Outdoor Recreation and Ecotourism*. New York and London: Chapman and Hall.

548. Likens, G. E., and F. H. Bormann. 1974. Linkages between terrestrial and aquatic ecosystems. *BioScience* 24:447–56.

550. Likens, G. E., F. H. Bormann, R. S. Pierce, J. S. Eaton, and N. M. Johnson. 1977. *Biogeochemistry of a Forest Ecosystem*. New York: Springer-Verlag.

551. Likens, G. E., C. T. Driscoll, and D. C. Buso. 1996. Long-term effects of acid rain: Response and recovery of a forest ecosystem. *Science* 272:244–46.

552. Lima, S. L., and P. A. Zollner. 1996. Towards a behavioral ecology of ecological landscapes. *Trends in Ecology and Evolution* 11:131–35.

553. Ling, P. 1990. *America and the Automobile: Technology Reform and Social Change*. Manchester, England: Manchester University Press.

554. Ling, R. W., J. P. Van Amberg, and J. K. Werner. 1986. Pond acidity and its relationship to larval development of *Ambystoma maculatum* and *Rana sylvatica* in Upper Michigan. *Journal of Herpetology* 20:230–36.

555. Linsdale, J. M. 1929. Roadways as they affect bird life. *Condor* 31:143–45.

556. Linsley, R. K., Jr., M. A. Kohler, and J. L. H. Paulkus. 1975. *Hydrology for Engineers*. New York: McGraw-Hill.

557. Little, S. J., R. G. Harcourt, and A. P. Clevenger. 2002. Do wildlife passages act as prey-traps? *Biological Conservation* 107:135–45.

558. Liu, B. 1997. Route finding by using knowledge about the road network. *IEEE Transactions on Systems, Man and Cybernetics, Part A—Systems and Humans* 27(4):436–48.

559. Liu, D. S., L. R. Iverson, and S. Brown. 1993. Rates and patterns of deforestation in the Philippines: Application of geographic information systems analysis. *Forest Ecology Management* 57:1–16.

560. Liu, J., and W. W. Taylor, eds. 2002. *Integrating Landscape Ecology into Natural Resource Management*. New York and Cambridge: Cambridge University Press.

561. Liu, M. 2002. Atmospheric Deposition of Phosphorus and Particles to Lake Tahoe, California–Nevada. Master's thesis, University of California, Davis.

562. Loehle, C., and B. L. Li. 1996. Habitat destruction and the extinction debt revised. *Ecological Applications* 6:784–89.

563. Lonsdale, W. M., and A. M. Lane. 1994. Tourist vehicles as vectors of weed seeds in Kakadu National Park, Northern Australia. *Biological Conservation* 69:277–83.

564. Lopez, B. 1998. *About This Life: Journeys on the Threshold of Memory.* New York: Vintage Books.

565. Lovallo, M. J., and E. M. Anderson. 1996. Bobcat movements and home ranges relative to roads in Wisconsin. *Wildlife Society Bulletin* 24:71–76.

566. Lovich, J. E., and D. Bainbridge. 1999. Anthropogenic degradation of the southern California desert ecosystem and prospects for natural recovery and restoration. *Environmental Management* 24:309–26.

567. Loving, B. L., K. M. Waddell, and C. W. Miller. 2000. *Water and Salt Balance of Great Salt Lake, Utah, and Simulation of Water and Salt Movement through the Causeway, 1987–98.* U.S.G.S. Water Resources Investigations Report 00-4221. Salt Lake City, Utah: U.S. Geological Survey.

568. Lowe, J. C., and S. Moryadas. 1975. *The Geography of Movement.* Boston: Houghton-Mifflin.

570. Ludwig, J., and T. Bremicker. 1983. Evaluation of 2.4 m fences and one-way gates for reducing deer-vehicle collisions in Minnesota. *Transportation Research Record* 913:12–22.

571. Ludwig, J., D. Tongway, D. Freudenberger, J. Noble, and K. Hodgkinson, eds. 1997. *Landscape Ecology Function and Management: Principles from Australia's Rangelands.* Collingwood, Victoria, Australia: CSIRO Publishing.

572. Luey, J. E., and I. R. Adelman. 1980. Downstream natural areas as refuges for fish in drainage development watersheds. *Transactions of the American Fisheries Society* 109:332–35.

573. Lugo, A. E., and H. Gucinski. 2000. Function, effects, and management of forest roads. *Forest Ecology and Management* 133:249–62.

574. Lyles, L. 1988. Basic wind erosion processes. *Agriculture, Ecosystems and Environment* 22/23:91–101. (Reprinted in J. R. Brandle, D. L. Hintz, and J. W. Sturrock, eds., *Windbreak Technology,* Amsterdam: Elsevier, 1988.)

575. Lynam, D. R., G. D. Pfeifer, B. F. Fort, and A. A. Gelbcke. 1990. Environmental assessment of MMT fuel additive. *Science and the Total Environment* 93:107–14.

576. Lynch, K., and G. Hack. 1996. *Site Planning.* Cambridge: MIT Press.

577. Lytle, C. M., B. N. Smith, and C. Z. McKinnon. 1995. Manganese accumulation along Utah roadways: A possible indication of motor vehicle exhaust pollution. *Science and the Total Environment* 162:105–9.

578. MacDonald, L. H., R. W. Sampson, and D. M. Anderson. 2001. Runoff and road erosion at the plot and road segment scales, St. John, U.S. Virgin Islands. *Earth Surface Processes and Landforms* 26:251–72.

579. Mace, R. D., J. S. Waller, T. L. Manley, L. J. Lyon, and H. Zuuring. 1996. Relationships among grizzly bears, roads and habitat in the Swan Mountains, Montana. *Journal of Applied Ecology* 33:1395–1404.

580. Macek-Rowland, K. M., M. J. Barr, and G. B. Mitton. 2001. *Peak Discharges and*

Flow Volumes for Streams in the Northern Plains 1996–97. U.S.G.S. Circular 1185-B. Washington, D.C.: U.S. Geological Survey.

581. Mack, R. N. 1981. Invasion of *Bromus tectorum* L. into western North America: An ecological chronicle. *Agro-Ecosystems* 7:145–65.
582. Madej, M. A. 2001. Erosion and sediment delivery following removal of forest roads. *Earth Surface Processes and Landforms.* 26:175–90.
583. Mader, H.-J. 1984. Animal habitat isolation by roads and agricultural fields. *Biological Conservation* 29:81–96.
584. Mader, H.-J., C. Schell, and P. Kornacker. 1990. Linear barriers to arthropod movements in the landscape. *Biological Conservation* 54:209–22.
585. Maehr, D. S., E. D. Land, and M. E. Roelke. 1991. Mortality patterns of panthers in Southwest Florida. *Proceedings of the Annual Conference of Southeastern Association Fish and Wildlife Agencies* 45:201–7.
586. Maki, S., R. Kalliola, and K. Vuorinen. 2001. Road construction in the Peruvian Amazon: Process, causes and consequences. *Environmental Conservation* 20:199–214.
587. Malanson, G. P. 1993. *Riparian Landscapes.* New York: Cambridge University Press.
588. Malingreau, J. P., and C. J. Tucker. 1988. Large-scale deforestation in the southeastern Amazon Basin of Brazil. *Ambio* 17(1):49–55.
590. Maltby, L. A., B. A. Boxall, D. M. Farrow, P. Calow, and C. I. Betton. 1995. The effects of motorway runoff on freshwater ecosystems. 2. Identifying major toxicants. *Environmental Toxicology and Chemistry* 14:1093–1101.
591. Mansergh, I. M., and D. J. Scotts. 1989. Habitat-continuity and social organization of the mountain pygmy-possum restored by tunnel. *Journal of Wildlife Management* 53:701–7.
592. Mansfield, T. M., and B. D. Miller. 1975. *Highway Deer-Kill District 02 Regional Study.* Internal report. Sacramento: California Department of Transportation.
593. Marino, F., A. Ligero, and D. J. Diaz Cosin. 1992. Heavy metals and earthworms on the border of a road next to Santiago (Galicia, Northwest of Spain): Initial results. *Soil Biology and Biochemistry* 24:1705–9.
594. Marland, G., R. J. Andres, T. Boden, C. Johnston, and A. Brenkert. 1999. *Global, Regional, and National CO_2 Emission Estimates from Fossil-Fuel Burning, Cement Production, and Gas Flaring: 1751–1996.* Report NDP-030. Oak Ridge, Tenn.: Carbon Dioxide Information Analysis Center, Oak Ridge National Laboratory.
595. Marland, G., T. A. Boden, and R. J. Andres. 2001. Global, regional, and national fossil fuel CO_2 emissions. In *Trends: A Compendium of Data on Global Change.* Oak Ridge, Tenn.: Carbon Dioxide Information Analysis Center, Oak Ridge National Laboratory.
596. Maron, J., and R. Jefferies. 2001. Restoring enriched grasslands: Effects of mowing on species richness, productivity, and nitrogen retention. *Ecological Applications* 11:1088–1100.
597. Marsh, P. C., and J. E. Luey. 1982. Oases for aquatic life within agricultural watersheds. *Fisheries* 7(6):16–19, 24.
598. Marshik, J., L. P. Renz, J. L. Sipes, D. Becker, and D. Paulson. 2002. Preserving a spirit of place: U.S. Highway 93 on the Flathead Indian Reservation. In *Interna-*

tional Conference on Ecology and Transportation 2001 Proceedings: A Time for Action, 244–56. Center for Transportation and the Environment. Raleigh: North Carolina State University.

599. Martin, R. J. 1939. Highway toll. *Michigan Conservation* 9:2.
600. Massachusetts EOEA (Executive Office of Environmental Affairs). 1995. *Phragmites—Controlling the All-too-common Reed.* Boston: Commonwealth of Massachusetts.
601. ———. 1998. *Neponset River Watershed Restoration Plan.* Boston: Commonwealth of Massachusetts.
602. Matson, P. A., W. H. McDowell, A. R. Townsend, and P. M. Vitousek. 1999. The globalization of N deposition: Ecosystem consequences in tropical environments. *Biogeochemistry* 46:67–83.
603. Maurer, M. E. 1999. Development of a community-based, landscape-level terrestrial mitigation decision support system for transportation planners. In *Proceedings of the Third International Conference on Wildlife Ecology and Transportation,* edited by G. L. Evink, P. Garrett, D. Zeigler, and J. Berry, 99–109. FL-ER-73-99. Tallahassee: Florida Department of Transportation.
604. May, S. A., and T. W. Norton. 1996. Influence of fragmentation and disturbance on the potential impact of feral predators on native fauna in Australian forest ecosystems. *Wildlife Research* 23:387–400.
605. McBean, E., and S. Al-Nassri. 1987. Migration pattern of de-icing salts from roads. *Journal of Environmental Management* 25:231–38.
606. McCarthy, J. J., O. F. Canziani, N. A. Leary, D. J. Dokken, and K. S. White, eds. 2001. *Climate Change 2001: Impacts, Adaptation, and Vulnerability.* Contribution of Working Group II to the Third Assessment Report of the Intergovernmental Panel on Climate Change. New York: Cambridge University Press.
607. McClaran, M. P., and T. T. Van Devender. 1995. *The Desert Grassland.* Tucson: University of Arizona Press.
608. McClure, H. L. 1951. An analysis of animal victims on Nebraska's highways. *Journal of Wildlife Management* 15:410–20.
610. McDonnell, M. J., and E. W. Stiles. 1983. The structural complexity of old field vegetation and the recruitment of bird-dispersed plant species. *Oecologia* 56:109–16.
611. McGuire, T. M., and J. F. Morrall. 2000. Strategic highway improvements to minimize environmental impacts within the Canadian Rocky Mountain national parks. *Canadian Journal of Civil Engineering* 27:523–32.
612. McHarg, I. 1969. *Design with Nature.* New York: Doubleday.
613. McIntyre, S., and S. Lavorel. 1994. Predicting richness of native, rare, and exotic plants in response to habitat and disturbance variables across a variegated landscape. *Conservation Biology* 8:521–31.
614. McKelvey, K. S., Y. K. Ortega, G. Koehler, K. Aubry, and D. Brittell. 1999. Canada lynx habitat and topographic use patterns in north central Washington: A reanalysis. In *Ecology and Conservation of Lynx in the United States,* edited by L. F. Ruggiero, K. B. Aubry, S. W. Buskirk, G. M. Koehler, C. J. Krebs, K. S. McKelvey, and J. R. Squires, 307–36. General Technical Report RMRS-GTR-30WWW. Fort Collins, Colo.: USDA Forest Service.

615. McLellan, B. N., and D. M. Shackleton. 1988. Grizzly bears and resource-extraction industries: Effects of roads on behavior, habitat use and demography. *Journal of Applied Ecology* 25:451–60.
616. McNaughton, K. G. 1988. Effects of windbreaks on turbulent transport and microclimate. *Agriculture, Ecosystems and Environment* 22/23:17–39. (Reprinted in J. R. Brandle, D. L. Hintz, and J. W. Sturrock, eds., *Windbreak Technology,* Amsterdam: Elsevier, 1988.)
617. Mech, L. D. 1970. *The Wolf: The Ecology and Behavior of an Endangered Species.* St. Paul: University of Minnesota Press.
618. Mech, L. D., S. H. Fritts, G. L. Raddle, and W. J. Paul. 1988. Wolf distribution and road density in Minnesota. *Wildlife Society Bulletin* 16:85–87.
619. Meehan, W. R., ed. 1991. *Influences of Forest and Rangeland Management on Salmonid Fishes and Their Habitats.* Special Publication 19. Bethesda, Md.: American Fisheries Society.
620. Meffe, G. K., C. R. Carroll, and contributors. 1997. *Principles of Conservation Biology.* Sunderland, Mass.: Sinauer Associates.
621. Megahan, W. F., M. Wilson, and S. B. Monson. 2001. Sediment production from granitic cutslopes on forest roads in Idaho, USA. *Earth Surface Processes and Landforms* 26:153–63.
622. Melman, P. J. M., and H. J. Verkaar. 1991. Layout and management of herbaceous vegetation in road verges. In *Nature Engineering and Civil Engineering Works,* edited by P. Aanen, W. Alberts, et al., 62–78. Wageningen, Netherlands: Pudoc.
623. Mena, I. 1980. Manganese. In *Metals in the Environment,* edited by H. A. Waldron, 199–220. New York: Academic Press.
624. Merriam, G. 1990. Ecological processes in the time and space of farmland mosaics. In *Changing Landscapes: An Ecological Perspective,* edited by I. S. Zonneveld and R. T. T. Forman, 121–33. New York: Springer-Verlag.
625. ———. 1991. Corridors and connectivity: Animal populations in heterogeneous environments. In *Nature Conservation 2: The Role of Corridors,* edited by D. A. Saunders and R. J. Hobbs, 133–42. Chipping Norton, Australia: Surrey Beatty.
627. ———. 1998. Important concepts from landscape ecology for game biologists. In *Twenty-Third Congress of the International Union of Game Biologists,* edited by P. Havet, E. Taran, and J. C. Berthos. *Gibier, Faune Sauvage* 15 (Hors série Tome 2): 525–31.
628. Merriam, G., and A. Lanoue. 1990. Corridor use by small mammals: Field measurement for three experimental types of *Peromyscus leucopus. Landscape Ecology* 4:123–31.
630. Merriam, G., K. Michal, E. Tsuchiya, and K. Hawley. 1989. Barriers as boundaries for metapopulations and demes of *Peromyscus leucopus* in farm landscapes. *Landscape Ecology* 29:227–35.
631. Merrill, L. 1998. Making the dust go away. *Erosion Control Magazine,* July/August, 26–32.
632. Messmer, T. A., C. W. Hendricks, and P. W. Klimack. 2000. Modifying human behavior to reduce wildlife-vehicle collisions using temporary signing. In *Wildlife and Highways: Seeking Solutions to an Ecological and Socio-economic*

Dilemma, edited by T. A. Messmer and B. West, 125–39. Nashville, Tenn.: The Wildlife Society.

633. Meunier, F. D., C. Verheyden, and P. Jouventin. 2000. Use of roadsides by diurnal raptors in agricultural landscapes. *Biological Conservation* 92:291–98.
634. Milberg, P., and B. B. Lamont. 1995. Fire enhances weed invasion of roadside vegetation in Southwestern Australia. *Biological Conservation* 73:45–49.
635. Milberg, P., and T. Persson. 1994. Soil seed bank and species recruitment in road verge grassland vegetation. *Annales Botanica Fennoscandia* 31:155–62.
636. Milchunas, D. G., K. A. Schulz, and R. B. Shaw. 1999. Plant and environmental interactions: Plant community responses to disturbance by mechanized military maneuvers. *Journal of Environmental Quality* 28:1533–47.
637. Ministry of Transport, Public Works and Water Management. 1994a. *Towards Sustainable Verge Management in The Netherlands.* No. 59. Delft, Netherlands: Ministerie van Verkaar en Waterstaat.
638. ———. 1994b. *Managing Roadside Flora in The Netherlands.* No. 60. Delft, Netherlands: Ministerie van Verkaar en Waterstaat.
639. ———. 1994c. *The Chemical Quality of Verge Grass in The Netherlands.* No. 62. Delft, Netherlands: Dienst Weg- en Waterbouwkunde, Ministerie van Verkaar en Waterstaat.
640. ———. 2000. *National Highway Verges. . . . National Treasures!* Delft, Netherlands: Ministerie van Verkaar en Waterstaat.
641. Mitch, W. J., and J. G. Gosselink. 2000. *Wetlands.* New York: Van Nostrand Reinhold.
642. Mitchell, J. H. 1984. *Ceremonial Time: Fifteen Thousand Years on One Square Mile.* New York: Addison-Wesley.
643. Mladenoff, D. J., T. A. Sickley, R. G. Haight, and A. P. Wydeven. 1995. A regional landscape analysis of favorable gray wolf habitat in the northern Great Lakes region. *Conservation Biology* 9:279–94.
644. Mladenoff, D. J., T. A. Sickley, and A. P. Wydeven. 1999. Predicting gray wolf landscape recolonization: Logistic regression models vs. new field data. *Ecological Applications* 9:37–44.
645. Mobley, C. D. 1994. *Light and Water: Radiative Transfer in Natural Waters.* New York: Academic Press.
646. Montgomery, D. R., K. Sullivan, and H. M. Greenberg. 1998. Regional test of a model of shallow landsliding. *Hydrological Processes* 12:943–55.
647. Montgomery, D. R. 1994. Road surface drainage, channel initiation, and slope stability. *Water Resources Research* 30:1925–32.
648. Mooney, H. A., and J. A. Drake, eds. 1986. *Ecology of Biological Invasions of North America and Hawaii.* New York: Springer-Verlag.
650. Moran, J. M., and M. D. Morgan. 1994. *Meteorology: The Atmosphere and the Science of Weather.* New York: Macmillan.
651. Morawska, L., S. Thomas, D. Gilbert, C. Greenaway, and E. Rijnders. 1999. A study of the horizontal and vertical profile of submicrometer particles in relation to a busy road. *Atmosphere and Environment* 33:1261–74.
652. Morel, G. A., and B. P. M. Specken. 1992. Versnippering van de ecologische hoofdstructuur door de natte infrastructuur. *Project Versnippering Deel 4* (H. Duel,

schrijver). Delft, Netherlands: Directoraat-Generaal Rijkswaterstaat, Dienst Weg- en Waterbouwkunde (Ministry of Transport, Public Works and Water Management).

653. Morin, P. J. 1999. *Community Ecology.* Malden, Mass.: Blackwell Science.
654. Morrall, J. F., and T. M. McGuire. 2000. Sustainable highway development in a national park. *Transportation Research Record* 1702:3–10.
655. Morrison, D. 1981. Utilization of prairie plants on disturbed sites. *Transportation Research Record* 822:10–17.
656. Mowat, G., K. G. Poole, and M. O'Donoghue. 1999. Ecology of lynx in northern Canada and Alaska. In *Ecology and Conservation of Lynx in the United States,* edited by L. F. Ruggiero, K. B. Aubry, S. W. Buskirk, G. M. Koehler, C. J. Krebs, K. S. McKelvey, and J. R. Squires, 265–306. General Technical Report RMRS-GTR-30WWW. Fort Collins, Colo.: USDA Forest Service.
657. Mueller, D. S. 2000. *National Bridge Scour Program—Measuring Scour of the Streambed at Highway Bridges.* U.S.G.S. Fact Sheet FS 107-00. Louisville, Ken.: U.S. Geological Survey.
658. Mueller-Navarra, D. C., M. T. Brett, A. M. Liston, and C. R. Goldman. 2000. A highly unsaturated fatty acid predicts carbon transfer between primary producers and consumers. *Nature* 403:74–77.
659. Muller, F. 1997. State-of-the-art in ecosystem theory. *Ecological Modelling* 100:135–61.
660. Mumme, R. L., S. J. Schoech, G. E. Woolfenden, and J. W. Fitzpatrick. 2000. Life and death in the fast lane: Demographic consequences of road mortality in the Florida scrub-jay. *Conservation Biology* 14:501–12.
661. Munguira, M. L., and J. A. Thomas. 1992. Use of road verges by butterfly and burnet populations, and the effect of roads on adult dispersal and mortality. *Journal of Applied Ecology* 29:316–29.
662. Murphy, S. M., and J. A. Curatolo. 1987. Activity budgets and movement rates of caribou encountering pipelines, roads and traffic in northern Alaska. *Canadian Journal of Zoology* 65:2483–90.
663. Murphy, T. J., and P. V. Doskey. 1976. Inputs of phosphorus from precipitation to Lake Michigan. *Journal of Great Lakes Research* 2:60–70.
664. Muskett, C. J., and M. P. Jones. 1980. The dispersal of lead, cadmium and nickel from motor vehicles and effects on roadside invertebrate macrofauna. *Environmental Pollution* (Series A) 23:231–42.
665. Nagler, A., W. Schmidt, and T. Stottele. 1989. Die Vegetation an Autobahnen und Strassen in Sudhessen. *Tuexenia* 9:151–82.
666. Naiman, R. J., T. J. Beechie, L. E. Benda, et al. 1992. Fundamental elements of ecologically healthy watershed in the Pacific Northwest Coastal region. In *Watershed Management: Balancing Sustainability and Environmental Chance,* edited by R. J. Naiman, 127–88. New York: Springer-Verlag.
667. Naiman, R. J., R. E. Bilby, and P. A. Bisson. 2000. Riparian ecology and management in the Pacific Coastal rain forest. *BioScience* 50:996–1011.
668. Nakamura, F., F. J. Swanson, and S. M. Wondzell. 2000. Disturbance regimes of stream and riparian systems—a disturbance-cascade perspective. *Hydrological Processes* 14:2849–60.

670. Nassauer, J. I., ed. 1997. *Placing Nature: Culture and Landscape Ecology.* Washington, D.C.: Island Press.
671. Nassauer, J. I., and R. Westmacott. 1987. Progressiveness among farmers as a factor in heterogeneity of farmed landscapes. In *Landscape Heterogeneity and Disturbance,* edited by M. G. Turner, 199–210. New York: Springer-Verlag.
672. National Council for Air and Stream Improvement. 1992. *Status of the NCASI Cumulative Watershed Effects Program and Methodology.* Technical Bulletin 634. Washington, D.C.
673. National Research Council. 1995. *Expanding Metropolitan Highways: Implications for Air Quality and Energy Use.* Washington, D.C.: National Academy Press.
674. ———. 1997. *Toward a Sustainable Future: Addressing the Long-Term Effects of Motor Vehicle Transportation on Climate and Ecology.* Washington, D.C.: National Academy Press.
675. ———. 2002. *Surface Transportation Environmental Research: A Long-Term Strategy.* Special Report 268. Washington, D.C.: Transportation Research Board.
676. Nellemann, C., and R. D. Cameron. 1996. Effects of petroleum development on terrain preferences of calving caribou. *Arctic* 49:23–28.
677. ———. 1998. Cumulative impacts of an evolving oil-field complex on the distribution of calving caribou. *Canadian Journal of Zoology* 76:1425–30.
678. Nellemann, C., I. Vistnes, P. Jordhoey, and O. Strand. 2001. Winter distribution of wild reindeer in relation to power lines, roads and resorts. *Biological Conservation* 101:351–60.
679. Nepstad, D., G. Carvalho, A. C. Barros, A. Alencar, J. Capobianco, J. Bishop, P. Moutinho, P. Lefebvre, and U. Silva Jr. 2001. Road paving, fire regime feedbacks, and the future of Amazon forests. *Forest Ecology and Management* 154:295–407.
680. Nepstad, D. C., A. Verissimo, A. Alencar, C. Nobre, E. Lima, P. Lefebre, P. Schlesinger, C. Potter, P. Moutinho, E. Mendoza, M. Cochrane, and V. Brooks. 1999. Large-scale impoverishment of Amazonian forests by logging and fire. *Nature* 398:505–08.
681. *New York Times.* 1998. *1999 Almanac.* New York: Penguin Reference Books.
682. Newton, I., I. Wyllie, and A. Asher. 1991. Mortality causes in British barn owls *Tyto alba,* with a discussion of aldrin-dieldrin poisoning. *Ibis* 133:162–69.
683. Nieuwenhuizen, W., and R. C. van Apeldoorn. 1995. *Mammal Use of Fauna Passages on National Road A1 near Oldenzaal.* Report P-DWW-95-737. Delft, Netherlands: Ministry of Transport, Public Works and Water Management.
684. Norton, D. A., R. J. Hobbs, and L. Atkins. 1995. Fragmentation, disturbance, and plant distribution: Mistletoes in woodland remnants in the Western Australian wheatbelt. *Conservation Biology* 9:426–38.
685. Noss, R. F. 1991. Landscape connectivity: Different functions at different scales. In *Landscape Linkages and Biodiversity,* edited by W. E. Hudson, 27–39. Washington, D.C.: Island Press.
686. Noss, R. F., H. B. Quigley, M. G. Hornocker, T. Merrill, and P. C. Paquet. 1996. Conservation biology and carnivore conservation in the Rocky Mountains. *Conservation Biology* 10:949–63.
687. Nowland, A. 1997. *Sustainable Management Strategy for Travelling Stock Routes and Reserves in Central Western New South Wales.* New South Wales, Australia: Con-

dobolin, Coonabarabran, Coonamble, Dubbo, Forbes, Molong, and Nyngan Rural Lands Protection Boards.

688. Oberts, G. L. 1986. Pollutants associated with sand and silt applied to roads in Minnesota. *Water Resources Bulletin* 22:479–83.
690. O'Neill, R.V., D. L. DeAngelis, J. B. Waide, and T. F. H. Allen, eds. 1986. *A Hierarchical Concept of Ecosystems.* Princeton, N.J.: Princeton University Press.
691. Odum, E. P. 1971. *Fundamentals of Ecology.* Philadelphia: Saunders.
692. Odum, H. T. 1983. *Systems Ecology: An Introduction.* New York: John Wiley.
693. Oetting, R. B., and J. F. Cassel. 1971. Waterfowl nesting on interstate highway right-of-way in North Dakota. *Journal of Wildlife Management* 35:774–81.
694. Ontario Ministry of Natural Resources. 1995. *Environmental Guidelines for Access Roads and Water Crossings.* Toronto: Ontario Ministry of Natural Resources.
695. Opdam, P. 1991. Metapopulation theory and habitat fragmentation: A review of holarctic breeding bird studies. *Landscape Ecology* 5:93–106.
696. Opdam, P. F. M. 1997. How to choose the right solution for the right fragmentation problem? In *Habitat Fragmentation & Infrastructure,* edited by K. Canters, 55–60. Delft, Netherlands: Ministry of Transport, Public Works and Water Management.
697. Opdam, P. 2002. Assessing the conservation potential of habitat networks. In *Applying Landscape Ecology in Biological Conservation,* edited by K. J. Gutzwiller, 381–404. New York: Springer.
698. Oris, J. T., A. Hatch, J. Weinstein, R. Findlay, P. McGinn, S. Diamond, R. Garrett, W. Jackson, G. A. Burton, and B. Allen. 1998. *Toxicity of Ambient Levels of Motorized Watercraft Emissions to Fish and Zooplankton in Lake Tahoe, California/Nevada, USA.* Poster 3E-P005. 8th annual meeting SETAC-Europe (European Society of Environmental Toxicology and Chemistry), Bordeaux, France.
699. Ortega, Y. K., and D. E. Capen. 1999. Effects of forest roads on habitat quality for ovenbirds in a forested landscape. *Auk* 116:937–46.
700. Oxley, D. J., M. B. Fenton, and G. R. Carmody. 1974. The effects of roads on populations of small mammals. *Journal of Applied Ecology* 11:51–59.
701. Panetta, F. D., and A. J. M. Hopkins. 1991. Weeds in corridors: Invasion and management. In *Nature Conservation 2: The Role of Corridors,* edited by D. A. Saunders and R. J. Hobbs, 341–51. Chipping Norton, Australia: Surrey Beatty.
702. Parendes, L. A. 1997. Spatial patterns of invasion by exotic plants in a forested landscape. Ph.D. dissertation, Oregon State University, Corvallis.
703. Parendes, L. A., and J. A. Jones. 2000. Role of light availability and dispersal in exotic plant invasion along roads and streams in the H. J. Andrews Experimental Forest, Oregon. *Conservation Biology* 14:64–75.
704. Park, G. 1995. *Nga Uruora: Ecology and History in a New Zealand Landscape.* Wellington, New Zealand: Victoria University Press.
705. Parker, G. R. 1981. Winter habitat use and hunting activities for lynx *(Lynx canadensis)* on Cape Breton Island, Nova Scotia. In *Worldwide Furbearer Conference Proceedings,* edited by J. A. Chapman and D. Pursley, 221–48. Frostburg: University of Maryland.
706. Parker River Clean Water Association. 1996. *Tidal Crossing Inventory and Assessment. Summary Report: Upper North Shore, Massachusetts.* Newburyport, Mass.: Parker River Clean Water Association.

707. Parr, T. W., and J. M. Way. 1988. Management of roadside vegetation: The long term effects of cutting. *Journal of Applied Ecology* 25:1073–87.
708. Parsons, R. 1995. Conflict between ecological sustainability and environmental aesthetics: Conundrum, canard or curiosity. *Landscape and Urban Planning* 32:227–44.
710. Paterson, A. M., B. F. Cumming, J. P. Smol, J. M. Blais, and R. L. France. 1998. Assessment of the effects of logging, forest fires and drought on lakes in northwestern Ontario: A 30-year paleolimnological perspective. *Canadian Journal of Forest Research* 28:1546–56.
711. ———. 2000. A paleolimnological assessment of the effects of logging on two lakes in northwestern Ontario, Canada. *Verh. Internat. Verein. Limnol.* 27:1214–19.
712. Patric, J. H., J. O. Evans, and J. D. Helvey. 1984. Summary of sediment yield data from forested land in the United States. *Journal of Forestry* 76:101–04.
713. Patten, B. C. 1992. Energy, emergy, and environs. *Ecological Modelling* 62:29–70.
714. Patterson, P. 2000. *Explaining VMT Growth.* Washington, D.C.: Office of Technology Transfer, U.S. Department of Energy.
715. Peiser, R. 2001. Decomposing urban sprawl. *Town Planning Review* 72:275–98.
716. Perdikaki, K., and C. F. Mason. 1999. Impact of road run-off on receiving streams in Eastern England. *Water Research* 33:1627–33.
717. Perring, F. H. 1967. Verges are vital: A botanist looks at our roadsides. *Journal of the Institute of Highway Engineers* 14:13–16.
718. Perrins, J., A. Fitter, and M. Williamson. 1993. Population biology and rates of invasion of three introduced Impatiens species in the British Isles. *Journal of Biogeography* 20:33–44.
719. Peterken, G. F. 1993. *Woodland Conservation and Management.* New York and London: Chapman and Hall.
720. ———. 1996. *Natural Woodland: Ecology and Conservation in Northern Temperate Regions.* New York: Cambridge University Press.
721. Peters, N. E., and J. T. Turk. 1981. Increases in sodium and chloride in the Mohawk River, New York, from the 1950s to the 1970s attributed to road salt. *Water Resources Bulletin* 17:586–98.
722. Pfister, H. P., V. Keller, H. Reck, and B. Georgii. 1997. *Bio-okologische Wirksamkeit von Grunbrucken uber Verkehrswege* (Bio-ecological effectiveness of wildlife overpasses or "green bridges" over roads and railway lines). Bonn-Bad Godesberg, Germany: Herausgegeben vom Bundesministerium fur Verkehr Abeteilung Strassenbau.
723. Pickett, S. T. A., and M. J. McDonnell. 1989. Changing perspectives in community dynamics: A theory of successional forces. *Trends in Ecology and Evolution* 4:241–45.
724. Pickett, S. T. A., and P. S. White, eds. 1985. *The Ecology of Natural Disturbance and Patch Dynamics.* New York: Academic Press.
725. Piepers, A. A. G., ed. 2002. *Infrastructure and nature: fragmentation and defragmentation.* Dutch State of the Art Report for COST Activity 341. Report P-DWW-2000-41. Delft, Netherlands: Ministry of Transport, Public Works and Water Management.

726. Pimentel, D., ed. 1993. *World Soil Erosion and Conservation*. New York and Cambridge: Cambridge University Press.

727. Pimm, S. L., J. H. Lawton, and J. E. Cohen. 1991. Food web patterns and their consequences. *Nature* 350:669–74.

728. Pisarski, A. E. 1996. *Commuting in America II: The Second National Report on Commuting Patterns and Trends.* Washington, D.C.: Eno Transportation Foundation.

730. ———. 2001. US roads. In *Millennium Book,* 52–79. Paris: International Road Federation.

731. Platt, D. D., ed. 1998. *Rim of the Gulf: Restoring Estuaries in the Gulf of Maine.* Rockland, Maine: Island Institute.

732. Platt, R. H., R. A. Rowntree, and P. C. Muick. 1994. *The Ecological City: Preserving and Restoring Urban Biodiversity.* Amherst: University of Massachusetts Press.

733. Poff, N. L., J. D. Allan, M. B. Bain, J. R. Karr, K. L. Prestegaard, B. D. Richter, R. E. Sparks, and J. C. Stromberg. 1997. The natural flow regime. *BioScience* 47:769–84.

734. Pollard, E., M. D. Hooper, and N. W. Moore. 1974. *Hedges.* London: W. Collins.

735. Pollock, S. 1990. *Mitigating Highway Deicing Salt Contamination of Private Water Supplies in Massachusetts.* Wellesley: Research and Materials Section, Massachusetts Department of Public Works (reprinted in C. Goldman and G. Malyj, 1990).

736. Pope, S. E., L. Fahrig, and H. G. Merriam. 2000. Landscape complementation and metapopulation effects on leopard frog populations. *Ecology* 81:2498–508.

737. Port, G. R., and J. R. Thompson. 1980. Outbreaks of insect herbivores on plants along motorways in the United Kingdom. *Journal of Applied Ecology* 17:649–56.

738. Pratt, D. W., R. A. Black, and B. A. Zamora. 1984. Buried viable seed in a ponderosa pine community. *Canadian Journal of Botany* 62:44–52.

739. President's Council on Sustainable Development. 1996. *Sustainable America: A New Consensus for Prosperity, Opportunity, and a Healthy Environment for the Future.* Washington, D.C.: President's Council on Sustainable Development.

740. Prince George's County. 1999. *Low-Impact Development Design Strategies: An Integrated Design Approach.* Prince George's County, Md.: Prince George's County Department of Environmental Resources.

741. Puglisi, M. J., J. S. Lindzey, and E. D. Bellis. 1974. Factors associated with highway mortality of white-tailed deer. *Journal of Wildlife Management* 38:799–807.

742. Pulliam, H. R. 1988. Sources, sinks, and population regulation. *American Naturalist* 132:652–61.

743. Purinton, T. 1998. Restoring tidal flow to salt marshes. *Coastal Monitor* 8:16–18.

744. Putman, R. J. 1997. Deer and road traffic accidents: Options for management. *Journal of Environmental Management* 51:43–57.

745. Quarles, H. D., R. B. Hanawalt, and W. E. Odum. 1974. Lead in small mammals, plants, and soil at varying distances from a highway. *Journal of Applied Ecology* 11:937–49.

746. Rabenold, K. N., P. T. Fauth, B. W. Goodner, J. A. Sadowski, and P. G. Parker. 1998. Response of avian communities to disturbance by an exotic insect in spruce-fir forests of the Southern Appalachians. *Conservation Biology* 12:177–89.

747. Race, M. S., and M. S. Fonseca. 1995. Fixing compensatory mitigation: What will it take? *Ecological Applications* 6:94–101.
748. Racin, J. A., R. B. Howell, G. R. Winters, and E. C. Shirley. 1982. *Estimating Highway Runoff Quality.* FHWA/CA/TL-82/11. Sacramento: Federal Highway Administration and California Department of Transportation.
750. Rajvanshi, A., V. B. Mathur, G. C. Teleki, and S. K. Mukherjee. 2001. *Roads, Sensitive Habitats and Wildlife: Environmental Guideline for India and South Asia.* Dehradun, India: Wildlife Institute of India.
751. Randall, J. M., and J. Marinelli, eds. 1996. *Invasive Plants: Weeds of the Global Garden.* New York: Brooklyn Botanic Garden.
752. Ranney, J. W., M. C. Bruner, and J. B. Levenson. 1981. The importance of edge in the structure and dynamics of forest islands. In *Forest Island Dynamics in Man-Dominated Landscapes,* edited by R. L. Burgess and D. M. Sharpe, 67–92. New York: Springer-Verlag.
753. Raty, M. 1979. Effect of highway traffic on tetraonid densities. *Ornis Fennica* 56:169–70.
754. Reck, H., and G. Kaule. 1992. *Strassen und Lebensraume: Ermittlung und Beurteilung strassenbedingter Auswirkungen auf Pflanzen, Tiere und ihre Lebensraume.* Bonn-Bad Godesberg, Germany: Forschung Strasssenbau und Strassenverkehrstechnik, Heft 654, Herausgegeben vom Bundesminister fur Verkehr.
755. Reed, D. F. 1981. Effectiveness of highway lighting in reducing deer-vehicle collisions. *Journal of Wildlife Management* 45:721–26.
756. Reed, D. F., and A. L. Ward. 1985. Efficacy of methods advocated to reduce deer-vehicle accidents: Research and rationale in the USA. In *Routes et faune sauvage* 285–93. Bagneaux, France: Service d'Etudes Techniques des Routes et Autoroutes.
757. Reed, D. F., T. M. Pojar, and T. N. Woodard. 1974. Use of one-way gates by mule deer. *Journal of Wildlife Management* 38:9–15.
758. Reed, D. F., T. N. Woodard, and T. M. Pojar. 1975. Behavioral response of mule deer to a highway underpass. *Journal of Wildlife Management* 39:361–67.
759. Reed, D. M., and J. A. Schwarzmeier. 1978. The prairie corridor concept: Possibilities for planning large-scale preservation and restoration. In *Proceedings of the Fifth Midwest Prairie Conference,* edited by Lewin and Landers, 158–65. Ames: Iowa State University.
760. Reed, R. A., J. Johnson-Barnard, and W. L. Baker. 1996. Contribution of roads to forest fragmentation in the Rocky Mountains. *Conservation Biology* 10:1098–106.
761. Reeve, A. F. 1988. *Vehicle-Related Mortality of Mule Deer in Nugget Canyon, Wyoming.* Laramie: Wyoming Cooperative Fisheries and Wildlife Research Unit.
762. Reeve, A. F., and S. H. Anderson. 1993. Ineffectiveness of Swareflex reflectors at reducing deer-vehicle collisions. *Wildlife Society Bulletin* 21:127–32.
763. Regional Ecosystem Office. 1995. *Ecosystem Analysis at the Watershed Scale, Version 2.2.* Washington, D.C.: U.S. Government Printing Office.
764. Reh, W. 1989. Investigations into the influence of roads in the genetic structure of populations of the common frog *Rana temporaria.* In *Amphibians and Roads,*

edited by T. E. S. Langton, 101–03. Shefford, Bedforshire, England: ACO Polymer Products.

765. Reh, W., and A. Seiz. 1990. The influence of land use on the genetic structure of populations of the common frog *Rana temporaria*. *Biological Conservation* 54:239–49.
766. Reid, L. M., and T. Dunne. 1984. Sediment production from forest road surfaces. *Water Resources Research* 20:1753–61.
767. Reijnen, M. J. S. M., G. Veenbaas, and R. P. B. Foppen. 1995. *Predicting the Effects of Motorway Traffic on Breeding Bird Populations.* Delft, Netherlands: Ministry of Transport, Public Works and Water Management.
768. Reijnen, R., and R. Foppen. 1994. The effects of car traffic on breeding bird populations in woodland. I. Evidence of reduced habitat quality for willow warblers *(Phylloscopus trochilus)* breeding close to a highway. *Journal of Applied Ecology* 31:85–94.
769. ———. 1995. The effects of car traffic on breeding bird populations in woodland. IV. Influence of population size on the reduction of density close to a highway. *Journal of Applied Ecology* 32:481–91.
770. Reijnen, R., R. Foppen, and H. Meeuwsen. 1996. The effects of car traffic on the density of breeding birds in Dutch agricultural grasslands. *Biological Conservation* 75:255–60.
771. Reijnen, R., R. Foppen, C. ter Braak, and J. Thissen. 1995. The effects of car traffic on breeding bird populations in woodland. III. Reduction of density in relation to the proximity of main roads. *Journal of Applied Ecology* 32:187–202.
772. Reiner, E. 1999. *Salt Marsh Restoration in Massachusetts.* Boston: Association of Conservation Commissions.
773. Richardson, C. J., and K. Nunnery. 2001. Ecological functional assessment (EFA): A new approach to determining wetland health. In *Transformations of Nutrients in Natural and Constructed Wetlands,* edited by J. Vymazal, 95–111. Leiden, Netherlands: Backhuys Publishers.
774. Richardson, J. H., R. F. Shore, and J. R. Treweek. 1997. Are major roads a barrier to small mammals? *Journal of Zoology* (London) 243:840–46.
775. Richman, T., J. Worth, P. Dawe, J. Aldrich, and B. Ferguson. 1997. *Start at the Source: Residential Site Planning and Design Guidance Manual for Stormwater Quality Protection.* Los Angeles: Bay Area Stormwater Management Agency Association.
776. Ricklefs, R. E., and G. L. Miller. 2000. *Ecology.* New York: Freeman.
777. Ries, L., D. M. Debinski, and M. L. Wieland. 2001. Conservation value of roadside prairie restoration to butterfly communities. *Conservation Biology* 15:401–11.
778. Riffell, S. K. 1999. Road mortality of dragonflies (Odonata) in a Great Lakes coastal wetland. *Great Lakes Entomologist* 32:63–73.
779. Roach, G. L., and R. D. Kirkpatrick. 1985. Wildlife use of roadside woody plantings in Indiana. *Transportation Research Record* 1016:11–15.
780. Roberston, S. B. 1989. Technical comments: Impacts of petroleum development in the Arctic. *Science* 25:764–65.
781. Rodríguez, A., G. Crema, and M. Delibes. 1996. Factors affecting crossing of red foxes and wildcats through non-wildlife passages across a high speed railway. *Ecography* 20:287–94.

782. ———. 1997. Use of non-wildlife passages across a high speed railway by terrestrial vertebrates. *Journal of Applied Ecology* 33:1527–40.
783. Rolley, R. E., and L. E. Lehman. 1992. Relationship among raccoon road-kill surveys, harvests, and traffic. *Wildlife Society Bulletin* 20:313–18.
784. Romin, L. A. 1994. Factors associated with the highway mortality of mule deer at Jordanelle Reservoir, Utah. Master's thesis, Utah State University, Logan.
785. Romin, L. A., and J. A. Bissonette. 1996a. Deer-vehicle collisions: Status of state monitoring activities and mitigation efforts. *Wildlife Society Bulletin* 24:276–83.
786. ———. 1996b. Temporal and spatial distribution of highway mortality of mule deer on newly constructed roads at Jordanelle Reservoir, Utah. *Great Basin Naturalist* 56:1–11.
787. Romin, L., and L. B. Dalton. 1992. Lack of response by mule deer to wildlife warning whistles. *Wildlife Society Bulletin* 20:382–84.
788. Romme, W. H. 1997. Creating pseudo-rural landscapes in the Mountain West. In *Placing Nature: Culture and Landscape Ecology,* edited by J. I. Nassauer, 139–61. Washington, D.C.: Island Press.
790. Roof, J., and J. Wooding. 1996. Evaluation of the S.R. 46 wildlife crossing in Lake County, Florida. In *Trends in Addressing Transportation Related Wildlife Mortality,* edited by G. L. Evink, P. Garrett, D. Zeigler, and J. Berry, 329–36. Publication FL-ER-58-96. Tallahassee: Florida Department of Transportation.
791. Rosell, C., J. Parpal, R. Campeny, S. Jove, A. Pasquina, and J. M. Velasco. 1997. Mitigation of barrier effect on linear infrastructures on wildlife. In *Habitat Fragmentation & Infrastructure,* edited by K. Canters, 367–72. Delft, Netherlands: Ministry of Transport, Public Works and Water Management.
792. Rosell Papes, C., and J. M. Velasco Rivas. 1999. Manual de prevencio i correccio dels impactes de les infraestructures viaries sobre la fauna. Departament de Medi Ambient, Numero 4. Barcelona, Spain: Generalitat de Catalunya.
793. Rosen, P. C., and C. H. Lowe. 1994. Highway mortality of snakes in the Sonoran Desert of southern Arizona. *Biological Conservation* 68:143–48.
794. Rosenberg, N. J., B. L. Blad, and S. B. Verma. 1983. *Microclimate: The Biological Environment.* New York: John Wiley.
795. Rosenberry, D. O., P. A. Bukaveckas, D. C. Buso, G. E. Likens, A. M. Shapiro, and T. C. Winter. 1999. Movement of road salt to a small New Hampshire lake. *Water, Air, and Soil Pollution* 109:179–206.
796. Ross, S. M. 1986. Vegetation change on highway verges in south-east Scotland. *Journal of Biogeography* 13:109–13.
797. Rost, G. R., and J. A. Bailey. 1979. Distribution of mule deer and elk in relation to roads. *Journal of Wildlife Management* 43:634–41.
798. Rothwell, R. L. 1983. Erosion and sediment control at road-stream crossings. *Forestry Chronicle* 59:62–66.
799. Royal Commission on Environmental Pollution. 1994. *Transport and the Environment.* London: Stationery Office.
800. Royal Commission on National Passenger Transportation. 1992. *Directions: The Final Report of the Royal Commission on National Passenger Transportation.* Ottawa: Supply and Services Canada.
801. Rudolph, C., S. Burgdorf, R. Conner, and R. Schaefer. 1999. Preliminary evalu-

ation of the impact of roads and associated vehicular traffic on snake populations in eastern Texas. In *Proceedings of the Third International Conference on Wildlife Ecology and Transportation,* edited by G. L. Evink, P. Garrett, and D. Zeigler, 129–36. Publication FL-ER-73-99. Tallahassee: Florida Department of Transportation.

802. Ruediger, B. 1996. The relationship between rare carnivores and highways. In *Trends in Addressing Transportation Related Wildlife Mortality,* edited by G. L. Evink, P. Garrett, D. Zeigler, and J. Berry, 24–38. Publication FL-ER-58-96. Tallahassee: Florida Department of Transportation.
803. Ruediger, B., J. Claar, and J. Gore. 1999. Restoration of carnivore habitat connectivity in the Northern Rockies. In *Proceedings of the Third International Conference on Wildlife Ecology and Transportation,* edited by G. L. Evink, P. Garrett, and D. Zeigler, 5–20. Publication FL-ER-73-99. Tallahassee: Florida Department of Transportation.
804. Russell, E. W. B. 2001. Applications of historical ecology to land-management decisions in the northeastern United States. In *Applying Ecological Principles to Land Management,* edited by V. H. Dale and R. A. Haeuber, 119–35. New York: Springer.
805. Russell, H. N., and D. Amadon. 1938. A note on highway mortality. *Wilson Bulletin* 50:205–06.
806. Russell, W. H., and C. Jones. 2001. The effects of timber harvesting on the structure and composition of adjacent old-growth coast redwood forest, California, USA. *Landscape Ecology* 16:731–41.
807. Ruthsatz, B., and A. Otte. 1987. Kleinstrukturen im Raum Ingolstadt: Schutz und Zeigerwert. Teil III. Feldwegrander und Ackerraine. *Tuexenia* 7:139–63.
808. Ryden, K. C. 2001. *Landscape with Figures: Nature & Culture in New England.* Iowa City: University of Iowa Press.
810. Safford, H. D., and S. P. Harrison. 2001. Grazing and substrate interact to affect native vs. exotic diversity in roadside grasslands. *Ecological Applications* 11:1112–22.
811. Salvesen, D. 1994. *Wetlands: Mitigating and Regulating Development Impacts.* Washington, D.C.: Urban Land Institute.
812. Samson, F. B., and F. L. Knopf, eds. 1996. *Prairie Conservation: Preserving North America's Most Endangered Ecosystem.* Washington, D.C.: Island Press.
813. Sandova, M. 1979. Indikationseigenschaften der Vegetation am Beisdpiel der Pflanzengesellschaften der Strasse Susice-Modrava (Bohmerwald). *Folia Museu Rer Natura Bohemia Occidentalis Botanica* 13:1–35.
814. Santelmann, M.V., and E. Gorham. 1988. The influence of airborne road dust on the chemistry of *Sphagnum* mosses. *Journal of Ecology* 76:1219–31.
815. Saunders, D. A., and R. J. Hobbs, eds. 1991. *Nature Conservation 2: The Role of Corridors.* Chipping Norton, Australia: Surrey Beatty.
816. Saunders, D. A., G. W. Arnold, A. A. Burbidge, and A. J. M. Hopkins, eds. 1987. *Nature Conservation: The Role of Remnants of Native Vegetation.* Chipping Norton, Australia: Surrey Beatty.
817. Saunders, S. C., M. R. Mislivets, J. Chen, and D. T. Cleland. 2002. Effects of roads on landscape structure within nested ecological units of the Northern Great Lakes Region, USA. *Biological Conservation* (in press).
818. Sawchuk, W. 2001. *Ice Road Technology in the Murphy Oil Chicken Creek*

B-94-B/94-G-6 Natural Gas Drilling Project. Chetwynd, B.C.: A review by the Chetwynd Environmental Society.

819. Schafer, J. A., and S. T. Penland. 1985. Effectiveness of Swareflex reflectors in reducing deer-vehicle accidents. *Journal of Wildlife Management* 49:774–76.
820. Schaffers, A. P., M. C. Vesseur, and K. V. Sykoia. 1998. Effects of delayed hay removal on the nutrient balance of roadside plant communities. *Journal of Applied Ecology* 35:349–64.
821. Schloss, J. A. 2002. GIS watershed mapping: Developing and implementing a watershed natural resources inventory (New Hampshire). In *Handbook of Water Sensitive Planning and Design,* edited by R. L. France. Boca Raton, Fla.: Lewis Publishers.
822. Schmidt, W. 1989. Plant dispersal by motor cars. *Vegetatio* 80:147–52.
823. Schonewald-Cox, C., and M. Buechner. 1992. Park protection and public roads. In *Conservation Biology: The Theory and Practice of Nature Conservation, Preservation and Management,* edited by P. L. Fiedler and S. L. Jain, 373–95. New York: Chapman and Hall.
824. Schroeder, H. W., and T. C. Daniel. 1980. Predicting the scenic quality of forest road corridors. *Environment and Behavior* 12:349–66.
825. Schueler, T. 1995. *Site Planning for Urban Stream Protection.* Ellicot City, Md.: Center for Watershed Protection.
826. Schulze, E. D., O. L. Lange, and R. Oren, eds. 1989. *Forest Decline and Air Pollution: A Study of Spruce* (Picea abies) *on Acid Soils.* New York: Springer-Verlag.
827. Schwartz, M. W., ed. 1997. *Conservation in Highly Fragmented Landscapes.* New York and London: Chapman & Hall.
828. Scott, N. E., and A. W. Davison. 1985. The distribution and ecology of coastal species on roadsides. *Vegetatio* 62:433–40.
830. Scott, T. G. 1938. Wildlife mortality on Iowa highways. *American Midland Naturalist* 20:527–39.
831. Scott, W. S., and N. P. Wylie. 1980. The environmental effects of snow dumping—a literature review. *Journal of Environmental Management* 10:219–40.
832. Scottish Natural Heritage. 1997. *Scotland's Wildlife: Otters.* Edinburgh, Scotland: Scottish Natural Heritage.
833. Seabrook, W. A., and E. B. Dettmann. 1996. Roads as activity corridors for cane toads in Australia. *Journal of Wildlife Management* 60:363–68.
834. Seddon, G. 1997. *Landprints: Reflections on Place and Landscape.* New York: Cambridge University Press.
835. Sedinger, J. S., and A. A. Stickney. 2000. Black brant. In *The Natural History of an Arctic Oil Field: Development and the Biota,* edited by J. C. Truett and S. R. Johnson, 221–32. San Diego, Calif.: Academic Press.
836. Semlitsch, R.D. 2000. Principles for management of aquatic-breeding amphibians. *Journal of Wildlife Management* 64:615–31.
837. Shaver, G. R., J. Canadell, F. S. Chapin III, J. Gurevitch, J. Harte, G. Henry, P. Ineson, S. Jonasson, J. Melillo, L. Pitelka, and L. Rustad. 2000. Global warming and terrestrial ecosystems: A conceptual framework for analysis. *BioScience* 50:871–82.
838. Shaw, D. L. 1988. The design and use of living snowfences in North America. *Agriculture, Ecosystems and Environment* 22/23:351–62. (Reprinted in J. R. Brandle,

D. L. Hintz, and J. W. Sturrock, eds., *Windbreak Technology,* Amsterdam: Elsevier, 1988.)

839. Shen, J. X. 1983. A behavioral study of vibrational sensitivity in the pigeon *(Columba livia). Journal of Comparative Physiology* 152:251–55.
840. Shiffer, M. 1994. *Taking Charge: The Electric Automobile in America.* Washington, D.C.: Smithsonian Institution Press.
841. Shroba, R. R., P. W. Schmidt, E. J. Crosby, and W. R. Hansen. 1979. Storm and flood of July 31–August 1, 1976, in the Big Thompson River and Cache la Poudre River basins, Larimer and Weld Counties, Colorado. In *Part B, Geologic and Geomorphic Effects in the Big Thompson Canyon Area, Larimer County,* 87–148. U.S. Geological Survey Professional Paper 1115. Washington, D.C.: U.S. Geological Survey.
842. Shutes, R. B. E., D. M. Revitt, I. M. Lagerberg, and V. C. E. Barraud. 1999. The design of vegetative constructed wetlands for the treatment of highway runoff. *Science and the Total Environment* 235:189–97.
843. Sidle, R. C., A. J. Pearce, and C. L. O'Loughlin. 1985. *Hillslope Stability and Land Use.* Water Resources Monograph No. 11. Washington, D.C.: American Geophysical Union.
844. Singer, F. J., and J. B. Beattie. 1986. The controlled traffic system and associated wildlife responses in Denali National Park. *Arctic* 39:195–203.
845. Singer, F. J., and J. L. Doherty. 1985. Managing mountain goats at a highway crossing. *Wildlife Society Bulletin* 13:469–77.
846. Singleton, P. H., and J. F. Lehmkuhl. 1999. Assessing wildlife habitat connectivity in the Interstate 90 Snoqualmie Pass corridor, Washington. In *Proceedings of the Third International Conference on Wildlife Ecology and Transportation,* edited by G. L. Evink, P. Garrett, and D. Zeigler, 75–84. Publication FL-ER-73-99. Tallahassee: Florida Department of Transportation.
847. Skidmore, E. L., and L. J. Hagen. 1977. Reducing wind erosion with barriers. *Transactions of the American Society of Agricultural Engineers* 20:911–15.
848. Skole, D., and C. J. Tucker. 1993. Tropical deforestation and habitat fragmentation in the Amazon: Satellite data from 1978 to 1988. *Science* 260:1905–10.
850. Smardon, R. C. 1988. Perception and aesthetics of the urban environment: A review of the role of vegetation. *Landscape and Urban Planning* 15:85–106.
851. Smith, D. 1999. Identification and prioritization of ecological interface zones on state highways in Florida. In *Proceedings of the Third International Conference on Wildlife Ecology and Transportation,* edited by G. L. Evink, P. Garrett, and D. Zeigler, 209–29. Publication FL-ER-73-99. Tallahassee: Florida Department of Transportation.
852. Smith, D. L. 1977. Wildlife considerations in managing highway rights-of-way. *Transportation Research Record* 647:23–25.
853. Smith, D. S., and P. C. Hellmund, eds. 1993. *Ecology of Greenways: Design and Function of Linear Conservation Areas.* Minneapolis: University of Minnesota Press.
854. Smith, J. 1999. Wetlands health assessments in Massachusetts. *Coastlines* (Spring 1999):6.
855. Smith, M. A., M. G. Turner, and D. H. Rusch. 2002. The effect of military training

activity on eastern lupine and the Karner blue butterfly at Fort McCoy, Wisconsin, USA. *Environmental Management* 29:102–15.

856. Smith, R. L. 1996. *Ecology and Field Biology.* New York: HarperCollins.
857. Smith, W. H. 1990. *Air Pollution and Forests: Interaction between Air Contaminants and Forest Ecosystems.* New York: Springer-Verlag.
858. Smith, W. T., and R. D. Cameron. 1985. Reactions of large groups of caribou to a pipeline corridor on the arctic coastal plain of Alaska. *Arctic* 38:53–57.
859. Snow, W. 1959. *The Highway and the Landscape.* New Brunswick, N.J.: Rutgers University Press.
860. Soil Conservation Service. 1975a. *Soil Taxonomy: A Basic System for Making and Interpreting Soil Surveys.* Agriculture Handbook 436. Washington, D.C.: U.S. Department of Agriculture.
861. ———. 1975b. *Urban Hydrology for Small Watersheds.* Technical Release 55. Washington, D.C.: U.S. Department of Agriculture.
862. Spellerberg, I. F. 1998. Ecological effects of roads and traffic: A literature review. *Global Ecology and Biogeography Letters* 7:317–33.
863. Spellerberg, I. F., and M. J. Gaywood. 1993. *Linear Features: Linear Habitats and Wildlife Corridors.* Research Report 63. Peterborough, England: English Nature.
864. Spencer, C. N., and C. L. Schelske. 1998. Impact of timber harvest on sediment deposition in surface waters in northwest Montana over the last 150 years: A paleolimnological study. In *Forest-Fish Conference: Land Management Practices Affecting Aquatic Ecosystems,* edited by M. K. Brewin and D. M. Monita, 187–201. Ottawa: Canadian Forest Service.
865. Sperling, D. 1995. *Future Drive: Electric Vehicles and Sustainable Transportation.* Washington, D.C.: Island Press.
866. Sperling, D. 1998. *New Transportation Fuels: A Strategic Approach to Technological Change.* Berkeley: University of California Press.
867. Sperling, D., and D. Salon. 2002. *Developing Countries and Global Climate Change: Transportation Strategies and Policies.* Arlington, Va.: Pew Center for Climate Change.
868. Spillios, L. C., and R. L. Rothwell. 1998. Freeze-core sampling for sediment intrusion from road stream crossings in Alberta's foothills: A preliminary discussion. In *Forest-Fish Conference: Land Management Practices Affecting Aquatic Ecosystems,* edited by M. K. Brewin and D. M. Monita, 445–50. Ottawa: Canadian Forest Service.
870. Spirn, A. W. 1984. *The Granite Garden: Urban Nature and Human Design.* New York: Basic Books.
871. Stamps, J. A., M. Buechner, and V. V. Krishman. 1987. The effects of edge permeability and habitat geometry on emigration from patches of habitat. *American Naturalist* 129:533–52.
872. Stanford, J. A., and J. V. Ward. 1988. The hyporheic habitat of river ecosystems. *Nature* 335:64–66.
873. Starrett, W. C. 1938. Highway casualties in Central Illinois during 1937. *Wilson Bulletin* 50:193–96.
874. Stedman, S.-M., and J. Hanson. 1997. *Wetlands, Fisheries, and Economics in the New England Coastal States: Habitat Connections.* Vol. 1.3. Boston: National Oceanic and Atmospheric Administration.

875. Steedman, R. J., and R. L. France. 2000. Origin and transport of aeolian sediment from new clearcuts into boreal lakes, northwestern Ontario, Canada. *Water, Air, and Soil Pollution* 122:139–52.
876. Steiner, F. 2002. *Human Ecology.* Washington, D.C.: Island Press.
877. Steinitz, C. 1990. Toward a sustainable landscape with high visual preference and high ecological integrity: The Loop Road in Acadia National Park, U.S.A. *Landscape and Urban Planning* 19:213–50.
878. Stevenson, D. 1991. Is lead's replacement really safer? *Canadian Consumer* 21:4–5.
879. Stiles, E. W. 1980. Patterns of fruit presentation and seed dispersal in bird-disseminated woody plants in the eastern deciduous forest. *American Naturalist* 116:670–88.
880. Stoker, Y. E. 1996. *Effectiveness of a Stormwater Collection and Detention System for Reducing Constituent Loads from Bridge Runoff in Pinellas County, Florida.* U.S.G.S. Open-File Report 96-484. Tampa, Fla.: U.S. Geological Survey.
881. Stoms, D. M. 2000. GAP management status and regional indicators of threats to biodiversity. *Landscape Ecology* 15:21–33.
882. Stoner, D. 1925. The toll of the automobile. *Science* 61:56–57.
883. ———. 1936. Wildlife casualties on the highways. *Wilson Bulletin* 48:276–83.
884. Stottele, T. 1995. *Vegetation und Flora am Strassennetz Westdeutschlands.* Stuttgart, Germany: J. Cramer.
885. Strahler, A. N. 1957. Quantitative analysis of watershed geomorphology. *American Geophysical Union Transactions* 38:913–20.
886. Straker, A. 1998. Management of roads as biolinks and habitat zones in Australia. In *Proceedings of the International Conference on Wildlife Ecology and Transportation,* edited by G. L. Evink, P. Garrett, D. Zeigler, and J. Berry, 181–88. Publication FL-ER-69-98. Tallahassee: Florida Department of Transportation.
888. Strategic Highway Research Program. 1991. *Snow Fence Guide.* Washington, D.C.: National Research Council.
890. Sukopp, H., N. Numata, and A. Huber, eds. 1995. *Urban Ecology as the Basis of Urban Planning.* The Hague, Netherlands: SPB Academic Publishing.
891. Swanson, F. J., and C. T. Dyrness. 1975. Impact of clear-cutting and road construction on soil erosion by landslides in the western Cascade Range, Oregon. *Geology* 3:393–96.
892. Swanson, F. J., S. L. Johnson, S. V. Gregory, and S. A. Acker. 1998. Flood disturbance in a forested mountain landscape. *BioScience* 48:681–89.
893. Swanson, G. A., T. C. Winter, V. A. Adomaitis, and J. W. LaBaugh. 1988. *Chemical Characteristics of Prairie Lakes in South-Central North Dakota, Their Potential for Influencing Use by Fish and Wildlife.* Technical Report 18. Washington, D.C.: U.S. Fish and Wildlife Service.
894. Swap, R., M. Garstang, and S. Greco. 1992. Saharan dust in the Amazon Basin. *Tellus* 44B:133–49.
895. Swift, T. J. 2001. Determinants of optical characteristics of Lake Tahoe, CA-NV. Ph.D. dissertation, University of California, Davis.
896. Swihart, R. K., and N. A. Slade. 1984. Road crossing in *Sigmodon hispidus* and *Microtus ochrogaster. Journal of Mammalogy* 65:357–60.

897. Sykora, K. V., L. J. de Nijs, and T. A. H. M. Pelsma. 1993. *Plantengemeenschappen van Nederlandse wegbermen*. Utrecht, Netherlands: Stichting Uitgeverij Koninklijke Nederlandse Natuurhistorische Vereniging.
898. Taaffe, E. J., and H. L. Gauthier Jr. 1973. *Geography of Transportation*. Englewood Cliffs, N.J.: Prentice-Hall.
899. Taaffe, E. J., R. L. Morrill, and P. R. Gould. 1963. Transport expansion in underdeveloped countries: A comparative analysis. *Geophysical Review* 53:503–29.
900. Tabler, R. D. 1974. New engineering criteria for snow fence systems. *Transportation Research Record* 506:65–84.
901. Tabor, R. 1974. Earthworms, crows, vibrations and motorways. *New Scientist* 62:482–83.
902. Tamm, C. O., and T. Troedsson. 1955. An example of the amounts of plant nutrients supplied to the ground in road dust. *Oikos* 6:61–70.
903. Tanghe, M. 1986. Approche floristique et phytosociologique des espaces verts autoroutiers de la moyenne Belgique (Brabant–Hainaut). *Bulletin de la Société Royale de Botanique de Belgique* 119:22–34.
904. Taskula, T. 1997. The moose ahead. *Traffic Technology International* 42:170–73.
905. Taylor, P. D., L. Fahrig, K. Henein, and G. Merriam. 1993. Connectivity is a vital element of landscape structure. *Oikos* 68:571–73.
906. Ter Haar, G.L., M. E. Griffing, M. Brandt, D. G. Oberding, and M. Kapron. 1975. Methylcyclopentadienyl manganese tricarbonyl as an antiknock: Composition and fate of manganese exhaust products. *Journal of Air Pollution Control Association* 25:858–60.
907. Terrene Institute. 1994. *Urbanization and Water Quality: A Guide to Protecting the Urban Environment*. Washington, D.C.: Terrene Institute.
908. Thiel, R. P. 1985. Relationship between road densities and wolf habitat suitability in Wisconsin. *American Midland Naturalist* 113:404–07.
910. Thompson, J. R., A. J. Rutter, and P. S. Ridout. 1986. The salinity of motorway soils. 2. Distance from the carriageway and other local sources of variation in salinity. *Journal of Applied Ecology* 23:269–80.
911. Thompson, K., and J. P. Grime. 1979. Seasonal variation in the seed banks of herbaceous species in ten contrasting habitats. *Journal of Ecology* 67:893–921.
912. Thoreau, H. D. 1849. *A Week on the Concord and Merrimack Rivers*. Sentry Edition. Boston: Houghton Mifflin.
913. Thurber, J. M., R. O. Peterson, T. D. Drummer, and S. A. Thomasma. 1994. Gray wolf response to refuge boundaries and roads in Alaska. *Wildlife Society Bulletin* 22:61–68.
914. Tibke, G. 1988. Basic principles of wind erosion control. *Agriculture, Ecosystems and Environment* 22/23:103–22. (Reprinted in J. R. Brandle, D. L. Hintz, and J. W. Sturrock, eds., *Windbreak Technology*, Amsterdam: Elsevier, 1988.)
915. Tiemann, K. H. 1971. Die Auswirkungen des Strassenverkehrs auf Boden, Pflanzen und Wasser. *Wasserbau TU Hannover* 21:157–225.
916. Tikka, P. M., H. Hogmander, and P. S. Koski. 2001. Road and railway verges serve as dispersal corridors for grassland plants. *Landscape Ecology* 16:659–66.
917. Tilman, D., R. M. May, C. L. Lehman, and M. A. Nowak. 1994. Habitat destruction and the extinction debt. *Nature* 371:65–66.

918. Tinker, D. B., C. A. C. Resor, G. P. Beauvais, K. F. Kipfmueller, C. I. Fernandes, and W. L. Baker. 1998. Watershed analysis of forest fragmentation by clearcuts and roads in a Wyoming forest. *Landscape Ecology* 13:149–65.
919. Tischendorf, L., and L. Fahrig. 2000a. On the usage and measurement of landscape connectivity. *Oikos* 90:7–19.
920. ———. 2000b. How should we measure landscape connectivity? *Landscape Ecology* 15:633–41.
921. Tivy, J. 1990. *Agricultural Ecology.* Harlow, Essex, England: Longman.
922. Tong, S. T. Y. 1990. Roadside dusts and soils contamination in Cincinnati, Ohio, USA. *Environmental Management* 14:107–14.
923. Townsend, C. R., J. L. Harper, and M. Begon. 2000. *Essentials of Ecology.* Malden, Mass.: Blackwell Science.
924. Transport Association of Canada. 1995. *Transportation in Canada: A Statistical Overview.* Ottawa: Transport Association of Canada.
925. Transportation Association of Canada. 2000. *Transportation in Canada 2000.* Ottawa: Transportation Association of Canada.
926. Transportation Research Board. 1991. *Highway Deicing: Comparing Salt and Calcium Magnesium Acetate.* Special Report 235. Washington, D.C.: National Research Council.
927. Transportation Research Board. 1993. *Stormwater Management for Transportation Facilities.* NCHRP Synthesis 174. Washington, D.C.: National Research Council.
928. Treweek, J., and N. Veitch. 1996. The potential application of GIS and remotely sensed data to the ecological assessment of proposed new road schemes. *Global Ecology and Biogeography Letters* 5:249–57.
930. Trombulak, S. C., and C. A. Frissell. 2000. Review of ecological effects of roads on terrestrial and aquatic communities. *Conservation Biology* 14:18–30.
931. Truett, J. C., and S. R. Johnson, eds. 2000. *The Natural History of an Arctic Oil Field: Development and the Biota.* San Diego, Calif.: Academic Press.
932. Turner, A. K., and R. L. Schuster, eds. 1996. *Landslides: Investigation and Mitigation.* Transportation Research Board Special Report 247. Washington, D.C.: National Academy Press.
934. Turner, M. G. 1989. Landscape ecology: The effect of pattern on process. *Annual Review of Ecology and Systematics* 20:171–97.
935. Turner, M. G., R. H. Gardner, and R. V. O'Neill. 2001. *Landscape Ecology in Theory and Practice: Pattern and Process.* New York: Springer-Verlag.
936. Turtle, S. L. 2000. Embryonic survivorship of the spotted salamander *(Ambystoma maculatum)* in roadside and woodland vernal pools in southeastern New Hampshire. *Journal of Herpetology* 34:60–67.
937. Tyser, J. W., and C. A. Worley. 1992. Alien flora in grasslands adjacent to road and trail corridors in Glacier National Park, Montana (USA). *Conservation Biology* 6:253–62.
938. Ujvari, M., H. J. Baagoe, and A. B. Madsen. 1998. Effectiveness of wildlife warning reflectors in reducing deer-vehicle collisions: A behavioral study. *Journal of Wildlife Management* 62:1094–99.
939. Ulanowicz, R. E. 1997. *Ecology, the Ascendent Perspective.* New York: Columbia University Press.

940. Ullmann, I., and B. Heindl. 1989. Geographical and ecological differentiation of roadside vegetation in temperate Europe. *Botanica Acta* 102:261–340.
941. Ullmann, I., P. Bannister, and J. B. Wilson. 1995. The vegetation of roadside verges with respect to environmental gradients in southern New Zealand. *Journal of Vegetation Science* 6:131–42.
942. ———. 1998. Lateral differentiation and the role of exotic species in roadside vegetation in southern New Zealand. *Flora* 193:149–64.
943. Ullmann, I., B. Heindl, and B. Schug. 1990. Naturraumliche Gliederung der Vegetation auf Strassenbegleit-flachen im westlichen Unterfranken. *Tuexenia* 10:197–222.
944. Underhill, J. E., and P. G. Angold. 2000. Effects of roads on wildlife in an intensively modified landscape. *Environmental Review* 8:21–39.
945. USDA Forest Service. 1999. *Roads Analysis: Informing Decisions about Managing the National Forest Transportation System.* Miscellaneous Report FS-643. Washington, D.C.: U.S. Department of Agriculture.
946. U.S. Department of Transportation. 1999. *Transportation Equity Act for the 21st Century.* Washington, D.C.: Federal Highway Administration.
947. U.S. Department of Transportation and Environmental Protection Agency Joint Report. 1993. *Clean Air through Transportation: Challenges in Meeting National Air Quality Standards.* Washington, D.C.
948. U.S. Fish and Wildlife Service. 1981. *Standards for the Development of Suitability Index Models.* Ecological Services Manual 103. Washington, D.C.: U.S. Department of Interior.
950. ———. 2001. *Karner Blue Butterfly* (Lycaeides melissa samuelis) *Recovery Plan.* Technical Agency Draft. Fort Snelling, Minn.: U.S. Fish and Wildlife Service.
951. van Bohemen, H., C. Padmos, and H. de Vries. 1994. Versnippering—ontsnippering: Beleid en onderzoek bij verkeer en waterstaat. *Landschap* 1994, no. 3: 15–25.
952. van Bohemen, H. D. 1996. Mitigation and compensation of habitat fragmentation caused by roads: Strategy, objectives and practical measures. *Transportation Research Record* 1475:133–37.
953. ———. 1998. Habitat fragmentation, infrastructure and ecological engineering. *Ecological Engineering* 11:199–207.
954. van der Sluijs, J., and P. J. M. Melman. 1991. Layout and management of planted road and canal verges. In *Nature Engineering and Civil Engineering Works,* edited by P. Aanen, W. Alberts, G. J. Bekker, et al., 79–85. Wageningen, Netherlands: Pudoc.
955. van der Sluijs, J., and H. D. van Bohemen. 1991. Green elements of civil engineering works and their (potential) ecological importance. In *Nature Engineering and Civil Engineering Works,* edited by P. Aanen, W. Alberts, G. J. Bekker, et al., 21–32. Wageningen, Netherlands: Pudoc.
956. van der Zande, A. N., W. J. Ter Keurs, and W. J. van der Weijden. 1980. The impact of roads on the densities of four bird species in an open field habitat: Evidence of a long-distance effect. *Biological Conservation* 18:299–321.
957. VanderZanden, M. J., G. Cabana, and J. B. Rasmussen. 1999. Stable isotope evi-

dence for food web consequences of species invasions in lakes. *Nature* 401:464–67.

958. van der Zee, F. F., J. Wiertz, C. J. F. Ter Braak, and R. C. Apeldoorn. 1992. Landscape change as a possible cause of the badger *Meles meles* L. decline in The Netherlands. *Biological Conservation* 61:17–22.
959. van Dorp, D., P. Schippers, and J. M. van Groenendael. 1997. Migration rates of grassland plants along corridors in grassland landscapes assessed with a cellular automation model. *Landscape Ecology* 12:39–50.
960. van Dyke, F. B., R. H. Brocke, H. G. Shaw, B. B. Ackerman, T. P. Hemker, and F. G. Lindzey. 1986. Reactions of mountain lions to logging and human activity. *Journal of Wildlife Management* 50:95–102.
961. van Eimern, J., R. Karschon, L. A. Razumova, and G. W. Robertson. 1964. *Windbreaks and Shelterbelts.* Technical Note 59. Geneva: World Meteorological Organization.
962. Vankat, J. L., and D. G. Roy. 2002. Landscape invasibility by exotic species. In *Applying Landscape Ecology in Biological Conservation,* edited by K. J. Gutzwiller, 170–91. New York: Springer.
963. Vannote, R. L., G. W. Minshall, K. W. Cummins, J. R. Sedell, and C. E. Cushing. 1980. The river continuum concept. *Canadian Journal of Fisheries and Aquatic Science* 37:130–37.
964. Vassant, J., S. Brandt, and J. M. Jullien. 1993a. Influence du passage de l'autoroute A5 sur les populations cert et sanglier du Massif d'Arc-en-Banois: 1èr partie. *Bulletin de l'Office National de la Chasse* 183:15–25.
965. ———. 1993b. Influence du passage de l'autoroute A5 sur les populations cert et sanglier du Massif d'Arc-en-Banois: 2ème partie. *Bulletin de l'Office National de la Chasse* 184: 24–33.
966. Vavrek, M. C., N. Fetcher, J. B. McGraw, G. R. Shaver, F. S. Chapin III, and B. Bovard. 1999. Recovery of productivity and species diversity in tussock tundra following disturbance. *Arctic, Antarctic, and Alpine Research* 31:254–58.
967. Veen, J. 1973. De verstoring van weidevogelpopulaties. *Stedeb. en Volkshuisv* 53:16–26.
968. Veenbaas, G., and G. J. Brandjes. 1999. The use of fauna passages along waterways under motorways. In *Key Concepts in Landscape Ecology,* edited by J. W. Dover and R. G. H. Bunce, 315–20. Preston, England: International Association for Landscape Ecology.
970. Venier, L., and L. Fahrig. 1996. Habitat availability causes the species abundance-distribution relationship. *Oikos* 76:564–70.
971. Verkaar, H. J., and G. J. Bekker. 1991. The significance of migration to the ecological quality of civil engineering works and their surroundings. In *Nature Engineering and Civil Engineering Works,* edited by P. Aanen, W. Alberts, G. J. Bekker, et al., 44–61. Wageningen, Netherlands: Pudoc.
972. Verkaar, H. J., P. Aanen, and C. F. van de Watering. 1991. Theoretical background to the application of nature engineering knowledge. In *Nature Engineering and Civil Engineering Works,* edited by P. Aanen, W. Alberts, G. J. Bekker, et al., 33–43. Wageningen, Netherlands: Pudoc.
973. Vermeulen, H. J. W. 1994. Corridor function of a road verge for dispersal of

stenotopic heathland ground beetles (Carabidae). *Biological Conservation* 69:339–49.

974. Vermeulen, H. J. W., and P. F. M. Opdam. 1995. Effectiveness of roadside verges as dispersal corridors for small ground-dwelling animals: A simulation study. *Landscape and Urban Planning* 31:233–48.
975. Vermuelen, J., and T. Whitten, eds. 1999. *Conservation of Biodiversity and Cultural Property in the Exploitation of Limestone: Lessons from East Asia.* World Bank Technical Paper (prepublication draft). Washington, D.C.: World Bank.
976. Vigier, F. 1987. Housing in Tunis. Cambridge, Mass.: Graduate School of Design, Harvard University.
977. Vileisis, A. 1997. *Discovering the Unknown Landscape: A History of America's Wetlands.* Washington, D.C.: Island Press.
978. Vitousek, P. M., and R. W. Howarth. 1991. Nitrogen limitation on land and in the sea: How can it occur? *Biogeochemistry* 13:87–115.
979. Vitousek, P. M., J. D. Aber, R. W. Howarth, G. E. Likens, P. A. Matson, D. W. Schindler, W. H. Schlesinger, and D. G. Tilman. 1997. Human alteration of the global nitrogen cycle: Sources and consequences. *Ecological Applications* 7:737–50.
980. Voorhees, A. M. 1956. *A General Theory of Traffic Flow.* 1955 Proceedings. New Haven, Conn.: Institute of Traffic Engineers.
981. Vos, C. C. 1997. Effects of road density: A case study of the moor frog. In *Habitat Fragmentation & Infrastructure,* edited by K. Canters, 93–97. Delft, Netherlands: Ministry of Transport, Public Works and Water Management.
982. Vos, C. C., and J. P. Chardon. 1998. Effects of habitat fragmentation and road density on the distribution pattern of the moor frog *Rana arvalis. Journal of Applied Ecology* 35:44–56.
983. Vos, C. C., H. Baveco, and C. J. Grashof-Bokdam. 2002. Corridors and species dispersal. In *Applying Landscape Ecology in Biological Conservation,* edited by K. J. Gutzwiller, 84–104. New York: Springer.
984. Wace, N. 1977. Assessment of dispersal of plant species: The car-borne flora in Canberra. *Proceedings of the Ecological Society of Australia* 10:167–86.
985. ———. 2000. The botany of the motor car. In *The Best of the Science Show,* edited by R. Williams, 70–79. Sydney, Australia: Nelson.
986. Wade, K. J., J. T. Flanagan, A. Currie, and D. J. Curtis. 1980. Roadside gradients of lead and zinc concentrations in surface-dwelling invertebrates. *Environmental Pollution* (Series B) 1:87–93.
987. Wales, B. A. 1972. Vegetation analysis of northern and southern edges in a mature oak-hickory forest. *Ecological Monographs* 42:451–71.
988. Walker, D. A. 1996. Disturbance and recovery of arctic Alaskan vegetation. In *Landscape Function and Disturbance in Arctic Tundra,* edited by J. F. Reynolds and J. D. Tenhunen, 35–71. New York and Berlin: Springer-Verlag.
990. Walker, D. A., and K. R. Everett. 1987. Road dust and its environmental impact on Alaskan taiga and tundra. *Arctic and Alpine Research* 19:479–89.
991. Walker, D. A., and M. D. Walker. 1991. History and pattern of disturbance in Alaskan arctic terrestrial ecosystems: A hierarchical approach to analyzing landscape change. *Journal of Applied Ecology* 28:244–76.
992. Walker, D. A., D. Cate, J. Brown, and C. Racine. 1987a. *Disturbance and Recovery*

of Arctic Alaskan Tundra Terrain: A Review of Recent Investigations. Report 87-11. Hanover, N.H.: Cold Regions Research and Engineering Laboratory.

993. Walker, D. A., P. J. Webber, E. F. Binnian, K. R. Everett, N. D. Lederer, E. A. Norstrand, and M. D. Walker. 1987b. Cumulative impacts of oil fields on northern Alaskan landscapes. *Science* 338:757–61.
994. Walling, E. 1985. *Country Roads: The Australian Roadside.* Lilydale, Victoria, Australia: Pioneer Design Studio.
995. Wallis, M. 2001. *Route 66: The Mother Road.* New York: St. Martin's Griffin.
996. Walter, K. 2000. Ecosystem effects of the invasion of Eurasian watermilfoil *(Myriophyllum spicatum)* at Lake Tahoe. Master's thesis, University of California, Davis.
997. Ward, A. L. 1982. Mule deer behavior in relation to fencing and underpasses on Interstate 80 in Wyoming. *Transportation Research Record* 859:8–13.
998. Ward, J. V. 1989. The four dimensional nature of lotic ecosystems. *Journal of the North American Benthological Society* 8:2–8.
999. Waring, G. H., J. L. Griffis, and M. E. Vaughn. 1991. White-tailed deer roadside behavior, wildlife warning reflectors, and highway mortality. *Applied Animal Behavior Science* 29:215–23.
1000. Warner, R. E., G. B. Joselyn, and S. L. Etter. 1987. Factors affecting roadside nesting by pheasants in Illinois. *Wildlife Society Bulletin* 15:221–28.
1001. Warren, E. R. 1936. Casualties among animals on mountain roads. *Science* 83:14.
1002. Warren, M. L., and M. G. Pardew. 1998. Road crossings as barriers to small-stream fish movement. *Transactions of the American Fisheries Society* 127:637–44.
1003. Washington Forest Practices Board. 1995. *Standard Methodology for Conducting Watershed Analysis under Chapter 222-22 WAC, Version 3.0.* Olympia, Wash.: Department of Natural Resources, Forest Practices Division.
1004. Wasser, S. K., K. Bevis, G. King, and E. Hanson. 1997. Noninvasive physiological measures of disturbance in the northern spotted owl. *Conservation Biology* 11:1019–22.
1005. Waters, D. 1988. *Monitoring Program Mitigative Measures: Trans-Canada Highway Twinning.* Final Report to Parks Canada. Alberta, Canada: Banff National Park.
1006. Watson, R. T., I. R. Noble, B. Bolin, N. H. Ravindranath, D. J. Verardo, and D. J. Dokken. 2000. *Land Use, Land-Use Change, and Forestry.* Special Report of the Intergovernmental Panel on Climate Change (IPCC). New York and Cambridge: Cambridge University Press.
1007. Watts, M. T. 1975. *Reading the Landscape of America.* New York: Macmillan.
1008. Way, J. M. 1977. Roadside verges and conservation in Britain: A review. *Biological Conservation* 12:65–74.
1009. Wayne, R. K., S. B. George, D. Gilbert, P. W. Collins, S. D. Kovach, D. Girman, and N. Lehman. 1991. A morphologic and genetic study of the island fox *Urocyon littoralis. Evolution* 45:1849–68.
1010. Weathers, K. C., M. L. Cadenasso, and S. T. A. Pickett. 2001. Forest edges as nutrient and pollutant concentrators: Potential synergisms between fragmentation, forest canopies, and the atmosphere. *Conservation Biology* 15:1506–14.
1011. Weaver, T., and D. Dale. 1978. Trampling effects of hikers, motorcycles and horses in meadows and forest. *Journal of Applied Ecology* 15:451–57.

1012. Webb, R. H. 1982. Off-road motorcycle effects on a desert soil. *Environmental Conservation* 9:197–208.

1013. Webb, R. H., and H. G. Wilshire, eds. 1983. *Environmental Effects of Off-Road Vehicles: Impact and Management in Arid Regions.* New York: Springer-Verlag.

1014. Webber, M. 1992. The joys of automobility. In *The Car and the City: The Automobile, the Built Environment and Daily Life,* edited by M. Wachs and M. Crawford, 274–84. Ann Arbor: University of Michigan Press.

1015. Webster's Third New International Dictionary: The English Language Unabridged. 1986. Springfield, Mass.: Merriam-Webster.

1016. Weiner, E. 1997. *Urban Transportation Planning in the United States: An Historical Overview.* Publication DOT-T-97-20. Washington, D.C.: U.S. Department of Transportation.

1017. Wemple, B. C., J. A. Jones, and G. E. Grant. 1996. Channel network extension by logging roads in two basins, western Cascades, Oregon. *Water Resources Bulletin* 32:1195–1207.

1018. Wemple, B. C., F. J. Swanson, and J. A. Jones. 2001. Forest roads and geomorphic process interactions, Cascade Range, Oregon. *Earth Surface Processes and Landforms* 26: 191–204.

1019. West, G. B., J. H. Brown, and B. J. Enquist. 1997. A general model for the origin of allometric scaling laws in biology. *Science* 276:122–26.

1020. Wester, L., and J. O. Juvik. 1983. Roadside plant communities on Mauna Loa, Hawaii. *Journal of Biogeography* 10:307–16.

1021. Westhoff, V., P. A. Bakker, C. G. van Leeuwen, and E. E. van der Maarel. 1970. *Wilde planten: Flora en vegetatie in onze natuurgebieden.* Amsterdam: Vereniging tot Behoud van Natuurmonumenten.

1022. Wetzel, R. G., and G. E. Likens. 1991. *Limnological Analysis.* Philadelphia: W. B. Saunders.

1023. White, F. B. 1927. Birds and motor cars. *Auk* 44:265–66.

1024. ———. 1929. Birds and motor cars. *Auk* 50:236.

1025. Whitney, G. G., and W. C. Davis. 1986. From primitive woods to cultivated woodlots: Thoreau and the forest history of Concord, Massachusetts. *Journal of Forest History* 30:70–81.

1026. Whittaker, R. H. 1975. *Communities and Ecosystems.* New York: Macmillan.

1027. Whitten, K., and R. D. Cameron. 1983. Movements of collared caribou, *Rangifer tarandus,* in relation to petroleum development on the Arctic Slope of Alaska. *Canadian Field-Naturalist* 97:143–46.

1028. Whyte, W. H. 1980. *The Social Life of Small Urban Spaces.* Washington, D.C.: Conservation Foundation.

1030. Wiche, G. J., R. M. Lent, W. F. Rannie, and A. V. Vecchia. 1997. A history of lake-level fluctuations for Devils Lake, North Dakota, since the early 1800s. In *Proceedings of the 89th Annual Meeting,* 34–39. Grand Forks: North Dakota Academy of Science.

1031. Wiche, G. J., A. V. Vecchia, L. Osborne, C. M. Wood, and J. T. Fay. 2000. *Climatology, Hydrology, and Simulation of an Emergency Outlet, Devils Lake Basin, North Dakota.* U.S.G.S. Water Resources Investigations Report 00-4174. Bismarck, N.D.: U.S. Geological Survey.

1032. Wiens, J. A. 1996. Wildlife in patchy environments: Metapopulations, mosaics, and management. In *Metapopulations and Wildlife Conservation,* edited by D. R. McCullough, 53–84. Washington, D.C.: Island Press.

1033. ———. 2002. Central concepts and issues of landscape ecology. In *Applying Landscape Ecology in Biological Conservation,* edited by K. J. Gutzwiller, 3–21. New York: Springer.

1034. Wiggins, G. B., R. J. Mackay, and I. M. Smith. 1980. Evolutionary and ecological strategies of animals in annual temporal pools. *Archives fur Hydrobiologia Supplement* 58:97–206.

1035. Wilcox, B. A., and D. D. Murphy. 1985. Conservation strategy: The effects of fragmentation on extinction. *American Naturalist* 125:879–87.

1036. Wilcox, D. A. 1989. Migration and control of purple loosestrife (*Lythrum salicaria* L.) along highway corridors. *Environmental Management* 13:365–70.

1037. Wilder, T. 1995. *Fort McCoy Karner Blue Butterfly Conservation Plan.* Fort McCoy, Wis.: Directorate of Public Works, Environmental and Natural Resources Division.

1038. Wilkins, K. T. 1982. Highways as barriers to rodent dispersal. *Southwestern Naturalist* 27:459–60.

1039. Williams, P. J. 1995. Permafrost and climate change: Geotechnical considerations. *Philosophical Transactions: Physical Sciences and Engineering* 3:56–60.

1040. Williamson, P., and P. R. Evans. 1972. Lead: Levels in roadside invertebrates and small mammals. *Bulletin of Environmental Contamination and Toxicology* 8:280–88.

1041. Wilson, E. O. 1984. *Biophilia.* Cambridge: Harvard University Press.

1042. ———. 1996. *The Diversity of Life.* Cambridge: Belknap Press of Harvard University Press.

1043. Wilson, J. B., G. L. Rapson, M. T. Sykes, A. J. Watkins, and P. A. Williams. 1992. Distributions and climatic correlations of some exotic species along roadsides in South Island, New Zealand. *Journal of Biogeography* 19:183–94.

1045. Winner, W. E. 1994. Mechanistic analysis of plant responses to air pollution. *Ecological Applications* 4:651–61.

1046. Winter, T. C. 2001. The concept of hydrologic landscapes. *Journal of the American Water Resources Association* 37:335–49.

1047. Winter, T. C., J. W. Harvey, O. L. Franke, and W. M. Alley. 1998. *Ground Water and Surface Water: A Single Resource.* Circular 1139. Reston, Va.: U.S. Geological Survey.

1048. Wisconsin Department of Transportation. 1995. FHWA/NEPA Project Development Process. Unpublished report. Madison: Wisconsin Department of Transportation.

1050. Wisdom, M. J., R. S. Holthausen, and B. K. Wales. 2000. *Source Habitats for Terrestrial Vertebrates of Focus in the Interior Columbia Basin: Broad-Scale Trends and Management Implications.* General Technical Report PNW GTR-485. Portland, Ore.: USDA Forest Service.

1051. With, K.A., and A.W. King. 1999. Extinction thresholds for species in fractal landscapes. *Conservation Biology* 13:314–26.

1052. Wohlgemuth, N. 1997. World transport energy demand modeling: Methodology and elasticities. *Energy Policy* 25:1109–19.

1053. Wolska, L., W. Wardencki, M. Wiergowski, et al. 1999. Evaluation of pollution degree of the Odra river basin with organic compounds after the 1997 summer flood: General comments. *Acta Hydrochimica et Hydrobiologica* 27:343–49.

1054. Woods, J. G. 1990. *Effectiveness of Fences and Underpasses on the Trans-Canada Highway and Their Impact on Ungulate Populations.* Report to Banff National Park Warden Service. Banff, Alberta.
1055. Woodward, S. M. 1990. Population density and home range characteristics of woodchucks, *Marmota monax,* at expressway interchanges. *Canadian Field-Naturalist* 104:421–28.
1056. World Commission on Environment and Development. 1987. *Our Common Future.* New York: Oxford University Press.
1057. World Health Organization. 1981. Manganese. In *Environmental Health Criteria* 17. Geneva: World Health Organization.
1058. Wyatt, R. 1996. More on the southward spread of common milkweed, *Asclepias syriaca* L. *Bulletin of the Torrey Botanical Club* 123:68–69.
1059. Wyatt, R., A. Stoneburner, S. B. Broyles, and J. R. Allison. 1993. Range extension southward in common milkweed, *Asclepias syriaca* L. *Bulletin of the Torrey Botanical Club* 120:177–79.
1060. Yanes, M., J. Velasco, and F. Suárez. 1995. Permeability of roads and railways to vertebrates: The importance of culverts. *Biological Conservation* 71:217–22.
1061. Yost, A. C., and R. G. Wright. 2001. Moose, caribou, and grizzly bear distribution in relation to road traffic in Denali National Park, Alaska. *Arctic* 54:41–48.
1062. Young, A., and N. Mitchell. 1994. Microclimate and vegetation edge effects in a fragmented podocarp-broadleaf forest in New Zealand. *Biological Conservation* 67:63–72.
1063. Young, G. K., S. Stein, P. Cole, T. Kammer, F. Graziano, and F. Bank. 1996. *Evaluation and Management of Highway Runoff Water Quality.* Publication FHWA-PD-96-032. Washington, D.C.: Federal Highway Administration, U.S. Department of Transportation.
1064. Young, W. C. 1968. Ecology of roadside treatment: To successfully treat a roadside with vegetation, one need only follow basic ecological principles and simulate nature's way of revegetating a disturbed site. *Journal of Soil and Water Conservation* 23:47–50.
1065. Zachar, D. 1982. *Soil Erosion.* Amsterdam: Elsevier.
1066. Zacks, J. L. 1986. Do white-tailed deer see red? Premise underlying the design of Swareflex wildlife reflectors. *Transportation Research Record* 1075:35–43.
1067. Zembrzuski, T. J., Jr., and M. L. Evans. 1989. *Flood of April 4–5, 1987, in Southeastern New York State, with Flood Profiles of Schoharie Creek.* U.S.G.S. Water Resources Investigations Report 89-4084. Albany, N.Y.: U.S. Geological Survey.
1068. Ziegler, A. D., R. A. Sutherland, and T. W. Giambelluca. 2001. Interstorm surface preparation and sediment detachment by vehicle traffic on unpaved mountain roads. *Earth Surface Processes and Landforms* 26:235–50.
1070. Zielinski, J. A. 2002. Open spaces and impervious surfaces: Model development principles and benefits. In *Handbook of Water Sensitive Planning and Design,* edited by R. L. France, 49–64. Boca Raton, Fla.: Lewis Publishers.
1071. Zimmerman, R. 2002. Global warming, infrastructure, and land use in the metropolitan New York area: Prevention and response. In *Global Climate Change and*

Transportation: Coming to Terms, 55–64. Washington, D.C.: Eno Transportation Foundation.

1072. Zimmerman, R., and M. Cusker. 2001. Institutional decision-making. In *Climate Change and a Global City: The Potential Consequences of Climate Variability and Change—Metro East Coast,* chapter 9. Report for the U.S. Global Change Research Program, National Assessment of the Potential Consequences of Climate Variability and Change for the United States. New York: Columbia Earth Institute.
1073. Zobel, D. B., L. F. Roth, and G. M. Hawk. 1985. *Ecology, Pathology and Management of Port-Orford Cedar* (Chamaecyparis lawsoniana). General Technical Report PNW-TER-184. Portland, Ore.: USDA Forest Service.
1074. Zonneveld, I. S., and R. T. T. Forman, eds. 1990. *Changing Landscapes: An Ecological Perspective.* New York: Springer-Verlag.
1075. Zug, L. S. 1997. Habitat, water quality, and wetland preservation: U.S. Route 220 (I-99) replacement wetlands, Blair County, Pennsylvania. *Wetland Journal* 9(4): 3–7.
1076. Zwaenepoel, A. 1996. Wegbermen in Vlaanderen: Een refugium voor botanische zeldzaamheden. *Dumortiera* 64/65:25–35.
1077. Zwaenepoel, A. 1997. Floristic impoverishment by changing unimproved roads into metalled roads. In *Habitat Fragmentation & Infrastructure,* edited by K. Canters, 127–137. Delft, Netherlands: Ministry of Transport, Public Works and Water Management.
1078. ———. 1998. *Werk aan de berm! Handboek botanishch bermbeheer.* Antwerp, Belgium: Stichting Leefmilieu.

About the Authors

RICHARD T. T. FORMAN is the PAES Professor of Landscape Ecology at Harvard University, where he teaches ecological courses in the Graduate School of Design and in Harvard College. His research and writing include landscape and regional ecology, road ecology, land-use planning and conservation, and spatially meshing nature and people in the land mosaic. Forman served on two National Academy of Sciences committees on surface transportation and the environment and began publishing road ecology articles in 1996. His books include *Land Mosaics: The Ecology of Landscapes and Regions* (Cambridge University Press, 1995) and *Landscape Ecology Principles for Landscape Architecture and Land-Use Planning* (Island Press, 1996). He is a fellow of the AAAS; served as vice president of the Ecological Society of America and the International Association for Landscape Ecology; has received medals and honors from Italy, Australia, France, the Czech Republic, China, and the United Kingdom; was named Distinguished Landscape Ecologist (USA); and received the Lindback Foundation Award for Excellence in Teaching. He received a Haverford College B.S., a University of Pennsylvania Ph.D., and honorary degrees from Miami University, Harvard University, and Florida International University, and has taught at the University of Wisconsin, Rutgers University, and in Central and South America. Address: Harvard University, Harvard Design School, Cambridge, Massachusetts, 02138, USA.

DANIEL SPERLING is professor of civil engineering and environmental science and policy and founding director of the Institute of Transportation Studies (ITS–Davis) at the University of California, Davis. Dr. Sperling is recognized as a leading international expert on transportation technology assessment, energy and environmental aspects of transportation, and transportation policy. He has authored or coauthored over 140 technical papers and six books and is associate editor of *Transportation Research D (Environment)*, a current or recent editorial board member of four other scholarly journals, a recent member of seven U.S. National Academy of Sciences committees on various transport and environment topics, and founding chair of the Alternative Fuels Committee of the U.S. Transportation Research Board. Dr. Sperling has testified numerous times before the U.S. Congress and various government agencies and is the recipient of the 1993 Gilbert F. White Fellowship from Resources for the Future, the 1996 Distinguished Public Service Award from the University of California, Davis, the 1997 Clean Air Award from the American Lung Association of Sacramento, and the 2002 Carl Moyer Award for Scientific Achievement from the Clean Air Coalition. Address: Institute of Transportation Studies, One Shields Avenue, University of California, Davis, California, 95616, USA.

JOHN A. BISSONETTE is a research scientist with the U.S. Geological Survey. He leads the Utah Cooperative Fish and Wildlife Research Unit and is a professor in the College of Natural Resources at Utah State University. His research interests include landscape effects on wildlife species. He is interested in the conceptual foundation for landscape ecology and how it might be used in real-life applications. Bissonette has published three other volumes: *Integrating People and Wildlife for a Sustainable Future* (The Wildlife Society, 1995), *Wildlife and Landscape Ecology: Effects of Pattern and Scale* (Springer, 1997), and *Landscape Ecology and Resource Management: Linking Theory with Practice* (Island Press, 2002). He has been invited to present keynote addresses in Australia, Germany, and Portugal, and was a Senior Fulbright Scholar at the Technic University of Munich in 2002. When not working or traveling, he rides his horse in the mountains of Utah. Address: USGS Utah Cooperative Fish and Wildlife Research Unit, College of Natural Resources, Utah State University, Logan, Utah, 84322, USA.

ANTHONY P. CLEVENGER is a wildlife research ecologist currently contracted by Parks Canada to study road effects on wildlife populations in the Banff-Bow Valley and the surrounding national and provincial parks. In that capacity, his research has focused primarily on the factors influencing mammal passage through drainage culverts and wildlife crossing structures, developing GIS-based modeling approaches to identify mitigation placement along roads, and investigating spatial patterns and factors influencing wildlife road mortality. He has worked as a wildlife biologist for the World Wide Fund for Nature–International (Gland, Switzerland), Ministry of Environment–France (Toulouse), U.S. Forest Service, and U.S. National Park Service. Dr. Clevenger is a graduate of the University of California, Berkeley, has a master's degree in wildlife ecology from the University of Tennessee, and earned a doctoral degree in zoology from the University of León, Spain. He has been an adjunct assistant professor at the University of Tennessee since 1989 and at the University of

Calgary since 1998. Address: 3-625 Fourth Street, Canmore, Alberta T1W 267, Canada.

CAROL D. CUTSHALL is director of the Bureau of Environment at the Wisconsin Department of Transportation. She has responsibility for approving environmental impact statements prepared by the department. Her staff provides technical assistance in the areas of history, archaeology, air and noise analysis, endangered species, water quality, wetlands, and land-use and socioeconomic factors. The department's cleanup program for petroleum-contaminated sites and its wetland mitigation banking program are also located in her office. Cutshall has overall responsibility for developing the department's environmental policies and administrative rules. She is very active in the American Association of State Highway and Transportation Officials' Standing Committee on the Environment. In addition, she serves as chairman of the Transportation Research Board's Committee on Environmental Analysis in Transportation. She holds a master's degree in urban and regional planning from the University of Wisconsin–Madison. Address: Bureau of Environment, Wisconsin Department of Transportation, P.O. Box 7965, Madison, Wisconsin, 53707, USA.

VIRGINIA H. DALE is an ecologist seeking solutions for issues dealing with land-use change, forest development after disturbances, landscape ecology, and environmental decision making. She obtained her Ph.D. in mathematical ecology from the University of Washington. She is a corporate fellow in the Environmental Sciences Division at Oak Ridge National Laboratory in Tennessee, where she has been a staff member since 1984. She has worked on developing modeling tools and indicators for resource management, vegetation recovery subsequent to disturbances, effects of air pollution and climate change on forests, tropical deforestation, and integrating socioeconomic and ecological models of land-use change. She has also worked on the effects of land-use change in tropical Latin America, the management of rare species on military reservations, and the ecological recovery of Mount St. Helens subsequent to the 1980 eruption. She serves on advisory boards for the Environmental Protection Agency, The Nature Conservancy, the National Academy of Sciences, the Scientific Committee on Problems of the Environment, and the Grand Canyon Monitoring and Research Center. She is also editor-in-chief of the journal *Environmental Management*. Address: Environmental Sciences Division, P.O. Box 2008, Oak Ridge National Laboratory, Oak Ridge, Tennessee, 37831, USA.

LENORE FAHRIG is professor of biology at Carleton University, Ottawa, Canada. Dr. Fahrig studies the effects of landscape structure on wildlife populations. She uses spatial simulation modeling to formulate predictions and tests those predictions using a wide range of organisms, including plants, insects, amphibians, mammals, and birds. Her current work on road system ecology includes empirical studies of road impacts on small mammal and amphibian populations and movements as well as generalized simulation modeling of population responses to road networks. Dr. Fahrig has published over 50 papers in landscape ecology, and many of her recent papers focus on ecological impacts of roads. She is currently a member of the U.S. National Academy of Sciences Committee on Ecological Impacts of Road Density. Address: Carleton University, Department of Biology, Ottawa, Ontario, K1S 5B6, Canada.

ROBERT FRANCE is associate professor at the Harvard Design School, Harvard University, where he teaches courses on watershed management, urban stormwater design, and environmental theory. He is president of W.D.N.R.G. Limnetics, a firm specializing in ecological restoration. He conducts research on the aquatic ecology (limnology) effects of riparian deforestation. France is author of over 100 scientific papers, with emphases on terrestrial-aquatic interactions, stable isotopes, biodiversity, and contaminants and polynyas in the Canadian arctic. His authored and edited books include *Handbook of Water Sensitive Planning and Design* (Lewis Publishing, 2002).He is the editor of a book series for Lewis Publishing entitled *Integrative Studies in Water Management and Land Development*. France received a BSc from the University of Manitoba, an MSc from the Freshwater Institute in Winnipeg, and a Ph.D. from the University of Toronto, and has conducted research and taught at McGill University. Address: Harvard University, Harvard Design School, Cambridge, Massachusetts, 02138, USA.

CHARLES R. GOLDMAN is professor of limnology in the Department of Environmental Science and Policy at the University of California, Davis. He has served as chair of the Division of Environmental Studies and as founding director of the Institute of Ecology at U.C.–Davis, and he is currently director of the Tahoe Research Group. His four decades of research on Lake Tahoe, California, have included lake dynamics, eutrophication, the development of artificial wetlands, de-icing agents for highways, and comparative analyses of Lake Baikal, Russia, and hydroelectric impoundments worldwide. Dr. Goldman received bachelor's and master's degrees from the University of Illinois and a Ph.D. in limnology-fisheries from the University of Michigan. His many awards include a Guggenheim Fellowship in Italy, an NSF Senior Postdoctoral Fellowship in Lapland, the Congressional Antarctic Service Medal, a Fulbright Professorship in Yugoslavia, and the Albert Einstein World Award of Science. He served as vice president of the Ecological Society of America and as president of the American Society of Limnology and Oceanography. He has published four books and over 400 research articles and has frequently translated his research findings to state, national, and international policy decisions, including the conservation and judicious use of aquatic resources on all continents. Address: University of California, Department of Environmental Science and Policy, Davis, California, 95616, USA.

KEVIN HEANUE, after a 40-year career with the U.S. Federal Highway Administration, consults on transportation planning, environmental analysis, organizational development, and related issues. His consulting practice has included projects for the Eno Transportation Foundation, American Highway Users Alliance, Transportation Research Board, and FHWA. During his federal career, he served for eight years as director of the Office of Environment and Planning, administering statewide, intermodal, urban planning, and environmental programs. He is active in the Transportation Research Board and currently serves as chair of the Task Force on Transportation in a Sustainable Environment. In 1970–71, he directed the Dublin, Ireland, Transportation Study for the United Nations Development Program. He has consulted with the World Bank in China and serves on the International Steering Group for the World Bank, Urban Transport Strategy Review. He holds a B.S. in civil engineering from Tufts University and an M.S. in civil engineering from Georgia Institute of Technology. Address: 610 Pullman Place, Alexandria, Virginia, 22305, USA.

JULIA A. JONES is a professor in the Department of Geosciences at Oregon State University, where she teaches graduate and undergraduate courses related to spatial statistics, landscape ecology, and geographical analysis of watershed dynamics. Her research interests include the hydrological effects of road networks in National Forest land, roadside plants, physical stream processes, and the spatio-temporal analysis of ecological and physical processes at landscape to regional scales. Dr. Jones received a B.A. in economic development from Hampshire College and an M.A. in international relations and a Ph.D. in geography and environmental engineering from Johns Hopkins University. She served as a research assistant at Resources for the Future, Washington, D.C., and as associate professor in geography and environmental studies at the University of California, Santa Barbara. Address: Oregon State University, Department of Geosciences, Corvallis, Oregon, 97331, USA.

FREDERICK J. SWANSON is a research geologist with the USDA Forest Service, Pacific Northwest Research Station, in Corvallis, Oregon, and a Forest Service lead scientist for the ecosystem research team based at the H. J. Andrews Experimental Forest in the Oregon Cascade Range. He has been a leader of the National Science Foundation–sponsored Long-Term Ecological Research program based at the Andrews Forest since 1980. Throughout a 30-year career,

Dr. Swanson's research has focused on interactions of geophysical processes with forest and stream ecosystems in mountain landscapes under both natural conditions and influences of land management, including roads. His interest in interactions of science and policy is reflected in part by his coeditorship of the book *Bioregional Assessments: Science at the Crossroads of Management and Policy* (Island Press, 1999). He holds a bachelor's degree from Pennsylvania State University and a Ph.D. from the University of Oregon, both in geology. Address: USDA Forest Service, 3200 Jefferson Way, Corvallis, Oregon, 97331, USA.

THOMAS TURRENTINE began his anthropology career studying processes of cultural change among villagers in the Peruvian Andes. For the past 12 years, he has been with the Institute of Transportation Studies at the University of California, Davis, studying consumer and citizen response to clean automotive technologies and policies. Dr. Turrentine has been spearheading the development and funding of an interdisciplinary research center at ITS–Davis, which will focus on understanding the problems of transportation in fragile environments. He is currently working on market strategies for fuel-cell vehicles. He received his Ph.D. in anthropology from the University of California, Davis, and held a postdoctoral fellowship at the University of Laval, Quebec. He has conducted field research in Peru, Bolivia, Chile, Canada, New Zealand, and the USA; served on an OECD panel on solutions to the negative impacts of car use; and works with the U.S. National Park Service on solutions to the environmental impacts of car use in national parks. Address: Institute of Transportation Studies, One Shields Avenue, University of California, Davis, California, 95616, USA.

THOMAS C. WINTER is a senior research hydrologist with the U.S. Geological Survey in Denver, Colorado. He earned B.A. and M.S. degrees in geology and a Ph.D. in hydrogeology at the University of Minnesota. From 1961 to 1972, he conducted geological and water resource studies in Minnesota and was in charge of USGS groundwater studies there from 1968 to 1972. Since 1973, he has conducted research on the hydrology of lakes and wetlands, with emphasis on their interaction with groundwater and evaporation. In the late 1970s, he helped establish, and has since been a principal investigator at, four long-term field research sites: the Mirror Lake watershed in New Hampshire, the Shingobee River headwaters area in Minnesota, the Cottonwood Lake wetland complex in North Dakota, and the Island Lake area of the Crescent Lake National Wildlife Refuge in Nebraska. He also has been involved with lake and wetland studies in Washington, California, Colorado, Wisconsin, Massachusetts, and Florida. He has received the Distinguished Service Award from the U.S. Department of the Interior, the M. King Hubbert Award from the National Ground Water Association, the W. R. Boggess Award from the American Water Resources Association, the Lifetime Achievement Award from the Society of Wetland Scientists, and the O. E. Meinzer Award from the Geological Society of America. Address: U.S. Geological Survey, Mail Stop 413, Denver Federal Center, Denver, Colorado, 80225, USA.

Index